POROUS PAVEMENTS

Integrative Studies in Water Management and Land Development

Series Editor
Robert L. France

Published Titles

Handbook of Water Sensitive Planning and Design
Edited by Robert L. France

Boreal Shield Watersheds: Lake Trout Ecosystems in a Changing Environment
Edited by J.M. Gunn, R.J. Steedman, and R.A. Ryder

Forests at the Wildland–Urban Interface: Conservation and Management
Edited by Susan W. Vince, Mary L. Duryea, Edward A. Macie, and L. Annie Hermansen

The Economics of Groundwater Remediation and Protection
Paul E. Hardisty and Ece Özdemiroğlu

Restoration of Boreal and Temperate Forests
Edited by John A. Stanturf and Palle Madsen

POROUS PAVEMENTS

Bruce K. Ferguson

Taylor & Francis Group

Boca Raton   London   New York   Singapore

A CRC title, part of the Taylor & Francis imprint, a member of the
Taylor & Francis Group, the academic division of T&F Informa plc.

Library of Congress Cataloging-in-Publication Data

Ferguson, Bruce K.
  Porous pavements / by Bruce K. Ferguson.
    p. cm. – (Integrative studies in water management and land development; 6)
    Includes bibliographical references and index.
    ISBN 0-8493-2670-2 (alk. paper)
    1. Pavements, Asphalt. 2. Asphalt. I. Title. II. Series.

TE270.F47 2005
625.8'5--dc22   2004019318

This book contains information obtained from authentic and highly regarded sources. Reprinted material is quoted with permission, and sources are indicated. A wide variety of references are listed. Reasonable efforts have been made to publish reliable data and information, but the author and the publisher cannot assume responsibility for the validity of all materials or for the consequences of their use.

Neither this book nor any part may be reproduced or transmitted in any form or by any means, electronic or mechanical, including photocopying, microfilming, and recording, or by any information storage or retrieval system, without prior permission in writing from the publisher.

All rights reserved. Authorization to photocopy items for internal or personal use, or the personal or internal use of specific clients, may be granted by CRC Press, provided that $1.50 per page photocopied is paid directly to Copyright Clearance Center, 222 Rosewood Drive, Danvers, MA 01923 USA. The fee code for users of the Transactional Reporting Service is ISBN 0-8493-2670-2/05/$0.00+$1.50. The fee is subject to change without notice. For organizations that have been granted a photocopy license by the CCC, a separate system of payment has been arranged.

The consent of CRC Press does not extend to copying for general distribution, for promotion, for creating new works, or for resale. Specific permission must be obtained in writing from CRC Press for such copying.

Direct all inquiries to CRC Press, 2000 N.W. Corporate Blvd., Boca Raton, Florida 33431.

Trademark Notice: Product or corporate names may be trademarks or registered trademarks, and are used only for identification and explanation, without intent to infringe.

Visit the CRC Press Web site at www.crcpress.com

© 2005 by CRC Press

No claim to original U.S. Government works
International Standard Book Number 0-8493-2670-2
Library of Congress Card Number 2004019318
Printed in the United States of America 2 3 4 5 6 7 8 9 0
Printed on acid-free paper

Series Statement: Integrative Studies in Water Management and Land Development

Ecological issues and environmental problems have become exceedingly complex. Today, it is hubris to suppose that any single discipline can provide all the solutions for protecting and restoring ecological integrity. We have entered an age where professional humility is the only operational means for approaching environmental understanding and prediction. As a result, socially acceptable and sustainable solutions must be both imaginative and integrative in scope; in other words, garnered through combining insights gleaned from various specialized disciplines, expressed and examined together.

The purpose of the CRC Press series Integrative Studies in Water Management and Land Development is to produce a set of books that transcends the disciplines of science and engineering alone. Instead, these efforts will be truly integrative in their incorporation of additional elements from landscape architecture, land-use planning, economics, education, environmental management, history, and art. The emphasis of the series will be on the breadth of study approach coupled with depth of intellectual vigor required for the investigations undertaken.

Robert L. France
Series Editor
Integrative Studies in Water Management
and Land Development
Associate Professor of Landscape Ecology
Science Director of the Center for
Technology and Environment
Harvard University
Principal, W.D.N.R.G. Limnetics
Founder, Green Frigate Books

Foreword by Series Editor: Targeting Causes Rather Than Treating Symptoms

Pavements are the most ubiquitous imprints left by humans upon the natural landscape; the extent of coverage, as described by Bruce K. Ferguson and his contributing authors Gregg A. Coyle, Ronald Sawhill, and Kim Sorvig in the present book, is truly shocking. A large body of literature that links impervious surfaces to a wide variety of environmental problems exists. Indeed, this relationship is so established that one can truncate the old adage and, regardless of inferred intentions, simply state that the road to Hell is paved…Period.

Not far from where I live in the Alewife Brook watershed of Cambridge, Massachusetts (described as the case study in my book *Facilitating Watershed Management: Fostering Awareness and Stewardship*) is a region whose premier characteristic is that it is a sea of pavement. Indeed, it is actually possible to walk for a kilometer linking up one sprawling parking lot with another, several of which are of a size large enough to accommodate the landing of a jet airplane! It comes as no surprise that the nearby stream is the most flood-prone and nonpoint source polluted river in the eastern part of the State. In contrast, only twenty kilometers away, near the waters of Walden Pond, which have been empowered by many environmentalists around the world with near-sacred status (see my edited volume *Profitably Soaked: Thoreau's Engagement With Water*), lies one of the nation's first successful porous pavement parking lots (see page 64 and 124–125 in the present book). I well remember the day, following the 2000 Harvard conference, which gave rise to the first book in this series (*Handbook of Water Sensitive Planning and Design*), when Bruce Ferguson held the interest of a group of hydrologists with his demonstration of pouring water onto and *into* the asphalt there.

The present book, the sixth in the series by CRC Press — Integrative Studies in Water Management and Land Development — is the long-awaited and eagerly sought comprehensive review of porous pavements. The seamless fusion of landscape architecture, structural engineering, and hydrology are a perfect fit to the aspirations of this series of books. Herein we learn from Ferguson and his colleagues not only of the role of porous pavements in reducing the "feast or famine" nature of urban stream hydrology (in terms of there being either too much or too little water due to rapid runoff and lack of groundwater replenishment), but also of the role that such surfaces play in promoting well-watered and healthy trees, microclimatic thermal regulation, quieter and safer streets, and also in creating beauty in our (sub)urban landscapes.

The breadth of study, exhaustive research, wealth of technical detail, illustrative and informative case studies, great photographs and clear figures, and diversity of references and web-pages cited, will ensure that this book will become the standard reference manual for practitioners. The detailed examination of the various porous pavement typologies, each given its own chapter, in which strengths and limitations, maintenance issues, and application suggestions are honestly and straightforwardly presented, will mean that what Ferguson describes at the start of the book as "the controversial and technically challenging field of porous pavements" may not be quite so in the years to come. There is much to learn from these pages, which provide a clarion call for the imaginative use of porous technologies to mimic natural landscape functionality and thus alleviate many of the environmental stresses that plague our developed watersheds. Such an approach that specifically targets the causes of environmental dysfunction rather than only dealing with the symptoms of the disease will go far toward promoting healthy watersheds.

And finally, this book makes the point that when consideration is given to related infrastructure costs needed to alleviate watershed disturbance, porous pavements may often be the less expensive option in the long run. One final example, again from the Boston area, illustrates how attention to issues of groundwater infiltration could have saved thousands of dollars. Many older cities that developed as a result of filling in their wetlands and coastal areas may look forward with apprehension to what Boston is now having to address. There, some of the historic buildings, anchored as in Amsterdam and elsewhere, to massive cribs of wooden timbers buried deep within the formerly moist ground, are now showing signs of instability. This is due to the increase in the extent of impervious coverage that has prevented the infiltration of water needed to preserve the structural integrity of the crib anchors. Engineers are now examining expensive methods of artificially injecting water into the ground to saturate the building foundations. How much simpler it would have been to have either left more open space free of impervious coverage or to have used any of the diversity of porous pavement options that Ferguson and his colleagues advance in these pages.

Robert L. France
Harvard University

Preface

Of all the structures built by human beings, pavements are the most ubiquitous. They occupy twice the area of buildings. And of all the physical features of contemporary cities they are the most influential. They dominate the quality of urban environments. In urban watersheds impervious pavements produce two thirds of the excess runoff. They are responsible for essentially all the hydrocarbon pollutants. They produce two thirds of the groundwater decline and the resulting local water shortages. They produce two thirds of the temperature increase in the urban "heat island". They determine whether urban trees extend their roots and live, or die.

The polluted quality of urban runoff, the overflowing of combined sewer systems, the diminishment of water supplies, the wasteful consumption of urban energy, and the decline and death of the "urban forest" force our attention on the reclamation of paved areas for the benefit of the biophysical environment and the human beings who live with it.

Porous pavements are those that have built-in networks of void spaces where water and air pass through. Although some porous paving materials are nearly indistinguishable from nonporous materials in construction and superficial appearance, their environmental effects are qualitatively different. They cause air, water and heat to enter different parts of the environment, there to undergo qualitatively different processes of storage, treatment, and flow.

Porous pavements can allow the oils from cars and trucks to biodegrade safely, the rainwater to infiltrate the soil, the heat of the sun to dissipate, the groundwater to be replenished, the roots of trees to breathe, and the streams to flow in dry summers. A large part of the solution to urban environmental problems is under our feet. By paying appropriate attention to the everyday materials on which we walk and drive, we can replenish renewable resources, restore regenerative processes, and produce a cleaner, healthier, safer, more sustainable world in which to live. Porous pavements are potentially the most important development in urban watersheds since the invention of the automobile.

But in most parts of North America porous pavements are outside the ordinary conventions of urban design and construction. Many people are curious about porous pavements, and many are skeptical.

PURPOSE AND NEED

This book's purpose is to give responsible professionals the information they need to put porous pavement materials into appropriate, informed, beneficial, and successful use. With factual knowledge of experience in the field and theoretical understanding of underlying mechanisms, individual designers can evaluate one kind of pavement material against another, participate in responsible professional debates, and competently and correctly adapt porous pavements to site-specific conditions.

This book is addressed to landscape architects, building architects, urban designers, civil engineers, urban foresters, construction contractors, construction product manufacturers, city planners, environmental policy-makers, and all others professionally concerned with urban construction and the urban environment. It supplements basic training in site design, site drainage, construction materials, and horticulture with the special concepts of porous pavements and their implications for the urban environment. This is a reference for practitioners who need to update and expand their applied skills, and a textbook for university classes in site construction, watershed protection, and sustainable development.

Previous guides to porous pavements have been published as the technology developed in the last 30 years. For porous asphalt several books emerged when the U.S. Environmental Protection Agency was supporting research (Thelen et al., 1972; Thelen and Howe, 1978; Diniz, 1979). For porous concrete the Florida Concrete and Products Association published fine guidelines based on its seminal experience (FCPA, no date; Wingerter and Paine, 1989; Paine, 1990). For paving blocks and grids, product licensing and manufacturing groups beginning with Uni-Group USA admirably invested in research and published the results (Rollings and Rollings, 1992 and 1999), and their work is now joined by a fine summary manual from the Interlocking Concrete Pavement Institute (Smith, 2001). This book leans gratefully on those earlier works. However each previous guide focused on an individual type of material without defining the field as a whole, and some of the early works are now out of date.

Professionals who have to design sites creatively and cost-effectively to meet combinations of criteria need an overview of the available materials. They need information that will allow them to choose and apply materials to meet site-specific conditions and objectives, and examples of how the materials have fared in a variety of settings. They need lines of thought for evaluating the feasibility and appropriateness of alternative pavement applications on specific sites.

This book fills the void in the compilation of porous pavement information. It is the first that has inventoried the range of available materials, arranged them to contrast their applications in different types of settings, and related them to the context of the general site environment. It defines and organizes the field for the first time.

The research for this book consumed seven years, during which I interviewed 170 experienced researchers, designers, and suppliers, read 800 technical articles and reports, and personally surveyed 270 installations of all kinds of porous pavements in all parts of North America. Near the beginning, Tom Richman of Catalyst in San Francisco clarified the image of a work that would be immediately usable by practitioners. In conceiving it and pursuing its completion I have been inspired as usual by the examples of Albert B. Ferguson and Louise E. Ferguson of the motivation and discipline to work intensely and joyfully for the good of the community, and by the example of Ian L. McHarg of moral will to seek new and better ways specifically in environmental design.

ARRANGEMENT AND CONTENT

This book begins with broad basics to establish a foundation for all porous pavement materials and applications. The first five chapters introduce the types of materials and arrangements and the roles they play in the urban environment, and outline the principles of pavement structure, hydrology, and rooting space.

Each of the remaining nine chapters is dedicated to one of the families of porous pavement materials (those families being defined and distinguished here for the first time): porous aggregate, porous turf, plastic geocells, open-jointed blocks, open-celled grids, porous concrete, porous asphalt, "soft" pavement materials, and decks. Each chapter outlines the nature of the material, the organization of the industry that supplies it, and its distinctive installation methods, performance levels, and appropriate applications.

This book emphasizes practice and experience in North America. North America has been a leader in some kinds of porous pavements, for example porous concrete, plastic geocells, the early development of porous asphalt, and now in this book the recognition of unbound aggregate as a valid and purposeful porous paving material; for the benefit of workers in all regions of the world this book reviews North American experience with those materials. In some other kinds of porous pavements, notably those of blocks and grids and recent developments in porous asphalt, North America has been behind countries in Europe and elsewhere; for the benefit of North American practitioners this book reviews the nature and availability of those materials and the growing experience with them.

This book is lengthy because it confronts practitioners' numerous, challenging, technical questions about porous pavement. As this book introduces the controversial field of porous pavements to the world for the first time, it is valid that those questions be asked, and necessary that they be answered.

This book emphasizes factual data from observed experience. Factual on-the-ground experience supercedes any degree of speculative theory. In the controversial and technically challenging field of porous pavements, factual evidence must be visible and accessible. Numerous case studies of specific materials in specific settings illustrate some features that are models for emulation, and others that are failures from which we can learn to do better in the future. Where the facts are simply not known, this book calls for further research. In addition this book cites numerous references because such references are the trail of recorded knowledge; they support specific statements and show where to go for further information. They lead readers to ongoing sources where they can update product information and obtain applicable industrial standards firsthand.

ADDITIONAL SOURCES OF INFORMATION

Site construction books such as those cited at the end of this preface give additional general background in pavement construction. The sources listed in Table P.1 provide updates on the general fields of urban construction and its use in environmental protection.

Specific paving products mentioned in this book were identified through searches on the web, exhibits at professional conferences, membership lists of industrial associations, and articles and advertisements in professional magazines. Practitioners must know what is available for their use. However, listing of proprietary products is for information only; it does not imply any recommendation or endorsement.

TABLE P.1

Examples of Sources of General Information on Sustainable and Environmentally Restorative Construction

Name	Contact Information
Environmental Building News	www.buildinggreen.com
Environmental Design + Construction	www.edcmag.com
GreenClips	www.greenclips.com
Low Impact Development Center	www.lowimpactdevelopment.org
American Recycler	www.americanrecycler.com
Rocky Mountain Institute	www.rmi.org
Smart Communities Network	www.sustainable.doe.gov
Southface Institute	www.southface.org
Sustainable Communities Network	www.sustainable.org
U.S. Green Building Council	www.usgbc.org
Green Builder	www.greenbuilder.com/sourcebook

TABLE P.2

Examples of Multi-industry Information Sources for Updating and Expanding Lists of Specific Porous Paving Materials

Name	Contact Information
CAD Details	www.caddetails.com
Erosion Control	www.forester.net/ec.html
LA Info Online	www.la-info.com
Landscape Architecture	www.asla.org/nonmembers/lam.cfm
Landscape Catalog	www.landscapecatalog.com
Landscape Online	www.landscapeonline.com
Material Connexion Library	www.materialconnexion.com
Stormwater	www.stormh2o.com
Sweets	www.sweets.com

Additional products and companies surely exist, or could exist in the future. Lists can be updated and enlarged at any time by re-searching the same types of sources, including the magazines and multi-industry "catalog" web sites listed in Table P.2. Industry-specific information sources are given in specific chapters of this book.

ROLES OF PRACTITIONERS

Every site-specific project presents a unique combination of conditions and objectives. Where porous pavements are used, they must be used right. Practitioners must apply porous pavements with the same degree of knowledge, selectivity, care, and ingenuity they would bring to any other aspect of any development project. Although pavements are mundane things, professionals must become accustomed to paying attention to them in ways they may never have done before.

No statement in this book constitutes a recommendation for any specific site. The information in this book is intended to be used by design professionals competent to evaluate its significance and limitations and who will accept the responsibility for its proper application. With knowledge and care, responsible designers can adapt pavement materials and configurations to satisfy specific performance criteria, write appropriate and precise specifications, compare one type of material with another, objectively evaluate the causes of failure when it occurs, and select and adapt new types of materials where they are appropriate.

This book does not advocate replacing one rigidly conventional technology with another, or provide fixed recommendations to be followed blindly into all project sites. Instead it advocates a complete "toolbox" from which designers can choose selectively and appropriately in their everyday work. No type of pavement, porous or nonporous, should be smeared thoughtlessly everywhere. Porous pavements do not, by themselves, solve all urban environmental problems. But pavements are so ubiquitous, and the potential effects of making them porous are so fundamental, that anyone who does not acquire the ability to use porous pavements is not working with a complete professional toolbox.

Every year the U.S. paves or repaves a quarter of a million acres of land. Today we are able to answer many of the technical questions that have in the past inhibited the adoption of porous pavements. It is time now for porous pavements to take their place alongside other paving materials as alternatives that practitioners can draw on selectively and knowledgeably in their everyday work.

The potential for porous pavements has built up like an overbalanced snowbank leaning over a mountain ridge and ready to fall. They say that, when a snowbank is like that, you can start an avalanche with a clap of your hands: the small sound makes the whole mountainside quiver and come tumbling down. Perhaps, for porous pavements, this book will be a clapping of hands.

References

Croney, David, and Paul Croney (1998). *Design and Performance of Road Pavements*, 3rd ed., New York: McGraw-Hill.

Diniz, Elvidio V. (1980). *Porous Pavement, Phase 1 — Design and Operational Criteria*, EPA-600/2-80-135, Cincinnati: U.S. Environmental Protection Agency, Municipal Environmental Research Laboratory.

Florida Concrete and Products Association (no date). *Construction of Portland Cement Pervious Pavement*, Orlando: Florida Concrete and Products Association.

Harris, Charles W., and Nicholas T. Dines (Eds.) (1998). *Time-Saver Standards for Landscape Architecture: Design and Construction Data*, 2nd ed., New York: McGraw-Hill.

Hopper, Len (editor in chief) (in press). *Landscape Architectural Graphic Standards*, Hoboken, New Jersey: John Wiley and Sons.

Keating, Janis (2001). Porous Pavement, *Stormwater* 2, 30–37.

Landphair, Harlow C., and Fred Klatt (1998). *Landscape Architecture Construction*, 3rd ed., New York: Prentice-Hall.

Nichols, David B. (1992). Paving, in *Materials,* Vol. 4 of *Handbook of Landscape Architectural Construction*, pp. 69–138, Scott S. Weinberg and Gregg A. Coyle, Eds., Washington: Landscape Architecture Foundation.

Nichols, David B. (1993). Fresh Paving Ideas for Special Challenges, *Landscape Design* 6, 17–20.
Paine, John E. (1990). *Stormwater Design Guide, Portland Cement Pervious Pavement*, Orlando: Florida Concrete and Products Association.
Rollings, Raymond S., and Marian P. Rollings (1992). *Applications for Concrete Paving Block in the United States Market*, Palm Beach Gardens, Florida: Uni-Group USA.
Rollings, Raymond S., and Marian P. Rollings (1999). *A Permeable Paving Stone System*, Mississauga, Ontario: SF Concrete Technology.
Smith, David R. (2001). *Permeable Interlocking Concrete Pavements: Selection, Design, Construction, Maintenance*, 2nd ed., Washington: Interlocking Concrete Pavement Institute.
Thelen, Edmund, and L. Fielding Howe (1978). *Porous Pavement*, Philadelphia: Franklin Institute Press.
Thelen, Edmund, Wilford C. Grover, Arnold J. Holberg, and Thomas I. Haigh (1972). *Investigation of Porous Pavements for Urban Runoff Control*, 11034 DUY, Washington: U.S. Environmental Protection Agency.
Thompson, J. William, and Kim Sorvig (2000). *Sustainable Landscape Construction, A Guide to Green Building Outdoors*, Washington: Island Press.
Walker, Theodore D. (1992). *Site Design and Construction Detailing*, 3rd ed., New York: Van Nostrand Reinhold.
Weinberg, Scott S., and Gregg A. Coyle, (Eds.) (1992). *Materials for Landscape Construction*, Vol. 4 of *Handbook of Landscape Architectural Construction*, Washington: Landscape Architecture Foundation.
Wingerter, Roger, and John E. Paine (1989). *Field Performance Investigation, Portland Cement Pervious Pavement*, Orlando: Florida Concrete and Products Association.

About the Author

BRUCE K. FERGUSON

Bruce K. Ferguson is a landscape architect who has specialized in environmental management of urban watersheds for 25 years.

Ferguson guided the development of award-winning guidelines for urban development to protect watersheds in the San Francisco Bay region. With the Rocky Mountain Institute, he conceived the restorative redevelopment of urban watersheds in Pittsburgh. As a member of interdisciplinary teams he has guided new development in the metropolitan regions of Atlanta, Miami, and Los Angeles, and the state of Georgia. He guided stormwater quality protection at the Goddard Space Flight Center, and the conservation of irrigation water on the lawn of the White House.

Harvard professor Robert France referred to Ferguson as "the world's expert in stormwater infiltration". Ferguson's 1994 book *Stormwater Infiltration* is considered a landmark in the integration of urban development with natural watershed processes. His 1998 book *Introduction to Stormwater* is the most frequently referenced book in the field. He is the author of 130 scientific and professional papers on environmental management of urban watersheds. Using the results of his research, he lectures at major universities and conducts continuing education courses for design practitioners at Georgia and Harvard.

Ferguson is a Fellow of the American Society of Landscape Architects and a past president of the Council of Educators in Landscape Architecture. He is a recipient of the Council's Outstanding Educator Award, the highest award for landscape architectural education in North America.

Ferguson is Professor of Landscape Architecture and Director of the School of Environmental Design at the University of Georgia. He obtained the B.A. degree at Dartmouth College and the M.L.A. under Ian McHarg at the University of Pennsylvania. He is a licensed landscape architect in Georgia and Pennsylvania.

Contributors

GREGG A. COYLE

Gregg A. Coyle is a landscape architect who designs estates and residential communities in the southeastern U.S. working with construction materials, decks, plantings, and lighting. He is overseeing the sustainable design and construction of EcoLodge San Luis, a University of Georgia residential and laboratory facility in the cloud-forest mountains of Costa Rica for research and study of ecology, anthropology, and landscape architecture. He coedited a volume of the Landscape Architecture Foundation's *Handbook of Landscape Architectural Construction: Materials for Landscape Construction* (volume 4, 1992), and contributed a chapter, "Contract Conditions and Specifications," to *Irrigation* (volume 3, 1988). He is associate professor of landscape architecture and director of the landscape architectural internship program at the University of Georgia. He earned the B.F.A. degree at Peru State College and the M.L.A. at Iowa State University. He is a licensed landscape architect in Georgia.

RONALD B. SAWHILL

Ronald B. Sawhill is a landscape architect who designs residential communities, recreational facilities, and commercial developments in the southeastern U.S. working with grading, planting, irrigation, and stormwater management. Using the results of his experience, he conducts continuing education courses for design practitioners. He is the author of the turf section of *Landscape Architectural Graphic Standards*. He is president of the Georgia Chapter of the American Society of Landscape Architects. He earned the B.L.A. and M.L.A. degrees at the University of Georgia. He is assistant professor of landscape architecture at the University of Georgia, and a licensed landscape architect in Georgia and South Carolina.

KIM SORVIG

Kim Sorvig is a landscape architect who specializes in sustainable design. He is the author of *To Heal Kent State: A Memorial Meditation* (1990). He is a coauthor of *Sustainable Landscape Construction: A Guide to Green Building Outdoors* (2000), which is organized around ten key principles of sustainability and which evaluates materials and methods of landscape construction using criteria such as energy savings, non-toxicity, and renewability. He is a contributing editor of *Landscape Architecture* magazine, where the American Society of Landscape Architects awarded him the Bradford Williams Medal for landscape architectural writing. He is research associate professor at the University of New Mexico. He was educated in plant ecology and horticulture at the Royal Botanic Gardens in London, and earned the M.L.A. degree at the University of Pennsylvania. He is a licensed landscape architect in New Mexico.

Table of Contents

Chapter 1
Why Make Pavements Porous?...1

Chapter 2
Dimensions of Porous Pavement Installations...35

Chapter 3
Porous Pavement Structure 69

Chapter 4
Porous Pavement Hydrology..119

Chapter 5
Porous Pavement Tree Rooting Media..171

Chapter 6
Porous Aggregate 199

Chapter 7
Porous Turf..241
Ronald B. Sawhill

Chapter 8
Plastic Geocells 285

Chapter 9
Open-Jointed Paving Blocks 323

Chapter 10
Open-Celled Paving Grids 381

Chapter 11
Porous Concrete 417

Chapter 12
Porous Asphalt 457

Chapter 13
Soft Porous Surfacing 513
Kim Sorvig

Chapter 14
Decks 539
Gregg A. Coyle

Index 567

1 Why Make Pavements Porous?

CONTENTS

The Magnitude of Pavements in America	1
Pavements in Alternative Patterns of New Development	5
Where not to Make Pavements Porous	6
The Promise of Clean Water	7
The Promise of Long-Lived Trees	10
The Promise of Cool Cities	15
The Promise of Quiet Streets	18
The Promise of Safe Driving	20
The Promise of Reducing Cost	22
The Promise of Meeting Development Regulations	23
The Promise of Preserving Native Ecosystems	24
The Promise of Beauty	26
Acknowledgments	29
References	29

Wherever pavements are built, porous pavements can improve the environment in vital ways. A pavement is any treatment or covering of the earth surface that bears traffic. A porous pavement is one with porosity and permeability high enough to significantly influence hydrology, rooting habitat, and other environmental effects. "Dense" pavements are those that are not porous. This chapter introduces the magnitude of pavements and the types of effects that porous pavements can achieve for water, air, living things, and human welfare, alone or in partnership with other aspects of urban design and construction.

THE MAGNITUDE OF PAVEMENTS IN AMERICA

Figure 1.1 shows the proportion of land covered by built structures in contemporary urban land-use districts. The dark portion of each column represents pavements; the white portion represents the roofs of buildings. The data are averages of measurements in the areas of Chesapeake Bay (Appendix D of Cappiella and Brown, 2001) and Puget Sound (Wells, 1994, p. 11).

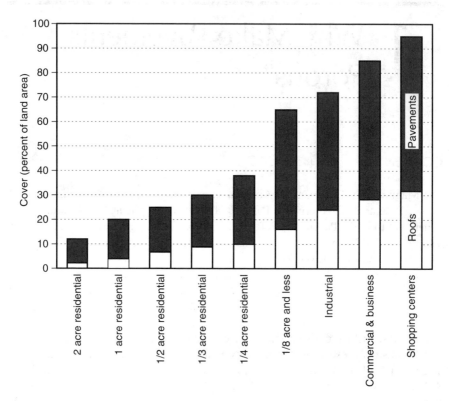

FIGURE 1.1 Built cover in contemporary urban land uses (total built cover from Arnold and Gibbons [1996]; distribution of pavements and roofs from an average of data in Appendix D of Cappiella and Brown [2001] and Wells [1994, p. 11]).

The left side of the chart shows dispersed, large-lot residential areas covering 12 percent or more of the land with built construction. Progressing toward the right of the chart, one finds increasingly intense residential, industrial, and commercial land uses producing a correspondingly greater built cover. Shopping centers are in a class by themselves, routinely covering more than 90 percent of the land with built structures. Pavements occupy 65 to 70 percent of the built cover. In intensely built-up areas, pavements cover more than half of all the land.

Figure 1.2 analyzes built cover by comparing the areas of building roofs and three categories of pavements. The height of each column represents the proportion of built cover occupied by each type of structure. The roof areas of buildings have white columns; the pavements have dark columns. The three charts show data for single-family residential, multifamily residential, and commercial land uses. In all three charts, the white columns for building roofs show that buildings occupy about one third of the built cover; pavements occupy the other two thirds.

In single-family residential districts the area of street pavements is large because long streets are necessary to connect the dispersed dwellings. Local streets occupy 69 percent of all road mileage in the U.S. (derived from data in Table No. 1019 of the U.S. Census Bureau, no date). The parking is residential driveways. In single-family

Why Make Pavements Porous?

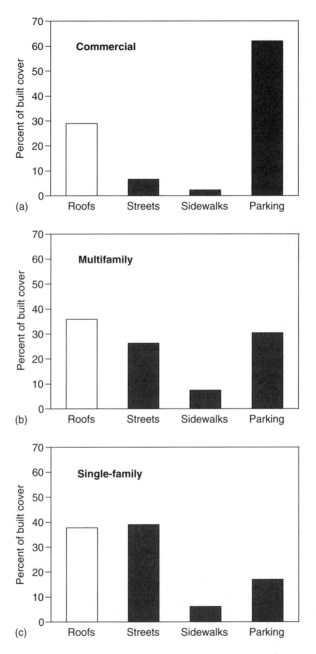

FIGURE 1.2 Types of built cover in three land uses (average of data from Wells [1994, p. 11] and Appendix D of Cappiella and Brown [2001]).

districts the driveways, local streets, and pedestrian sidewalks all have low traffic loads, so they are all eligible for consideration as porous pavement materials without structural conflicts.

In multifamily residential districts the areas of both streets and on-site parking are substantial. The sidewalks and most of the parking lots in such districts have low traffic loads, so they are eligible for consideration as porous pavement materials. The traffic load in the streets could vary from place to place.

In commercial districts on-site parking lots dominate the built cover. Although public highways in commercial districts are wide, they occupy a small area compared with nearby parking lots. Large portions of parking lots have low or moderate traffic loads, including most of the parking stalls and all the outer, less-used portions of parking lots. The low- and moderate-traffic areas are eligible for selective porous pavement construction.

In summary, these figures show that it is possible to select porous pavement materials for approximately half of the built cover in most urban land uses.

In a large region such as a county with diverse interacting land uses, the total amount of pavement depends on the number of people living and working there. Figure 1.3 shows built cover in relation to population density. The total height of the curve represents the total built cover in a region. The dark portion represents the area of pavements; the white portion represents building roofs. These regional values are lower than most of those for individual urban land uses because they average in a region's parks and undeveloped lands along with built-up urban districts. With increasing population density the amount of built cover increases as the intensity of streets, buildings, and parking lots increases to support the people.

Figure 1.4 shows the regional amounts of pavement and roofs per person. This curve goes in the direction opposite from that for coverage of the land: as population density increases, the amount of built cover per person declines. This is because at high densities people live and work in multistory buildings that are close together and require fewer connecting streets, and the population uses public transportation and parking garages that require less pavement space for the storage of cars.

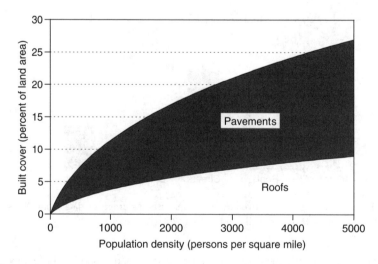

FIGURE 1.3 Built cover per land area in large areas such as counties (based on the equation for "medium" total built cover in Stankowski [1972] and pavements = 2/3 of total).

The amount of pavement in America is rapidly increasing as urban and suburban areas expand with the growing population. In recent years the U.S. population has been growing at a rate of 3.27 million persons per year (derived from data for 1990 and 2000 in *American Factfinder*, http://factfinder.census.gov). Using an arbitrary value of 0.05 acres of pavement per person from Figure 1.4, one can conclude that the country's paved area is growing at a rate of approximately 250 square miles per year.

PAVEMENTS IN ALTERNATIVE PATTERNS OF NEW DEVELOPMENT

When one is planning the future development of a site or a region, the opposite directions of the curves in Figures 1.3 and 1.4 present opposing choices between dispersed, low-density development and concentrated, high-density development. Table 1.1 shows that the choices present contrary arrays of pavement per acre and per person. The same type of choice applies both within an individual development site and

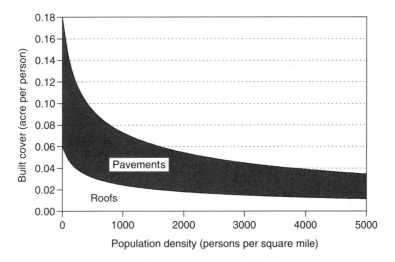

FIGURE 1.4 Built cover per person in large areas such as counties (based on the equation for "medium" total built cover in Stankowski [1972] and pavements = 2/3 of total).

TABLE 1.1
Implications of Choices in Urban Land-Use Pattern in a Site or Region

	Concentrated, High-Density Development	Dispersed, Low-Density Development
Local concentration of pavement where development is built	High	Low
Total quantity of pavement to serve a given population	Low	High

Summarized from Center for Watershed Protection, 1998; University of Georgia School of Environmental Design, 1997; Richman and Associates, 1997.

across an urbanizing region. For application in a given specific locale, each development pattern has a combination of advantages and disadvantages (Center for Watershed Protection, 1998; University of Georgia School of Environmental Design, 1997; Richman and Associates, 1997). The selection of each is likely to depend on site-specific conditions.

In dispersed development including large-lot single-family residences, the quantity of pavement is high for a given number of residents because the area paved for automobiles to connect to their dispersed buildings is large. However, the effect is diffuse, and its intensity at any one location is low.

On the other hand, dense development concentrates a given unit of development on only a portion of the available land. It generates high local concentrations of vehicles and people. But it uses a relatively small amount of pavement to support a given unit of development while leaving other areas pristine. Within an individual development site, a concentrated layout features a "clustering" of dwellings on small lots with correspondingly short streets and driveways. On a regional scale, concentrated development is done with compact mixtures of land uses where everyday needs can be met within small distances, nonautomotive transportation, and high residential and commercial densities. The total and per capita pavement areas are low. The total runoff and pollution from a site or a region as a whole are lower than they would be with dispersed development. A densely developed area that absorbs a given population growth is in effect a sacrificial area to preserve the quality of pristine lands elsewhere.

Within any given land-use pattern, the dimensions of necessary pavements can be minimized within the functional requirements of site-specific traffic and land use. In commercial districts the required amount of parking is that needed for actual utilization by a specific land use in a specific location; some jurisdictions could reduce their requirement by 30 percent (Albanese and Matlack, 1999; Willson, 1995). In residential districts the required street width is that needed for actual utilization by traffic and on-street parking; some municipalities could reduce their pavement widths by one third. Half the residential driveway pavements could be eliminated by reducing the driveways to separate wheel treads, as shown in Figure 1.5.

WHERE NOT TO MAKE PAVEMENTS POROUS

On certain special sites, pavements should remain dense and impervious for the sake of resource conservation and environmental protection.

On some sites the surface runoff from dense pavements is a resource that can be "harvested" into special swales or cisterns and used for irrigation or other productive purposes. Harvesting reduces the amount of freshwater to be imported from municipal supplies. However, the expense of collecting, storing, and perhaps treating the harvested water is worth the benefit only in certain limited climatic and site circumstances. Harvesting can be done only where the surface is correctly pitched toward a point of collection or use. The potential uses of the water tend to be limited to those that are tolerant of low water quality unless a treatment system is added.

On many old industrial "brownfield" sites, dense pavements prevent rainwater from percolating through old toxic deposits in the soil. This protects aquifers and streams by preventing the leaching of pollutants out into the environment.

Why Make Pavements Porous? 7

FIGURE 1.5 A residential driveway reduced to two wheel tracks.

In portions of many sites the necessary provisions for porous pavements that will be described in Chapter 2 cannot easily be met. The site layout directs clogging sediment onto the pavement surface, or the slope is excessively steep, or the traffic loading is too great. In these areas porous pavements may not be feasible.

Apart from these exceptions, in all land uses, in all patterns of development, on all types of sites, large areas of pavements are eligible for construction with porous paving materials. The environmental effects they promise are geographically widespread and functionally multifaceted.

THE PROMISE OF CLEAN WATER

The scene in Figure 1.6 exemplifies the problem that urban watersheds present when they are developed with impervious structures. It shows a culvert in the densely built-up Nine Mile Run watershed in Pittsburgh discharging water during a rainstorm. Before the storm these surfaces, like those in any built-up area, had been accumulating pollutants deposited from the atmosphere, dripped from vehicles, leached from metal gutters, and defecated by animals. When the first rain fell, the watershed's impervious pavements and roofs turned essentially all of the pollutants into surface runoff that flushed the pollutants into the stream. As the rain continued, even though the culvert was big enough to walk through, it flowed nearly full. Growing volumes of runoff eroded stream banks, destroying habitats and producing further sediment pollution. Bed materials shifted; banks sloughed in; biota were flushed out of the chute-like channel. In Pittsburgh and other old cities the floods got into sanitary sewers, adding overflows of raw sewage to the stream flow.

FIGURE 1.6 Discharge during a storm from the main Nine Mile Run culvert in Pittsburgh, Pennsylvania (photo courtesy of STUDIO for Creative Inquiry, Carnegie-Mellon University).

FIGURE 1.7 Discharge from the Nine Mile Run culvert after rainfall has stopped (photo courtesy of STUDIO for Creative Inquiry, Carnegie-Mellon University).

Figure 1.7 shows the discharge from the same culvert when the rain stopped. Little flow remained in the stream because there was no water left in the watershed: it was all flushed out during the storm. Groundwater levels were low. Fish were gasping for oxygen in the shallow, warm, sluggish water. Some cities were left with local water shortages.

Why Make Pavements Porous? 9

Impervious pavements and roofs such as those in the Nine Mile Run watershed are collection pans that propel runoff and pollutants into streams without conservation or treatment. The large area that pavements cover, and the automobiles that use them, make impervious pavements the most significant generators of urban runoff and pollutants (Arnold and Gibbons, 1996). The water that dense pavements spoil and discard would, if it were conserved, be capable of supporting the future population growth of millions of people (Otto et al., 2002).

Too often the response has been to construct detention basins like the one shown in Figure 1.8, which illustrates the culvert bringing pulses of surface runoff from the

FIGURE 1.8 Detention basin in the corner of a commercial site.

shopping center's impervious roofs and pavements into a reservoir. The basin stores the runoff briefly so that it discharges slower and slightly later than it would otherwise. All single-purpose stormwater basins like this one cost money to construct. The land dedicated to them is lost to the local economy and the life of urban residents. Detention basins have failed to prevent downstream flooding and erosion, and have never done anything for water quality, ground water replenishment, or urban water supplies (Ferguson, 1998, p. 164). Paradoxically, we have specified impermeable pavements that flush away runoff, then paid for detention basins to counteract the pavements' runoff and pollution, and then paid again to import water supplies to replace the naturally occurring rainwater we have spoiled and thrown away.

Figure 1.9 shows how porous pavements can protect urban watersheds and aquifers before off-pavement stormwater basins are necessary. Some water has been poured on the surface of a porous concrete parking lot. The circular stain indicates that the water has gone down through the pavement's pores, and not across the surface. A porous pavement infiltrates and treats rainwater where it falls. Its pore space stores water like a detention basin. Almost every porous pavement reduces runoff and restores infiltration during small, frequent, numerous storms; some reduce runoff also during rare large storms when downstream flooding would be a severe concern. Infiltrating water recharges aquifers and sustains stream base flow. The pores house a microecosystem that filters and biodegrades the pollutants that occurs generically on residential, commercial, and office pavements; the underlying soil ecosystem is a backup treatment system that assures high treatment levels. Spreading out stormwater infiltration and treatment systems over a development site with porous pavements makes full use of the land's ability to infiltrate, treat, and store subsurface water. Porous pavements cure the diseases of urban watersheds and aquifers at the source, reducing or eliminating symptoms to be treated downstream.

Figure 1.10 contrasts the hydrologic effects of porous and dense pavements. In newly developing areas porous pavements protect the pristine resources of watersheds and aquifers. In old cities, renovating old pavements with porous paving materials compensates for the inadequacy of old combined sewer systems. Wherever paving must be done, porous pavement materials bring rainwater back into contact with the underlying soil. By controlling the fate of precipitation where it falls, they unify stormwater management and the fulfillment of practical urban needs efficiently in single structures.

THE PROMISE OF LONG-LIVED TREES

In the U.S., over half a million trees are planted every year in densely built-up urban settings (Arnold, 1993, p. 121). The scene in Figure 1.11 exemplifies the problem that trees present where they are surrounded by impervious pavements. Trees that could live for 100 years or more, when planted in narrow pits surrounded by dense pavement, are found to be dead or dying only seven years after planting (Moll, 1989). Almost all are diminutive in size for trees of their age (Grabosky and Gilman, 2004; Quigley 2004). In the background of the figure are trees planted at the same time outside the pavement; they have grown into large, healthy trees while those in the pavement have failed.

Why Make Pavements Porous? 11

FIGURE 1.9 Infiltration into the surface of the porous concrete parking lot at the Florida Aquarium in Tampa, Florida.

Where an "urban forest" lives, it replaces carbon dioxide in the air with oxygen and improves air quality by removing sulfur dioxide, nitrogen dioxide, carbon monoxide, ozone, and particulate matter (American Forests, 1999; Nowak et al., 2002; Robinette, 1972; Urban, 2000). Trees cool the air by shading and transpiring; the cooling may further reduce the air pollution from parked vehicles in parking lots (Scott et al., 1999; Greg McPherson, personal communication, 2003). Trees reduce glare, and attenuate noise. They house natural birds and insects. To a city they add color and gentle movement, and symbolize the presence of nature. Their arrangements frame vistas, screen objectionable views, and define spatial units in "outdoor architecture" (Arnold, 1993). They enhance worker productivity, reduce stress, attract

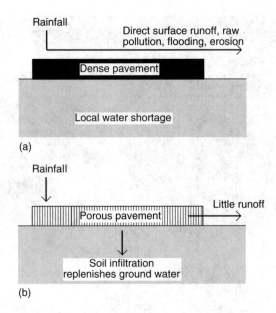

FIGURE 1.10 Contrasting hydrologic effects of dense (impervious) and fully permeable porous pavements.

customers to commercial districts, and add economic value to property (Wolf, 2003). The benefits for which trees are planted are fully achieved only when trees grow to full size and live long lives. American Forests' *City Green* software evaluates the effects of tree cover in individual urban developments and districts. Updates on the environmental effects of urban trees are available from the U.S. Department of Agriculture's Center for Urban Forest Research (http://cufr.ucdavis.edu).

Tree survival and growth require a large rooting zone with free exchange of air, water, and nutrients. The zone is ordinarily within 24 to 36 inches of the surface. Tree roots grow by tentatively exploring in all directions with numerous slender absorbing roots and extending in the directions where they find oxygen and moisture most abundant (MacDonald et al., 1993).

Tree-planting pits only a few feet wide surrounded by impervious pavements and compacted soil provide too little volume of aerated, penetrable soil for roots to grow as trees require. Roots that do penetrate beyond the pit into the soil below an impervious pavement quickly exhaust the soil's air because there is no exchange with the atmosphere; in anoxic conditions the roots fail to function and die. As a tree's root system fills the pit's rooting space to capacity, the growth of the crown slows and the tree becomes small in stature for its age. But as the crown continues to grow slowly, it becomes large in proportion to the confined root system that supplies water to it. With a small rooting volume supplying water to the leaves and branches, the tree becomes drought-stressed and increasingly susceptible to disease and insect infestations. Confined root space ultimately limits the size and lifetime of the tree (Watson and Himelick, 1997, pp. 10, 43–44).

Why Make Pavements Porous?   13

FIGURE 1.11 Dwarfed, declining, and dead sugar maples planted in a densely paved parking lot, seven years after planting.

For a tree in a root space that is only marginally constricted, frequent watering may for a time compensate for the soil's small native moisture reservoir (Watson and Himelick, 1997, pp. 12). Selection of tree species relatively adapted to constricted and compacted soil can further assist tree health and longevity within the ultimate constraint of rooting space. References such as those of Arnold (1993), Dirr (1998), Hightshoe (1988), Trowbridge and Bassuk (2004), Wyman (1965), Watson and Himelick (1997, p. 19–26), and Zion (1968) evaluate numerous tree species for tolerance to conditions such as these, as well as for the heat and air pollution that are likely to be present in urban districts.

To aerate the soil, a layer of porous aggregate has sometimes been placed under a dense paved surface; if the aggregate is exposed to the air at intervals, then air might move laterally from the uncovered spots to areas under the pavement. Networks of perforated pipes are intended similarly to distribute air into the soil zone (Arnold, 1993, pp. 128–130). But these systems are crutches added to overcome the natural barrier of dense surface pavement.

A porous pavement is a complete and vital way to allow air and water into rooting media in densely built-up areas. It allows the exchange of air and moisture through the pavement surface similar to that in a healthy natural soil surface. The soil's moisture regime fluctuates like that in natural soils, with rapid wetting during rain or snowmelt, followed by evapotranspirative drying and re-aeration, while the continuous exchange of air with the atmosphere maintains high soil oxygen levels.

Under a porous pavement, it is possible today to construct load-bearing rooting media made of open-graded aggregate in which the networks of pore spaces are partly filled with soil for root growth and are partly open for the exchange of water and air (Watson and Himelick, 1997, p. 44). "Structural soils" like those that will be described in Chapter 5 combine stone aggregate for the structural support of load-bearing pavements and porous aerated soil for tree roots. Beneath a porous surface and a porous structural-soil base, the subgrade soil is an additional reservoir of potential rooting media. For all subsurface rooting media, porous pavement surfacing is essential for the continuous exchange of air and water.

Trees have thrived where they have been given viable rooting zones under porous pavement surfaces, growing to the full size for which they were intended. Figure 1.12 shows healthy trees rooted in a heavily used park called The Commons

FIGURE 1.12 Large, healthy honey locust trees rooted in "structural soil" beneath a porous aggregate pavement in the Metrotech Business Improvement District in Brooklyn, New York, 14 years after installation.

Why Make Pavements Porous?

in the Metrotech Business Improvement District in downtown Brooklyn, New York. The surface pavement is porous aggregate through which air and water penetrate. Below the surface aggregate layer is a structural soil combining porous aggregate and soil. The structural soil extends under the entire paved surface, making a large rooting zone for the trees while supporting the load of thousands of pedestrians per day. Porous pavements transform the "urban wasteland" into a thriving habitat for people and trees together.

THE PROMISE OF COOL CITIES

Built-up areas in the U.S. are typically 2 to 8°F higher than the surrounding countryside (Akbari et al., 1992, p. 16). Figure 1.13 shows an example at Woodfield Mall, a shopping center in Schaumburg, Illinois, on a cold, clear, windless evening in 1972 (Norwine, 1973). The large multistory building is surrounded by a dense asphalt parking lot big enough to hold 10,000 cars. In 1972 the mall was newly built; the area around the mall was still mostly farmland, only beginning its transition to a suburban commercial district. The contours of temperature show that the built-up area is 2 to 4°F warmer than the unpaved surroundings. The maximum temperature is near the center where both the pavement and the building contribute to the temperature effect. The effect is called the urban "heat-island" because on a map like that of Woodfield Mall, the built-up area appears as an island of warmth; on larger maps entire cities appear as islands in a sea of cooler rural temperatures.

The heat-island effect is greatest in the late afternoon and evening, and particularly in clear, calm weather. Over 90 percent of the increase in temperature is due to urban construction materials that absorb and store solar heat without evapotranspirative cooling; only the remaining 1 to 10 percent comes from the active emissions of vehicles, buildings, and factories (Rosenfeld et al., 1997). A solid structure

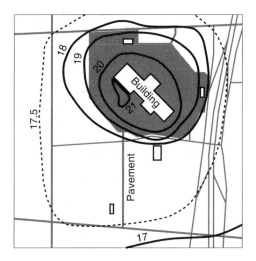

FIGURE 1.13 Contour map of temperature (°F) in the area of Woodfield Mall, Schaumburg, Illinois, on the evening of March 9, 1972 (after Norwine, 1973).

absorbs solar heat and conducts it into the depth of the material, making the structure into a thermal "storage battery." Late in the day, when the sun's heat is not so intense, solid construction materials re-emit their stored heat to the air, raising the urban air temperature even after the sun has set (Asaeda et al., 1996). Pavements contribute at least as much as buildings to heat-island formation because pavements have high thermal inertia at the ground surface (Goward, 1981).

Excess heat has a combination of advantages and disadvantages for cities. In many cities in temperate parts of the U.S., the heat island reduces the demand for winter heating by about 8 percent, as indicated by the decrease in "heating degree days," a measure of the climatic requirement for heating (Akbari et al., 1992, pp. 16–17; Landsberg, 1981, pp. 119–121). In cities with winter temperatures that hover near freezing, warmer temperatures reduce the frequency of snowfall and the necessity of snow removal.

However, in the same cities during the summer, the heat-island increases the climatic demand for air conditioning by about 12 percent (Landsberg, 1981, p. 120). The greater energy consumption needed in cities for cooling than for heating is magnified by air-conditioning technology, which requires more energy to produce a given amount of cooling than to produce an equivalent amount of heating. Three to eight percent of today's urban electric demand is used to compensate for the heat-island effect alone. Americans spend about one billion dollars per year for that extra energy (Akbari et al., 1992, p. 16). With more energy being used, power-plant generators run faster, polluting the atmosphere with increased carbon dioxide emissions.

Higher urban temperature aggravates air pollution in the city itself. Heat accelerates chemical reactions in the atmosphere that transform emissions from cars and smokestacks into ozone, an irritating gas that is the main ingredient of smog. For a 5°F increase in temperature, the number of ozone-polluted days increases by 10 percent (Akbari et al., 1992, p. 21; Rosenfeld et al., 1997).

City heat also produces a kind of water pollution. The runoff that drains off hot urban surfaces is correspondingly warm, raising the temperatures of nearby streams compared with those where the water has passed through cool porous soil. Figure 1.14 shows this effect in watersheds in Maryland. As stream temperature rises, the water's capacity to hold dissolved oxygen to support aquatic life declines.

For people outdoors, excessively high urban temperatures are associated with decreased comfort and are implicated in heat-related health problems including some deaths of persons with heart conditions (Huang, 1996; Landsberg, 1969, pp. 59–60).

The heat-island effect has a subtle influence on rainfall that could be considered either an advantage or a disadvantage. During summer thunderstorm conditions, city heat enhances convective rainfall downwind of city centers. Seasonal rainfall increases of 9 to 17 percent are possible (Changnon and Westcott, 2002; Changnon et al., 1991; Huff and Changnon, 1973; Landsberg, 1970).

Limiting the amount of pavement to serve a given unit of urban development would limit the opportunity for the heat-island effect to occur. Choices in patterns of development that influence the amount of pavement were discussed earlier in this chapter. Consideration of the heat-island effect complicates the choice of development pattern because a dense conurbation with canyon-like complexes of buildings and streets tends to absorb and store more solar energy than does an isolated complex like Woodfield Mall as it existed in 1972 (Goward, 1981).

Why Make Pavements Porous?

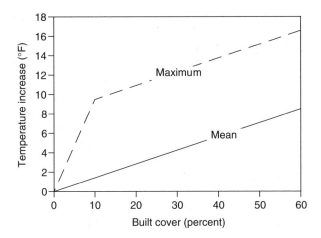

FIGURE 1.14 Increase in urban stream temperature over background temperature of rural streams, in the Maryland Piedmont (after Schueler, 1994).

For a given amount of construction, the use of light-colored construction materials may reduce the buildup of urban heat (Rosenfeld et al., 1997). Light-colored materials such as concrete absorb less solar heat than do dark-colored materials such as asphalt.

Shading by canopy trees is a powerful and certain way to limit the heat-island effect (Akbari et al., 1992, p. 21). Tree canopies intercept solar heat before it enters any "storage battery" on the ground and actively cool themselves with evapotranspiration. As described earlier in this chapter, in densely built-up places porous pavements are a prerequisite for the growing of large, long-lived shade trees for this purpose.

Porous "grass pavements" actively cool the ground surface with their natural evapotranspiration. This was demonstrated in Japan, where Asaeda and Ca (2000) monitored the surface temperature of grass on a warm sunny day in August. At noon, the grass surface was 18°F cooler than a nearby dense asphalt surface; at 6:00 pm it was still 14° cooler; at midnight it was 9° cooler. The grass was cooler even at depths of several feet below the surface. Porous pavements with grass components — whether grass alone or grass reinforced by geocells or concrete grids — are eligible for selective use in areas with infrequent traffic. The eligible areas are small and scattered, but together all the fragments can add up to a significant portion of an urban district. In some areas the maintenance of living grass would be inhibited by a requirement of water for irrigation.

In Asaeda and Ca's study, the 42 percent porosity of the soil in which the grass was growing may have added a small insulating effect, suppressing the material's storage battery effect. However, the cooling effect of the grass was due mostly or entirely to evapotranspiration of water; in the same study, a nonliving porous pavement material did not have the same cooling effect. The researchers simultaneously monitored a concrete block's surface with 30 percent porosity, and found its surface temperature to be essentially identical to that of dense asphalt all day long.

The thermal similarity of porous concrete and dense asphalt was a surprising result of the Japanese study because of the concrete's light color and high porosity.

Porous materials have less thermal conductivity and thermal capacity than corresponding dense materials (ASHRAE, 1993, pp. 22.6–22.9; Malhotra, 1976, p. Table 13; CRC Press, 2000, pp. 12–204; Geiger, 1965, pp. 29, 145–146; Livet, 1994, cited in Huber, 2000, p. 24), so they ought to conduct daytime heat downward and hold it in an internal storage battery less effectively than dense materials. Perhaps on the clear sunny day of the Japanese study, when radiation was the dominant means of heat transfer, the dark-colored asphalt was able to radiate its accumulated heat outward at a rate proportional to its absorption of incoming solar radiation, ending up with the same net temperature as that of the concrete. In these conditions the insulating effect of porous concrete's air-filled pores might have had no significance. Or perhaps the advection (movement of air) through the pores of a porous material counteracts its low thermal conductivity and capacity: in one day a sandy soil can "breathe" through its surface a volume of air equal to a column 70 feet high, transferring heat between surface and subsurface (Geiger, 1965, p. 27).

Slightly different results were observed in Ontario (James and Thompson, 1996), where during clear days the surface of a porous pavement of open-jointed concrete blocks with aggregate joint fill was cooler than that of a nearby dense asphalt pavement, and at night it was warmer; on average the temperature was the same. The researchers attributed the difference in temperature between the materials to the difference in color (albedo). Daytime rain cooled the dense asphalt surface markedly, but had little influence on the porous concrete–aggregate surface.

Research comparing corresponding porous and nonporous pavement materials, for example, porous concrete and dense concrete, is called for. Research to confirm the Japanese result and extend it into other types of weather conditions is needed. Table 1.2 lists examples of web sites where information on urban heat islands may be updated in the future.

THE PROMISE OF QUIET STREETS

Traffic noise is objectionable where residential areas adjoin highways and busy streets. Most people consider traffic noise problematic within 100 or 200 feet of moderately traveled roads and 500 feet of heavily trafficked freeways (United States Federal Highway Administration, 1980). The noise of a moving vehicle originates in the engine exhaust, the flexing of rolling tires, the rumbling of tires that pass over a rough pavement surface, and the splashing of tires on a wet surface. Engine exhaust noise is reduced by vehicular provisions such as mufflers; the other noise factors depend at least partly on the pavement.

TABLE 1.2
Agencies That May Update Information on Urban Heat Islands

Agency	Contact Information
National Aeronautics and Space Administration	http://science.msfc.nasa.gov/; at that address use the "Search" command to find "heat island"
Lawrence Berkeley Laboratory	http://Eetd.LBL.gov/HeatIsland/

Why Make Pavements Porous?

The bel (B) is a unit for expressing the intensity of sound energy (Webster, 2000). In application the units are recorded in decibels (dB); one decibel is one tenth of a bel. A decibel compares the intensity of a sound to that of a reference sound on a logarithmic scale (Truax, 1999). The internationally agreed-upon reference is the threshold of human hearing, which is assigned a value of 0 dB. One decibel is approximately equal to the smallest difference in sound energy detectable by the human ear. The scale extends to the loudest sound the human ear can tolerate without pain at about 120 to 140 dB.

To the human ear, the subjective impression of loudness is modified by a sound's frequency or "pitch" (Truax, 1999). The ear perceives a sound with high pitch as having greater loudness than a sound of objectively similar intensity but lower pitch. For a measure that simulates the overall impression of loudness perceived by the ear, the objective sound intensity (dB) is weighted according to frequency, and assigned the symbol dBA. Figure 1.15 shows examples of dBA for some common sounds.

Other scales of noise have been developed to take additional variables into account, such as the Traffic Noise Index developed in Britain, which takes into account both the peak noise levels and the general ambient noise level over a 24-hour period. Another is the Community Noise Equivalent Level, developed in California, which weights noises according to social factors such as time of day (on the assumption that evening noises are most annoying), season, type of residential area where the noise is heard, and previous community experience with similar noises.

One way to reduce the traffic noise that reaches sensitive communities is the construction of noise barriers in the form of earth mounds or masonry walls. Properly constructed barriers can reduce noise by 10 to 15 dB. Because the decibel scale is logarithmic, a reduction of 10 dB amounts to cutting the loudness in half

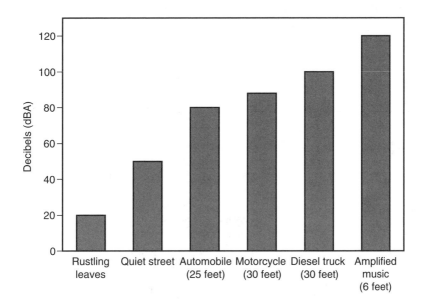

FIGURE 1.15 Typical average noise levels for some common sounds (data from Truax, 1999).

(United States Federal Highway Administration, 1980). However, walls and mounds require space and funds for construction and are appropriate and feasible only along certain stretches of freeways.

Porous pavements reduce traffic noise at the source, particularly the noise from tires. A porous surface both absorbs sound energy and allows some of the air around tires to be pressed into the voids, dissipating air pressure before any noise is generated. Noise reduction is particularly effective for high-frequency (high-pitched) sounds which are perceived relatively loudly. This means that the tire noise from a porous pavement is both lower in loudness and lower in pitch than that from a corresponding dense pavement. Recently installed porous asphalt reduces noise compared with dense asphalt by 3 dBA or more (Huber, 2000, pp. 6–7 and 9; Kuennen, 1996; Bendtsen and Larsen, 1999).

The intensity of traffic noise tends to rise in wet conditions because the tires force water noisily across the pavement surface (Shackel and Pearson, 1997). At these times porous pavements have an additional advantage over nonporous pavements because the surface of porous pavements is better drained in wet weather and any puddled water is squeezed through the pores as much as across the noisy surface.

THE PROMISE OF SAFE DRIVING

In wet weather, driving is difficult and dangerous where the pavement is slippery. The wheels separate from the pavement with hydroplaning, sheets of water obscure pavement markings, and moving vehicles throw up curtains of blinding mist.

Table 1.3 lists the types of street settings where pavement skid resistance is most important to safety, based on tests of skid resistance in places where accidents were reported in Britain. In the "most critical" category of sites are those urban streets where vehicles turn rapidly around sharp corners or need to stop suddenly as signals change and traffic backs up.

TABLE 1.3
Relative Importance of Skid Resistance to Driving Safety in Various Urban Settings

Category of Street	Examples
Most critical sites (pavement skid resistance is most critical)	Roundabouts (traffic circles)
	Streets with sharp bends (radius less than 500 feet)
	Steep gradients of greater than 5 percent, or longer than 300 feet
	Approaches to traffic signals
	Approaches to pedestrian crossings
Intermediate sites	Freeways and other roads designed for high speeds
	Urban streets with high traffic volume
	Other principal roads
Other sites (pavement skid resistance is of only ordinary importance)	Straight roads with low gradients
	Curves without intersections
	Streets with passenger-car traffic only

Sabey, 1968, cited in Croney and Croney, 1998, pp. 470–471 and 483.

Why Make Pavements Porous?

The resistance of almost any pavement surface to skidding of vehicles comes mostly from the numerous small edges of aggregate particles in the pavement material (Croney and Croney, 1998, pp. 471–478). Both dense pavements and corresponding porous pavements possess that kind of friction when they are dry. However, in wet weather a film of water over a dense pavement's surface inhibits the firm contact of tires with the surface. A dense pavement can be particularly slippery in the first minutes of a rain event (Croney and Croney, 1998, pp. 474–475) because the lubricants that vehicles drop onto pavement surfaces in the days or weeks before the rain combine with water in a sheet of water and oil that simultaneously relaxes the tires' contact with the pavement and adds a layer of lubricant (the lubricants are the same petroleum products that show up as "first-flush" pollutants when surface runoff carries them into streams).

Porous pavements remove water and oil from the surface directly downward through their pores, preventing surface accumulation. The same pores are pressure-relief channels where any ponded water escapes from beneath vehicle tires, keeping the tires in contact with the surface (Diniz, 1980, p. 5). Figure 1.16 shows the resulting contrast in friction between porous and dense materials. When wet, the friction coefficient of dense asphalt collapses to only one fourth of its dry-weather value, while porous asphalt retains its dry-weather friction value.

In wet conditions porous pavements also improve driving visibility. With no layer of water over the pavement, vehicles do not kick up plumes of mist from their wheels. The pavement itself is more visible because of the absence of puddled water. At night, well-drained pavements produce little glare from vehicle lights.

For these reasons the state highway departments in Georgia, Oregon, and other states place porous asphalt overlays on their major highways. In wet weather the porous surfaces improve driving comfort and reduce accidents. In effect, by improving traffic flow, they increase the capacity of highways without the expense of widening the roads.

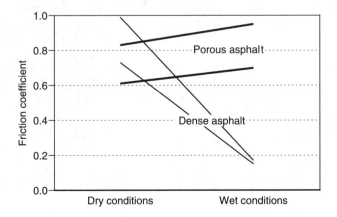

FIGURE 1.16 Friction coefficients of pavement surface materials (data from Diniz, 1980, p. 27).

THE PROMISE OF REDUCING COST

The selection of porous pavements in place of dense ones has directly reduced construction costs in some developments. In most regions of North America, porous aggregate used alone without binding or reinforcement is the least expensive of all paving materials, including conventional dense asphalt. Although the low-traffic places where aggregate can be used are individually small and isolated, they occur in pockets throughout urban districts, and together represent a large area of pavement. Where aggregate is used in place of other paving materials, the construction cost per square yard of pavement is immediately reduced.

Where other porous materials are used, they are more expensive, but they are still capable of reducing construction cost because they perform necessary stormwater functions that would otherwise have to be accomplished by additional pipes and reservoirs. Because porous pavements absorb, store, and treat water within the pavement structure, they reduce or eliminate the need for drainage inlets, storm drainage pipes, and stormwater detention areas. A porous pavement with little or no drainage structures is commonly less expensive than a dense pavement with the large drainage and treatment systems it requires.

Even more expensive porous paving materials can reduce the total development cost by avoiding the larger cost of land acquisition for off-pavement stormwater management facilities. Figure 1.17 shows an example where a shopping center's porous asphalt parking lot protects stream water quality. The stream is visible just beyond the curb in the background; the shopping center had no additional land for single-purpose stormwater facilities. The porous asphalt absorbs rainwater and

FIGURE 1.17 Porous asphalt parking lot protecting stream water quality at Exton Square Mall in Exton, Pennsylvania.

biodegrades automotive pollutants within the pavement structure; only rarely does excess stormwater overflow through the surface grate. In valuable, densely developed locations like this, the selection of porous pavement makes sustainable development economically feasible.

Porous pavements have also helped reduce the long-term costs of taxes and fees required from urban properties. Many municipal stormwater "utilities" and stormwater management departments impose taxes or fees based on impervious coverage (http://stormwaterfinance.urbancenter.iupui.edu). Converting pavements to porous, pervious materials reduces the basis for the tax or fee. Some other agencies reduce the fees for properties where runoff controls such as porous pavements are installed.

THE PROMISE OF MEETING DEVELOPMENT REGULATIONS

Municipal jurisdictions impose requirements on new developments for their effects on stormwater, tree preservation, and impervious coverage, all of which can be partly or wholly satisfied by the selective and appropriate use of porous pavements. Where developers do not perceive a direct interest for themselves in the effects of porous pavements, they still have a vital interest in obtaining permission to build. Information on methods of development regulation is available from the American Planning Association (www.planning.org).

Governments regulate stormwater quality and quantity during small "first-flush" storms and large flood-hazard storms. A New York State law specifies soil infiltration as the preferred approach to stormwater control and "pervious surfaces" as one of the specific techniques that would fulfill the infiltration goal. The town of Hilton Head Island, South Carolina, requires that 1 inch of runoff from all impervious surfaces be dissipated by percolation into the soil, and expects routinely that part of that requirement will be met by porous pavements. In Washington State, the Puget Sound Water Quality Management Plan aims for developments to make no net detrimental change in natural surface runoff and infiltration and requires municipalities to adopt stormwater ordinances that make infiltration "the first consideration in stormwater management." Porous pavements assist in meeting these regulations by reducing and detaining runoff, increasing infiltration, and treating water quality.

Some municipalities require preservation of trees over a certain size or planting of a certain quantity of new trees as part of new developments. In Savannah, Georgia, this type of requirement has been a major motivation for constructing porous pavements of blocks, grids, geocells, and porous concrete. Figure 1.18 shows a grid pavement at a Savannah restaurant, satisfying the city's requirement for tree preservation while letting commercial activities move forward.

Some jurisdictions directly limit impervious cover to protect water quality and the environment in general. Figure 1.19 shows a plastic geocell grid being installed in a parking lot at a fraternity in Athens, Georgia. The geocell is to reinforce porous aggregate which will be placed in the grid's cells. When the fraternity moved into its new building, a local ordinance required it either to reduce its impervious cover or to control the excess runoff from impervious surfaces. There was no room for stormwater control basins on this small in-town lot. So the fraternity made its parking lot pervious by surfacing it with geocell-reinforced aggregate. This provision

FIGURE 1.18 Open-celled concrete grids filled with porous aggregate preserving live-oak tree roots at a restaurant in Savannah, Georgia.

satisfied the city's requirements and permitted the new use to go forward, while providing the number of parking spaces required for the property's new use.

THE PROMISE OF PRESERVING NATIVE ECOSYSTEMS

Figure 1.20 shows a boardwalk crossing a fragile marsh ecosystem. A boardwalk is a porous surrogate for a pavement. It is supported by footings that touch the soil only at discrete points or lines. Because the decking surface is isolated from the ground, it preserves the dispersed flows of surface and subsurface waters. Water percolates through the decking without concentration and infiltrates native soil below. The continuous flows of water and sediment build ecological equilibrium, rebuild it with

FIGURE 1.19 Plastic geocell being installed to reinforce a porous aggregate parking lot at an office building in Athens, Georgia.

changing circumstances, and maintain reservoirs of soil, water, and propagules. During the construction of a deck the only necessary disturbance of soil and roots is that for the discrete footings. A finished deck floats suspended through a functioning ecosystem.

Figure 1.21 shows a porous concrete road meandering through a pine-forest preserve. Rainwater that falls on the pavement infiltrates the underlying sandy soil as it did before the road was built. An on-the-ground porous pavement like this is one degree more intrusive on an ecosystem than a boardwalk, but in return it carries the weight of cars. For the road in the picture there are no curbs, no gutters, no drainage inlets, not even any drainage swales: all drainage is immediately downward through the pavement to the soil, wherever the rain falls, as it is through the adjacent forest

FIGURE 1.20 Boardwalk over wetland preserve at Huntley Meadows Park near Alexandria, Virginia.

floor. The trees have a continuous rooting medium with a natural regime of moisture and aeration. Far below, excess soil moisture replenishes a natural limestone aquifer whence it discharges as the base flow of streams and the habitat of the preserve's aquatic organisms.

THE PROMISE OF BEAUTY

The selection of pavement material sets the stage for the character of an urban place. Figures 1.22 and 1.23 illustrate pavements of porous grass that "soften" the character of the paved areas, making them "green" and consonant with residential communities and relaxed pedestrian activity.

Aesthetics integrates values in which symbolism and functional information are as important as neatness and attractiveness (Nassauer, 1995). Design is capable of revealing and integrating. It can embed the solution to environmental problems in land use, transportation, and the urban way of life. The characteristics of a place can make the processes through which hydrologic and ecological restoration take place visible and comprehensible. What a system looks like, how it functions ecologically and socially, and what it symbolizes in the way of stewardship can be congruent. One can lay hands on the details of construction materials to bring restorative processes to every inch of an inhabited place. Permeable materials are visibly distinctive with their open voids and, in some cases, their living vegetation. Wherever we go in cities, they can make us conscious of the careful return of water and air to the soil.

Why Make Pavements Porous?

FIGURE 1.21 Porous concrete road amid native longleaf pine and wiregrass at Jones Ecological Research Center near Newton, Georgia.

Some observers believe that the American landscape is a battleground between opposing values: those embodied by the machine (control, architecture, technology, human dominance) and those represented by wild nature (Marx, 1964). Urban pavements are a locus in the struggle to resolve this opposition of values: wherever land is paved, technological materials displace vegetated soil to exert force and impose character on urban environments. One type of resolution of this conflict is the pastoral landscape (Marx, 1964), in which machine-like urban features are dispersed through nature in a park-like, garden-like "middle landscape." This compromise, which avoids the excesses of both sides, has been criticized as losing the full and best aspects of both (MacElroy and Winterbottom, 1997): dispersed, fragmented, garden-like landscapes are neither intensely used by people, nor rioting in natural regeneration.

FIGURE 1.22 Emergency access lane of grass reinforced by concrete grids at the Saddleback Church in Lake Forest, California.

Hough (1995) proposed another type of resolution: "living machines" that represent symbioses between technology and natural process. "Grass pavements" and all other porous pavements are examples of living machines. They are functional components of cities that support the natural processes of the environments they change; they merge environmental process with urban infrastructure. Percolation through pavements every time rain falls makes the natural process visible to the people who live and work in cities. As porous pavements restore natural functions, they also restore the perceptual connection between environment and society. Unlike stormwater detention basins and treatment wetlands which are added to development sites without changing the developments themselves, porous pavements are under people's feet all the time. Porous pavements make no distinction between the quality of the

FIGURE 1.23 Parking of grass reinforced by plastic geocells at the Orange Bowl Stadium in Miami, Florida, as seen from an adjacent residential street.

environment and the quality of the human life. They give everyone who uses them a role in restorative environmental processes.

ACKNOWLEDGMENTS

The following persons generously provided information used in this chapter: Alan Lell of the U.S. Army Corps of Engineers; Joe Martin of the U.S. National Park Service; E. Gregory McPherson of the U.S. Forest Service (http://cufr.ucdavis.edu); Richard Pinkham of the Rocky Mountain Institute (www.rmi.org); Klaus Scott of the California Air Resources Board; and William Siler, a former landscape architectural student at the University of Georgia now in practice in North Carolina.

The following persons constructively critiqued early drafts of this chapter: Lucia Athens of the City of Seattle Green Building Team (www.cityofseattle.net/sustainablebuilding/), E. Gregory McPherson of the U.S. Forest Service, Richard Pinkham of the Rocky Mountain Institute, and Neil Weinstein of the Low Impact Development Center (www.lowimpactdevelopment.org).

REFERENCES

Akbari, Hashem, Susan Davis, Sofia Dorsano, Joe Huang, and Steven Winnett, Eds. (1992). *Cooling Our Communities, A Guidebook on Tree Planting and Light Colored Surfacing*, Washington: U.S. Environmental Protection Agency.

Albanese, Brett, and Glenn Matlack (1999). Utilization of Parking Lots in Hattiesburg, Mississippi, USA, and Impacts on Local Streams, *Environmental Management* 24, 265–271.
American Forests (1999). *Urban Ecosystem Analysis, The District of Columbia*, Washington: American Forests.
Arnold, Chester L. and C. James Gibbons (1996). Impervious Surface Coverage, the Emergence of a Key Environmental Indicator, *Journal of the American Planning Association* 62, 247–258.
Arnold, Henry F. (1993). *Trees in Urban Design*, 2nd ed., New York: Van Nostrand Reinhold.
Asaeda, Takashi and Vu Thanh Ca (2000). Characteristics of Permeable Pavement During Hot Summer Weather and Impact on the Thermal Environment, *Building and Environment* 35, 363–375.
Asaeda, Takashi, Vu Than Ca, and A. Wake (1996). Heat Storage of Pavement and its Effect on the Lower Atmosphere, *Atmospheric Environment* 30, 413–427.
ASHRAE (1993). *Fundamentals, 1993 ASHRAE Handbook*, Atlanta: American Society of Heating, Refrigerating, and Air Conditioning Engineers.
Bendtsen H. (1997). Noise Reduction by Drainage Asphalt, *Nordic Road & Transport Research* 1997, No. 1, 6–8.
Bendtsen, H., and L. Larsen (1999). Noise-Reducing Pavements for Urban Roads, in *Nordic Road & Transport Research* 1999, No. 3 (www.vti.se/Nordic).
Cappiella, Karen and Kenneth Brown (2001). *Impervious Cover and Land Use in the Chesapeake Bay Watershed*, Ellicott City, MD: Center for Watershed Protection.
Center for Watershed Protection (1998). *Consensus Agreement on Model Development Principles to Protect our Streams, Lakes, and Wetlands*, Ellicott City, MD: Center for Watershed Protection.
Changnon, Stanley A. and Nancy E. Westcott (2002). Heavy Rainstorms in Chicago: Increasing Frequency, Altered Impacts, and Future Implications, *Journal of the American Water Resources Association* 38, 1467–1475.
Changnon, Stanley A., Robin T. Shealy, and Robert W. Scott (1991). Precipitation Changes in Fall, Winter and Spring Caused by St. Louis, *Journal of Applied Meteorology* 30, 126–134.
CRC Press (2000). *CRC Handbook of Chemistry and Physics*, 81st ed., Cleveland: CRC Press.
Croney, David, and Paul Croney (1998). *Design and Performance of Road Pavements*, 3rd ed., New York: McGraw-Hill.
Day, Susan D. and Nina L. Bassuk (1994). A Review of the Effects of Soil Compaction and Amelioration Treatments on Landscape Trees, *Journal of Arboriculture* 20, 9–17.
Diniz, Elvidio V. (1980). *Porous Pavement, Phase 1 — Design and Operational Criteria*, EPA-600/2-80-135, Cincinnati: U.S. Environmental Protection Agency Municipal Environmental Research Laboratory; downloadable at www.epa.gov/ednnrmrl/repository/abstrac2/abstra2.htm.
Dirr, Michael (1998). *Manual of Woody Landscape Plants: Their Identification, Ornamental Characteristics, Culture, Propagation and Uses*, 5th ed., Champaign, IL: Stipes Publishing.
Ferguson, Bruce K. (1998). *Introduction to Stormwater: Concept, Purpose, Design*, New York: John Wiley and Sons.
Ferguson, Bruce K. (2002). Stormwater Management and Stormwater Restoration, in *Handbook of Water-Sensitive Planning and Design*, Robert France, Ed., Boca Raton: CRC Press.
Geiger, Rudolf (1965). *The Climate Near the Ground*, Cambridge: Harvard University Press.
Girling, Cynthia L. and Ronald Kellett (2001). Comparing Stormwater Impacts and Costs on Three Neighborhood Plan Types, *Landscape Journal* 21, 100–109.

Goward, Samuel N. (1981). Thermal Behavior of Urban Landscapes and the Urban Heat Island, *Physical Geography* 2, 19–33.

Grabosky, Jason and Edward Gilman (2004). Measurement and Prediction of Tree Growth Reduction from Tree Planting Space Design in Established Parking Lots, *Journal of Arboriculture* 30, 154–164.

Hightshoe, Gary L. (1988). *Native Trees, Shrubs and Vines for Urban and Rural America: A Planting Design Manual for Environmental Designers*, New York: Van Nostrand Reinhold.

Hough, Michael (1995). *Cities and Natural Process*, New York: Routledge.

Huang, Joe (1996). Urban Heat Catastrophes: The Summer 1995 Chicago Heat Wave, *Center for Building Science News* (Lawrence Berkeley Laboratory) No. 12, p. 5.

Huber, Gerald (2000). *Performance Survey on Open-Graded Friction Course Mixes*, National Cooperative Highway Research Program Synthesis of Highway Practice 284, Washington: Transportation Research Board.

Huff, F.A., and S.A. Changnon (1973). Precipitation Modification by Major Urban Areas, *Bulletin American Meteorological Society* 54, 1220–1232.

James, William, and Michael K. Thompson (1996). Contaminants from Four New Pervious and Impervious Pavements in a Parking Lot, in *Advances in Modeling the Management of Stormwater Impacts*, Vol. 5, Guelph, Ontario: Computational Hydraulics International.

Kopinga, Jitze (1994). Aspects of the Damage to Asphalt Road Pavings Caused by Tree Roots, in *The Landscape Below Ground: Proceedings of an International Workshop on Tree Root Development in Urban Soils*, Gary W. Watson and Dan Neely, Eds., pp. 165–178, Savoy, IL: International Society of Arboriculture.

Kuennen, Tom (1996). Open-Graded Mixes: Better the Second Time Around, *American City & County* August 1996, pp. 40–53.

Landsberg, Helmut E. (1969). *Weather and Health, An Introduction to Biometeorology*, New York: Doubleday.

Landsberg, Helmut E. (1970). Man-Made Climatic Changes, *Science* 170, 1265–1274.

Landsberg, Helmut E. (1981). *The Urban Climate*, New York: Academic Press.

Livet, J. (1994). The Specific Winter Behavior of Porous Asphalt: The Situation as it Stands in France, in *Proceedings of the 9th PIARC International Winter Road Congress*, Seefield, Austria, March 1994.

MacDonald, J.D., L.R. Costello, and T. Berger (1993). An Evaluation of Soil Aeration Status around Healthy and Declining Oaks in an Urban Environment in California, *Journal of Arboriculture* 19, 209–219.

MacElroy, William P., and Daniel M. Winterbottom (1997). Toward a New Garden: A Model for an Emerging Twenty-First Century Middle Landscape, in *Critiques of Built Works of Landscape Architecture*, Vol. 4, pp. 10–14, Baton Rouge: Louisiana State University School of Landscape Architecture.

Malhotra, V.M. (1976). No-Fines Concrete — Its Properties and Applications, *Journal of the American Concrete Institute* November 1976, pp. 628–644.

Marx, Leo, (1964), *The Machine in the Garden: Technology and the Pastoral Ideal in America*, New York: Oxford University Press.

McPherson, E. Gregory (1994). Benefits and Costs of Tree Planting and Care in Chicago, in *Chicago's Forest Ecosystem: Results of the Chicago Urban Forest Climate Project*, pp. 115–128, E. Gregory McPherson, David J. Nowak, and Rowan A. Rowntree, Eds., General Technical Report NE-186, Radnor, Pennsylvania: U.S. Forest Service Northeastern Forest Experiement Station.

Moll, Gary (1989). The State of Our Urban Forest, *American Forests* 95, 61–64.

Nassauer, Joan Iverson (1995). Messy Ecosystems, Orderly Frames, *Landscape Journal* 14, 161–170.
Norwine, James R. (1973). Heat Island Properties of an Enclosed Multi-Level Suburban Shopping Center, *Bulletin of the American Meteorological Society* 54, 637–641.
Nowak, David J., Jack C. Stevens, Susan M. Sisinni, and Christopher J. Luley (2002). Effects of Urban Tree Management and Species Selection on Atmospheric Carbon Dioxide, *Journal of Arboriculture* 28, 113–120.
Otto, Betsy, Katherine Ransel, Jason Todd, Deron Lovaas, Hannah Stutzman, and John Bailey (2002). *Paving Our Way to Water Shortages: How Sprawl Aggravates Drought*, Washington: American Rivers, Natural Resources Defense Council, and Smart Growth America; downloadable from www.americanrivers.org.
Quigley, Martin F. (2004). Street Trees and Conspecifics: Will Long-Live Trees Reach Full Size in Urban Conditions? *Urban Ecosystems* 7, 29–39.
Richman, Tom and Associates (1997). *Start at the Source, Residential Site Planning and Design Guidance Manual for Stormwater Quality Protection*, Oakland, CA: Bay Area Stormwater Management Agencies Association.
Robinette, Gary O. (1972). *Plants, People, and Environmental Quality*, Washington: U.S. National Park Service.
Rosenfeld, A.H., H. Akbari, J.J. Romm, and M. Pomerantz (1998). Cool Communities: Strategies for Heat Island Mitigation and Smog Reduction, *Energy and Buildings* 28, 51–62.
Rosenfeld, Arthur H., Joseph J. Romm, Hasem Akbari, and Alan C. Lloyd (1997). Painting the Town White — and Green, *Technology Review* 100, 52–59.
Sabey, B.E. (1968). The Road Surface and Safety of Vehicles, in *Symposium on Vehicle and Road Design for Safety*, Cranfield 3–4 July 1968, London: Institution of Mechanical Engineers.
Schueler, T. (1994). The importance of imperviousness, *Watershed Protection Techniques* 1, 100–111.
Scott, Klaus I., James R. Simpson, and E. Gregory McPherson (1999). Effects of Tree Cover on Parking Lot Microclimate and Vehicle Emissions, *Journal of Arboriculture* 25, 129–142.
Shackel, Brian, and Alan R. Pearson (1997). Concrete Segmental Pavements, *Constructional Review*, Vol. 70 February 42–47.
Stankowski, Stephen J. (1972). *Population Density as an Indirect Indicator of Urban and Suburban Land-Surface Modification*, Professional Paper 800-B, pp. B219–B224, Washington: U.S. Geological Survey.
Thayer, Robert L. (1994). *Gray World, Green Heart: Technology, Nature, and the Sustainable Landscape*, New York: John Wiley and Sons.
Trowbridge, Peter J. and Nina L. Bassuk (2004). *Trees in the Urban Landscape: Site Assessment, Design, and Installation*, New York: John Wiley and Sons.
Truax, Barry, Ed. (1999). *Handbook for Acoustic Ecology*, 2nd ed., Cambridge Street Publishing (www.sfu.ca/sonic-studio/index.html).
United States Census Bureau (no date). *Statistical Abstract of the United States*, downloadable from www.census.gov/prod/www/statistical-abstract-us.html.
United States Federal Highway Administration (1980). *Highway Traffic Noise*, Washington: U.S. Federal Highway Administration.
University of Georgia School of Environmental Design (1997). *Land Development Provisions to Protect Georgia Water Quality*, Atlanta: Georgia Department of Natural Resources, Environmental Protection Division; downloadable from the *Technical Guidance — Water Quality Engineering & Design* page at www.dnr.state.ga.us/dnr/environ/.
Urban, James (1996). Room to Grow, *Landscape Architecture* 86, 74–79, 96–97.

Urban, James (2000). Environmental Effects of Trees, in *Architectural Graphic Standards*, 10th ed., p. 177, John Ray Hoke, editor in chief, New York: John Wiley and Sons.

Wagar, J. Alan and Albert L. Franklin (1994). Sidewalk Effects on Soil Moisture and Temperature, *Journal of Arboriculture* 20, 237–238.

Watson, Gary W. and E. B. Himelick (1997). *Principles and Practice of Planting Trees and Shrubs*, Savoy, IL: International Society of Arboriculture.

Webster Len F. (2000). *The Wiley Dictionary of Civil Engineering and Construction*, New York: John Wiley & Sons.

Wells, Cedar (1994). *Impervious Surface Reduction Study*, Olympia, Washington: City of Olympia Public Works Department.

Willson, Richard S. (1995). Suburban Parking Requirements and the Shaping of Suburbia, A Tacit Policy for Automobile Use and Sprawl, *Journal of the American Planning Association* 61, 29–42.

Wolf, Kathleen L. (2003). Public Response to the Urban Forest in Inner-City Business Districts, *Journal of Arboriculture* 29 (http://joa.isa-arbor.com).

Wyman, Donald (1965). *Trees for American Gardens*, revised and enlarged edition, New York: Macmillan.

Zion, Robert (1968). *Trees for Architecture and the Landscape*, New York: Reinhold.

2 Dimensions of Porous Pavement Installations

CONTENTS

Porous Pavement Components 36
    Surface and Base Courses 36
    Overlay 37
    Reservoir 38
    Lateral Outlet... 40
    Filter Layers... 42
    Liners 42
Porous Paving Materials 42
    Porous Aggregate... 42
    Porous Turf 43
    Plastic Geocells 45
    Open-Jointed Paving Blocks 45
    Open-Celled Paving Grids... 47
    Porous Concrete... 47
    Porous Asphalt... 48
    Soft Paving Materials 49
    Decks 51
Examples of Selective Application 51
    Office Parking, Medford Village, New Jersey....................................... 52
    Cardinal Ridge, Medford, New Jersey 52
    Pier A Park, Hoboken, New Jersey 55
    Mitchell Center Arena, Mobile, Alabama 57
Provisions for All Porous Pavement Applications.. 58
    Selective Application 58
    Drainage at Pavement Edges 59
    Protection of Pavement during Further Construction 61
    Appropriate Specification 63
    Contractor Qualifications and Communications 65
    Appropriate Maintenance and Signage 65
Trademarks 66
Acknowledgments 67

References 67

Porous pavements are of various types. Their materials and configurations must be selected to suit the context and requirements of each specific project. This chapter introduces the major alternatives in pavement components and materials, and identifies provisions that should be included in almost any porous pavement application.

POROUS PAVEMENT COMPONENTS

Pavements, porous and dense alike, are assembled from several types of components. Table 2.1 defines some potential types of components. Few pavements contain all of the listed components; instead, each pavement has a specific combination of components to meet its own requirements.

SURFACE AND BASE COURSES

The construction of a pavement in two or more courses (layers) is common. Differentiating the courses allows each layer to be optimized for the special purpose it serves in the pavement, and the structure as a whole to be built with the least possible expense. Figure 2.1 shows a pavement with distinct surface and base courses.

The surface course directly receives the traffic load and the disintegrating effects of traffic abrasion. It is likely to be made of special, relatively expensive material to

TABLE 2.1
Terms with Particular Application to Porous Pavement Components

Term	Definition
Base course	Layer placed below a surface course to extend pavement thickness; may be called simply base
Course	Layer in a pavement structure
Filter layer	Any layer inserted between two other layers, or between a pavement layer and the subgrade, to prevent particles of one from migrating into the void space of the other
Geomembrane	Impermeable manufactured fabric; sometimes called liner
Geotextile	Permeable manufactured fabric; sometimes called filter fabric
Pavement	Any treatment or covering of the earth surface to bear traffic
Overlay	Layer applied on top of a preexisting or otherwise complete pavement
Pavement structure	A combination of courses of material placed on a subgrade to make a pavement
Reservoir	Any portion of a pavement that stores or transmits water; a reservoir may overlap or be combined with other pavement layers such as base and subbase; sometimes called reservoir base, drainage layer, or drainage blanket
Subbase	Layer of material placed below a base course to further extend pavement thickness
Subgrade	The soil underlying a pavement structure and bearing its ultimate load
Surface course	Pavement layer that directly receives the traffic load; this layer presents a pavement's surface qualities such as accessibility, travel quality, appearance, and resistance to direct traffic abrasion

Dimensions of Porous Pavement Installations

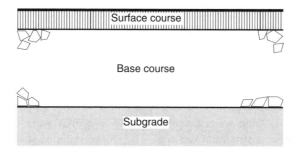

FIGURE 2.1 Section through a pavement with distinct surface and base courses.

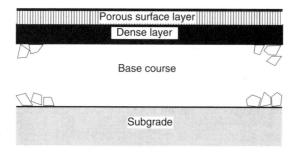

FIGURE 2.2 Pavement with a porous overlay.

resist abrasion and provide qualities such as appearance and accessibility. A wide variety of materials are available to meet the special requirements of surface courses.

A distinct base course builds up the thickness of a pavement with comparatively inexpensive material in order to spread out the traffic load over the subgrade or to protect the subgrade from frost penetration. If necessary, a subbase can be added to further thicken the pavement structure or to store more water as it discharges to a pipe or infiltrates the soil.

In a dense-surfaced pavement, air and water do not penetrate into any part of the pavement or soil, although in many dense-surfaced installations the base course is made of porous aggregate. An impervious surface prevents the pavement as a whole from functioning as a porous, permeable structure.

OVERLAY

An overlay is any layer applied on top of a preexisting or otherwise complete pavement. Figure 2.2 shows a common type of overlay, consisting of a porous layer over an otherwise dense-surfaced pavement. Highway departments use this type of overlay to enhance a pavement's surface benefits. The porous layer drains water away from the surface, improves visibility, increases traction, and reduces noise and glare. It makes highway driving safer and increases a highway's capacity to carry large amounts of traffic without the cost of widening the road. The dense lower portions of the pavement remain an inexpensive way to support a highway's heavy traffic loads. The characteristic thinness of the porous layer, however, limits its treatment and the storage of water, and the underlying impervious layer prevents the aeration of tree roots and the recharge of ground water.

Reservoir

A reservoir is any portion of a pavement that stores or conveys water while it exits through a drainage pipe or into the soil. A reservoir includes all pavement materials where stored or flowing water occurs with any frequency, even though the same materials also have a structural function. The storage volume is in the void space between particles of material. Portions of pavement with this type of function are familiar in pavement construction (AASHTO, 1986, p. AA-32; 1993, p. I-18; Cedergren, 1989, p. 350; Mathis, 1989; Moulton, 1980, pp. 87–98; Moulton, 1991, pp. 12-32–12-43; Nichols, 1991). The reservoir has also been called a drainage layer or drainage blanket.

Where water stored in a pavement reservoir discharges relatively slowly via lateral pipes, the in-pavement storage reduces downstream flooding and erosion and the required size of downstream drainage systems. Where water infiltrates from the reservoir into the subgrade, it reduces downstream storm flows and in addition maintains ground water aquifers and stream base flow. As water passes through almost

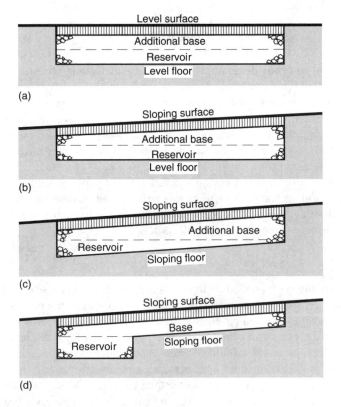

FIGURE 2.3 Reservoir in level and sloping pavements.

Dimensions of Porous Pavement Installations   39

any porous paving material, it is treated by filtration and the biochemical activity of microorganisms.

Commonly, the hydrologic and structural functions of pavement material are merged into a single uniform layer of material called a base reservoir. A base reservoir may be drained by a pipe at a certain elevation, which divides the base into a rapidly drained segment above the pipe, and a more frequently and persistently wet segment below where water remains for infiltration. The upper, dry segment is free of frost hazard and has a primarily structural function. The lower, wetter segment has equal structural and hydrologic functions.

Figure 2.3 shows how the configuration of a reservoir can vary under level and sloping pavement surfaces. Under a level surface the reservoir can be a horizontal layer at the bottom of the pavement. Under a sloping surface the reservoir can remain

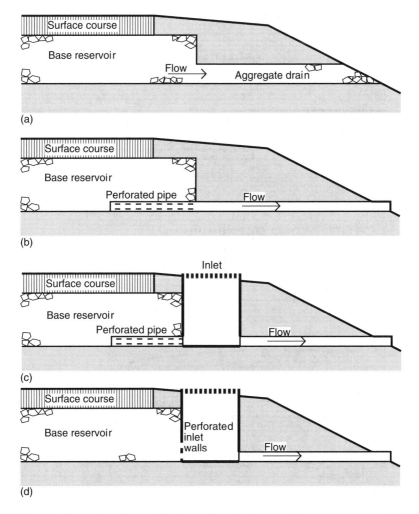

FIGURE 2.4 Examples of lateral discharge from the bottom of a pavement reservoir.

on a level floor, so long as the additional base material required to take up the difference in slopes is not prohibitively expensive. On a sloping pavement floor, water flows down to the low point, so the reservoir is limited to the pavement's low edge. The base near the low edge may be specially shaped to enlarge the reservoir.

Lateral Outlet

A pipe or any other lateral outlet can discharge excess water from a pavement reservoir safely and limit the depth and duration of ponding in the upper segment of the pavement. The capacity of the outlet controls the rate of discharge. Specific outlet configurations may be chosen for reasons of maintenance or cost.

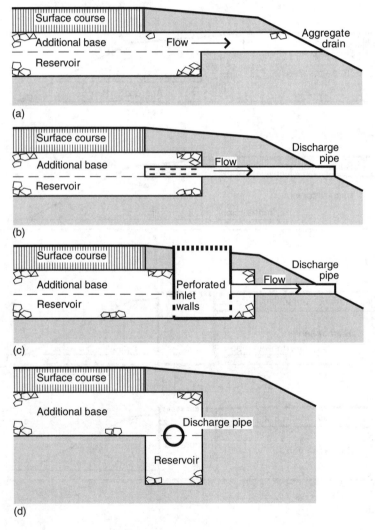

FIGURE 2.5 Examples of lateral discharge from the top of a pavement reservoir.

Dimensions of Porous Pavement Installations

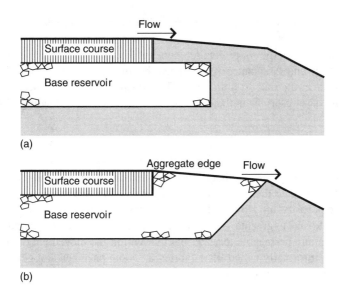

FIGURE 2.6 Arrangements for reservoir overflow at a pavement surface.

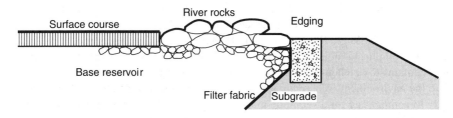

FIGURE 2.7 Typical construction of a Cahill edge drain.

Figure 2.4 shows possible outlet arrangements located at the bottom of the reservoir. The outlet drains potentially all of the water laterally, leaving little or none to infiltrate the subgrade. Water enters a lateral pipe through perforations in either the pipe or the walls of an inlet structure. An inlet structure permits monitoring and maintenance access.

Figure 2.5 shows outlets for discharge that overflows at the top of the reservoir. Amounts of water that exceed the storage capacity of the reservoir and the infiltration rate of the subgrade discharge laterally through the outlet. The reservoir below the lateral discharge area retains water for infiltration into the subgrade. A trench constructed at the edge of a pavement could enlarge the reservoir economically.

Figure 2.6 shows outlets that overflow at the pavement surface. Water rises in the reservoir and discharges at the pavement edge when the quantity of water exceeds the infiltration rate of the soil and the storage capacity of the reservoir. This arrangement is appropriate in regions where there is limited frost hazard, on sites with limited traffic load to stress-saturated pavement material, or in settings of climate and soil where the water would be expected to rise to the surface only seldom.

The provision of an aggregate edge at the side of a pavement keeps overflows off the pavement surface.

Figure 2.7 shows the more detailed construction of an aggregate edge as practiced by Cahill and Associates (www.thcahill.com) in their many installations of porous asphalt in the mid-Atlantic area. This construction is also used as a backup inlet into the base reservoir in the event of any clogging or overloading of the pavement surface.

Filter Layers

Filter layers are layers inserted between two other layers, or between a layer and the subgrade, to segregate their materials. Segregation is needed in some pavements to maintain the porosity and structural integrity of each layer. Filter layers can be made of intermediate-size aggregate. Alternatively, they can be made of geotextiles, which are fabrics that are permeable to water but that inhibit the movement of small particles; they are informally called filter fabrics. In some pavements, a geotextile adds tensile strength to the pavement.

Liners

Some porous pavements are lined at the bottom to prevent the infiltration of water into the subgrade. The technical term for plastic lining sheets is "geomembranes", which refers to manufactured impermeable fabrics (ASTM D 4439; Koerner, 1998; Rollings and Rollings, 1996, pp. 428–429). Where the soil under a porous pavement would swell during moisture fluctuations, a liner stabilizes the structure. Where the soil contains toxic chemicals, a liner prevents infiltrating water from leaching them into the environment. Where great thickness of pavement structure would be necessary to compensate for a subgrade made soft by infiltrating water, a liner could be a cost-saving alternative. In some projects a liner turns the base course into a reservoir of harvested water. The stormwater management functions of a lined pavement are limited to detention and water-quality treatment, not soil infiltration.

POROUS PAVING MATERIALS

This book arranges porous paving materials in nine general types or families. A careful choice of material is essential to fit a pavement to a project's functional, environmental, appearance, and cost requirements. The nine families are introduced here with summaries of their advantages and disadvantages. Later chapters will review their properties and applications in detail. Table 2.2 lists some terms with particular application to pavement materials.

Porous Aggregate

Aggregate is any mass of particulate material such as gravel, crushed stone, crushed recycled brick, or decomposed granite. Single-size particles create an aggregate mass with 30 to 40 percent void space; such "open-graded" material can be extremely permeable to air and water.

Dimensions of Porous Pavement Installations

TABLE 2.2
Terms with Particular Application to Pavement Materials

Term	Definition
Aggregate	Any mass of particulate material
Asphalt (asphalt concrete)	Aggregate bound by asphalt cement
Block	Solid piece of material used as a construction unit
Concrete (Portland cement concrete)	Aggregate bound by Portland cement; or generically, aggregate bound by any type of cement
Geocell	Plastic lattice that forms a web of cells
Open-celled grid	Block or slab containing openings through its entire thickness

Aggregate is by far the most ubiquitous material in pavement construction. It is the most common material in pavement base courses; in that role alone it is a major component in most kinds of pavements. Single-size aggregate is also the principal component of porous asphalt and porous concrete, and is used as porous fill in the open cells and joints of paving blocks, grids, and geocells.

Unbound open-graded aggregate is used directly as a surface course in very low-traffic settings such as residential driveways, lightly used portions of parking lots, and lightly used pedestrian walkways. Locations for it must be selected carefully in order to avoid displacement by traffic. Figure 2.8 shows an appropriate application in the parking lot for a riding stable, where the traffic is intense only on event days.

Aggregate surfaces are distinctively applicable in settings with freezing or swelling soil, because deflections from heaving are not noticeable on the aggregate's irregular surface. In some installations, maintenance is required from time to time to level the surface and replace lost material.

In appropriate settings, unbound aggregate has important advantages simultaneously for economics and the environment. In most regions, aggregate is the least expensive of all firm surfacing materials. At the same time, the high porosity and permeability of single-size aggregate make it the most favorable of all pavement materials for restoring watershed hydrology and tree rooting habitat. Aggregate materials are available from natural and recycled sources at low energy cost.

POROUS TURF

A turf surface is a "green" open space. It is also in effect a pavement that supports pedestrian or vehicular traffic. Figure 2.9 shows turf used for overflow parking at a large retail store. Turf's permeability is positive as long as it is not compacted by excessive traffic; healthy turf actively maintains soil permeability by building soil aggregation. The transpiration of living grass actively counteracts potential urban heat islands.

Locations for turf use must be selected to avoid compaction by frequent traffic. It is particularly easy to damage where the topsoil is plastic clay, since wheels and heels can dig into it in wet weather. On the other hand it is distinctively applicable in settings with swelling soil or frost heave, because subgrade deflections are barely noticeable on its irregular surface. Turf has been used without reinforcement for

FIGURE 2.8 Unbound aggregate parking at Bar Gee Farm in Allison Park, Pennsylvania.

FIGURE 2.9 Turf surface for overflow parking at a Home Depot store in Brandon, Florida.

lightly used garden walks and parking at frequencies of up to once per week. With reinforcement by geocells, grids, or blocks, it adds its living "green" look and flexible surface to settings with heavier or more frequent traffic loads.

All turf must be regularly maintained with mowing and some degree of fertilization and irrigation; this is the inherent cost of a living "green" surface. Because maintenance must be scheduled, turf should be used only where traffic can be controlled or is predictably scheduled, such as in an office, church, or in event parking; it should not be used in a multifamily residential parking area that could be in use at any hour of any day. A particular turf variety must be selected for the climate, soil, and shade in each specific setting.

Plastic Geocells

Plastic geocells are manufactured lattice-like products that hold aggregate or topsoil in their cells, inhibiting displacement and compaction. Geocells extend the use of aggregate and turf into more demanding traffic settings than they could bear alone, including portions of commercial and institutional parking lots and emergency access lanes. Figure 2.10 shows an appropriate example in the parking lot of a small office.

Most plastic geocells are flexible, so they are adaptable to sites with swelling or freezing soil. In most models the plastic ribs occupy a very small portion of the surface area, so the surface permeability, temperature, and visual appearance are essentially those of the grass or aggregate fill. Many geocell models are made partly or wholly from recycled materials.

Open-Jointed Paving Blocks

Paving blocks are solid units of concrete, brick, or stone laid side by side to bear traffic loads. The models that can be used to make porous pavements are shaped

FIGURE 2.10 Plastic geocell with turf fill at the Southface Institute in Atlanta, Georgia.

FIGURE 2.11 Open-jointed block pavement at sidewalk tree plantings at the Harbourfront Community Centre in Toronto, Ontario.

to produce open joints between adjacent units. Porous aggregate or turf in the joints gives the pavement its porosity and permeability. Figure 2.11 shows an example.

Many block products are remarkably durable, giving their installations long lifetimes and low life-cycle costs. They can bear very heavy traffic. A block surface with correctly open-graded aggregate in the joints can be highly permeable. However, block pavements are more expensive to construct than some other types of pavement. They are sensitive to deformation in the base or subgrade, so in a setting with deep frost they may require a thick base course to prevent heaving; over swelling soil they may require a lining to prevent water from infiltrating into the soil.

OPEN-CELLED PAVING GRIDS

Open-celled paving grids are units of concrete or brick, which are designed with open cells that can be filled with porous aggregate or turf. The units are laid side by side like blocks. The resulting surface is a gridwork of solid ribs or pedestals commonly an inch or more wide, alternating with cells of aggregate or grass. Figure 2.12 shows an example.

The type and condition of the aggregate or grass in the cells strongly influence the surface appearance of the grids. Where infrequent traffic and good maintenance allow grass to spread over the ribs, the pavement can have the impression of a "green" open space and be smooth enough for walking comfortably. Where grass is poorly maintained or the cells are filled with aggregate, the irregular surface can make walking difficult.

Although paving grids require relatively expensive paving materials, they play a valuable role in selected settings. Many models are durable and long-lived. Many can bear the loads of heavy vehicles. In emergency access lanes and seldom-used parking stalls, the grass in the cells provides permeability and a lawn-like look while the strong ribs support heavy vehicles on the rare occasions when the pavements are called into use. Grid pavements, like those of paving blocks, should be protected from heaving or swelling soil.

POROUS CONCRETE

Porous concrete is made of single-size aggregate bound together by Portland cement, cast in place to form a rigid pavement slab. It is a subtle variation of conventional

FIGURE 2.12 Concrete open-celled grid draining the edge of a courtyard pavement at Olympia City Hall in Olympia, Washington.

FIGURE 2.13 Porous concrete parking stalls at Finley Stadium in Chattanooga, Tennessee.

dense concrete, requiring a special specification and an experienced installer. Figure 2.13 shows an example.

Porous concrete is moderately high in initial cost but the long life of properly installed material can make its life-cycle cost low. Porous concrete's durability in cold climates has not yet been proven; it can probably be enhanced by treatments such as "air entrainment" and polymer fiber reinforcing. It is intolerant of swelling soil because it is susceptible to cracking.

Properly installed porous concrete is appropriate for both the low traffic loads of driveways and walkways, and moderate traffic loads such as those of commercial parking lots and residential streets. It can be made to accommodate remarkably heavy traffic loads. The finished material can have substantial porosity and permeability. It forms a firm level surface that is universally accessible.

Porous Asphalt

Porous asphalt is made of single-size aggregate bound together by bituminous asphalt binder. Asphalt is a familiar and inexpensive paving material, but porous asphalt is a subtle variation of conventional dense asphalt, requiring a special specification. Figure 2.14 shows an example of the material.

Numerous installations have proven that porous asphalt's permeability can be high. However, some installations have suffered from clogging by the asphalt binder. Where the binder is too fluid or the bond between the binder and the aggregate is weak, the binder can drain gradually from the pavement surface downward through the pavement's pores, accumulating into a clogging layer inside the structure, and leaving the surface particles unbound.

Dimensions of Porous Pavement Installations 49

FIGURE 2.14 Porous asphalt parking lot at the Centre County-Penn State Visitor Center in State College, Pennsylvania.

In Europe and the U.S. several highway departments have developed technology to stabilize the binder and assure that porous asphalt has a long, useful life. Their motivation is the unusually safe and continuous travel that is possible on highways with well-drained, high-traction porous asphalt overlays. Their positive results benefit applications of porous pavement everywhere. As updated technology becomes more widely available, it makes porous asphalt a reliably durable choice in porous paving material for a wide range of common urban settings.

Soft Paving Materials

The category of "soft" paving materials includes any granular material from an organic or recycled source such as bark mulch, crushed shells, or rubber granules. These

FIGURE 2.15 Wood mulch walkways in the Melvin Hazen Community Garden in Washington, District of Columbia.

materials are exclusively suited for areas of very light traffic, such as pedestrian walkways, lightly used residential driveways, equestrian ways, and very lightly used parking stalls. Figure 2.15 shows an example in the walkways between garden plots.

"Soft" materials can be highly appropriate in naturalistic, historic, or informal settings. Where they are suited to the local environment and traffic load, they can bring gentle beauty and integration with the organic life of the soil. Under excessive traffic the particles could be displaced, or crushed and compacted into an impervious mass. Some soft materials are unstable in windy open places or under concentrated surface runoff.

The cost of specific materials varies with their regional availability. Their embedded energy costs vary with their extraction and recycling histories. The infiltration rates of installed materials vary greatly; durable single-size particles and resilient organic materials have the greatest permeability.

DECKS

Decks and boardwalks are surrogates for pavements. They are bridge-like structures built on footings that suspend them over the soil surface. They leave the soil below almost entirely free for rooting and water infiltration. They are completely permeable to air and water as long as their decking components are perforated or spaced apart from each other. Figure 2.16 shows an example.

Decks and boardwalks can bear substantial pedestrian traffic without compacting the underlying soil. In certain instances they can bear vehicular traffic. They are suited to all kinds of freezing and swelling soil conditions, because their footings can be seated below the active soil zone with the structure spanning above. They are uniquely suited to steeply sloping sites and to sites where the native soil, tree roots, or ecosystem dynamics are to be very conscientiously protected.

Decks and grates can be made from a variety of natural, manufactured, and recycled materials. Their durability varies with the material and its preservative treatment. Decks made of wood can have an informal outdoor character; those of metal and reinforced concrete appear more urban and structured. The principal disadvantage of decks and grates is their cost, which can be high compared to some common pavement materials.

EXAMPLES OF SELECTIVE APPLICATION

Each material and configuration of porous pavement can be successful only in certain types of locations within a project site. In order to fully achieve the advantages of cost, permeability, appearance, and stability, it is imperative to analyze a site's

FIGURE 2.16 Boardwalk over the restored Spring Peeper Meadow wetland in the Minnesota Landscape Arboretum, Chanhassen, Minnesota.

traffic patterns, topography, use schedule, and expected maintenance provisions, to identify the microsettings where each type of material would be completely appropriate.

Almost every paving installation described in this book exemplifies the selective use of materials, providing lessons in how selection could be improved in the future. The following case studies pointedly exemplify sites where selective choices have been made using diverse paving materials for specific purposes. Based on these studies, pavements of different types have been logically applied in settings defined by traffic and use.

Office Parking, Medford Village, New Jersey

A clear and simple example of selective application is a parking lot in Medford Village, New Jersey, made partly of porous aggregate and partly of conventional dense asphalt. This parking lot is one of many that crowd the alleys and back lots of the village's historic center to serve offices, storefront shops, and upstairs apartments. Almost all of them are partly or wholly of porous aggregate to satisfy with an economical material the objectives established in an environmental study of Medford Township by Ian McHarg and his colleagues (Juneja, 1974).

Figure 2.17 shows the parking lot in 1996 (it was still extant in 2004). It bears the moderate daily traffic load of a small office. The traveling lane is of dense asphalt, which is a well-known material and a structurally reliable response to the traffic load; it successfully resists the abrasion of many vehicles moving and turning per day. In contrast, the parking stalls are of open-graded crushed stone to infiltrate rain water into the soil. The particles are stable under the stalls' limited, slow-moving traffic. The aggregate is at once highly permeable, and less expensive than the nearby dense asphalt. The 3/4 to 1 inch particles make an adequately walkable surface. A concrete wheel stop at each stall organizes the parking and keeps vehicle wheels a foot or two from the aggregate edge, protecting the edge from displacement.

The aggregate absorbs both the direct rainfall on the parking stalls and the runoff from the traveling lane. The parking lot as a whole infiltrates rainwater as if the entire surface were permeable. Surface water essentially never occurs in the parking stalls due to the aggregate's very high permeability.

The vegetated swale shown in Figure 2.18 receives the parking lot's overflow during occasional large storms. Water that reaches the swale has already been reduced, detained, and cleaned in the pavement structure. In the swale it has a further chance to slow down, infiltrate, and be treated.

Cardinal Ridge, Medford, New Jersey

A more complex example of selective application is in the residential setting of Cardinal Ridge, a community of 77 single-family homes in an outlying part of Medford Township. The community as a whole uses an aggressive combination of porous surfaces to satisfy the township's objectives to infiltrate stormwater into the soil and to preserve the native oak-pine forest (Juneja, 1974; Palmer, 1981, pp. 97–119; Ferguson, 1994, pp. 228–231).

Dimensions of Porous Pavement Installations 53

FIGURE 2.17 A parking lot for a small office in Medford Village, New Jersey, selectively combining dense asphalt and porous aggregate.

Each home has a driveway of porous crushed-stone aggregate like the one shown in Figure 2.19. The use of porous aggregate in place of conventional dense asphalt reduced the cost of home construction. In a visit to Cardinal Ridge about 20 years after the community was built, compaction of the wheel tracks was visible on the driveway surfaces, but not displacement of aggregate particles.

Outside the driveways' timber edging the native forest floor is scrupulously preserved, supporting a woodland canopy that shades and cools the pavements. In turn, the infiltration of air and water through the porous pavements supports the plants and trees. Encroachment of vegetation from the sides of the driveways has been very limited, because the only species in the native wooded setting are slow-growing forest species, and the deep forest shade suppresses germination of weed seeds.

FIGURE 2.18 A vegetated swale draining aggregate parking stalls in Medford Village, New Jersey.

FIGURE 2.19 Porous aggregate driveway in the Cardinal Ridge residential community in Medford, New Jersey.

Dimensions of Porous Pavement Installations

The homes' outdoor patios are pervious wooden decks. Residential walkways are of porous wood mulch, well-spaced paving stones, or stone aggregate.

The public streets are of conventional dense asphalt, a structurally reliable response to the streets' moderate traffic load. The street pavements however, are minimized in width to preserve adjacent vegetated soil and the woodland canopy it supports. The length of the impervious streets, and thus the area of their pavement, was minimized by the arrangement of houses in clusters of small lots. Natural swales draining the small amount of runoff from streets and roofs contain small check dams to give the water one last chance to slow down and infiltrate into the woodland soil.

PIER A PARK, HOBOKEN, NEW JERSEY

Pier A Park in Hoboken, New Jersey, exemplifies an aggressive combination of porous pavements in a very heavily used pedestrian setting. The park was constructed on a former shipping pier projecting into the Hudson River, where spectacular views of Manhattan attract thousands of walkers and joggers daily. The concrete pier is the park's impervious "bedrock," upon which growth of the park's turf and 97 London plane trees is vital to the park's public success. To provide a rooting zone, the pier is overlaid with a layer of "structural soil" rooting medium combining aggregate and soil (Arnold, 1993, pp. 128–129; Arnold, 2001; Arnold/Wilday, 1997; Henry Arnold, personal communication, 2001; Thompson, 2001). The layer is 18 inches in thickness under a lawn and increases to 48 inches under trees.

Above the rooting zone, the entire surface of the 5.1-acre park is a mosaic of porous surfacing materials to support different levels of pedestrian traffic. Figure 2.20 shows the park's combination of surfaces: aggregate in the foreground, turf in the middle ground, and blocks in the background. The porous pavements and their structural-soil base give air and water to tree roots while allowing park users to walk directly under the trees without compacting the root-zone soil.

Waterfront views attract heavy pedestrian traffic to a broad perimeter walkway made of blocks. The blocks have 1/4-inch-wide joints filled with single-size aggregate. The blocks are placed up to the tree trunks without tree grates; trees are seemingly planted in the block pavement itself. As the tree bases grow the blocks can be lifted out. Blocks of selected colors and sizes form panels and patterns in the walkway surface.

In the center of the park is a large panel of single-size aggregate with a canopy of closely spaced trees. The aggregate's small, angular particles make a level, firm, smooth surface that supports moderate to heavy foot traffic during summer events in the shade of the trees where grass would not grow well. Benches, public restrooms, and play equipment support the use of the space. The surface permeability is high.

The remainder of the park is a large lawn for creative play. Recreation brings moderate foot traffic to the grass at all times of year. A few trees grow at the lawn's northern edge; the remainder of the lawn is open so that sunlight can reach the growing grass. Trees and grass together extend their roots naturally through the topsoil and the underlying structural soil.

FIGURE 2.20 Porous aggregate, porous turf, and porous open-jointed blocks at Pier A Park in Hoboken, New Jersey.

MITCHELL CENTER ARENA, MOBILE, ALABAMA

The parking lot at Mitchell Center Arena in Mobile, Alabama, exemplifies an aggressive combination of porous paving materials in a large parking lot. The arena houses events such as basketball games and concerts about once a week in season. It is located on the campus of the University of South Alabama, in a part of the city with poor storm drainage and low-lying neighborhoods subject to urban flooding (Rabb, 1999). At one particularly busy intersection, fatal auto accidents had been attributed to flooding by runoff from the campus. When the arena was constructed, the limitation of runoff at the source in the new 438-car parking lot was an economical alternative to reconstruction of the city's streets and storm sewers. Consequently, the entire parking lot is constructed of selectively located porous aggregate and porous turf, all reinforced by appropriate models of plastic geocells.

The parking bays are of porous Bermuda turf reinforced with "Grasspave^2" geocells. Bermuda grass is used even for the accessible (handicapped) parking stalls, where, in the limited traffic of those stalls, it is in excellent condition and makes a uniform surface for walking. Figure 2.21 shows an example. The other parking spaces close to the building are used daily by about 25 staff and 100 students. In those spaces, frequent traffic and shade have weakened the sod and have required it to be replaced twice in four years. In the remaining spaces the grass has grown thick and healthy under traffic that is only occasionally intense (Andy Lindsey, University of South Alabama, cited in project profile at www.invisiblestructures.com).

FIGURE 2.21 Parking bay of turf reinforced with plastic geocells at Mitchell Center Arena, University of South Alabama, Mobile, Alabama.

The parking is organized primarily by the alignment of the bays. Wheel stops were added to internal bays to prevent vehicles from cutting across the grass. Each bay is a shallow swale that drains any overflow runoff to a grate inlet at one end.

The turning lanes are of single-size crushed limestone aggregate 3/8 inch in size reinforced with "Gravelpave^2" geocells. The lanes carry moving, turning traffic that would be too frequent for grass. Heavily used walkways are of similar reinforced aggregate. Some loose aggregate particles have migrated onto the edges of the grass parking stalls, but not enough to alter the character or use of the grass surface. The aggregate is noticeably displaced out of its geocells only at the one spot where almost all vehicular traffic turns to enter the parking lot from a dense asphalt access road. A more detailed analysis of the layout might have identified these few square yards as requiring firmer surfacing, whether porous or dense, to bear the large amount of traffic.

A 12-inch-wide concrete band defines the edges between grass bays and aggregate lanes and sets aside "islands" for plantings at the corners. It expands into a raised curb at the parking lot's perimeter.

The construction of Mitchell's porous paving system costs about twice as much as conventional dense asphalt (Rabb, 1999). However, the total cost of site development was less than it would have been with dense asphalt, because no off-pavement drainage structures were necessary. The porous grass and aggregate infiltrate rainwater into the parking lot's sandy, permeable subgrade. During storms, the parking lot produces only a small fraction of the runoff that would be produced by conventional dense asphalt, accommodating the capacity of city storm sewers and protecting nearby streets.

PROVISIONS FOR ALL POROUS PAVEMENT APPLICATIONS

Almost any porous pavement application requires certain provisions which would not necessarily be anticipated in applications of conventional dense pavements. The following provisions are not complicated, but they are special for porous pavements and vital to their success.

SELECTIVE APPLICATION

By far the most vital and universal requirement is to analyze a site to identify the settings in which specific paving materials and configurations would be beneficial. No kind of pavement — porous or nonporous — should be built thoughtlessly. The purpose is to put the right type of pavement in each type of place. The case studies described above are the outcome of making this kind of distinction.

A site can be analyzed as soon as a draft site plan is available showing the locations of streets, parking, walkways, patios, and other areas that require paving to bear traffic. Those areas' types and the frequencies of traffic define microenvironments where different paving materials and configurations would be appropriate. One can distinguish lightly trafficked streets from heavily trafficked ones. From within individual streets, one can distinguish driving lanes from parking lanes, within parking lots, one can distinguish parking stalls from driving lanes, and frequently used areas

Dimensions of Porous Pavement Installations

from those used only for overflow or event parking; and from among walkways one can distinguish accessible routes from general routes. In all paved areas, one can distinguish different needs for tree rooting, appearance, cost, and hydrology.

One can distinguish pavements that are steeply sloping (perhaps over 5 percent) from those that are gently sloping. On steep slopes, surfaces of grass and unbound aggregate are susceptible to erosion by traffic or runoff. Beneath the pavement surface, water flowing rapidly through an open-graded base may erode the subgrade soil.

One can distinguish the special maintenance regimes that exist on different sites. On sites with established grounds maintenance crews such as sports stadiums, parks, campuses, and golf courses, grass-based paving is a practical possibility. In municipalities where road crews spread sand or cinders for winter traction, an all-porous pavement may be impractical for public streets under the load of clogging material.

DRAINAGE AT PAVEMENT EDGES

Where earthen slopes abut porous pavement edges, the slopes must be drained and downward away from the pavement in every possible direction. The purpose is to preserve the pavement's porosity and permeability by preventing sediment from being washed onto the surface. A porous pavement should ordinarily not be used to absorb drainage running toward it from off-pavement areas; it should be used to infiltrate only the rainwater that falls on it.

Figure 2.22 illustrates this provision along the edge of a porous concrete road in a new park. The picture was taken soon after the road was installed, and while the surrounding soil was still being graded for sports fields. The soil was eroding freely;

FIGURE 2.22 Edge of a porous concrete road in Webb Bridge Park, Alpharetta, Georgia, in 1998.

flowing sediment was visible through a roadside drainage swale. If the sediment were to flow onto the pavement, it would clog the surface before the road was even put into service. The road edge is properly shaped to prevent that. Pavement drainage is downward into a swale on the side of the road shown in the picture and down the hill slope on the other side. In every direction, sediment drains down eroding slopes without entering the pavement's pores.

The only exception is a limited amount of runoff from clean impervious roofs and pavements. Where such additional runoff is brought into a pavement for storage or infiltration, the amount must be added to the pavement's inflow in any hydrologic model.

Sediment, debris, and overflow runoff that get onto a pavement surface should be allowed to wash freely off the pavement edge. Figure 2.23 shows an installation

FIGURE 2.23 Sediment on a porous concrete parking area surrounded by raised curbs at an office building in Hobe Sound, Florida.

Dimensions of Porous Pavement Installations

of porous concrete where a raised curb was added, it was supposed, to pond up water and force it into infiltration through the pavement. But vehicles tracked in sand and overhanging vegetation contributed organic debris. Wind and water carried the sediment and debris to the foot of the curb where the ponding of water forced them to accumulate into a solid layer on the pavement, clogging the pores and reducing infiltration. To allow debris to be washed off the pavement, raised curbs should be avoided as far as possible, particularly on the downslope edge of a pavement. Where raised curbs are necessary they should be constructed with numerous cuts or gaps to let drainage through. Paradoxically, in order to preserve and maximize infiltration, a porous pavement must be designed to overflow freely.

A porous pavement needs safe provision for occasional overflow as does any other component of a drainage system, either from the surface or via a pipe above the reservoir.

Where a porous pavement abuts a dense pavement, drainage from the base of the porous pavement into that of the dense pavement must usually be prevented. Dense pavements are seldom designed in anticipation of water in their bases; excess water could soften the subgrade or pose a freezing hazard, leading to the settlement or heaving of the dense pavement structure. Where a porous pavement slopes toward a dense pavement, it should be separated by a strip of concrete or other material like that illustrated in Figure 2.24.

PROTECTION OF PAVEMENT DURING FURTHER CONSTRUCTION

It is a common occurrence in site development that after a pavement is installed, the grading, landscaping, and construction of buildings continue for some time. During this time construction materials are stored and moved about, rainfall erodes unvege-

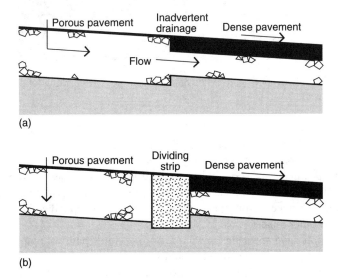

FIGURE 2.24 Alternative configurations at an edge between a porous pavement and a dense pavement.

FIGURE 2.25 Construction sand spilled onto a porous concrete parking lot during erection of an adjacent wall.

tated soil, and construction vehicles and personnel track mud all about. Until all construction is completed and the site is stabilized, mobile sediment poses the hazard of clogging a porous pavement surface.

Project specifications should direct construction traffic away from installed porous pavements to eliminate the tracking of sediment. When sediment is deposited on a pavement surface despite precautions, it should be immediately removed by vacuuming or other means.

Construction sand must be prevented from spilling onto a completed porous pavement. Sand-producing materials could be stored away from porous pavement surfaces, or on protective sheets placed over a pavement surface. Figure 2.25 shows how important a provision like this can be. In the instance shown in the picture, the sand was effectively removed from the pavement pores by vacuuming.

Where earthen slopes drain down toward a porous pavement surface, the eroding soil must be prevented from washing onto the surface using sediment barriers such as silt fences or compost berms. The barriers should be kept in place until the ground is stabilized by thoroughly established vegetation.

Some potential sources of trackable soil can be eliminated by placing pavement subbase material over all planned traffic areas immediately upon initiating a project. This keeps traffic out of the mud during construction and prevents compaction of the underlying soil. Pavement base and surface materials can be added after all other construction is complete, the site is stabilized, and construction equipment is withdrawn from the site.

APPROPRIATE SPECIFICATION

Project specifications describe the types and qualities of materials and configurations desired for a pavement. They guide contractors in the submission of competitive bids and regarding compliance with requirements. Large public projects subject to competitive bidding require full formal specifications for bidding to be uniform and fair. Small private projects subject to little regulation may get by with less formality or thoroughness, leaving contractors to work out many details on the ground. Various approaches to the writing of specifications are a combination of advantages and disadvantages (Beery, 2002; Nichols, 1991, pp. 15-2–15-3). Different approaches are appropriate for different projects and for different parts of a single project.

"Descriptive" (or "design" or "method") specifications dictate materials the contractors must use and the procedures they must follow. With them the writer of the specifications definitely controls a project and takes responsibility for project performance. Commitment to this type of specification requires a thorough knowledge of the relevant materials and methods, and of the project site. A "proprietary" specification is a particularly narrow descriptive specification that designates a manufactured material in terms of its brand name or proprietary source. A proprietary specification may be forbidden in a public project because it precludes competition from other products, but it can be written to allow substitutions subject to approval by the designer or the client.

"Performance" specifications describe desired qualities in measurable terms such as permeability and strength, leaving methods and even materials up to the contractor. They can be enforced through testing at specified stages in the project, using specified test methods. The specifications invite innovation from contractors in ways that produce the desired outcomes and to deal with site-specific problems. They place responsibility on the contractor for the project's performance. Their implementation requires the availability of competent contractors who are capable of problem-solving.

A "reference" specification makes reference to an established and publicly accessible industry or jurisdictional standard. The reference incorporates the standard in the project specification as if it had been written in its entirety. It expedites project completion where local suppliers are organized to deliver standard products and local contractors understand their uses. It assures compliance with a legitimate benchmark and is legally required, where a local jurisdiction will accept only those specifications it has already adopted as standard.

A given standard, however, is not necessarily a universal measure of quality; it is ultimately applicable only in specific types of application. Under no conditions should standards be blindly perpetuated from one project or jurisdiction to another. Standards must be skeptically critiqued and appropriately modified for application to each specific project; each designer is ultimately responsible for the specifications for his own project. On the other hand, on diverging from an accepted standard, one should be able to state exactly the circumstances that require the divergence. One should not reinvent a new "wheel" for every project as if there were no standards.

Table 2.3 lists examples of sources of standard specifications for porous pavement. Libraries of specifications are accessible through web sites such as CADdetails (www.caddetails.com), LA-Info (www.la-info.com), Landscape OnLine (www.landscapeoneline.com), and Landscape Catalog (www.landscapecatalog.com). ASTM and AASHTO standards may be available on-line from third-party services such as Normas (www.normas.com) and IHS Engineering (www.ihserc.com).

The Construction Specifications Institute (www.csinet.org) promotes uniformity in specification format. The institute's *CSI Format for Construction Specifications* (Construction Specifications Institute, 1996) arranges information in a hierarchy of categories and subcategories. The objective is to eliminate repetitions and omissions by placing a given piece of information in a given standard location. CSI's section 02795 is Porous Paving. The hierarchy's numerical system allows the addition of special specifications for specific jobs.

The American Institute of Architects (www.aia.org/masterspec) and the American Society of Landscape Architects (www.asla.org/nonmembers/masterspec.html) have implemented the CSI format in the Masterspec guideline specification system. For a given project the system suggests options for wording each required CSI section in accordance with contemporary technology and provides links to information about products and standards.

TABLE 2.3
Examples of Sources of Standard or Guideline Specifications for Porous Pavement

Pavement material	Source of specification	Contact information
Open-graded aggregate	State highway agencies	Varies from state to state
Open-graded aggregate	ASTM International	www.astm.org
Open-graded aggregate	American Association of State Highway and Transportation Officials (AASHTO)	www.transportation.org
Porous turf	Turfgrass Producers International	www.turfgrasssod.org
Plastic geocells	Individual manufacturers	Listed in Chapter 8
Paving blocks and grids	Interlocking Concrete Pavement Institute	www.icpi.org
Paving blocks and grids	National Concrete Masonry Association	www.ncma.org
Porous concrete	Florida Concrete and Products Association	www.fcpa.org
Porous concrete	Georgia Concrete and Products Association	www.gcpa.org
Porous asphalt	Oregon Department of Transportation	www.odot.state.or.us

CONTRACTOR QUALIFICATIONS AND COMMUNICATIONS

It is advantageous, and even essential for qualified and experienced contractors to supervise the installation of the different types of porous paving. Contractors can prove specialized competence by possessing certification from an authoritative industry association or by placing a test panel before proceeding with the rest of a project. Criteria such as these can be used to prequalify contractors before they are allowed to bid on a project.

Thorough communication with the contractor is essential. Contractors must understand what is to be done and why compliance with specifications is essential. Owing to the lack of communication and oversight, shoddy contractors can circumvent project objectives and even well-meaning contractors can err in astonishingly unexpected ways.

Cahill (1993, p. 22; 1994) has emphasized the following practices based on his experience as an engineer for numerous porous asphalt installations:

State prominently and repeatedly in plans and specifications the special nature and purpose of each distinctively porous material.

Meet with the contractors in person to review the specifications and make sure the contractors understand the objectives.

Oversee the contractors and any testing laboratory to make sure the objectives are carried out; the necessary oversight must be budgeted into the design project.

Maintain a written record documenting review and approval at critical project stages such as excavation of the subgrade and quality checks of base and surface materials.

Inspect the site to make sure construction vehicles are not allowed to traverse the excavated subgrade or pavement structure at any inappropriate stage.

Forbid construction traffic from tracking soil onto the finished pavement surface.

The introduction of soil or debris not included in the specifications to bring the pavement up to level or for any other reason must be explicitly forbidden.

APPROPRIATE MAINTENANCE AND SIGNAGE

Written and verbal communication to a porous pavement's owner or manager should make clear the pavement's special characteristics, its purposes, and the special maintenance practices that may be required.

In cold locations the maintenance regime must prohibit sanding. Application of sand or cinders can be a gross source of clogging sediment. A maintenance program that does not include sanding must be agreed to by the owner or manager even before the decision to design a porous pavement is made.

"Seal coats" must be absolutely forbidden. Applications of semifluid materials clog surface pores.

FIGURE 2.26 Informational sign at a porous asphalt parking lot at Walden Pond State Reservation in Concord, Massachusetts.

TABLE 2.4
Holders of Registered Trademarks Mentioned in This Chapter

Registered Trademark	Holder	Headquarters Location
Grasspave^2	Invisible Structures	Golden, Colorado
Gravelpave^2	Invisible Structures	Golden, Colorado

Permanent signs or stenciled messages can identify a porous pavement and its purposes. They can educate the public about what such potentially unfamiliar materials are for and discourage improper disposal of oil and other fluids. The signs can remind maintenance crews about the special restraint that needs to be taken in maintaining them, specifically warning them away from sanding or applying seal coats.

Figure 2.26 shows an informative sign at a porous asphalt parking lot at Walden Pond State Reservation in Massachusetts. It educates the public about what the distinctive material is for. It could be made even more useful by cautioning about the special restraint that needs to be taken in using and caring for this pavement's surface.

TRADEMARKS

Table 2.4 lists the holders of registered trademarks mentioned in this chapter.

ACKNOWLEDGMENTS

Near the beginning of this book project Shelly Cannady, then a graduate student in landscape architecture at the University of Georgia, made an exhaustively documented survey of porous pavement experiences in North America which contributed importantly to this chapter.

The following additional persons generously provided information used in this chapter: Soenke Borgwardt of Borgwardt + Wissenshaftliche Beratung, Norderstedt, Germany; Henry Arnold of Arnold Associates (www.arnoldassociates.org); Michael Grossman, Capitol Ornamental Concrete Specialties (www.capitolconcrete.com); and William Stoop, Planning Administrator of Medford Township, New Jersey.

The following persons constructively critiqued early drafts of this chapter: Harlow C. Landphair of Texas A&M University, Tom Richman of Catalyst (www.catalystworks.net), David Spooner of the University of Georgia School of Environmental Design, and Neil Weinstein of the Low Impact Development Center (www.lowimpactdevelopment.org).

REFERENCES

American Association of State Highway and Transportation Officials (1986). *AASHTO Guide for Design of Pavement Structure,* Vol. 2, Washington: American Association of State Highway and Transportation Officials.

American Association of State Highway and Transportation Officials (1993). *AASHTO Guide for Design of Pavement Structures 1993*, Washington: American Association of State Highway and Transportation Officials.

Arnold, Henry F. (1993). *Trees in Urban Design*, 2nd ed., New York: Van Nostrand Reinhold.

Arnold, Henry F. (2001). The Down and Dirty on Structural Soil, letter to the editor, *Landscape Architecture* 91, 9–11.

Arnold/Wilday (1997). *Section 02515, Precast Concrete Unit Pavement over Air Entrained Soil, Specifications for Pier A Park*, on file at Capitol Concrete, South Amboy, NJ.

Beery, William E. (2002). Specifications, in *Business Law for Landscape Architects,* pp. 187–200, Boston: Pearson Custom Publishing.

Cahill, Thomas (1993). *Porous Pavement with Underground Recharge Beds*, West Chester, Pennsylvania: Cahill Associates.

Cahill, Thomas (1994). A Second Look at Porous Pavement/Underground Recharge, *Watershed Protection Techniques* 1, 76–78.

Cedergren, Harry R. (1989). *Seepage, Drainage, and Flow Nets*, 3rd ed., New York: John Wiley and Sons.

Cedergren, Harry R., K.H. O'Brien, and J.A. Arman (1972). *Guidelines for the Design of Subsurface Drainage Systems for Highway Structural Sections*, FHWA-RD-72-30, Washington: Federal Highway Administration.

Construction Specifications Institute (1996). *Manual of Practice*, Alexandria: Construction Specifications Institute.

Ferguson, Bruce K. (1994). *Stormwater Infiltration*, Boca Raton: Lewis Publishers.

Juneja, Narendra (1974). *Medford: Performance Requirements for the Maintenance of Social Values Represented by the Natural Environment of Medford Township, N.J.*, Philadelphia: University of Pennsylvania Department of Landscape Architecture and Regional Planning.

Koerner, Robert M. (1998). *Designing with Geosynthetics*, 4th ed., Upper Saddle River, NJ: Prentice-Hall.
Mathis, Daniel M. (1989). Permeable Base Design and Construction, *Stone Review* August 1989, pp. 12–14.
Moulton, Lyle K. (1980). *Highway Subdrainage Design*, Report No. FHWA-TS-80-224, Washington: Federal Highway Administration.
Moulton, Lyle K. (1991). Aggregate for Drainage, Filtration, and Erosion Control, in *The Aggregate Handbook*, Richard D. Barksdale, Ed., Washington: National Stone Association, chap. 12.
Nichols, Frank P. (1991). Specifications, Standards, and Guidelines for Aggregate Base Course and Pavement Construction, in *The Aggregate Handbook*, Richard D. Barksdale, Ed., Washington: National Stone Association, chap. 15.
Palmer, Arthur E. (1981). *Toward Eden*, Winterville, NC: Creative Resource Systems.
Rabb, William (1999). Paved with Good Intentions, *Mobile Register,* March 7, 1999.
Rollings, Marian P. and Raymond S. Rollings, Jr. (1996). *Geotechnical Materials in Construction*, New York: McGraw-Hill.
Thompson, J. William (2001). Roots Over the River, *Landscape Architecture* 91, 62–69.

3 Porous Pavement Structure

CONTENTS

The Load of Traffic on Pavement 70
The Bearing of Load by Subgrade 71
    General Characteristics of Subgrade Soils 73
    Soil Characteristics That Influence Bearing Value 76
    California Bearing Ratio 77
    Unified Soil Classification... 77
    Sources of Information about Subgrade Soils...................................... 79
The Structural Roles of Pavement Materials... 81
    Required Pavement Thickness... 84
    Filter Layers and Geotextiles... 86
Compaction and Its Alternatives 89
    The Proctor Reference for Compaction 89
    Construction with Little or No Compaction 92
Pavement Adaptation to Freezing 93
    Frost Depth 93
    Protection of Subgrade 96
    Protection of Pavement Reservoir 98
    Experiences in Cold Climates 99
Pavement Adaptation to Special Subgrades 103
    Adaptation to Wet Subgrade 103
    Adaptation to Swelling Subgrade 103
    Adaptation to Plastic Subgrade 105
Pavement Adaptation at Edges 106
    Support of Pavement Edges... 106
    Restraint of Flexible Pavement Edges... 107
    Protection of Pavement Edges... 109
Trademarks 111
Acknowledgments 113
References 115

All pavements, porous and dense alike, must bear the traffic loads imposed on them in the soil and weather conditions where they are located. This chapter introduces important considerations in pavement structural design and the kinds of treatments

TABLE 3.1
Examples of Types of Distress That Pavement May Experience

Mode of Distress	Contributing Factors
Displacement or wide cracking	Repeated loading or braking by vehicles; temperature and moisture changes
Distortion	Repeated loading; swelling soil; frost heave; differential settlement; temperature and moisture changes
Disintegration	Traffic abrasion; freeze-thaw effects on material; loss of binder; chemical reactions; weathering

Adapted from U.S. Departments of the Army and Air Force, 1992, p. 18-2, and ASTM D 6433.

that can be used to respond to them. Later chapters will describe the requirements of specific porous paving materials.

Table 3.1 lists some of the types of distress that a pavement may experience, and the traffic and environmental factors that can contribute to them. The condition of a pavement tends to decline over time as the surface displaces, deforms, or disintegrates (AASHTO, 1993, pp. I-7–I-10; Rollings and Rollings, 1991). After a while it may require some sort of rehabilitation. If a pavement were to fail prematurely it would not be catastrophic like the collapse of a building or the failure of a dam, but it would represent a financial loss and a nuisance to the public. A failure of even a sidewalk or parking lot could translate into financial liabilities such as personal injury, vehicle damage, increased maintenance, or premature replacement costs. The objective of pavement structural design is to produce a structure that will maintain a desired condition for an acceptably long time. It does so by fitting pavement materials to the traffic load imposed from above, and to the bearing value of the soil below.

THE LOAD OF TRAFFIC ON PAVEMENT

The cumulative total traffic load on a pavement comes from the intensity of individual load events (pedestrian, automobile, heavy truck, or other load), and the quantity of events over time. Table 3.2 defines some terms with distinctive application to traffic load.

A combination of different loads is commonly expressed in terms of a standard reference vehicle. Each vehicle is related to the reference vehicle by its equivalent wheel load (EWL) or equivalent single-axle load (ESAL). A reference unit commonly used in the United States is ESAL of 18,000 pounds (18 kip); this is the load from one axle of a rather heavy truck (AASHTO, 1993, pp. I-10–I-11). For comparison, many passenger cars place only about 1500 pounds on each axle.

The effect of traffic load on pavement deterioration increases with ESAL raised to the fourth power (AASHTO, 1993, pp. I-10–I-11). Thus, one axle of a large truck bearing 18,000 pounds has more than 20,000 times the pavement-deteriorating effect of one axle of a passenger car bearing 1500 pounds. Most experimental observations of pavement performance have been under heavy traffic loads, including numerous trucks, for guiding the design of busy streets and highways. On the scale of such

TABLE 3.2
Some Terms with Distinctive Application to Traffic Load

Term	Definition
ADT	Average daily traffic (number of vehicles): ADT is most precise when it is specified separately for each of the two travel directions on a street; when only a single value is given it is assumed to represent the sum of traffic in both directions
ADV	Average daily volume, the same as ADT
ESAL	Equivalent single-axle load; in the U.S., this is commonly specified as the traffic load on a pavement equivalent to a single-axle load of 18,000 pounds
EWL	Equivalent wheel load; analogous to ESAL
kip	1000 pounds (one kilo-pound)
single-axle load	The total load transmitted to a pavement by all wheels of a single axle extending the full width of a vehicle

AASHTO (1993); U.S. Department of the Army and Air Force, 1992.

observations, the loading of automobiles on ordinary parking lots, driveways, and residential streets is almost zero (Croney and Croney, 1998, p. 52).

Stress due to traffic is cumulative over time. In the pavement design procedure of the American Association of State Highway and Transportation Officials (AASHTO, 1993), the number of predicted ESALs is summed over the pavement's expected lifetime, for example 20 years. When the design lifetime expires, the pavement is assumed to require some sort of rehabilitation.

Designing for high cumulative traffic load (high ESAL, high ADT, and long pavement life) produces a pavement that will stand up for a long life of rugged service, but in return the cost of the construction of a thick, strong pavement structure must be borne. Designing for low cumulative load yields a pavement that will be inexpensive to build, but if its design assumptions are unrealistic it will begin to show wear within a relatively short period of time. The degree of willingness to bear a high initial cost for the purpose of prolonging pavement life is specific to each project's client and setting.

THE BEARING OF LOAD BY SUBGRADE

Subgrade is the soil underneath any pavement structure, bearing the ultimate load of the pavement and its traffic. Subgrade originates as naturally occurring earth, previously disturbed urban soil, or artificially placed fill. This section briefly summarizes some soil characteristics that are relevant to pavement design. Further information is available in technical references such as those listed at the end of this chapter, including the concise *PCA Soil Primer* (Portland Cement Association, 1992) and the elucidative book by Rollings and Rollings, *Geotechnical Materials in Construction* (Mc Graw-Hill, 1996). Table 3.3 defines some terms with distinctive application to subgrade.

Although the subgrade is a pavement's ultimate underlying support, it is also ordinarily the weakest structural component in or around a pavement. One of the major purposes of a pavement structure is in effect to protect the subgrade from

TABLE 3.3
Some Terms with Distinctive Application to Pavement Subgrade

AASHTO	American Association of State Highway and Transportation Officials (www.aashto.org)
ASTM	American Society for Testing and Materials (www.astm.org)
Atterberg limits	The moisture contents at which a soil changes physical state or consistency, including the shrinkage limit, the plastic limit, and the liquid limit, together named for the scientist who developed this system of testing and characterization
Bulk density	The density of an entire granular material, counting the volume of both the solid material and the voids
CBR	California Bearing Ratio, a measure of bearing value in soil or other material
Cohesion	The tendency of some soil particles to bind together
Density	Weight per volume
Fines	Small ("fine") soil particles such as clay and silt
Fraction	Portion of a granular material that passes a given sieve size
Gradation	The combination of mineral particle sizes making up a soil; soil texture
Horizon	Distinctive layer in a soil profile
Internal friction	The resistance to sliding of one soil particle against another
Liquid limit	The moisture content at which a soil changes from a plastic state to a liquid state
Moisture content	The amount of moisture in a soil, expressed as a percentage by weight of the soil's oven-dry weight
NRCS	Natural Resources Conservation Service, a branch of the U.S. Department of Agriculture; until 1992 this agency was called the Soil Conservation Service (SCS)
Optimum moisture content	The moisture content of a given soil at which a given compactive effort produces the maximum compacted density
Plastic limit	The moisture content at which a soil changes from a semisolid state to a plastic state
Plasticity	A material's tendency, when subjected to pressure, to deform as a body rather than cracking or crumbling
Plasticity index	The numerical difference between the liquid limit and the plastic limit; the range of moisture content at which a soil is in a plastic state
Proctor	Test of soil density achieved by a given compactive effort at varying soil moisture contents, named for the scientist who developed this type of test
Profile	Vertical cross section of soil layers
SCS	Soil Conservation Service, the former name of the Natural Resources Conservation Service (NRCS); many local soil surveys were published by this agency while it was called by this old name
Shrinkage limit	The soil moisture content at which, during drying, a soil ceases shrinking while further drying continues
Sieve number (gage number)	Size of opening in a sieve for measuring the size of particles; ASTM designates a standard size for each sieve number
Soil structure	Aggregation of soil particles into composite patterns such as blocky, platy, or granular; this term should not be confused with texture or strength characteristics
Subgrade	Surface of earth or rock that receives the foundation of a pavement structure

TABLE 3.3 Continued

Swelling soil	Soil that expands with increase of moisture content
Texture (soil texture)	The combination of mineral particle sizes making up a soil; soil gradation
Unified classification	The soil classification system developed jointly by the U.S. Army Corps of Engineers and U.S. Bureau of Reclamation, and subsequently standardized by ASTM
USDA	United States Department of Agriculture

Portland Cement Association, 1992; Soil Survey Staff, 1993.

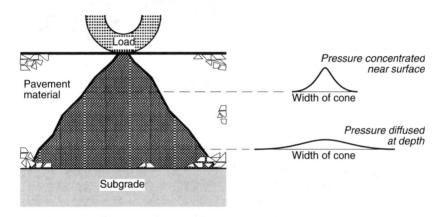

FIGURE 3.1 Diffusion of traffic load with depth in pavement.

the load. It does this by spreading the load over the subgrade to the extent that the soil can bear the load without deforming. Conceptually, from a point at the pavement surface where a load is applied, the load spreads out in a cone downward through the thickness of the pavement. As shown in Figure 3.1, with increasing depth, the load diffuses across the cone's width and the maximum pressure at the center declines, thereby protecting the subgrade. In an area with substantial winter frost or swelling soil, the thickness of a pavement also supplies insulation and weight that counteracts the soil's tendency to heave.

GENERAL CHARACTERISTICS OF SUBGRADE SOILS

Natural soils vary from place to place, so the unique characteristics of each site's subgrade must be understood as a basis for that site's design.

The individual mineral particles in soils vary greatly in size. Figure 3.2 shows that the terms clay, silt, sand, and gravel are used consistently to refer to the same relative size ranges. However, their exact definitions vary among classification schemes established by various organizations. The general term "fines" applies to the smallest particles such as silt and clay.

Physical and chemical differences accompany differences in particle size. Clay's extremely small particle size gives it a correspondingly large surface area per volume of soil. Natural clay mineralogy is that of complex aluminum silicates. The particles

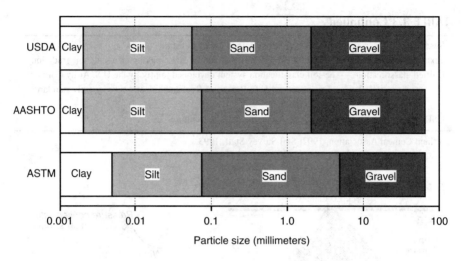

FIGURE 3.2 Particle size classes defined by three organizations; note that the horizontal scale is logarithmic (after Portland Cement Association, 1992, p. 6).

are commonly mica-like in shape. They bear surface charges to which ions and water are attracted, and which can bind particles together into larger soil aggregates.

In contrast, natural silt, sand, and gravel particles tend to be composed mostly of inert minerals such as quartz. Silt particles are irregular in shape; they often occur surrounded by an adhering film of clay. Sand and gravel particles are large in size and low in surface area per volume; they are variously angular or rounded, depending on their history of fracturing and weathering and they tend to function as separate particles.

The combination of mineral particle sizes in a soil is its texture or gradation. Table 3.4 defines some general types of soil gradations.

Figure 3.3 illustrates the triangular graph that USDA uses to categorize soil textures based on proportions by weight of sand, silt, and clay. Fine-grained categories are shown near the clay and silt corners, coarse-grained ones toward the sand corner. Loam appears near the center of the graph. USDA uses the presence of larger stone sizes such as gravel only as a modifier of the basic classification given by the three main particle sizes.

An additional constituent of many natural soils is organic matter. The proportion ranges from essentially zero in subsoils and some arid-region topsoils to almost 100 percent in bogs. Organic matter is characterized, like clay, by small particles, large surface area per volume, and surface charges to which ions and water are attracted.

Moisture content is ordinarily expressed as a percentage of a soil's oven-dry weight (Portland Cement Association, 1992, p. 13). It is determined by weighing the soil in its moist condition, then weighing it again after drying with standard drying heat and time. The difference between the two weights is the content of water in the original, moist soil. That weight divided by the soil's oven-dry weight is the moisture content.

Bulk density is the overall density of a soil, counting voids and solid particles together. It can be expressed either as wet density (including some water in the void space) or dry density (oven-dried with only air in the voids).

TABLE 3.4
Some General Types of Soil Gradations

Term	Definition
Fine-grained (fine-textured)	Continuous grading of sizes, with predominance of fines; such a soil can be plastic, cohesive, and slowly permeable
Dense-graded (well graded)	Continuous grading over a wide range of sizes including a su stantial quantity of fines; such a soil can be dense, slowly pe meable, and highly stable
Coarse-graded (coarse-grained, coarse-textured)	Continuous grading of sizes, with predominance of coarse particles such as sand; such a soil can be stable
Loam	A mixture of all three mineral particle sizes (sand, silt and clay) in more or less balanced proportions
Single-sized (open-graded, poorly graded)	Graded to a narrow range of particle size with few, if any, fines, leaving open void spaces between the coarse particles; dominance by gravel and sand can make such a soil highly porous and pe meable, well-drained, nonplastic, and loosely granular

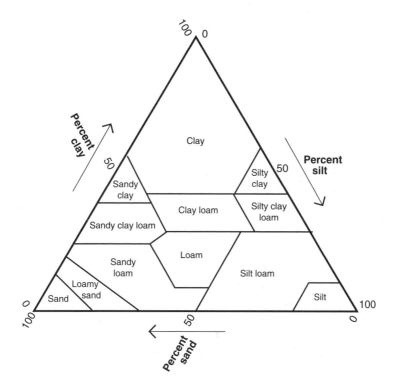

FIGURE 3.3 Texture classes defined by USDA (after Portland Cement Association, 1992, p. 7).

Soil particles are aggregated together in patterns that vary with texture, mineralogy, climate, and history of compaction. The term "soil structure" is used technically in this sense and should not be confused with texture or strength characteristics. A platy soil structure can be inherited from an underlying parent material. Prismatic and columnar structures can occur in arid and semiarid climates. Block structures can occur in humid regions and granular structures in extremely sandy subsoils.

Soil Characteristics That Influence Bearing Value

The stability or bearing value of a soil under a load is a composite of behavioral patterns governed by the physical and chemical attributes of individual particles and the overall gradation in the presence of varying amounts of moisture. Among the many behavioral patterns that interact to influence bearing value are internal friction, cohesion, and plasticity.

Internal friction is the resistance to the sliding of one particle against another. In gravel and sand internal friction tends to be high, no matter what the moisture content. In clay internal friction is usually low, but can vary greatly with moisture content.

Cohesion is the tendency of particles to bind together due to attractive molecular forces and tensile moisture films. The cohesion of sands free of clay and other fines can be close to zero. Cohesion in clays varies in complex ways with moisture content. It is low in pulverized, powder-dry clay. It increases as clay's moisture content is increased until the material becomes plastic; then it declines with the further addition of moisture. But if wet clay is oven-dried, the last remaining tight films of moisture can hold the clay grains together so firmly that a hammer is required to break the particles apart.

Plasticity is a material's tendency, when subjected to pressure, to deform as a body rather than cracking or crumbling. Moisture increases plasticity and reduces strength by suspending particles away from each other. A soil with substantial clay content exhibits plasticity because of the large number of water films surrounding the numerous small clay particles. The particles slide over each other, shearing the soil mass while the interconnecting water films stay intact. Plasticity can be experienced firsthand when a handful of moist fine-textured soil is "worked" in the hand. The quantitative level of moisture at which a soil changes physical state from solid or granular to plastic is the "plastic limit." Some soils containing no clay have no plastic limit and are called "nonplastic."

At high moisture contents clay deforms under its own weight; it flows when jarred and behaves like a liquid. The "liquid limit" is the moisture content at which the change occurs. High liquid limit indicates low load-carrying capacity.

The "plasticity index" (PI) is the liquid limit minus the plastic limit, in other words the range of moisture content over which a soil is in a plastic state. High PI indicates low bearing value because it suggests the tendency of water films to adhere to soil particles, reducing friction and strength.

Figure 3.4 summarizes the relationship between moisture level and soil state. Threshold moisture levels are referred to collectively as the Atterberg limits, after the Swedish scientist who defined them in 1911. ASTM (D 4318) and AASHTO (T89 and T90) standardize the laboratory tests for Atterberg limits.

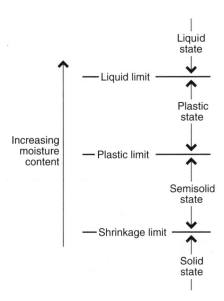

FIGURE 3.4 The Atterberg limits of soil state (adapted from Portland Cement Association, 1992, p. 14).

CALIFORNIA BEARING RATIO

A relatively straightforward rating of bearing value is the California Bearing Ratio (CBR), developed by the California Highway Department in the 1920s (Rollings and Rollings, 1996, pp. 138–140). In the CBR test (ASTM D 1883, AASHTO T 193), a piston is forced into a water-soaked sample of material, and the amount of pressure and rate of piston movement are monitored. The result is expressed as a percentage in relation to the same piston's penetration into a standard dense-graded, moderately compacted crushed-stone aggregate. Fine clayey soils can have CBR values lower than 5; a few dense-graded crushed-stone aggregates can have CBR values slightly higher than 100. Table 3.5 lists some typical CBR values.

Table 3.6 lists relative ratings of CBR value in pavement structural layers. High bearing value is of greatest value in the upper layers of a pavement, where the traffic load is most concentrated. In the subgrade very modest values of CBR are acceptable because the subgrade is protected from traffic load by the overlying layers.

CBR is useful for many aspects of pavement design because it is a simple and fixed measure. However, to reflect the full range of structural considerations in a pavement design — swelling, freezing, changing moisture content, and so on — CBR must be supplemented by additional, more complex information.

UNIFIED SOIL CLASSIFICATION

The Unified system of soil classification was developed to help interpret the composite of behavioral characteristics that a soil can exhibit under a range of environmental conditions. The system originated with the Army Corps of Engineers during

TABLE 3.5
Typical CBR Values

Material	CBR, percent
Dense-graded crushed aggregates typically used for pavement base	100
Dense-graded natural gravels typically used for pavement subbase	80
Limerock	80
Shell and sand mixtures	50 to 80
Gravelly sands	20 to 50
Fine clean sands	10 to 20
Wet clay soil	<1 to >3

Rollings and Rollings, 1996, p. 143.

TABLE 3.6
Relative Ratings of CBR Values in Pavement Structural Layers

	CBR in Base Course	CBR in Subbase Course	CBR in Subgrade
Excellent	100	50	—
Good	80	40	12+
Fair	—	30	9 to 12
Poor	50	—	4 to 8
Very poor	—	—	< 4

Rollings and Rollings, 1992, p. 17.

World War II. Subsequently, the Corps and the U.S. Bureau of Reclamation cooperatively made great improvements in the system, so it came to be called the "Unified" system. ASTM maintains the criteria for the system in ASTM D 2487.

Table 3.7 lists the major Unified soil categories. The system places soils in three basic groups: coarse-grained, fine-grained, and organic. In the coarse-grained group, categories are based only on textural characteristics. In the fine-grained group, the amount of fines is large enough to affect a soil's structural behavior. Categories in this group are based largely on the plasticity given by the fines. Each category is given initials that indicate its principal characteristics: GW for well-graded (dense-graded) gravels, CL for clays with low liquid limit, and so on. Intermediate categories are given hyphenated designations such as SP-SM or CL-ML. At the top of the table are soils that are coarse in texture and inorganic; they have relatively free drainage, little compressibility, and high stability. Materials dominated by fines or organic matter (in the lower part of the table, or at the lower end of each group of classes) have relatively slow drainage, great compressibility, and low stability. In applying a Unified classification to a specific project, local conditions that may not have been taken into account in the classification such as groundwater, degree of consolidation of the soil, and potential for swelling or frost movement, must be taken into account.

Figure 3.5 shows the frequency of occurrence of some USDA soil textures in Unified categories. It illustrates that only certain coarse USDA textures such as loamy sand or narrowly defined ones such as pure silt indicate a single Unified classification. In many texture categories, other variable characteristics such as plasticity broaden the Unified designation.

TABLE 3.7
Major Categories in the Unified Soil Classification System

Coarse-grained Soils (more than 50 percent retained on No. 200 sieve):

	Gravels (50% or more of fraction retained on No. 200 sieve is also retained on No. 4 sieve):
GW	Well-graded gravels and gravel–sand mixtures, little or no fines
GP	Poorly graded gravels and gravel–sand mixtures, little or no fines
GM	Silty gravels and gravel–sand–silt mixtures
GC	Clayey gravels and gravel–sand–clay mixtures
	Sands (50% or more of fraction retained on No. 200 sieve passes No. 4 sieve):
SW	Well-graded sands and gravelly sands, little or no fines
SP	Poorly graded sands and gravelly sands, little or no fines
SM	Silty sands and sand–silt mixtures
SC	Clayey sands and sand–clay mixtures

Fine-grained Soils (50 percent or more passes No. 200 sieve):

	Silts and clays with liquid limit less than 50%:
ML	Inorganic silts, very fine sands, silty or clayey fine sands
CL	Inorganic clays of low to medium plasticity, gravelly clays, sandy clays, silty clays, clays
OL	Organic silts and organic silty clays of low plasticity
	Silts and clays with liquid limit 50% or more:
MH	Inorganic silts, micaceous or diatomaceous fine sands or silts, inorganic silts, elastic silts
CH	Inorganic clays of high plasticity
OH	Organic clays of medium to high plasticity

Highly Organic Soils:

PT	Peat, muck and other highly organic soils

Portland Cement Association, 1992, p. 20.

Figure 3.6 illustrates the correlation of the Unified classification system with the California Bearing Ratio. The chart shows that soils with CBR of 15 and below tend to be in the clayey, silty, and organic categories; higher CBRs are in the coarse-textured soil categories. However, most Unified soil categories are associated with a range of possible CBR values. Consequently, it is vital to take into account site-specific conditions and, where subgrade stability is critical or soil conditions are questionable, to supplement general classifications with laboratory test data.

SOURCES OF INFORMATION ABOUT SUBGRADE SOILS

Information about the character and bearing value of soils on a specific site can come from a variety of potential sources of varying degrees of precision and cost. Different sources of information may be valuable at different stages of planning and design.

An immensely helpful source of general information for initial site planning studies and the planning of further, more detailed soil studies are the local soil surveys published by the U.S. Natural Resources Conservation Service (NRCS, http://soils.usda.gov). Each mapping unit in a survey map integrates the general soil environment at that place, from the surface to several feet below. It encompasses a soil profile with a unique range of important properties, arrayed in their various

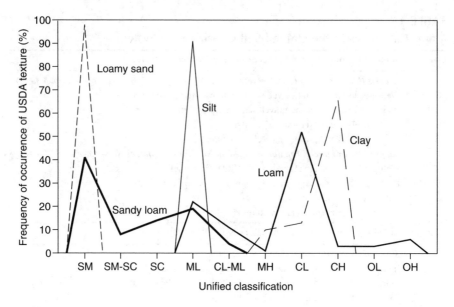

FIGURE 3.5 Frequency of occurrence of some USDA soil textures in Unified soil categories (data from Rollings and Rollings, 1996, p. 44).

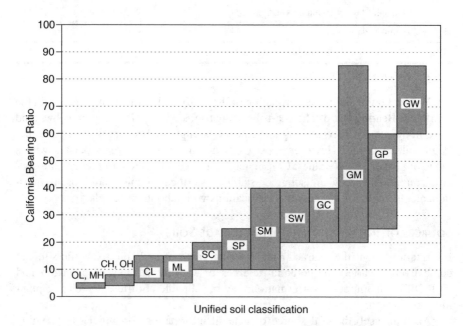

FIGURE 3.6 Relationship between California Bearing Ratio (CBR) and Unified soil classification (after Portland Cement Association, 1992, p. 29).

TABLE 3.8
Some Commonly Occurring Soil Horizons Designated in NRCS Soil Surveys

Horizon	Typical Character
A	Commonly high in organic matter mixed with mineral material
E	Mineral material commonly leached of colloidal material, leaving a concentration of sand and silt particles
B	A zone of accumulation of colloidal material carried in suspension from overlying horizons
C	Weathered or uncemented parent material such as sediment, saprolite, or unconsolidated bedrock
R	Hard bedrock

Soil Survey Staff, 1993.

layers or horizons. Table 3.8 lists commonly occurring types of soil horizons in order from the surface downwards. Figure 3.7 shows examples of soil profiles from different places in the U.S. Descriptions of representative profiles of the approximately 3000 soil series in the U.S. are maintained in the "Official Soil Descriptions" section of http://soils.usda.gov.

Other information sources can supplement soil surveys in anticipating general subgrade conditions. Where a good local geologic map is available, for example from a state geological survey, it is worth obtaining just because of the additional insight it provides from scientists who look at the ground from a viewpoint different from that of NRCS. On large sites, aerial photographs can help characterize general earth conditions and distribution, when analyzed using a basic knowledge of geology, landforms, and remote sensing (Way, 1978). A knowledgeable designer can roughly, but quickly and inexpensively, confirm some basic soil characteristics using the look and feel of the material in the field: texture can be estimated by the feel of moist soil when rubbed and ribboned in the hand; color can indicate the history of drainage and aeration (Brady and Weil, 2000).

By far the best — and most costly — source of soil information is a planned professional sampling and quantitative testing of site-specific soils. The investigation should be planned to focus on the areas where particular kinds of construction might be done, the kinds of soil problems that might be encountered, and the kinds of data that will be needed for specific design decisions. Investigations that physically penetrate the surface with pits, trenches, or borings can determine detailed stratigraphy, groundwater conditions, CBR, and Unified soil classifications with the greatest possible reliability. Less precisely, geophysical measurements that do not penetrate the surface can be used quickly over a large site to map irregularities in the bedrock surface or an interface between important strata; examples are seismic, electrical, and gravitational tests, calibrated with selected test borings.

THE STRUCTURAL ROLES OF PAVEMENT MATERIALS

An individual pavement material must be structurally sufficient to withstand the traffic load and the conditions brought to it by its environment. A number of different materials may be placed in a sequence of pavement layers or "courses" to respond, at

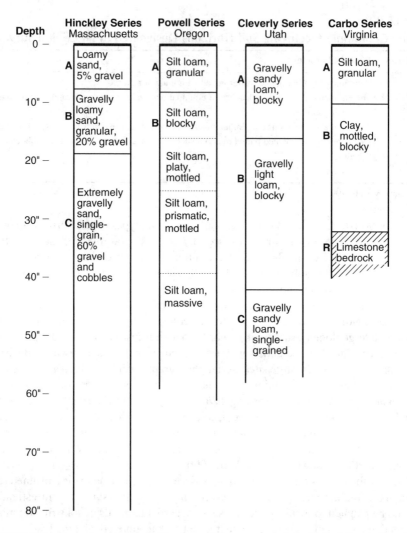

FIGURE 3.7 Examples of soil profiles from different parts of the United States (based on descriptions at http://soils.usda.gov).

minimum necessary cost, to the specific combination of forces that occur at each level, and together to adapt the traffic load to the bearing capacity of the subgrade (Croney and Croney, 1998, p. 73). Table 3.9 defines some terms with particular application to structural roles of pavement materials.

Different design criteria are used for so-called "flexible" and "rigid" pavement structures. A concrete pavement is rigid; almost all other pavements are flexible, including those of asphalt, aggregate, geocells, grids, and blocks. The two types distribute traffic load to the subgrade at different rates, as represented conceptually in Figure 3.8. A rigid structure resists bending, so it spreads a point load quickly over a wide soil area. In a flexible structure such as an asphalt surface course over an aggregate base, aggregate

TABLE 3.9
Some Terms with Particular Application to Structural Roles of Pavement Materials

Composite pavement	Pavement composed of layers of variously rigid and flexible materials, such as a flexible wearing course over a rigid base
Curb	Any relatively rigid unit at the edge of a pavement or between two different pavement areas
Flexible pavement	A pavement structure that distributes loads to the subgrade through particle interlock, friction, and cohesion
Pavement structure	Any combination of pavement layers placed on a subgrade to bear a traffic load
Rigid pavement	A pavement structure that distributes loads to the subgrade through the bening resistance of a slab

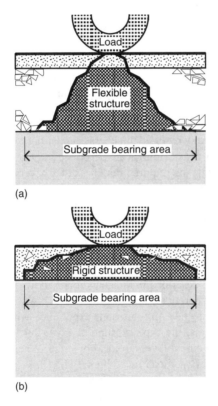

FIGURE 3.8 Transfer of traffic load to the subgrade in rigid and flexible pavement structures.

interlock spreads the load through the pavement only gradually, so a relatively thick pavement structure is required to protect the subgrade to the same degree.

A pavement surface layer must distinctively withstand the abrading force of traffic, while providing the surface appearance, accessibility, skid resistance, and other characteristics required for its setting and use. A surface course also contributes to the total thickness of a pavement in distributing the traffic load to the subgrade. To minimize construction cost, a surface layer is ordinarily only thick enough to maintain its own integrity under the direct traffic load. The rest of the pavement thickness is made up of base layers of less expensive material, most commonly aggregate, protected from abrasion by the surface layer.

Porous paving materials differ from corresponding nonporous materials, whether rigid or flexible, in having open voids between the materials' particles or units. The porosity sacrifices a degree of strength. Properly installed porous materials are nevertheless adequately strong for many of the ordinary traffic loads to which dense materials are also suited. Later chapters will describe the strengths and appropriate applications of specific porous materials.

Required Pavement Thickness

For a given type of paving material, the adequate distribution of traffic load over the subgrade requires a certain pavement thickness. In freezing conditions the protection of the pavement from frost may require a different thickness, which must be separately determined. In a porous pavement with a specific stormwater management function, the storage of an adequate amount of water may require some further thickness, which must be determined by hydrologic calculations. The greatest of the alternative thicknesses must be adopted in order for the pavement to meet all of its requirements. This section describes structurally required thicknesses under nonfreezing conditions; thicknesses required for other considerations are described in other sections of this book.

Procedures to estimate the required thickness of a pavement and the various layers within it have been formalized in manuals such as those of the American Association of State Highway and Transportation Officials (1993), the U.S. Departments of the Army and Air Force (1992), and the U.S. Federal Aviation Administration (1995). All of them take into account the bearing value of the subgrade at least in nonfreezing conditions. Some of them have been adapted and simplified for applying specific types of paving materials to local streets and parking lots; those adaptations will be described in the chapters on individual pavement materials.

Over and above theoretical design procedures, AASHTO (1993 p. II-17) acknowledges the need to take into account experience with local factors such as locally available materials, the type of support given by local subgrade, and the thickness of specific pavement layers in local practice. Where local experience conflicts with AASHTO's *Guide*, the *Guide* defers: "The *Guide* attempts to provide procedures for evaluating materials and environment; however, in cases where the *Guide* is at variance with proven and documented local experience, the proven experience should prevail" (AASHTO, 1993, p. I-5).

The U.S. Departments of the Army and Air Force manual (1992, pp. 3-1–3-2, 6-1, and 8-1–8-5) has illustrated the relationships between subgrade CBR, traffic load, and required pavement thickness. Figure 3.9 shows their results for flexible pavements in nonfreezing conditions. Great thickness is required to spread out traffic load on soft soils with low CBR. With increasing CBR the required thickness declines rapidly; for CBR of 15 and higher only the minimum thickness of 6 inches is required. At any CBR thicker pavements may be required for heavier traffic loads or frost protection.

Porous Pavement Structure

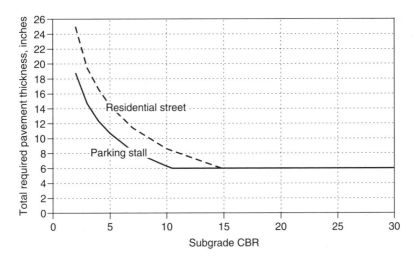

FIGURE 3.9 Minimum required total thickness of flexible vehicular pavement in nonfrost conditions for two different traffic loads (U.S. Departments of the Army and Air Force, 1992, p. 8-2; residential street data is Design Index 3; parking stall data is Design Index 1).

Rada et al. (1990; NCMA 1993; ICPI 1995) illustrated the relationship between required pavement thickness and subgrade moisture level, which can be very relevant to porous pavements that admit water into the subgrade. They applied the AASHTO design method specifically to block pavements assuming that the block surface is as strong as a layer of dense asphalt of equal thickness, that the base course is made of durable, well-drained, crushed aggregate at 98 percent compaction, and that the subgrade is compacted.

Table 3.10 lists Rada et al.'s (1990) categories of subgrade drainage conditions. On the left is the time necessary to remove water from the base and subgrade, in other words the ponding time, the determination of which will be discussed in Chapter 4. On the top is the percent of time the soil is exposed to saturation, which in a porous pavement comes from a combination of the ponding time and the precipitation regime where a pavement is located: exposure to saturation ≤1 percent of the time could occur in warm arid locales; exposure >25 percent of the time could occur where precipitation is common year-round. A subgrade drainage category of "dry," "medium" or "wet" can be read from the table.

Figure 3.10 shows the resulting minimum base thickness in non-frost conditions required for traffic loads that might occur on local residential streets. Different lines in the figure are for the "dry," "medium," and "wet" drainage categories. The curves show that, for base materials with CBR 80 or more, 4 inches is the minimum thickness required for any base, over any subgrade. Greater thicknesses, up to 8 inches, are required for soft soils or prolonged ponding in the base course. For base material with CBR down to 30, thicker bases are required, with 7 inches being the minimum in any condition. The thickness of the surface course is in addition to that of the base course; in block pavements the thickness might be 3 inches or more.

TABLE 3.10
Subgrade Drainage Categories in the Rada Method

Time to Remove Water	Exposure of Soil to Saturation			
	≤1% of Time	≤5% of Time	≤25% of Time	>25% of Time
2 h	Dry	Dry	Dry	Medium
1 d	Dry	Dry	Medium	Medium
7 d	Dry	Medium	Medium	Wet
30 d	Medium	Medium	Wet	Wet
Water will not drain	Medium	Wet	Wet	Wet

Rada et al. (1990, Tables 1 and 7).

Many porous pavements have a reservoir zone (defined in Chapter 2) where water is stored or transmitted. The presence of water may reduce the strength of certain reservoir construction materials. Above a reservoir there may be additional base or surface course material. By definition water does not rise above the reservoir except perhaps for a few hours following extreme, rare rainfall or snowmelt events; because the upper material is well-drained, it is of full bearing value (Cedergren, 1989, p. 342). Where there is doubt about a reservoir material's bearing value under saturated conditions, a prudent design might increase the thickness as shown in the curves for different base CBR values in Figure 3.10, or might count only the material above the reservoir in the pavement's structural thickness.

Software is available to help determine structurally required thicknesses of pavement layers in diverse conditions. An easily operable program is *Lockpave Pro*, distributed by Uni-Group USA (www.uni-groupusa.org). It applies established procedures to derive the required thickness of the base course for porous pavements surfaced with Uni-Group's open-jointed block products. It takes into consideration various traffic levels, various soil types, frost depth, and the possibility of subbase material being different from the base material. Another example of design software is that supplied by the U.S. Departments of the Army and Air Force (1992, p. 1-1–1-2) to implement their design calculations for thickness of rigid and flexible pavements.

FILTER LAYERS AND GEOTEXTILES

The open-graded aggregates used in the base courses of many porous pavements are commonly much coarser (of much larger particle size) than either the subgrade below the base, or the surface layer above. Filter layers are installed in some pavements to separate the various layers, and thereby to maintain the porosity and structural integrity of each layer. Filter layers are made of intermediate-size aggregate, or of permeable geotextile. Potential locations of filter layers are illustrated in Figure 3.11.

A filter layer over a base course prevents small particles of surface material from collapsing into the base's open pores. Examples of surface materials that may require this type of layer are turf rootzones, the aggregate setting beds of blocks and grids, the aggregate or soil fill of most geocells, and porous asphalt made with relatively small aggregate. Where a filter layer in this position is made of aggregate it is often referred to as a "choke layer" because its intermediate-size particles more or

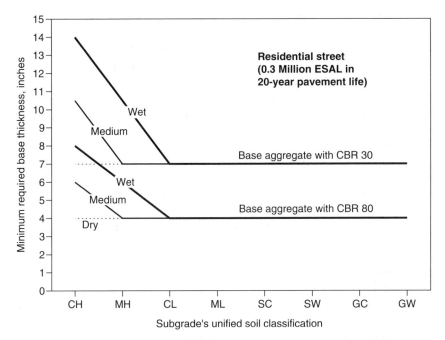

FIGURE 3.10 Aggregate base course thickness required in nonfrost conditions for block installations on local residential streets with traffic of 0.3 million equivalent single-axle loads in a 20-year life (Rada et al., 1990, Tables 5 and 7 and Figure 4; NCMA 1993, Table 3 and Figure 2).

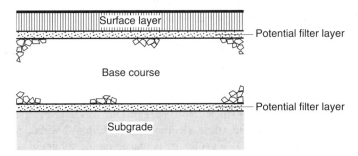

FIGURE 3.11 Potential locations of filter layers around an aggregate base course.

less fill or cover the large voids of the base aggregate. As described in Chapter 6, the selection of appropriate aggregate gradations for the base and surface courses can avoid the need for a separate filter layer between them.

A filter layer under a base course prevents plastic subgrade soil from flowing into the base's voids from below, preserving the structural and hydraulic capacity of the base at its original design thickness.

Geotextiles are an alternative to filter layers made of aggregate. Geotextiles are polymer fabrics that are permeable to water but which can inhibit movement of small particles (ASTM D 4439; Koerner, 1998; Rollings and Rollings, 1996, pp. 412–413).

They are informally called "filter fabrics" and are a variety of geosynthetic, which is a generic term for synthetic planar materials used in earth and pavement reinforcement. Woven geotextiles tend to have both permeability and tensile strength; nonwoven models can have high permeability but little tensile strength. The size of the openings (pores) can vary from sieve no. 30 (0.02 inch) to sieve no. 300 (0.002 inch), with the portion of the area in open pores varying up to about one third of the surface. The permeability (from one side of the fabric to the other) can range from 10 inches per hour to thousands of inches per hour. AASHTO M 288, *Geotextile Specification for Highway Applications*, sets standards for geotextile strength, permeability, opening size, and stability in ultraviolet light. A defining requirement is that the material be able to withstand stresses incurred during construction. Continued research is called for on the efficacy of geotextiles as separators (Soriano, 2004).

In some pavements a woven geotextile under the base course can contribute to distributing traffic load over soft subgrade. As the subgrade attempts to deform under a load the fabric is placed in tension, and its tensile strength adds support for the load. Whether the presence of a woven geotextile allows a reduction in the required base thickness however, is controversial.

Table 3.11 lists alternative guidelines for identifying subgrade conditions requiring some sort of geotextile at the bottom of the pavement structure. All the guidelines point to soft, plastic, fine-textured soils, which can flow easily into pavement voids and deflect enough under load to awaken the tensile reinforcement of a woven geotextile. For stronger subgrades a geotextile would act only as a separator. Most of the guidelines also point to very wet soil conditions, which means that porous pavements that admit water into the subgrade are likely to require geotextiles more often than pavements that leave the subgrade dry.

Table 3.12 lists examples of geotextile manufacturers or distributors. Most manufacturers publish guidelines for the selection and installation of their products. Further information can be found at web sites such as www.fibersource.com and www.geosource.com, at Drexel University's Geosynthetic Research Institute (www.drexel.edu.gri), at Geosynthetica (www.geosynthetica.net), and in review articles such as that of Duffy (1997).

During geotextile installation, careful storage and handling are necessary to prevent puncturing and undue stress (Rollings and Rollings, 1996, p. 426). In Cahill's practice (Brown, 1996), a geotextile is placed on the subgrade according to

TABLE 3.11

Alternative Guidelines for Identifying Subgrade Conditions Requiring Geotextile, According to Various Sources

Source	Subgrade Condition Requiring Geotextile
Burak, 2002	Plasticity index (PI) greater than 35, or soil expected to be saturated more than 50 percent of the time
Duffy, 1997	CBR no greater than 3
National Concrete Masonry Association, 1996	CBR no greater than 3, or high clay or silt content, or shallow water table, or soil subject to flooding

TABLE 3.12
Examples of Geotextile Manufacturers or Distributors

Name	Contact Information
BP Geotextiles	www.geotextile.com
Carthage Mills	www.carthagemills.com
Maccaferri	www.maccaferri-usa.com
Mirafi	www.tcmirafi.com
SI Geosolutions	www.fixsoil.com
Tensar	www.tensarcorp.com
Terratex	www.terratex.com
Trevira	www.trevira.com
Webtec	www.webtecgeos.com

the manufacturer's recommendations, with fabric sheets overlapping at the joints. It is extended beyond the edge of the base and securely positioned to prevent any sediment from washing into the base from adjacent soil. Base aggregate is placed on the fabric from one side, with vehicles riding on aggregate that has already been placed and not on the bare fabric. When the base aggregate is in place, the edges of the fabric are folded over its edges to further prevent eroding soil from entering the base voids. When the surrounding soil has been stabilized by established vegetation, the fabric may be trimmed and the excess removed.

COMPACTION AND ITS ALTERNATIVES

Compaction of a soil or aggregate during construction limits future settlement under traffic load and inhibits changes in volume from soil swelling or frost heave (U.S. Departments of the Army and Air Force, 1992, p. 1-1; Bader, 2001; Rollings and Rollings, 1996, pp. 134, 195–199). Various compaction machines work by rolling, tamping, or vibrating. Compaction increases a material's bulk density by reorienting particles and pushing them into closer contact. Where the particles are of various sizes, small particles are pushed into the voids between large particles.

No other treatment can produce so marked an increase in bearing value, at so low a cost, as does compaction (Portland Cement Association, 1992, p. 23). Increasing the subgrade bearing value decreases the cost required to build a stable overlying structure. In some areas, subgrade compaction is practiced as a low-cost "safety factor" even where no specific data indicate a need for greater strength to bear a given traffic load.

Pavement surfaces that would be least tolerant of the settlement of uncompacted soil are those surfaced with asphalt, concrete, blocks, and grids. For these pavement types compaction of subgrade is often required, and the associated reduction in soil infiltration rate must usually be accepted.

THE PROCTOR REFERENCE FOR COMPACTION

The degree of compaction of a soil or aggregate is indicated by its bulk density after compaction. For a given compactive effort (for example, three passes with a certain type of compacting machine), the resulting density depends on the material's moisture content at the time of compaction. Some moisture is necessary to allow particles to move about with respect to one another; too much moisture can suspend particles away from each other.

The Proctor test determines the maximum density that can be obtained in a material under standard compactive conditions, and the moisture content at which that density can be achieved (Portland Cement Association 1992, p. 23; Rollings and Rollings 1996, pp. 135–136). The more or less original version of the test is standardized in ASTM D 698. It is based on a level of compaction common in ordinary pavement construction projects; the result is commonly referred to as "standard density" or "standard Proctor." A modified version has been developed to take into account heavier compaction. It is standardized in ASTM D 1557; its result is known as "modified Proctor."

Both standard and modified Proctor tests determine "compaction curves" like those shown in Figure 3.12. The curves show that for a given material and compactive effort, the greatest compaction is achieved when the material is at an "optimum moisture content." For example, in the figure, the optimum moisture for a certain plastic clay is 20 percent, and the maximum density is 103 pounds per cubic foot. The maximum density is informally called "Proctor," "maximum Proctor," or "100% compaction."

The maximum Proctor density is a reference in comparison to which the compaction produced during construction can be evaluated. The bulk density of a compacted material can be measured and compared with the target density. In less formal projects a confirming measurement may not be made; instead, it is assumed

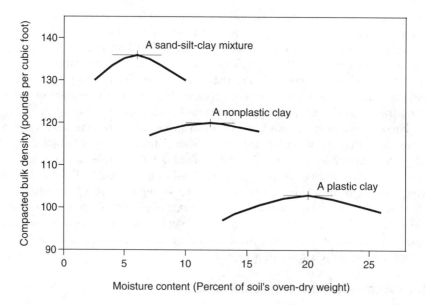

FIGURE 3.12 Examples of compaction curves produced by the Proctor test (adapted from Terzaghi and Peck, 1968, p. 447).

Porous Pavement Structure

that compaction by a well-known compaction method produces adequate density for the project.

In open-graded aggregates, such as those used in the bases of many porous pavements, it is almost impossible to measure in-place compacted density in the field accurately. (ASTM D 698; Rollings and Rollings, 1999, p. 10). Consequently, for the compaction of these materials it is more common to specify a method of compaction, for example a certain number of passes with a vibratory roller of a certain weight, than to specify a certain percentage of Proctor density.

Figure 3.13 illustrates the concept that compaction is increasingly necessary with proximity to a pavement's surface (U.S. Departments of the Army and Air Force, 1992, p. 4-1). Near the surface, traffic load is most concentrated, and protection from densification under future traffic load is most necessary. The curves for "cohesive soils" apply to soils such as plastic clay (PI > 5 and LL > 25); those for "noncohesive materials" are for materials such as open-graded aggregate (PI $\leq$ 5 and LL $\leq$ 25). Open-graded aggregate such as that used in the base of many porous pavements requires less compaction than plastic clay subgrades at the same depth. Parking stalls require less compaction than streets because of their lower traffic load. The compaction requirement declines with depth, eventually merging with natural ambient bulk densities; few natural soils have densities lower than 80 percent of modified Proctor. The curves in the figure are for materials with CBR $\leq$ 20; many pavement base materials have higher CBR and require correspondingly less compaction.

There are four ways for subgrade material to meet a compaction requirement (U.S. Departments of the Army and Air Force, 1992, p. 4-1): 1) a natural soil could

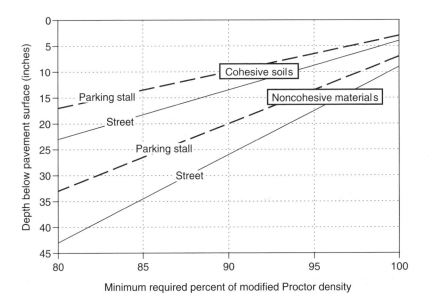

FIGURE 3.13 Minimum required compaction of subgrade and any pavement materials with CBR≤20 (U.S. Departments of the Army and Air Force, 1992, p. 4-1; residential street data is Design Index 3; parking stall data is Design Index 1).

have in-place density equal to or greater than the required value, so that no treatment is required during construction; 2) a subgrade with insufficient in-place density can be compacted to bring it to its required density; 3) the subgrade material can be removed and replaced with different material, and the new material compacted to its required density; or 4) a subgrade can be covered with sufficient additional material so that the uncompacted subgrade is at a depth where its in-place density is satisfactory.

CONSTRUCTION WITH LITTLE OR NO COMPACTION

In some porous pavement projects, subgrade compaction is deliberately avoided to preserve the soil's natural permeability. Compaction is most justifiably omitted in native cut (ICPI, 1997), or where the proposed surfacing material could tolerate undulations due to settlement as could turf, unbound aggregate, or geocells filled with turf or aggregate. Omission is also justified where a thick pavement structure is planned to compensate for the relatively low bearing value of uncompacted soil (AASHTO, 1993, p. I-14–I-15; Yoder and Witczak, 1975, p. 328). A thick pavement structure is ordinarily more expensive than subgrade compaction would be, but a permeable subgrade is a functioning part of a site's drainage system which may reduce off-pavement drainage costs.

The site-specific strength and permeability characteristics needed in the soil are the proper basis for subgrade compaction specifications (Rollings and Rollings, 1996, pp. 236–238). Natural in-place subgrade can be tested for its Proctor compaction and CBR and thereby its need for further strength and compaction during construction. Carefully limited compaction can maintain some subgrade permeability (Gray, 2002). Local experience may be a valuable guide to compaction requirements and to the likely outcome of proposed compaction practices.

In Cahill's installations of porous asphalt (Brown, 1996), the excavated subgrade is an infiltration "bed" that is not to be compacted, nor subjected to construction vehicle traffic. If the excavated surface is exposed even to rainfall, it is immediately scarified with a light rake to eliminate the thin crust of fines that forms under the erosive energy of falling raindrops. The engineer inspects the surface to authorize placement of geotextile on the bed. Base aggregate is placed over the fabric in 8-inch lifts (layers), starting at one edge of the infiltration bed. Aggregate trucks proceed across aggregate that has already been placed, and not across the fabric or any unprotected part of the subgrade. Each lift of base aggregate, unlike the subgrade, is lightly compacted, following the principle that compaction is increasingly necessary with proximity to the surface. The pavement thickness is ordinarily 18 inches or more. The great thickness compensates for the uncompacted subgrade by spreading out the traffic load. Some of these projects have been in place for more than 10 years without objectionable settlement. Specific examples will be illustrated in Chapter 12.

In some projects, staging of construction materials and equipment is needed in a spot where uncompacted subgrade is also desired for long-term drainage of a proposed porous pavement. Using pavement subbase material as a staging platform could protect the soil to a degree. Before any other construction, the pavement subgrade is excavated and a thick layer of subbase aggregate is placed. The subbase material is then a platform for construction equipment. An open-graded aggregate platform can be clean and stable; as a "rock blanket" it helps meet requirements for erosion and sediment control during construction. After completing other construction, clean base and surface course materials are placed on the subbase to finish the porous pavement.

New earth fill must always be compacted unless the proposed surface is of a type that can bear undulations from future settlement. Having been lifted and handled, the fill is in a loose condition subject to variable settlement, and can be made structurally reliable only by processing during construction. A porous concrete pavement that failed where it was placed on uncompacted fill will be discussed in Chapter 11. Fill material should be spread evenly in level lifts of limited thickness, and each lift compacted before the next is placed.

PAVEMENT ADAPTATION TO FREEZING

Upon freezing, water expands in volume by about 10 percent. The growing ice crystals exert pressure on anything nearby. Heaving movement takes place where the pressure of the growing crystals exceeds the load of the overlying structure.

Heaving can be followed by further damage caused by the loss of soil strength when the ice thaws. Ice's expansion pushes soil particles away from each other. Then as the ice melts, it simultaneously withdraws its support from the particles and adds liquid water to the soil. Under the pressure of subsiding soil particles, the water suspends the particles like those in quicksand. The material is left with very low strength and great susceptibility to deformation.

In regions with freezing winters it is vital either to select pavement surface materials that can tolerate undulations from frost-related movements, or to protect pavements from frost damage. Porous paving materials that can tolerate surface undulations are aggregate, turf, and geocells filled with aggregate or turf. Deck structures avoid frost damage by suspending their floors above heaving soil, concentrating their loads on footings placed below the frost penetration depth.

For pavement types that are sensitive to frost-related movement, several alternative design approaches are available for limiting frost damage (U.S. Departments of the Army and Air Force, 1992, p. 18-6; FAA, 1995, pp. 27–30; Goodings et al., no date). The approaches described here are based on placing within the frost penetration zone only those pavement components that are well-drained and nonsusceptible to frost damage, and limiting frost penetration into underlying components that hold water or that are susceptible to damage. A thick layer of paving material insulates underlying material from freezing temperatures, and its weight resists heaving pressure from frozen material below (AASHTO, 1993, p. I-14; Yoder and Witczak, 1975, p. 192). Protection from frost damage may require a thicker pavement than that required for bearing value in nonfrozen conditions.

As in all aspects of pavement design, "local experience should be given strong consideration for frost conditions" in addition to any approaches coming from general books and manuals (FAA, 1995, p. 29).

FROST DEPTH

The depth of frost penetration into a pavement and its subgrade is a function of temperature, the nature and moisture content of the material, and whether snow is left on the surface as insulation (FAA, 1995, p. 19). In many areas, local construction practices dictate a fixed nominal frost depth for the design of a proposed pavement, based on local experience. However, the depth of frost penetration varies from year to year with the severity and duration of freezing temperatures and amounts of moisture and snow.

Figure 3.14 shows the progress of frost during the winter of 1978–1979 in the porous asphalt parking lot at Walden Pond State Reservation in Concord, Massachusetts. The pavement had an aggregate base course to 38.5 inches below the surface, and sandy subgrade below. The curve shows where the temperature was exactly 32°F. Just inside the curve the temperature was just barely below freezing; the coldest temperature was near the pavement surface. Freezing progressed gradually into the ground as the winter continued. By the end of winter, the frost depth exceeded the depth of the entire pavement structure. At this extreme depth the frost remained only a few weeks; the ground was soon thawing from both above and below.

Potential frost depth can be estimated from above-ground air temperature recorded during the winter. The cumulative freezing temperature is expressed in freezing degree-days, using 31°F as the beginning of freezing (FAA, 1995, p. 16). By this measure, one day with an average temperature of 31°F represents one degree-day; one day at 22°F has 10 degree-days. The number of freezing degree-days accumulates during the winter season.

Recurrence interval, or frequency, is the average time between freezing events of a given magnitude; it is a way to express the probability that a given number of freezing degree-days and the corresponding frost depth might occur in a given year. A ten-year event is intense enough that it has recurred, on the average, in only one of every ten years in the local weather record. On average, the ten-year event has a 10 percent chance of occurring in any one year. There is always a risk that, in any one year, a freezing event

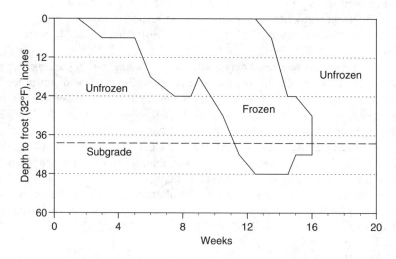

FIGURE 3.14 Frost depth in porous asphalt pavement at Walden Pond State Reservation, Concord, Massachusetts, December 1978 through April 1979 (after Wei 1986, pp. 5–15a, temperature probe 1).

selected for design could be exceeded. Designing for a deep, intense, infrequent frost limits the risk because it yields thick structures where frost seldom penetrates sensitive material. But thick structures are expensive to construct. A balance between risk and cost must be made that is appropriate to each pavement in its context.

The Federal Aviation Administration (1995, pp. 16–17) and the U.S. Departments of the Army and Air Force (1992, p. 18-7) have established the convention of designing pavements for the ten-year freezing event: the coldest winter observed in a ten-year period, or the average of the three coldest winters in a thirty-year period. In most years, frost penetration is not as deep as that during the ten-year event. On an average, every 11 years cumulative freezing days are greater, and frost penetration is correspondingly deeper. Figure 3.15 shows the ten-year freezing degree-days in the U.S. (this small-scale map should be used cautiously in mountainous areas, where temperature can vary greatly over a short distance).

Figure 3.16 shows the resulting penetration of frost into certain types of pavement and subgrade materials estimated by two different agencies. In making these estimates, both agencies assumed typical types of concrete, asphalt, or aggregate pavement materials in the first foot or so of depth below the surface. The assumed base materials have void space of 30 percent, like the open-graded materials commonly used in porous pavements. The curve labeled "FAA" is derived from the Federal Aviation Administration (1995, pp. 19–20). The other two curves are derived from the U.S. Departments of the Army and Air Force (1992, p. 18-11), and take into account the amount of moisture in the base and subgrade at the beginning of the freezing period. The various estimates are rather consistent: the FAA estimate is between the two Army-Air Force curves for most of the range of freezing degree-days shown in the chart.

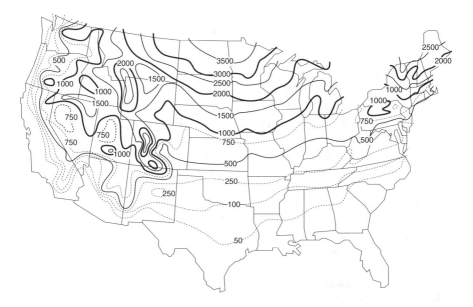

FIGURE 3.15 Ten-year freezing degree-days (°F-days; after FAA, 1995, p. 17).

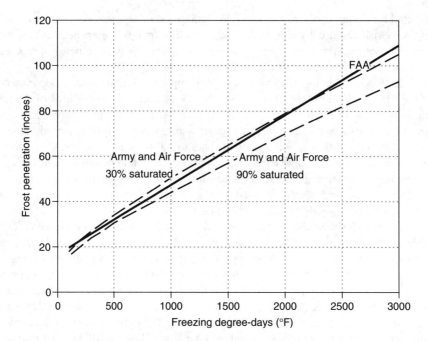

FIGURE 3.16 Depth of frost penetration below a pavement surface kept clear of snow and ice, estimated by two different agencies (data derived from FAA 1995, p. 20, and U.S. Departments of the Army and Air Force, 1992, p. 18-11, assuming base aggregate and subgrade have solid particle density of 165 pounds per cubic foot and void space of 30 percent).

Protection of Subgrade

Moisture suspended in subgrade soil can heave upon freezing, especially where growing ice is supplied with moisture, as it may be beneath a porous pavement. Fine-grained soil is particularly susceptible because moisture suspended in numerous small pores can migrate toward growing ice crystals, feeding and enlarging them (AASHTO, 1993, p. I-23; U.S. Departments of the Army and Air Force, 1992, p. 18-3). The formation of the ice out of the surrounding liquid moisture is called "ice segregation." A growing ice mass frequently takes the form of a horizontal vein or lens.

The same type of ice formation does not occur in well drained, coarse-grained soils such as gravel or coarse sand, because their large open pores suspend insufficient water to feed a growing ice mass (U.S. Departments of the Army and Air Force, 1992, p. 18-3). Instead, water freezes in isolated grains surrounded by open pore space without heaving the soil. Tests for the susceptibility of soils to frost damage are standardized in ASTM D 5918.

Table 3.13 lists the relative frost-damage susceptibilities of different soil and aggregate types (U.S. Departments of the Army and Air Force, 1992, pp. 18-4–18-5). The rating is based principally on Unified soil classification and the percentage of particles smaller than 0.02 millimeter. At the top of the table, open-graded crushed stone free of fine particles is non-frost-susceptible. The S1 and S2 groups are relatively

TABLE 3.13
Classification of Soil Frost Susceptibility

Frost Group	Frost Susceptibility	General Character	Percentage by Weight Finer than 0.02 mm	Examples of Unified Classification
Non-frost-susceptible (NFS)	Negligible to low	Crushed stone	0 to 1.5	GW, GP
		Sands	0 to 3	SW, SP
Potentially frost-susceptible (PFS)	Negligible to medium	Crushed stone	1.5 to 3	GW, GP
		Sands	3 to 10	SW, SP
S1	Very low to medium	Gravelly soils	3 to 6	GW, GP
S2	Very low to medium	Sandy soils	3 to 6	SW, SP
F1	Very low to high	Gravelly soils	6 to 10	GM
F2	Low to high	Gravelly soils	10 to 20	GM
		Sands	6 to 15	SM
F3	Medium to very high	Gravelly soils	>20	GM, GC
		Sands	>15	SM, SC
		Clays with PI>12	—	CL, CH
F4	Medium to very high	All silts	—	ML, MH
		Very fine silty sands	>15	SM
		Clays with PI<12	—	CL, CL-ML
		Varved clays	—	CL, CH, ML, SM

PI is Plasticity Index (U.S. Departments of the Army and Air Force (1992, p. 18-5); Rollings and Rollings (1996, p. 396)).

nonsusceptible to frost, to the extent that they are eligible for use as pavement subbase material. The F1 through F4 groups must generally be protected from frost where heave would be damaging. At the very bottom of the table, silty soils are among the most susceptible, because their void spaces are small enough to support networks of growing ice crystals, but not small enough to inhibit the suction of additional water into a growing crystal mass. Materials in the "potentially frost-susceptible" group — sands and crushed aggregates with moderate fines content — require laboratory determination to be certain of their susceptibility to frost. The result of that determination might likely put the material into the NFS, S1, or S2 category.

Complete protection of the subgrade from frost is accomplished by providing a sufficient thickness of non-frost-susceptible base and subbase material to eliminate frost penetration into the subgrade (FAA, 1995, p. 27). It is the most reliable method of subgrade protection. It is also the most costly method because of the great thickness of pavement structure that may be required. The U.S. Departments of the Army and Air Force manual (1992, p. 18-6) considers it "nearly always uneconomical and unnecessary." However, it may be justified for highly frost-susceptible subgrades and pavements that are extraordinarily intolerant of slight deformations (U.S. Departments of the Army and Air Force, 1992, p. 18-16; FAA, 1995, p. 29).

A less expensive alternative is to allow limited subgrade frost penetration. This approach permits a small and infrequent amount of frost penetration into any subgrade, including frost-susceptible ones, during the ten-year freezing event (U.S. Departments of the Army and Air Force 1992, pp. 18-6–18-7). In the standards of the FAA (1995, p. 27), non-frost-susceptible material is required for the upper 65 percent of the ten-year frost penetration depth; the bottom 35 percent may be in frost-susceptible subgrade. FAA's tolerance of the resulting small amount of frost damage is based on its experience. Some other agencies have required non-frost-susceptible material only in the upper 50 percent of the frost penetration depth (AASHTO, 1993, p. I-8). The limited-penetration design approach is implicitly followed in many cold regions, where the thickness of pavement structure is routinely less than the frost penetration depth. The degree to which this approach increases long-term maintenance costs depends on pavement type, traffic type, general setting, and user expectations.

To choose between the complete and limited protection approaches, one could first design a pavement section for nonfrozen conditions, then determine what modification would be needed to bring the pavement into compliance with each of the two approaches. An informed choice can then be made in consideration of cost and degree of protection. For example, it may be found that a proposed pavement would satisfy the limited-protection approach without modification, while it would require additional material and expense to satisfy the complete-protection approach.

Protection of Pavement Reservoir

If water ponded in a pavement's reservoir freezes, its expansion has some potential to heave the pavement. However, the possibility of a reservoir of water ponded in a porous pavement was not anticipated in pavement design manuals such as those of AASHTO (1993), FAA (1995, pp. 16–19) and U.S. Departments of the Army and Air Force (1992, chapter 18). All that can be presented here is an analysis of the potential problem and empirical experiences in the field.

Reservoirs that are rapidly drained by highly permeable subgrades or lateral drainage pipes face relatively little danger from frost because they seldom hold water long enough for it to freeze in place. Reservoirs that are slowly drained by slowly permeable subgrades, without any other drainage outlet, may display a relatively high susceptibility to frost heave because they hold some water relatively frequently, and, in addition, tend to maintain high moisture levels in their subgrades. A slow-draining reservoir could be made less susceptible to frost damage by the addition of a lateral drainage pipe at some elevation in the reservoir. The addition might convert the hydrologic function of the reservoir from primarily soil infiltration to primarily detention and lateral drainage.

For any reservoir, an extremely conservative approach to frost protection would completely prevent ten-year frost penetration into the top of the reservoir. A complete-protection approach would be reliable, but could be costly because of the great thickness of pavement material required above the reservoir. This approach might be justified for very slowly draining reservoirs that are expected to be full frequently or for significantly long periods, under pavement surfaces that would be intolerant of heaving distortion. Above the reservoir, where frost would penetrate, paving material can be made unsusceptible to frost damage by assuring that it is well drained and by using open-graded material that does not suspend pore water (U.S. Departments of the Army and Air Force, 1992, pp. 18-4–18-5).

Porous Pavement Structure 99

A less conservative approach would allow the ten-year frost to penetrate some depth into the reservoir. This approach is less costly than the complete-protection approach because it does not require such a thick pavement. It might be justified in reservoirs that are quickly drained and under pavement surfaces that are tolerant of distortion.

EXPERIENCES IN COLD CLIMATES

The following examples of installed porous asphalt and block pavements in cold climates have been successful from the point of view of the expectations of their users. They exemplify a partial-protection approach for the subgrade, and less protection for the reservoir. They drain rapidly. Those pavements located in the coldest places have

FIGURE 3.17 Porous pavement of open-jointed Eco-Stone blocks at the Thornbrough Building at the University of Guelph, Guelph, Ontario.

FIGURE 3.18 Porous pavement of open-jointed Eco-Stone blocks at Sunset Beach Park, Richmond Hill, Ontario.

allowed the frost to penetrate most deeply below the reservoir and into the subgrade, perhaps because surface distortions resulting from frost are most tolerated in the coldest climates. These successful case studies encourage optimism that constructing reservoirs in cold climates without the expense of excessively thick pavements may be possible. Further examples of porous pavements in cold climates will be presented in later chapters. Unfortunately, no porous pavement located in a notably cold climate has been identified that failed due to frost damage; studying a combination of failures and successes is necessary to determine the exact limits of prudent practice.

Figure 3.17 shows a porous pavement of open-jointed "Eco-Stone" block at the University of Guelph in Ontario, where the nominal frost depth is about 72 inches

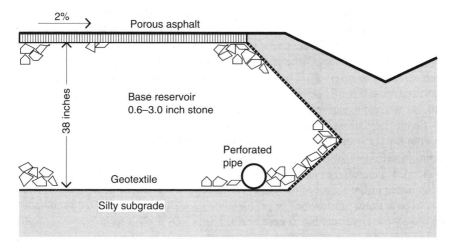

FIGURE 3.19 Construction of porous asphalt street in Luleå, Sweden (adapted from Stenmark, 1995).

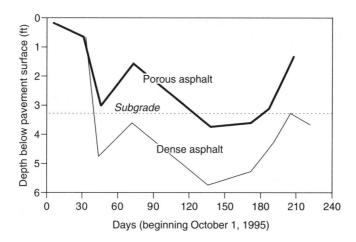

FIGURE 3.20 Depth of freezing temperature beneath comparative pavements in Luleå, Sweden (data from Backstrom, 2000).

(William James, personal communication, 2002; Thompson and James, 1994). In the ten years to the time of the photo the pavement experienced no objectionable distortion from frost heave. The pavement's total thickness is 22 inches, so the entire reservoir is well within the frost penetration depth. The reservoir is drained rapidly by a lateral drainage pipe and sandy, well-drained subgrade; the accumulation of water in the reservoir never lasts for more than a few hours.

Figure 3.18 shows an Eco-Stone pavement at Sunset Beach Park adjacent to Lake Wilcox in Richmond Hill, Ontario. The nominal frost depth here is about 72 inches. The total pavement thickness is 18 inches, so the entire reservoir is well within the frost depth. Although the subgrade soil is sandy, the reservoir could in a sense be characterized as slowly draining due to shallow groundwater, which presumably parallels the adjacent lake surface. The water table is seasonally near the bottom of the base course, where it can inhibit reservoir drainage; groundwater may encroach into the reservoir at times of high water. The photo was taken at the end of the installation's third winter. The structural condition at that time was excellent, with no noticeable surface displacement.

Figure 3.19 shows a section through a porous asphalt pavement on a residential street in Luleå, Sweden, within 1° of the Arctic Circle (Stenmark, 1995). Luleå's subarctic climate is roughly analogous to that of central Canada (Espenshade and Morrison, 1984, pp. 8–11). A large part of the annual precipitation comes in the form of snow, and accumulates in the winter snow pack. The greatest occurrences of liquid stormwater tend to be at times of snowmelt or precipitation on melting snow. The silty subgrade at the project site is susceptible to frost heave. The porous asphalt surface course overlies a 38-inch thick base reservoir of coarse aggregate with 35 to 40 percent porosity. A geotextile envelopes the aggregate to prevent intrusion of fine material. A perforated pipe at the bottom drains to a ditch off the road. Water that infiltrates through the porous surface into the base reservoir either percolates into the subgrade, or discharges through the pipe. The water table is more than four feet below the surface year-round.

The Luleå pavement has not been measurably distorted by frost. A large part of the explanation is illustrated in Figure 3.20, which shows the depth of freezing temperature during the winter of 1995–1996 as monitored by temperature probes. The porous pavement's base and subgrade froze less readily than those under nearby dense asphalt because the porous pavement's higher water content, constantly supplied by infiltration, increased latent heat (Backstrom, 2000). They thawed more rapidly because they were assisted by infiltrating meltwater. On the whole, frost penetration was shallower under the porous asphalt than under dense asphalt, and its duration was shorter. Consequently, the porous pavement had less cause for frost damage. The pavement and its context will be described further in Chapter 12.

In cold climates, protection of the reservoir has seldom been practiced to the same degree as protection of the subgrade. Perhaps protection of the reservoir is a less critical requirement. Ponded water is in a sense self-insulating (Penner, 1962). As water freezes, it delays freezing temperature from penetrating farther into unfrozen water below because of the heat made available from the water's change of state. Freezing a pound of water requires the withdrawal of 700 times more thermal energy than to change the temperature of a pound of dry soil by 1°F. In addition, the open-graded materials used for base reservoirs in many porous pavements may be tolerant of some freezing of water ponded in them: if a shallow floating lens of water freezes, it may expand into the open voids immediately above without heaving the pavement. When a layer of ice develops sufficient thickness so that further freezing water no longer has the ability to expand into open void space, the characteristic uniformity of pavement base material may limit differential heaving movement (Penner, 1962). Research is called for on the relationships between ponding, temperature, change of state, and heaving in porous pavement reservoirs. Long-term experience with aggressive installations in cold climates will be an appropriate part of the basis of firm reservoir frost-protection standards that may be developed in the future.

PAVEMENT ADAPTATION TO SPECIAL SUBGRADES

Moisture modifies the bearing value of soils. Fluctuations in moisture induce some soils to move of their own accord; some pockets of soil are extraordinarily weak. Every project-specific condition requires special adaptation in pavement design.

ADAPTATION TO WET SUBGRADE

The presence of water reduces the bearing value of almost any subgrade soil (AASHTO, 1993, pp. I-27–I-28). Porous pavements that allow moisture into the subgrade must compensate for it with a correspondingly thick pavement structure to spread out the load. A thicker pavement structure requires greater construction cost, but a porous pavement has a hydrologic function that could reduce the cost of off-pavement stormwater management structures.

The Rada et al., thickness requirements, illustrated in Figure 3.10 for residential streets, exemplify how much thicker a porous pavement may have to be. In the figure, with base material of CBR 30, and in soils of low bearing value (CH), moving from a "dry" or "medium" soil drainage category to a "wet" one requires increasing the base thickness by several inches, from 7 or 10 inches to 14 inches For soils of moderate (ML or SC) to high (GW) bearing value, no increase in thickness is necessary to compensate for increased moisture.

In the mid-Atlantic area, many of the porous asphalt parking lots by Cahill Associates (www.thcahill.com) have had aggregate base courses at least 18 inches thick; the total pavement is over 20 inches thick. This substantial thickness has successfully compensated for both the wetness of the subgrade under the porous surface, and Cahill's routine specification against subgrade compaction. This experience confirms the theoretical prediction of the Rada et al. method that great thickness compensates for wet subgrade. It proves empirically that, in the conditions of the mid-Atlantic region, an 18 inch base is sufficient or more than sufficient for stability over wet, uncompacted subgrade in moderately trafficked parking lots. Many case studies presented later in this book illustrate experiences with other pavement thicknesses.

Where the subgrade soil would be incapable of supporting a porous pavement's expected traffic load if saturated by infiltrating water, an impermeable liner at the bottom of the base reservoir may be called for. A liner's prevention of soil infiltration would limit the reservoir's hydrologic function to detention and water quality treatment.

ADAPTATION TO SWELLING SUBGRADE

Some soils swell when their moisture content increases, and shrink again when dry (Nelson and Miller, 1992; Rollings and Rollings, 1996, pp. 381–384). The moisture changes that cause swelling and shrinking can occur from season to season, between wet and dry years, or during extraordinary individual weather events. The most active soils are in semiarid environments such as Colorado and central Texas, where cycles of wetting and complete drying are relatively frequent; swelling activity is also known in many other areas. Where insufficient overlying weight counteracts the swelling pressure, the soil is capable of lifting, cracking, and displacing pavements and curbs. Figure 3.21 shows modest damage to a pavement and its curb built on swelling soil in Texas.

The potential of a soil to swell most commonly depends upon from the amount of montmorillonite clay in the soil (Nelson and Miller, 1992, p. 9). Montmorillonite chemically and physically attracts water molecules into the spaces between its tiny clay plates. As water is pulled in, the plates are pushed apart. Many montmorillonitic soils can expand up to 10 to 20 percent by volume.

USDA soil surveys, geologic maps, and experienced local professionals can indicate the general possibility of swelling soil in a locale. In some areas, swelling is associated with highly specific geologic strata or soil series, such as the Pierre formation along the Front Range in Colorado, the Yazoo clay of Alabama and

FIGURE 3.21 Cracking of a concrete curb and dense asphalt pavement resulting from swelling of the Houston black clay in Arlington, Texas.

Mississippi, certain elevations in the Hawthorne formation in northern Florida, the Whitehouse series of North Carolina, and some soils in the central valley of California (Nelson and Miller, 1992, p. 25).

The presence of a swelling soil can sometimes be confirmed in the field. When a swelling soil is dry it exhibits either a puffy structure like popcorn, or wide deep cracks; when it is wet, it is very sticky. It may be associated with undulations ("gilgai") in the soil surface (Gustavson, 1975). In profile, many swelling soils lack distinct horizons, because during dry periods surface sediment falls into open cracks, after which soil churning mixes all the materials together (Mitchell, 1999). Soil swelling can be further confirmed with drilling and testing (ASTM D 4829; Nelson and Miller, 1992, pp. 25–27, 51). Common tests are those of soil classification, mineralogy, swelling, and Atterberg limits. Expansive clays are commonly in the CH Unified classification.

Several types of porous paving materials are tolerant of surface undulations that could result from soil swelling: unbound aggregate, turf, and geocells with turf or aggregate fill. Although these materials are limited to applications with low traffic loads, in those applications they infiltrate rainwater into the soil while deforming flexibly with subgrade expansion and contraction as the moisture level fluctuates. Installations of tolerant surfaces on swelling soils will be illustrated in later chapters.

Deck structures can be used over swelling soils so long as the footings are placed below the actively swelling portion of the soil. A deck's beams must be at least a few inches above the soil surface to isolate them from upward-pressing soil and to concentrate the weight of the structure on the piers.

Surfaces of asphalt, concrete, blocks, and grids are intolerant of subgrade swelling. For pavements of these kinds, one way to inhibit subgrade swelling is to lay the pavement structure on a thick subbase of nonheaving material. The weight of the material counteracts the swelling pressure in the underlying soil. This approach is similar to that used for counteracting frost heave in cold climates.

Alternatively, the subgrade can be isolated from changes in moisture that would cause swelling. This approach allows at most the upper layers of a pavement structure to be porous and permeable, there is an impermeable layer at some level in the pavement (AASHTO, 1993, p. I-14; Yoder and Witczak, 1975, p. 342). Figure 3.22 shows the use of a geomembrane to isolate a substantial thickness of subgrade soil from surface moisture and from adjacent zones where changes in moisture could occur.

ADAPTATION TO PLASTIC SUBGRADE

Soft pockets of soil are easily recognized during earthwork operations as they deform ("pump") when heavy equipment drives over them. They present weak subgrade which could deform under future traffic load. Though not necessarily problematic for proposed pavement types that can tolerate gentle surface undulations such as grass, aggregate, and geocells filled with grass or aggregate, to support a less tolerant pavement such as asphalt, concrete, or blocks, however, pockets of soft plastic or organic subgrade must be corrected or compensated for before a pavement structure is placed on them.

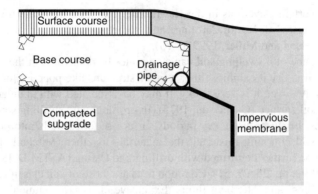

FIGURE 3.22 Isolation of potentially swelling subgrade from moisture changes.

The addition of woven geotextile to the bottom of a pavement structure can help support the pavement on soft subgrade by adding tensile strength.

The soil itself can be stabilized by mixing aggregate or cement into it and by compacting (AASHTO, 1993; U.S. Departments of the Army and Air Force, 1995; Rollings and Rollings, 1996, p. 249). However, binders and compaction tend to reduce soil permeability.

Alternatively, small areas of soft soil might be economically removed and replaced with more stable soil or aggregate. The thickness of the replacement material spreads out the traffic load on the remaining subgrade, making deformation less likely. Increasing the thickness of already-planned open-graded base material increases stability, hydraulic capacity, and protection from frost, without necessarily reducing soil permeability.

PAVEMENT ADAPTATION AT EDGES

The edges of pavements are potentially their weakest parts. Here the base and surface courses come to an end, ending the lateral spreading of load and resistance to deformation that exist elsewhere in the pavement. An unsupported or unreinforced edge might crack or subside soon after installation. Several forms of edge support and restraint are available.

SUPPORT OF PAVEMENT EDGES

In almost any pavement with multiple layers or courses, each successive lower layer should extend beyond the edge of the layer above, as exemplified in Figure 3.23. The outward extension provides lateral support for the upper layers equivalent to that at the center of the pavement. A guideline used by some practitioners is that the extension for a given layer should be at least equal to the thickness of the layer.

A rigid surface course may require strengthening at the edge to prevent cracking where vehicles may travel close to or across the edge. Thickening the slab is a common response to this requirement. Examples are shown in Figure 3.24.

Porous Pavement Structure

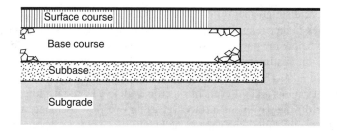

FIGURE 3.23 Outward extension of base and subbase courses.

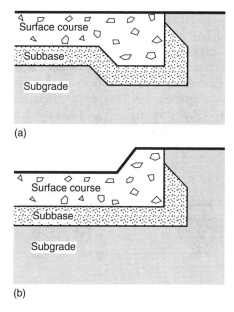

FIGURE 3.24 Alternative approaches to thickening a rigid pavement's edge.

RESTRAINT OF FLEXIBLE PAVEMENT EDGES

Flexible pavements require restraint against lateral deformation at the edge, where traffic load would otherwise push out the flexible material. Almost any block or grid pavement requires restraint on all sides (Burton, 2000; ICPI, 2000; NCMA, 1994). Asphalt, aggregate, and mulch require edge restraint where their settings require crisp, firm, continuous edges.

Curbs are any relatively rigid construction units used for edge restraint. Curbs can be made of almost any rigid material including poured-in-place concrete, units of stone or precast concrete large enough to anchor themselves in the base course and resist lateral forces, recycled plastic or timber pinned to the base course, or mortared masonry. It is common for curbs to be installed first, and then for the surface material to be placed between the curbs, an exception being certain installations of blocks in which a concrete curb is at least partly troweled into place after the blocks are laid.

Figure 3.25 distinguishes different curb configurations. Raised curbs are usually about 6 inches higher than the adjacent pavement; flush and submerged curbs are level with or below the pavement surface. All configurations perform the structural function of edge restraint. However they have the distinctive nonstructural effects listed in Table 3.14. They can be selected for specific projects according to the particular considerations and the costs of specific available materials. Flush curbs allow free drainage off the pavement surface, releasing overflow drainage and potentially clogging debris; they mark the edge visually but do not physically confine vehicular traffic or inhibit pedestrian accessibility. Raised curbs confine drainage and debris to the pavement surface and obstruct pedestrian accessibility, but they can be notched at intervals to improve surface drainage or to give access to pedestrian ramps.

Another approach to edge restraint for flexible surface courses is a thin strip of metal, plastic, or wood, held in place by stakes or pins. Many types of thin strips are less costly than curbs. They can be used in settings such as residential driveways and pedestrian walkways where they will not have to withstand the weight of numerous vehicles. Figure 3.26 shows two possible configurations. Table 3.15 lists some manufacturers of metal and plastic edgings. The ability of most models to resist lateral pavement deformation depends mostly on the stakes that are driven into the base course, so it is vital for the stakes to be long, strong, closely spaced, and firmly attached to the edging strip. Some models of edging are designed specifically to

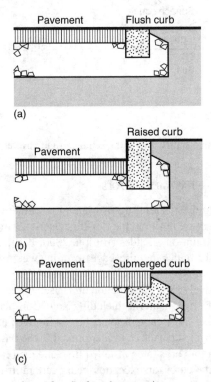

FIGURE 3.25 Configurations of curbs for edge restraint.

TABLE 3.14
Some Nonstructural Effects of Curb Configurations

	Flush and Submerged Curbs	Raised Curb
Drainage	Free movement of runoff off the pavement surface	Confinement of runoff to the pavement surface
Debris removal	Free movement of debris off the pavement surface	Confinement of debris to the pavement surface
Pedestrian accessibility	Continuous accessibility across pavement edge	Obstacle to movement at pavement edge
Control of vehicular traffic	Visual marking of pavement edge	Visual marking of pavement edge and physical confinement of traffic

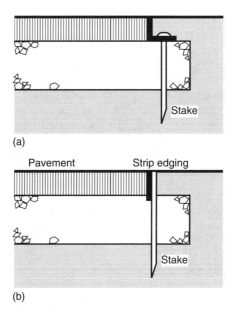

FIGURE 3.26 Examples of thin edgings.

restrain blocks, grids, or asphalt. Structural edging for pavements must be distinguished from "landscape edging" that is intended only to define planting areas.

PROTECTION OF PAVEMENT EDGES

A supplement or alternative to structural edge restraint is traffic control to keep vehicles away from pavement edges. This approach is useful at pavement curves and corners (where vehicles may veer over the pavement edge) and at the ends of parking stalls (where vehicles could roll beyond the pavement edge). Among the devices to control traffic are wheel stops, bumper fences, and bollards.

TABLE 3.15
Examples of Manufacturers of Pavement Curbing and Edge Restraints

Company	Contact Information
Barco Products	www.barcoproducts.com
Border Concepts	www.borderconcepts.com
Brickstop	www.brickstopcorportion.com
Carderock Stone	www.carderock.com
Collier Metal Specialties	www.colmet.com
Country Stone	www.countrystone.com
Curv-Rite	www.curv-rite.com
Dimex	www.edgepro.com
J. D. Russell Co.	www.jdrussellco.com
Endicott Clay Products	www.endicott.com
Gappsi	www.gappsi.com
Granicor	www.granicor.com
Polycor	www.polycor.com
Integrated Paving Concepts	www.streetprint.com
J.D. Russell	www.jdrussellco.com
Oly-Ola	www.olyola.com
Nitterhouse Concrete Products	www.nitterhouse.com
Park Structures	www.parkstructures.com
Pave Tech	www.pavetech.com
Permaloc	www.permaloc.com
Petersen Mfg. Co.	www.petersenmfg.com
Pine Hall Brick	www.pinehallbrick.com
Selectech	www.selectechinc.com
Snap Edge	www.snapedgeusa.com
Sure-loc Edging Corp.	www.surelocedging.com
Surfacing Systems	www.surfacingsystems.com
Wausau Tile	www.wausautile.com
Whitacre Greer	www.wgpaver.com

Figure 3.27 shows an example of a wheel stop. The open intervals between wheel stops release overflow drainage and surface debris from the pavement surface. Wheel stops manufactured from plastic or concrete can be purchased or fashioned on-site from timber. All wheel stops need to be fixed to the pavement, usually with metal pins driven into the pavement's base course.

Bumper fences are low fences at the ends of parking stalls. The footings may be located under the pavement, with the fence posts protruding from the pavement surface or just off the edge of the pavement, making the fence free-standing. Figure 3.28 shows an example. Solid construction is necessary to prevent damage from moving cars.

Bollards are standing posts as shown in Figure 3.29. They confine vehicular traffic without confining drainage. Bollards are manufactured variously of concrete, metal, recycled plastic, stone, and other materials. They can also be fashioned from timber, masonry, or other materials. Many bollards are carefully shaped and colored; some

Porous Pavement Structure

FIGURE 3.27 Concrete wheel stops at the Castaic Lake Water Conservatory in Santa Clarita, California, held by metal pins to a cast-in-place concrete edge; the parking lot pavement is Eco-Stone open-jointed block.

come with integral light fixtures (Elber, 1994). They can be located either integrally with a pavement, or just off the edge of the pavement. Table 3.16 lists some manufacturers of these devices.

Painted lines and decorative objects such as shrubs and boulders can be placed at a pavement edge to warn traffic. Their effect in controlling traffic is visual rather than physical.

TRADEMARKS

Table 3.17 lists the holders of registered trademarks mentioned in this chapter.

FIGURE 3.28 Timber bumper fence at the porous asphalt parking lot at Walden Pond State Reservation in Concord, Massachusetts.

TABLE 3.16
Examples of Manufacturers of Wheel Stops, Bumper Fences, and Bollards

Supplier	Contact Information
Ameron International	www.ameron-intl.com
Barco Products	www.barcoproducts.com
BRP Enterprises	www.brponline.com
Canterbury International	www.canterburyintl.com
Dura Art Stone	www.duraartstone.com
Polycor	www.polycor.com
LSI Greenlee Lighting	www.lsi-industries.com
Interlock Concrete Products	www.interlock-concrete.com
Ironsmith	www.ironsmith.cc
Keystone Ridge	www.keystoneridgedesigns.com
Kim Lighting	www.kimlighting.com
Louis Poulsen & Company	www.louispoulsen.com
Nitterhouse Masonry Products	www.nitterhouse.com
Petersen Mfg. Co., Concrete Leisure Division	www.petersenmfg.com
Plastic Lumber	www.plasticlumber.com
Hammer's Plastic Recycling	www.hammersplastic.com
Prosec	www.prostop.com
Quick Crete Products	www.quickcrete.com
SPJ Lighting	www.spjlighting.com
Spring City Electrical Manufacturing Co.	www.springcity.com
Stewart Iron Works	www.stewartironworks.com
W. J. Whatley, Inc.	www.whatley.com
Wausau Tile, Inc.	www.wausautile.com

FIGURE 3.29 Metal bollards confining vehicles to a concrete pavement and away from a pedestrian block pavement.

ACKNOWLEDGMENTS

The following persons generously provided information used in this chapter: Donald Bright and Gary Sander of Tensar Corporation (www.tensarcorp.com); Jack Clausen of the University of Connecticut; Craig Ditzler, National Soil Survey Center, Lincoln, Nebraska; William James, University of Guelph; Charles Machemehl of the Georgia Crushed Stone Association; Charles Pryor of the National Stone, Sand and Gravel Association (www.aggregates.com); Harry E. Stewart of Cornell University; and Robert Sykes of the University of Minnesota.

The following persons constructively critiqued early drafts of this chapter: Benjamin K. Ferguson, an engineering student at Cornell University; William James of

TABLE 3.17
Holders of Registered Trademarks Mentioned in This Chapter.

Registered Trademark	Holder	Headquarters Location
Ameron	Ameron International	Pasadena, California
Barco	Barco Products	Batavia, Illinois
Border Concepts	Border Concepts	Charlotte, North Carolina
BP	BP p.l.c.	London, UK
Brickstop	Brickstop	Toronto, Ontario
BRP	BRP Enterprises	Lincoln, Nebraska
Canterbury	Canterbury International	Los Angeles, California
Carderock	Carderock Stone	Bethesda, Maryland
Carthage Mills	Carthage Mills	Cincinnati, Ohio
Collier Metal Specialties	Collier Metal Specialties	Garland, Texas
Country Stone	Country Stone	Rock Island, Illinois
Curv-Rite	Curv-Rite	Wayland, Michigan
Dimex	Dimex	Marietta, Ohio
Dura Art Stone	Dura Art Stone	Fontana, California
Eco-Stone	F. von Langsdorff Lic. Ltd.	Mississauga, Ontario
Endicott	Endicott Clay Products	Fairbury, Nebraska
Gappsi	Gappsi	Commack, New York
Granicor	Granicor	St. Augustin, Quebec
Grasspave^2	Invisible Structures	Golden, Colorado
Gravelpave^2	Invisible Structures	Golden, Colorado
Hammer's Plastic	Hammer's Plastic Recycling	Iowa Falls, Iowa
Integrated Paving Concepts	Integrated Paving Concepts	Surrey, British Columbia
Interlock Concrete Products	Interlock Concrete Products	Jordan, Minnesota
Ironsmith	Ironsmith	Palm Desert, California
J.D. Russell	J.D. Russell	Tucson, Arizona
Kim Lighting	Kim Lighting	City of Industry, California
Louis Poulsen	Louis Poulsen & Company	Ft. Lauderdale, Florida
LSI Industries	Greenlee Lighting	Cincinnati, Ohio
Maccaferri	Maccaferri, S.p.A.	Williamsport, Maryland
Mirafi	Mirafi Construction Products	Pendergrass, Georgia
Nitterhouse	Nitterhouse Masonry Products	Chambersburg, Pennsylvania
Oly-Ola	Oly-Ola	Villa Park, Illinois
Park Structures	Park Structures	Coral Springs, Florida
Pave Tech	Pave Tech	Prior Lake, Minnesota
Permaloc	Permaloc	Holland, Michigan
Petersen Mfg.	Petersen Mfg. Co.	Denison, Iowa
Pine Hall Brick	Pine Hall Brick	Winston-Salem, North Carolina
Plastic Lumber	Plastic Lumber	Akron, Ohio
Polycor	Polycor	Quebec City, Quebec
Prosec	Prosec	Bensalem, Pennsylvania
Quik Crete	Quick Crete Products	Norco, California
Selectech	Selectech	Taunton, Massachusetts
SI	SI Geosolutions	Chattanooga, Tennessee
Snap Edge	Snap Edge	St. Charles, Illinois
SPJ	SPJ lighting	South El Monte, California

TABLE 3.17 Continued

Registered trademark	Holder	Headquarters Location
Spring City	Spring City Electrical Manufacturing Co.	Spring City, Pennsylvania
Stewart Iron Works	Stewart Iron Works	Covington, Kentucky
Sure-Loc	Sure-loc Edging Corp.	Holland, Michigan
Surfacing Systems	Surfacing Systems	Hicksville, New York
Tensar	Tensar	Atlanta, Georgia
Terratex	Interface Fabrics Group	Atlanta, Georgia
Trevira	Trevira	Frankfurt, Germany
W.J. Whatley	W. J. Whatley	Commerce City, Colorado
Wausau Tile	Wausau Tile	Wausaw, Wisconsin
Webtec	Webtec	Charlotte, North Carolina
Whitacre Greer	Whitacre Greer	Alliance, Ohio

the University of Guelph and Computational Hydraulics International (www.chi.on.ca); Raymond Rollings of the Army Corps of Engineers (www.crrel.usace.army.mil); and David R. Smith of the Interlocking Concrete Pavement Institute (www.icpi.org).

REFERENCES

American Association of State Highway and Transportation Officials, 1993, *AASHTO Guide for Design of Pavement Structures 1993*, Washington: American Association of State Highway and Transportation Officials.
Backstrom, Magnus (2000). Ground Temperature in Porous Pavement During Freezing and Thawing, *Journal of Transportation Engineering* 126, 375–381.
Bader, Charles D. (2001). Soil Compaction, *Grading & Excavation Contractor* 3, 8–23.
Brady, Nyle C., and Ray R. Weil. (2000). *Elements of the Nature and Properties of Soils*, Upper Saddle River, New Jersey: Prentice-Hall.
Brown, Daniel C. (1996). Porous Asphalt Pavement Rescues Parking Lots, *Asphalt Contractor* Jan. 1996, 70–77.
Burak, Rob (2001). Bases for Interlocking Concrete Pavements — The Foundation of Your Business, *Interlocking Concrete Pavement Magazine* November 2001, 22–28.
Burak, Rob (2002). Bases for Interlocking Concrete Pavements — The Foundation of Your Business: Part II, *Interlocking Concrete Pavement Magazine* February 2002, 16–18.
Burton, Brian (2000). Edge Restraints for Interlocking Concrete Pavements, *Landscape Architect and Specifier News* 18–20.
Cedergren, Harry R. (1989). *Seepage, Drainage, and Flow Nets*, 3rd ed., New York: John Wiley and Sons.
Croney, David, and Paul Croney. (1998). *Design and Performance of Road Pavements*, 3rd ed., New York: McGraw-Hill.
Duffy, Daniel P. (1997). Structural Reinforcement of Roadway Pavements with Geosynthetics, *Erosion Control* 4, 26–37.
Elber, Gail. (1994). Bollards, *Landscape Architecture* 84, 36–39.

Espenshade, Edward B., Jr., and Joel L. Morrison (1984). *Goode's World Atlas,* 16th ed., New York: Rand McNally.

Goodings, Deborah J., David Van Deusen, Maureen Kestler, and Billy Connor. (no date). *Frost Action*, A2L04, Washington: Transportation Research Board Committee on Frost Action (www.nationalacademies.org/trb).

Gray, Donald H. (2002). Optimizing Soil Compaction, *Erosion Control* 9, 34–41.

Gustavson, Thomas C. (1975). *Microrelief (Gilgai) Structures on Expansive Clays of the Texas Coastal Plain — Their Recognition and Significance in Engineering Construction*, Circular 75-7, Austin: University of Texas Bureau of Economic Geology.

Interlocking Concrete Pavement Institute (ICPI) (1995). *Structural Design of Interlocking Concrete Pavement for Roads and Parking Lots*, Tech Spec No. 4, Washington: Interlocking Concrete Pavement Institute.

Interlocking Concrete Pavement Institute (ICPI) (1997). *Concrete Grid Pavements*, Tech Spec No. 8, Washington: Interlocking Concrete Pavement Institute.

Interlocking Concrete Pavement Institute (ICPI) (2000). *Edge Restraints for Interlocking Concrete Pavements*, Tech Spec No. 3, Washington: Interlocking Concrete Pavement Institute.

Jewell, Linda. (1982). Granite Curbing, *Landscape Architecture* 72, 97–100.

Jewell, Linda. (1986). Stone and Concrete Bollards, *Landscape Architecture* 76, 93–96 and 112.

Koerner, Robert M. (1998). *Designing with Geosynthetics*, 4th ed., Upper Saddle River, New Jersey: Prentice-Hall.

Mitchell, Martha S. (1999). Dynamic Soils, Working with Earth that Heaves, Turns, Churns, Slips, and Slides, *Erosion Control* 6, 68–71.

Morse, Aric A., and Roger L. Green. (2003). Pavement Design and Rehabilitation, chapter 3 of *Highway Engineering Handbook*, 2nd ed., Roger L. Brockenbrough and Kenneth J. Boedecker, Eds., New York: McGraw-Hill.

National Concrete Masonry Association, (NCMA). (1993) *Structural Design of Interlocking Concrete Pavements for Roads and Parking Lots*, TEK 11-4, Herndon, Virginia: National Concrete Masonry Association.

National Concrete Masonry Association, (NCMA) (1994). *Edge Restraints for Concrete Pavers*, TEK 11-1, Herndon, Virginia: National Concrete Masonry Association.

National Concrete Masonry Association, (NCMA) (1996). *Concrete Grid Pavements*, TEK 11-3, Herndon, Virginia: National Concrete Masonry Association.

Nelson, John D., and Debora J. Miller. (1992). *Expansive Soils, Problems and Practice in Foundation and Pavement Engineering*, New York: John Wiley and Sons.

Pasko, Thomas J., Jr. (1998). Concrete Pavements: Past, Present, and Future, in *Public Roads* July–August 1998 (www.tfhrc.gov).

Penner, E. (1962). Ground Freezing and Frost Heaving, *Canadian Building Digest* CBD-26 (www.nrc.ca/irc/cbd/cbd026e.html).

Pereira, A. Taboza. (1977). *Procedures for Development of CBR Design Curves*, Instruction Report S-77-1, Vicksburg: U.S. Army Engineer Waterways Experiment Station.

Portland Cement Association. (1992). *PCA Soil Primer*, Skokie, Illinois: Portland Cement Association.

Rada, Gonzalo R., David R. Smith, John S. Miller, and Matthew W. Witczak (1990). Structural Design of Concrete Block Pavements, *Journal of Transportation Engineering* 116, 615–635.

Rollings, Marian P., and Raymond S. Rollings, Jr. (1996). *Geotechnical Materials in Construction*, New York: McGraw-Hill.

Rollings, Raymond S., and Marian P. Rollings. (1991). Pavement Failures: Oversights, Omissions and Wishful Thinking, *Journal of Performance of Constructed Facilities* 5, 271–286.

Rollings, Raymond S., and Marian P. Rollings. (1992). *Applications for Concrete Paving Block in the United States Market*, Palm Beach Gardens, Florida: Uni-Group USA.

Rollings, Raymond S., and Marian P. Rollings. (1999). *SF-Rima, A Permeable Paving Stone System*, Mississauga, Ontario: SF Concrete Technology.

Soil Survey Staff (1993). *Soil Survey Manual*, Handbook 18, Washington: U.S. Department of Agriculture (http://soils.usda.gov).

Soriano, Ariel (2004). Geosynthetics: The Benefits and Savings Behind the Technology, *Grading & Excavation Contractor* 6, 10–22.

Stenmark, C (1995). An Alternative Road Construction for Stormwater Management, *Water Science and Technology* 32, 79–84.

Terzaghi, Karl, and Ralph B. Peck (1968). *Soil Mechanics in Engineering Practice*, 2nd ed., New York: John Wiley and Sons.

Thompson, Michael K., and William James. (1994). Provision of Parking-Lot Pavements for Surface Water Pollution Control Studies, 381–398 of *Modern Methods for Modeling the Management of Stormwater Impacts*, Guelph, Ontario: Computational Hydraulics International.

U.S. Departments of the Army and the Air Force. (1992). *Pavement Design for Roads, Streets, Walks and Open Storage Areas,* Army TM-5-822-5 and Air Force AFM 88-7 Chap. 1, Washington: Headquarters, Departments of the Army and the Air Force (www.usace.army.mil/inet/usace-docs/armytm).

U.S. Federal Aviation Administration. (1995). *Airport Pavement Design and Evaluation*, Advisory Circular No. 150/5320-6D, Washington: Federal Aviation Administration (www.faa.gov/arp/150acs.htm#Design).

Way, Douglas S. (1978). *Terrain Analysis*, 2nd ed., New York: Van Nostrand Reinhold.

Wei, Irvine W. (1986) *Installation and Evaluation of Permeable Pavement at Walden Pond State Reservation*, Report to the Commonwealth of Massachusetts Division of Water Pollution Control, Boston: Northeastern University Department of Civil Engineering.

Yoder, E.J., and M.W. Witczak. (1975). *Principles of Pavement Design*, 2nd ed., New York: John Wiley & Sons.

4 Porous Pavement Hydrology

CONTENTS

Surface Infiltration and Runoff 122
    Surface Infiltration Rate 122
    Runoff Coefficient 125
    Runoff Velocity and Travel Time 127
    Runoff Observations at Symphony Square, Austin, Texas 128
Disposition of Water below Pavement Surface 129
Storage in Pavement Reservoir 131
    Storage Capacity 131
    Reservoir Configuration for Effective Storage 132
    Allowable Ponding Time 135
Reservoir Discharge through Perforated Pipe 136
    Reservoir-Limited Discharge 137
    Observations in Nottingham and Wheatley, England 140
Infiltration into Subgrade 142
    Soil Infiltration Rate 143
    Infiltration Ponding Time and Reservoir Thickness 145
    Lateral Discharge Following Partial Infiltration 146
    Subsurface Disposition of Infiltrated Water 148
    Watershed Discharge of Infiltrated Water 150
    Where Not to Infiltrate Stormwater 152
Water-Quality Treatment 152
    Constituents in Pavement Stormwater 154
    Capture of Solids and Attached Metals 155
        Observations in Rezé, France 156
        Observations in Nottingham, England 157
        Observation in Weinsberg, Germany 158
        Observations in the Netherlands 158
    Biodegradation of Oil 159
    Supplemental Treatment by Subgrade Infiltration 160

The Importance of Small Storms	161
Frequency of Small Storms	161
Cumulative Rainwater Infiltration	161
Trademarks	164
Acknowledgments	165
References	165

Given the large area that pavements occupy, the volume of drainage water that pavements control is vast. What happens to that water is a question vital to the welfare of people, and is one of the driving forces behind the increasing application of porous pavement materials. This chapter reviews the types of hydrologic effects that porous pavements can have and the characteristics they can be given to bring about those effects.

Those requiring a basic background in urban hydrology, its importance, its management alternatives, and its quantitative modeling are referred to Ferguson's *Stormwater Infiltration* (1994) and *Introduction to Stormwater* (1998), and Debo and Reese's (2002) *Municipal Storm Water Management*. Updates in evolving stormwater management technologies are available on the web sites of the Center for Watershed Protection (www.cwp.org), the Low Impact Development Center (http://lowimpactdevelopment.org), Nonpoint Education for Municipal Officials (www.nemo.uconn.edu) and *Stormwater* magazine (www.stormh2o.com). Table 4.1 defines some terms with particular application to porous pavement hydrology.

TABLE 4.1
Some Terms with Particular Application to Porous Pavement Hydrology

Term	Definition
Best management practice (BMP)	Any method believed to be effective in preventing or reducing pollution or otherwise protecting the environment
Design storm (design condition)	Particular rainfall or runoff condition for which a porous pavement or an overall drainage system is designed, or in terms of which its performance is characterized
Evaporation	Movement of water into the atmosphere as a vapor
Evapotranspiration	Movement of water into the atmosphere by a combination of evaporation and transpiration
Hydrology	Flow and storage of water
Infiltration	Movement of a fluid into the surface of a porous substance
Permeability	The rate at which a fluid flows through a porous substance, under given conditions
Porosity (void space)	The portion of a volume of material that is not solid
Reservoir routing	Mathematical model which relates a reservoir's inflow, outflow, and change in storage over successive time increments
Storm flow	Relatively high stream flow during storm events
Stormwater	Water that occurs during storms
Snowmelt	Liquid water that occurs during thawing of snow and ice
Transpiration	Emission of water vapor by plants

Porous Pavement Hydrology 121

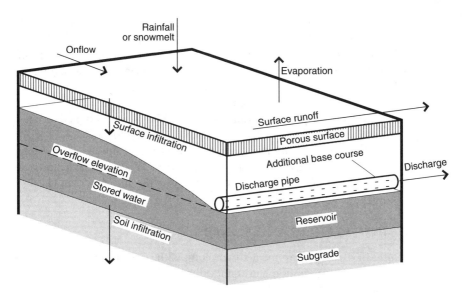

FIGURE 4.1 Hydrologic features and processes that could occur in a porous pavement (adapted from Diniz and Espey 1979, p. 50; Goforth et al., 1983, p. 3; Jackson and Ragan 1974; Smith, 2001, p. 16).

Figure 4.1 summarizes major hydrologic features and processes that could occur in a porous pavement. A pavement's hydrology begins at the surface with rainfall or snowmelt, at least some of which infiltrates the pavement's surface. Some additional water may arrive as runoff from off-pavement areas. Water infiltrating the pavement surface may further infiltrate the subgrade soil; the remainder overflows at the elevation provided for in the pavement's construction. Storage in the pavement's reservoir takes up temporary differences between rates of inflow and outflow. Evaporation pulls water back into the atmosphere at every stage.

Hydrologic models mathematically relate hydrologic flows and storages over time. They are useful because they allow the rapid comparison of clearly defined design alternatives. Some sort of modeling is almost inescapable in the design of porous pavements that must meet quantitative hydrologic performance criteria. Various combination of the hydrologic element of porous pavements can be written into ad hoc spreadsheets or large watershed models. They have been included in models by Jackson and Ragan (1974) Goforth et al. (1983); Debo (1994); Kipkie and James (2000); and James et al. (2001). Some models treat porous pavements merely as modifications to the land surface, having distinctive runoff or "depression storage" parameters. Other models take into account further processes in a pavement such as reservoir storage and soil infiltration.

The James model was developed relatively recently. It is distributed by Computational Hydraulics International (www.chi.on.ca) and Uni-Group USA (www.uni-groupusa.org) under the name *PC-SWMM for Permeable Pavements*. For a designated design storm it estimates surface runoff, depth of ponding in the base reservoir, and discharge from the base. It functions as a module of the U.S. Environmental Protection Agency's *Storm Water Management Model*. It is built upon earlier work by Diniz and others (Diniz, 1976; Diniz and Espey, 1979, pp. 41–62; Goforth et al., 1983; Huber and Dickinson, 1998; James et al., 1998).

Not all hydrologic modeling is completely accurate. The relative accuracy of different hydrologic modeling approaches has been the subject of long arguments. Many of those arguments have in fact been unresolvable, because they have taken place without the benefit of actual measured on-site hydrologic data. Relatively accurate results can be obtained only from a model that has been calibrated to actual local conditions. Too frequently — in fact, routinely — practitioners have to design facilities that are going to be built right away, in locales where similar features have not been monitored. In these cases, models based on general knowledge must inevitably be used to estimate hydrologic processes, with the understanding that site-specific accuracy is not a definable issue.

SURFACE INFILTRATION AND RUNOFF

The beginning of a porous pavement's hydrology is the partitioning of rainfall or snowmelt into surface infiltration and runoff. Table 4.2 lists some terms with distinctive application to surface infiltration and runoff.

SURFACE INFILTRATION RATE

A pavement's surface infiltration rate is measured with water ponded on top of the pavement (Rollings and Rollings, 1996, p. 184). In the laboratory, samples of pavement material are enclosed in boxes or cylinders to hold the water. In the field, small areas of pavement are diked or walled off to hold the water. Table 4.3 lists some test procedures for infiltration rate and permeability that have been standardized by ASTM (www.astm.org); other valid test procedures also exist but have not yet been standardized at this level.

Sustained ponding of the water yields the measurement of what is technically known as saturated hydraulic conductivity K, the permeability in saturated conditions under a unit pressure gradient. In the ponded test condition, all the available pore space is continuously used to pass water into and through the pavement's surface layer.

After the diking is removed and a pavement is put into service, surface ponding, if it ever exists, can seldom exceed a fraction of an inch in depth. Under such shallow ponding the value of the pressure gradient approaches unity, and the infiltration rate into the saturated pavement surface is equal to K (Ferguson, 1994, p. 92). Consequently, for saturated porous pavement surfaces the terms "infiltration rate" and "saturated hydraulic conductivity" can be used almost interchangeably.

Table 4.4 lists the saturated infiltration rates of some porous pavement surfaces, with those of some other surfaces for comparison. Further observations of the infiltration rates of specific paving materials are described in later chapters.

Open-jointed blocks, open-celled grids and plastic geocells form composite pavement surfaces, of which part is the solid material of the blocks or lattices and part is the grass or aggregate fill. The overall infiltration rate $I_{composite}$ of a composite surface is given by,

$$I_{composite} = I_{solid} \times (\text{solid area / total area}) + I_{fill} \times (\text{fill area / total area}),$$

TABLE 4.2
Some Terms with Particular Application to Surface Infiltration and Runoff

Term	Definition
Antecedent moisture	Quantity of moisture present in a porous material before a rainfall or snowmelt event begins
Catchment (catchment area)	Drainage area; watershed
Cfs	Cubic feet per second
Curve number	In the SCS hydrologic model, a summary characteristic of watershed soil and cover which quantifies the relationship between rainfall and runoff
Darcy's law	Relationship between flow rate through a porous medium, pressure gradient, and the medium's hydraulic conductivity
Drainage area (drainage basin)	Catchment area; watershed
Depression storage	Surface water retained in ground surface depressions, and eventually evaporated or infiltrated rather than contributing to runoff
Head	Difference in elevation between two points in a fluid, producing pressure
Hydraulic conductivity	Permeability under a given head or pressure of water
Runoff (direct runoff, surface runoff)	Water that flows on or near the ground surface during storms
Manning's equation	Relationship between flow rate and the slope, shape, and roughness of the surface over which water flows
Runoff coefficient	Ratio of surface runoff to rainfall
Saturated hydraulic conductivity (K)	Hydraulic conductivity under saturated conditions
SCS	Soil Conservation Service, a branch of the U.S. Department of Agriculture, which developed a hydrologic model using the curve number (the model still bears the SCS name although the agency's name was later changed to the Natural Resources Conservation Service)
Time of concentration	The time runoff takes to flow to a drainage area's discharge point from the most distant point in the area
Watershed	Area having a topographic surface that drains to a given discharge point

TABLE 4.3
Examples of ASTM Standard Tests for Infiltration Rate and Permeability

Number	Title
D 2434	Standard Test Method for Permeability of Granular Soils (Constant Head)
D 3385	Standard Test Method for Infiltration Rate of Soils in Field Using Double-Ring Infiltrometer
D 5093	Standard Test Method for Field Measurement of Infiltration Rate Using a Double-Ring Infiltrometer with a Sealed Inner Ring

where I_{solid} is the infiltration rate of the solid block or lattice material, and I_{fill} is the infiltration rate of the fill material. Where the solid material is essentially impervious, the value of I_{solid} approaches zero, and the infiltration rate of the composite surface reduces to,

TABLE 4.4
Surface Infiltration Rates under Saturated Conditions (Saturated Hydraulic Conductivity K)

Surface Type	Infiltration Rate (inch/hour)	Reference
Porous pavement surfaces		
Unbound aggregate		
1" uniform size	50,000	AASHTO, 1986, p. AA-18
1/2" uniform size	15,000	AASHTO, 1986, p. AA-18
1/4" uniform size	2500	AASHTO, 1986, p. AA-18
0.09" to 0.75" in size	1500	AASHTO, 1993, p. I-19
0.08" to 0.75" in size	1300	AASHTO, 1993, p. I-19
Turf on loamy soil, 2–10 years old		
Earth disturbed during construction	0.2 to 1.1	Hamilton, 1990, pp. 47–54
Earth not disturbed during construction	1.9 to 3.9	Hamilton, 1990, pp. 47–54
Open-jointed blocks with 0.08" to 0.20" aggregate fill		
Initially built	9.2	Borgwardt, 1999
6 years after construction	4.1	Borgwardt, 1999
Open-celled grids with cells in 10%+ of surface area		
With 0.1" to 0.2" aggregate fill	40+	Pratt et al., 1995
With 2/5 mm aggregate fill		
0.1 year after installation	9.4	Borgwardt, 1997a, 1997b
2 years after installation	6.1	Borgwardt, 1997a, 1997b
4.5 years after installation	4.8	Borgwardt, 1997a, 1997b
8 years after installation	4.12	Borgwardt, 1997a, 1997b
Two years after installation		
With 2/5 mm aggregate fill	6.1	Borgwardt, 1997b
With 1/3 mm aggregate fill	4.0	Borgwardt, 1997b
With 0/2 mm aggregate fill	2.8	Borgwardt, 1997b
Porous concrete		
Properly constructed	670 to 900	Wingerter and Paine, 1989, App. P-1 and P-3
Over-vibrated during construction	1.25 to 24	Wingerter and Paine, 1989, App. P-1 and P-3
Porous asphalt		
Immediately after construction	170 to 500+	St. John and Horner, 1997, p. xvi; Thelen and Howe, 1978, p. 13; Wei 1986, pp. 6–11
After 3 to 4 years	15 to 39	Wei, 1986, pp. 6–28 and 7–28
After 4 years of winter sanding	1.4	St. John and Horner, 1997
Other surfaces		
Forest soil	8 to 60	Lull and Rinehart, 1972
Clay loam desert soil with partial shrub cover	0.9 to 1.9	Smith and Leopold, 1942
Dense-graded aggregate	0.5 to 10	AASHTO, 1986, p. AA-18
Coarse sand	0.39 to 100	AASHTO, 1986, p. AA-18; van der Leeden et al., 1990, p. 284

TABLE 4.4 Continued

Surface type	Infiltration Rate (inch/hour)	Reference
Dense concrete	<0.00002	Rollings and Rollings, 1996, p. 149; and derived from Kosmatka and Paranese, 1988, p. 8
Dense asphalt	0.00006 to 6	Rollings and Rollings, 1996, p. 149

Values for open-jointed blocks and open-celled grids apply to the entire composite surfaces (0/2, 1/3, and 2/5 are German specifications for aggregate respectively 0 to 2, 1 to 3, and 2 to 5 millimeters in size).

$$I_{\text{composite}} = I_{\text{fill}} \times (\text{fill area} / \text{total area}).$$

The surface infiltration rates of some porous pavements change over time, due to processes such as compaction, sedimentation, migration of pavement binder, and aggregation of soil particles by growing vegetation. The infiltration rate of a specific pavement surface at a given moment in time depends on the quality of materials and construction, and the history of compaction and sedimentation. The infiltration rate assigned for the design of a surface should be a typical value which will apply for the life of the pavement. Surface infiltration rate can be protected as described in Chapter 2, and maintained and restored as described for specific materials in later chapters of this book.

RUNOFF COEFFICIENT

A runoff coefficient is a ratio of surface runoff to rainfall. Its value can range from 0 to 1. It can be used to estimate surface runoff in hydrologic models.

A pavement's runoff coefficient is measured in the field with natural rainfall, or in the field or laboratory with "artificial rainfall" sprinkled from above. Rain falling in drops tends not to saturate a pavement's surface pores as does ponded water during measurement of infiltration rate. Instead, the drops of rainwater, after first bouncing off the pavement's surface, wend their way randomly into the pavement's openings, forming a constantly shifting mosaic of water and air in the material's pores.

A porous pavement's runoff coefficient varies from storm to storm. Its value is relatively low during small, soft storms, when most of the rainfall infiltrates the surface. It is relatively high during intense storms, when the surface becomes more flooded and a greater proportion of the rainwater becomes surface runoff. On turf and fine aggregate surfaces the coefficient can be increased by antecedent moisture retained from previous storms in the numerous small pores.

Table 4.5 lists runoff coefficients reported for porous pavement materials, with those of some dense materials for comparison. A value for porous concrete has not yet been established; its value would presumably be comparable to that of porous asphalt. The coefficients for most porous pavements are below 0.5, which means that they are hydrologically more similar to grass than to dense pavements. In the absence of a specific reported value for a proposed porous pavement installation, one

TABLE 4.5
Runoff Coefficients of Pavement Surfaces

Surface Type	Runoff Coefficient	Reference
Porous pavement surfaces		
Aggregate		
Range of gradations	0.30 to 0.70	USFAA, 1965, Appendix 1, p. 1:
Turf, grass cover greater than 50 percent	0.05 to 0.53	Chow et al., 1988, p. 498; van der Leeden et al., 1990, p. 76
Open-jointed blocks		
With 0.80" to 0.20" aggregate fill	0.30 to 0.50	Borgwardt 1999, p. 69, and 1997b, p. 10
Open-celled Checkerblock and Monoslab grids		
With topsoil and Kentucky bluegrass	0.00 to 0.27	Day et al., 1981, p. 30
Open-celled Turfstone (Turfblock) grids		
With sandy loam and Bermuda grass	0.18 to 0.36	Goforth et al., 1983, p. 65
With topsoil and Kentucky bluegrass	d0.00 to 0.56	Day et al., 1981, p. 30
Porous asphalt		
Newly installed	0.12 to 0.40	St. John and Horner 1997, p. xvi
3 to 4 years after installation	0.18 to 0.29	Wei 1986, pp. 7–34
Dense pavement surfaces		
Dense asphalt	0.73 to 0.95	Chow et al., 1988, p. 498; van der Leeden et al., 1990, p. 76; St. John and Horner 1997, p. xvi
Dense concrete	0.75 to 0.97	Chow et al., 1988, p. 498; van der Leeden et al., 1990, p. 76

Where a range of values is given the higher values are for steep slopes, intense storms, and relatively impermeable surfaces.

might justifiably assign the pavement the same runoff coefficient that would be used for grass on the same site.

Although the runoff coefficients of porous pavements are generally low, they are higher than would be predicted from the materials' saturated hydraulic conductivities. Table 4.6 illustrates this paradox directly in an observation of porous asphalt at Walden Pond State Reservation in Massachusetts. Based on Walden's K values, one might predict a runoff coefficient of zero during any rainfall lower than 17 inches per hour. But during rainfall of only 3.5 inches per hour, approximately one quarter of the rainwater became runoff. Because the rainwater did not completely saturate and utilize the surface pores, it infiltrated at only a fraction of the pavements' saturated hydraulic conductivities. This means that as indicators of a pavement's runoff production during rainstorms, it is correct only to use the results of runoff tests and not those of infiltration tests that artificially saturate the surface.

Another way to describe a pavement's runoff production is the curve number in the SCS hydrologic model (U.S. Soil Conservation Service, 1986). Research has not yet been done to measure curve-number values for porous pavement surfaces. Until it is done, values could be estimated by analogy with the measured runoff coefficients listed above.

TABLE 4.6
Contrasting Infiltration and Runoff Rates in Porous Asphalt Pavements at Walden Pond State Reservation, Massachusetts, 3 to 4 Years after Installation

	Pavement Mixture J3	Pavement Mixture K	Reference
Saturated hydraulic conductivity K	17 inch/h	39 inch/h	Wei 1986, pp. 6–28
Runoff coefficient during rainfall of 3.5 inch/h	0.29	0.18	Wei 1986, pp. 7–34
Infiltration rate during rain rainfall	2.5 inch/h	2.9 inch/h	Calculated
Infiltration rate during rainfall as proportion of K	15 percent	7 percent	Calculated

RUNOFF VELOCITY AND TRAVEL TIME

During a given storm, the peak runoff rate reaching the low edge of a pavement or the bottom of a watershed is influenced by the runoff's velocity of travel across the surface. In many hydrologic models the velocity is reflected in the "time of concentration": the amount of time runoff takes to travel to the discharge point from the most distant point on the surface. Peak runoff rate increases with increasing velocity and thus with decreasing time of concentration.

Flow velocity across porous pavement surfaces has not been measured in actual observations. Until such measurements are made, velocity and travel time must be estimated by applying general theoretical knowledge. A theoretical basis for estimating sheet-flow travel time across a surface is the following version of Manning's equation (James, James and von Langsdorff, 2001; Overton and Meadows, 1976, pp. 58–88, cited in U.S. Soil Conservation Service, 1986, p. 3-3),

$$T_t = 0.007 \, (n L)^{0.8} / [(P_2)^{0.5} \, s^{0.4}],$$

where:

- T_t = travel time (hour),
- n = Manning's roughness coefficient,
- L = length of flow path (feet),
- P_2 = 2-year, 24-hour rainfall (inches),
- s = slope of surface along flow path (feet per feet).

The equation can be applied to estimate travel time for a proposed pavement. In addition, the factors in the equation can be analyzed to determine the general type of effect that porous pavements should have on travel time, when compared with dense pavements.

One relevant factor is surface roughness n, some values for which are listed in Table 4.7. The table shows that complex grass surfaces have n values many times higher than those of dense, smooth pavements. Many porous pavement surfaces presumably have values intermediate to those of dense pavements and grass, like those reported for open-jointed paving block. According to Manning's equation, porous pavements' high n values increase runoff travel time as compared with that on a dense pavement of the same length and slope.

TABLE 4.7
Manning's Roughness Factor *n* for Sheet Flow

Surface Type	Manning's *n*
Porous pavement surfaces	
Bermuda grass	0.41
Bluegrass and other "dense grasses"	0.24
Eco-Stone open-jointed paving block	0.03
Dense pavement surfaces	
Smooth surfaces (dense concrete, dense asphalt)	0.011

James et al. (2001); U.S. Soil Conservation Service (1986, p. 3-3).

A second relevant factor is the depth of runoff on the surface, which is reflected in the equation in the precipitation P that generates the runoff. As runoff depth decreases, a surface's roughness has on average more contact with the flowing water and impedes its velocity. The runoff coefficients listed earlier in this chapter indicate that porous pavements produce much less runoff than dense surfaces do; the shallow flows are in intimate contact with the rough pavement surfaces. According to Manning's equation, porous pavements' shallow runoff has a longer travel time than does the comparatively deep runoff on dense pavements.

Together, porous pavements' great roughness and shallow runoff multiply into the equation to produce longer travel time than that on otherwise similar dense pavements. For example, if an open-jointed paving block's runoff coefficient is one-half that of dense pavement, and its roughness is three times greater, then Manning's equation would predict that the travel time is 3.4 times longer. Its peak runoff rate would be correspondingly low. The degree of attenuation varies with each surface's roughness and the amount of runoff in specific storm conditions.

RUNOFF OBSERVATIONS AT SYMPHONY SQUARE, AUSTIN, TEXAS

Several aspects of a porous pavement's runoff were directly observed in Austin, Texas. Goforth et al. (1983, pp. 14–15, 18–20, 65) monitored runoff from a parking lot surfaced with Turfstone open-celled grids located at the intersection of Symphony Square and Red River Street. The lot was 0.14 acre in area and held 14 parking spaces; its surface sloped at 4.0 percent. The grids were laid directly on natural clay subgrade. The open cells of the 3-inch-thick grids were filled with sandy loam and planted with Bermuda grass which was in healthy, uniform condition at the time of the study. Infiltration into the subgrade was considered insignificant.

The researchers sprinkled artificial "rainfall" onto the pavement and measured the resulting runoff. During each event they maintained a constant rainfall rate until the pavement's runoff reached a stable peak rate. Figure 4.2 shows that the peak runoff rate was only a fraction of the rate of precipitation. Figure 4.3 shows that the time to peak was 20 minutes or more after the start of precipitation, even when rainfall intensity approached 2 inches per hour. The porous surface reduced and attenuated runoff even though infiltration a few inches below the surface was made insignificant by the impermeable subgrade.

Porous Pavement Hydrology

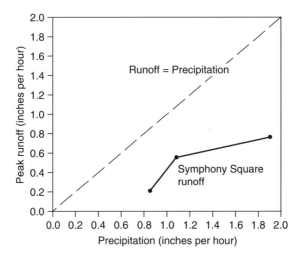

FIGURE 4.2 Peak runoff rate at a Turfstone parking lot on Symphony Square in Austin, Texas (data derived from Goforth et al., 1983, p. 65).

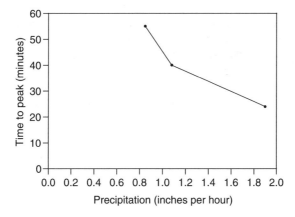

FIGURE 4.3 Time (after beginning of uniform rainfall) to peak runoff at the Symphony Square parking lot (data from Goforth et al., 1983, p. 65).

DISPOSITION OF WATER BELOW PAVEMENT SURFACE

After water infiltrates a pavement surface, some of it is retained in the pavement's pores, where it is available for evaporation. The remainder drains downward to the pavement's reservoir, perhaps after a delay due to slow movement through the pavement's pores. Two research teams have observed this combination of processes.

Andersen et al. (1999), in their laboratory in England, constructed porous pavement samples and simulated rainfall onto them. The pavement surfaces were open-jointed CeePy concrete blocks (CeePy is a model available in Britain). The blocks were laid on a variety of aggregate setting-bed materials; their joints were filled with aggregate. The samples had no base reservoirs; water that drained through the blocks and bedding fell out through a screen.

Figure 4.4 illustrates the drainage through one of the laboratory pavement structures during an experimental "storm." For the first 25 minutes there was no throughflow as the pavement structure was being wetted. Once throughflow started, it was lower than the rainfall, and continued at a low rate for more than 60 minutes after rainfall ceased.

In general, throughflow was smallest and delay was greatest in pavements with relatively small bedding particles; evidently the large internal surface area of small aggregate particles offered great opportunity for clinging, and absorption of water. During a one-hour event of 0.6 inch per hour, with small aggregate in the setting bed, 55 percent of the water could be retained in an initially air-dry structure, and 30 percent in an initially moist structure. The water that did not drain through evaporated. The rate of evaporation was greatest in the hours immediately following the rain event, when the structure was wettest.

Brattebo and Booth (2003) observed drainage through several types of porous and dense pavement surfaces at a parking lot in Renton, Washington. They monitored flow during one winter rainy season when the rainfall was typically gentle and there were no long periods of freezing weather or snow accumulation. Essentially all precipitation infiltrated the porous pavement surfaces.

Figure 4.5 shows the drainage during one storm event 4 inches below a surface of Turfstone concrete grids with the cells filled with soil and grass. Initially there was no throughflow, as rainwater merely wetted the pavement's grids and soil. After throughflow started, it lagged behind the rainfall due to its travel time through the soil and then through a pipe to the recording device. The storm's total throughflow volume was almost equal to that of the rainfall: evaporation was small in the cold

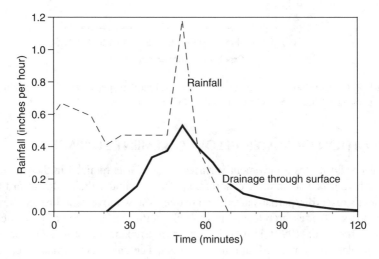

FIGURE 4.4 Rainfall on and drainage through a surface of open-jointed concrete blocks on a one-inch-thick bed of 0.2 to 0.4 inch aggregate (after Andersen et al., 1999, Figure 7).

Porous Pavement Hydrology

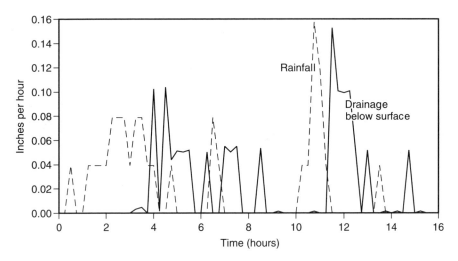

FIGURE 4.5 Rainfall on and drainage below a parking lot surface of Turfstone with soil and grass in Renton, Washington, during the storm of November 20–21, 2001 (after Brattebo and Booth, 2003, Figure 3).

winter night. In contrast, surface runoff on nearby dense asphalt closely mimicked the rainfall rate; the dense material's runoff coefficient was close to 1.0.

STORAGE IN PAVEMENT RESERVOIR

A pavement's reservoir is any portion of the pavement that stores or conveys water while the water discharges. Reservoir storage takes up any temporary difference between drainage from the pavement surface and outflow to soil infiltration and lateral discharge. In any increment of time, storage follows the principle of the basic water balance or "reservoir equation,"

$$\Delta \text{Storage} = \text{inflow} - \text{outflow},$$

where Δ Storage is the change in storage (the Greek letter D, delta, is a symbol for difference or change). Given the characteristics of a reservoir and its outlets, storage and discharge during a given storm event can be modeled quantitatively by "reservoir routing," which relates the reservoir's inflow, outflow, and storage over successive time increments (Ferguson, 1994, pp. 157–159).

STORAGE CAPACITY

A pavement reservoir's storage capacity is equal to the void space in the reservoir material. It can be expressed in units such as cubic feet. It can also be expressed as a proportion of the material's total volume,

$$V_d = \text{void volume} / \text{total material volume},$$

where V_d is void space or porosity, and the total material volume includes both void space and solid material.

The same value of V_d relates equivalent depth of water in the voids to the reservoir material's total thickness. With equivalent void thickness and total material thickness in inches,

$$V_d = \text{equivalent void thickness} / \text{total material thickness}.$$

For example, when 1 inch of water enters reservoir material with 33 percent porosity, the water rises 3 inches in the material. In this case, 3 inches is the total thickness of reservoir material, and 1 inch is the equivalent depth of water occupying the void space.

If a reservoir is to hold a predetermined volume of water for stormwater management, the total volume of reservoir material required to produce the necessary storage capacity is given by

$$\text{Total material volume} = \text{water storage volume} / V_d.$$

If the reservoir is to hold a predetermined depth of water such as a depth of precipitation, the total thickness of reservoir material required to produce the necessary storage capacity is given by

$$\text{Total material thickness} = \text{water storage depth} / V_d.$$

The most common material in base reservoir construction is open-graded (single-size) aggregate such as ASTM No. 57, which tends to have a porosity of 30 to 40 percent. Because a reservoir's storage capacity depends completely on the material's porosity, open-graded aggregate must be specified explicitly. The aggregate particles need not be large to produce this amount of void space; as a percentage the void space in all single-size aggregates is in the same range. Within that range, the exact amount depends on the particles' angularity: void space increases from about 30 percent for rounded particles to 40 percent and even higher for very angular particles (Shergold, 1953). The void space in a particular aggregate can be determined by a simple laboratory test (ASTM C 29). Many aggregate producers have void-space test data on file for their standard products. In the absence of project-specific aggregate data, specifying open-graded material and then assuming a value from the lower part of the 30 to 40 percent range would prudently take into account uncertainties in the material.

Perforated pipes and manufactured chambers can supplement a reservoir's storage capacity. They generate hydraulic capacity efficiently: the void space in many such devices is over 90 percent. Some of them allow access for monitoring and maintenance. Table 4.8 lists examples of stormwater chamber suppliers. Most suppliers provide guidelines for installation of their products. The installation of pipes should follow local practices applying to drainage pipes in general, including a minimum diameter for maintenance access a specified amount of backfill cover, and elevation of the flow line in relation to frost depth.

RESERVOIR CONFIGURATION FOR EFFECTIVE STORAGE

To store water, a reservoir must be configured to accommodate the elevations to which water at rest will naturally conform. The outlet elevation sets a reservoir's

TABLE 4.8
Examples of Suppliers of Stormwater Storage Chambers

Product	Company	Contact Information
Contech	Contech	www.contech-cpi.com
LandMax and EnviroChamber	Hancor	www.hancor.com
Rain Store	Invisible Structures	www.invisiblestructures.com
Recharger	Cultec	www.cultec.com
Single Trap and Double Trap	StormTrap	www.stormtrap.com
Storm Compressor	Advanced Drainage Systems	www.ads-pipe.com
StormChamber	Hydrologic Solutions	www.hydrologicsolutions.com
Stormtech	Stormtech	www.stormtech.com
StormVault	Jensen Precast	www.stormvault.com

upper storage boundary. Only the portion of a reservoir below that elevation stores water. Reservoir capacity constructed on a slope above that elevation cannot function for storage except when there are temporary and modest rises when there are storms. Two experiences in the field illustrate the importance of accommodating water level.

A porous asphalt parking lot in Austin, Texas exemplifies a hydrologically successful and efficient reservoir configuration. Although the parking lot was temporary and experimental, its installation and performance were thoroughly documented (Goforth et al., 1983, p. 42). It was built and monitored in 1981 at the city Public Works Department's truck maintenance yard on Kramer Lane. It provided 20 parking spaces on about 0.2 acres. Figure 4.6 shows the construction.

The subgrade was limestone bedrock and dense residual soil, both of which were essentially impervious. The firm subgrade and low frost hazard eliminated structural demands for a thick base course, so the base thickness was determined primarily by stormwater storage capacity in the base material's 35 percent void space. A perforated pipe discharged water from the low end of the reservoir.

The objective was to hold the total precipitation from a storm of 5.2 inches while releasing it through the perforated pipe. If the reservoir had been level, a uniform 13 inch thickness would have provided the required storage. However, the rock subgrade sloped at 1.5 to 2.0 percent. Consequently, the upslope end of the base was made thick enough to handle the mechanical traffic load only, which established the elevation of the top of the reservoir. At that elevation the surface was made level, producing a thickness of 24 to 30 inches at the downslope end and an average thickness of 15 or 16 inches, which produced total storage volume sufficient for the design storm.

As water filled the reservoir's void space, the water surface would rise. The level water surface paralleled the level pavement surface, making efficient use of the constructed reservoir volume. Although a level pavement surface has the disadvantage of possible surface puddling in the event of an extraordinarily large storm or the clogging of some part of the pavement, the Kramer Lane configuration made the base reservoir's entire volume functionally available for water storage, and confined pavement discharge to its intended route through the perforated pipe.

In contrast, a porous asphalt parking lot in Warrenton, Virginia illustrates how a sloping base reservoir can create inadvertent overflows and fail to store the intended

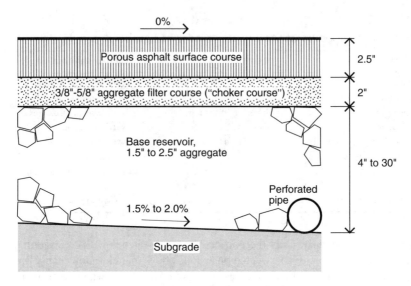

FIGURE 4.6 Construction of the Kramer Lane porous asphalt parking lot in Austin, Texas (based on data in Goforth et al., 1983, p. 42).

water volume. The lot was built at the intersection of Routes 29 and 605 by the Virginia Department of Transportation for commuters transferring between cars and busses (Wyant, 1992, pp. 4, 10, 41, 46). The lot opened for use in 1987 and was still extant in 2000. The porous asphalt area of 2.0 acres holds the parking stalls and turning lanes for 213 cars.

Figure 4.7 shows the construction. The sloping base reservoir is of uniform thickness. Perforated pipes located throughout the reservoir discharge to a storm sewer. Perforated monitoring wells are distributed throughout the lot.

The objective was to detain rainwater from the two-year storm while it discharged to the perforated pipes; little infiltration was possible into the compacted, fine-textured subgrade. It was expected that the uniform 6-inch reservoir thickness below the outlet pipes' perforations would produce sufficient storage capacity in the aggregate's 40 percent void space.

The Virginia Transportation Research Council monitored the lot's performance. The researchers observed during storms that water never rose to a visible level in the monitoring wells at the pavement's higher elevations, that outlet-pipe discharge started at the structure's low end earlier than that from anywhere else in the parking lot, and that large amounts of water overflowed from the pavement surface at the low edge. All of these results were unexpected and inadvertent.

The Warrenton parking lot did not retain water the way it was intended to because water drained down to the sump at the pavement's low edge; it was not held in the large uphill portions of the sloping base reservoir. The reservoir volume that actually stored water was only 21 percent of the volume that had been built.

To make a flat reservoir on a sloping site, the subgrade surface can be terraced, with terraces compartmentalized by berms (Cahill, 1993, p. 26). Figure 4.8 shows

Porous Pavement Hydrology

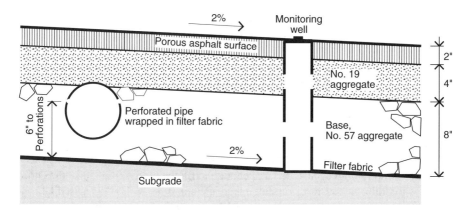

FIGURE 4.7 Construction of a porous asphalt parking lot in Warrenton, Virginia (based on data in Wyant, 1992, p. 4 and 10).

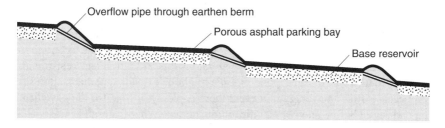

FIGURE 4.8 Terraced porous asphalt parking bays at the Siemens office in Great Valley, Pennsylvania (after project drawing by Andropogon Associates).

terracing at the Siemens porous asphalt parking lot in Pennsylvania, where a series of parking bays step down a long slope. Each terrace slopes at no more than 3 percent, to minimize unused storage space. This installation will be described further in Chapter 12.

Allowable Ponding Time

Ponding time is the length of time water occupies a reservoir following a design storm or snowmelt event. It is sometimes called drawdown time because it is the length of time for a completely filled reservoir to be drained out by perforated pipes or soil infiltration.

Ponding time in a base reservoir following a storm or snowmelt event must be limited in order to restore storage capacity for a subsequent event, to aerate the reservoir and subgrade for biodegradation of pollutants, and to reduce the hazard of freezing. If a certain ponding time is required, then the required base thickness can be derived from

$$\text{Total material thickness} = (\text{discharge} / \text{ponding time}) / V_d,$$

where discharge is the sum of all infiltration and pipe outflow rates in inches per hour.

AASHTO (1986, p. AA-21) suggested the following logic for determining allowable ponding time. Set the average proportion of time the base reservoir is permitted to be wet, for example 10 percent. Find the average length of time between rainfall events, for example five days. The base may then take 10 percent of the average time between events to drain, for example

$$0.10 \times 5 = 0.5 \text{ day},$$
or 12 hours.

This simple procedure might overly restrict allowable ponding time because it assumes that all rainfall events are large enough to fill the reservoir and take the entire 12 hours to drain. In fact most rainfall events are small, as will be discussed later in this chapter. An average rainfall event would only partly fill a reservoir designed to hold a large design storm and would take only a fraction of the designated ponding time to drain; storms large enough to fill the reservoir for the designated ponding time would be comparatively rare.

The designation of an allowable ponding time could take into account the pattern of rainfall frequencies where a pavement is located. Where rain can be frequent year-round, as in much of eastern U.S., a short ponding time of one to three days after filling of the reservoir may be appropriate to restore and aerate the reservoir and subgrade before the next rain occurs. Where rain is low and infrequent year-round, as in the arid southwestern states, ponding time might be extended, perhaps up to seven days, because after an average rain event there is plenty of time for the pavement to drain and aerate before another significant rain comes. Where rainfall is seasonal and moderate in total, as in parts of the Pacific states, intermediate ponding times might be used, because thorough aeration and biodegradation during the dry season may make up for prolonged saturation during the wet season. Further research and experience are necessary to refine ponding-time limits in specific locales.

RESERVOIR DISCHARGE THROUGH PERFORATED PIPE

Reservoir outflow through a perforated pipe contributes (with soil infiltration) to reservoir drainage and downstream flow rates. Pipe discharge is zero (reservoir outflow is only by soil infiltration) when the water level is at or below the pipe's elevation. When the water rises above the pipe's elevation, soil infiltration continues while additional discharge is through the pipe. The discharge rate reflects both the drainage of water from the surface into the reservoir, and detention in the reservoir while discharging.

Figure 4.9 shows a reservoir where water is ponded above a pipe while discharging. The pipe's capacity is small enough to limit the reservoir's discharge rate. Flow is determined by the head of water above the pipe and the sizes of the pipe and its perforations. In this condition the discharge rate can be manipulated by applying well-known principles of orifice and pipe hydraulics to the discharge pipe (Ferguson, 1998, pp. 127–133).

Figure 4.10 shows an alternative condition in which water is not ponded over the pipe because the pipe's capacity is large enough not to limit discharge out of the

Porous Pavement Hydrology

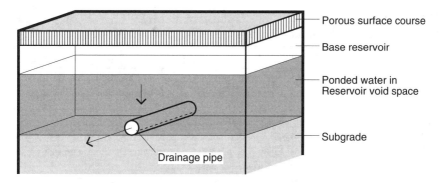

FIGURE 4.9 Water ponded in a reservoir with pipe-limited discharge.

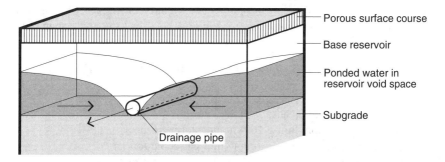

FIGURE 4.10 Water ponded in a reservoir with reservoir-limited discharge (adapted from Jackson and Ragan, 1974).

reservoir. Instead discharge is limited by the lateral flow rate through the reservoir aggregate; the pipe and its perforations carry water away as fast as the reservoir delivers it. In this condition the discharge rate is determined by the reservoir's ponding depth, porosity, and hydraulic conductivity. It can be manipulated by controlling the number of pipes and the reservoir's hydraulics, as described below.

If it is uncertain which condition will exist in a proposed reservoir, one should calculate the outflow both ways, and use the most limiting result. Some pavements may have the pipe-limited condition during large storms, and the reservoir-limited condition during small storms.

RESERVOIR-LIMITED DISCHARGE

Jackson and Ragan (1974) modeled reservoir discharge where water is not ponded over the pipe, and discharge is limited by the reservoir's hydraulics. In their models they assumed that all rainwater drained from a pavement's surface into the reservoir without delay. All reservoir discharge was through the pipe; there was no subgrade infiltration. The reservoir's floor was level, and ponding depth was measured from the floor elevation; the drainage pipe was in effect in a trench below the floor. Table 4.9 lists further assumptions in their model.

TABLE 4.9
Assumptions in Jackson and Ragan's (1974) Hydrologic Model

Factor	Assumption
Base reservoir material	Aggregate with 13 percent porosity and 100 in/hr hydraulic conductivity, to 40 percent porosity and 20,000 in/hr hydraulic conductivity
Drainage pipe spacing	60 ft to 360 ft
Storm type	160 minute rainfall with peak rate in the middle of the event
Precipitation amount	2.74 inches total with 8.3 in/hr peak intensity, to 3.86 inches total with 10.3 in/hr peak intensity (in Washington DC these storms have 5- to 25-year recurrence intervals)
Subgrade	Impervious; no infiltration

The researchers found that the ponded water surface was essentially flat everywhere in the reservoir except in the immediate vicinity of a pipe. The following equation described the maximum depth during a storm (Jackson and Ragan, 1974, equation 9):

$$H = (P/n) \exp[-0.459 + 0.217 \ln(n^2 s^2/4k) - 0.020(\ln(n^2 s^2/4 k))^2],$$

where:
 H = ponding depth in inches,
 P = total precipitation during 160-minute storm (inches),
 n = porosity of base reservoir (cubic foot per cubic foot),
 k = base reservoir hydraulic conductivity (inches per minute),
 s = pipe spacing in feet.

Figure 4.11 shows the maximum ponding depths at five pipe spacings in aggregate with 35 percent porosity. The ponding depth is seen to be only a few inches in most of the chart. For example, in a 3-inch rain, in a pavement with 90-foot pipe spacing, the water would rise 6.8 inches. To hold this amount of water the reservoir aggregate should be at least of equal thickness.

The following equation predicted peak pipe discharge rate (Jackson and Ragan, 1974, equation 8):

$$q_p = (2 P/n) \exp[-11.199 + 1 \; 0.499\ln(k)],$$

where q_p = peak pipe discharge (cfs per foot of perforated pipe).

Figure 4.12 shows the peak discharge at three reservoir hydraulic conductivities. The chart shows that the discharge rates are generally low, in accord with the shallow ponding depths. For example, the peak outflow from a typical parking lot drained by pipes 65 feet long would be only about 0.3 cfs from each pipe during a substantial storm. Each pipe's peak discharge does not change significantly with varying pipe-to-pipe spacings; instead, different spacings are taken up by different ponding depths as shown in the previous figure.

The researchers assumed that all precipitation immediately entered the base reservoir. However, later researchers observed that the precipitation of rainwater is slightly delayed and reduced while draining down through pavement material, as

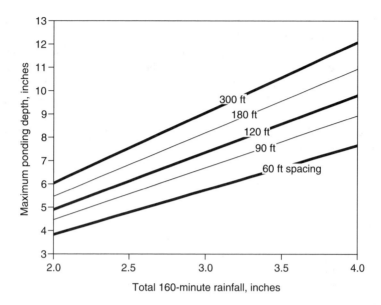

FIGURE 4.11 Maximum ponding depth in a base reservoir at different spacings of discharge pipe (data derived from Jackson and Ragan, 1974, equation 9, for base-reservoir porosity of 35 percent and permeability of 5000 inch/hour).

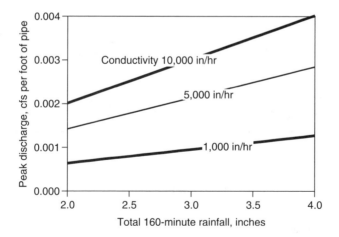

FIGURE 4.12 Peak discharge from a reservoir drainage pipe, at different reservoir aggregate hydraulic conductivities (data derived from Jackson and Ragan, 1974, equation 8, for base-reservoir porosity of 35 percent).

described earlier in this chapter. Consequently, Jackson and Ragan's equations may reflect to a slight degree an unrealistically rapid rise of water in the base, and a correspondingly rapid discharge. This could perhaps be corrected by substituting into their equations in place of precipitation P, the somewhat smaller amount of rainwater expected to reach the reservoir.

To use Jackson and Ragan's approach to discharge water at a predetermined rate, the number of pipes that will discharge at the total desired rate must be determined, and a reservoir thickness to hold the associated ponding depth must be established. It is then necessary to specify a pipe that can convey the given discharge without ponding (the head at the reservoir floor is zero). Conventional orifice and culvert calculations can be used to configure the pipe and its perforations to carry that discharge.

OBSERVATIONS IN NOTTINGHAM AND WHEATLEY, ENGLAND

Many aspects of reservoir discharge were directly observed in two parking lots in England.

In Nottingham, Pratt et al. (1989, 1995) monitored discharge from a parking lot surfaced with concrete open-jointed blocks (the Aquaflow block from the British company Formpave, www.formpave.co.uk). The specially constructed parking lot was located on the Clifton Campus of Nottingham Trent University (formerly called Trent Polytechnic); it held 16 cars. Figure 4.13 shows the construction. Impervious partitions separated the base reservoir into four cells, each of which was filled with a different type of open-graded aggregate material: gravel, slag, crushed granite, and crushed limestone. An impermeable liner directed each cell's drainage to a perforated pipe for monitoring.

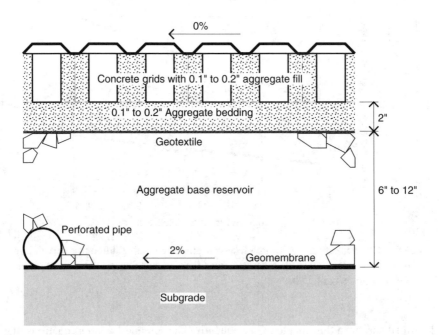

FIGURE 4.13 Parking lot construction at Nottingham Trent University in England (after Pratt et al., 1989).

Porous Pavement Hydrology

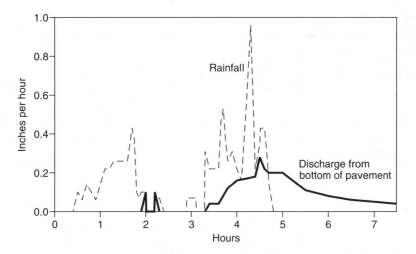

FIGURE 4.14 Base-reservoir discharge from the Nottingham Trent University parking lot during a day in September, 1987 with natural rainfall of 0.87 inch (after Pratt et al., 1989, Figure 9).

Figure 4.14 shows discharge from the slag reservoir during a day with natural rainfall of 0.87 inch. The first rainfall produced no discharge; it was entirely absorbed by the pavement materials. After discharge began, its peak rate was much lower and later than the peak rainfall. The discharge continued after the rain had stopped as the reservoir slowly drained out. Only 37 percent of the rainwater discharged from the pavement during the storm; only 51 percent discharged within one hour after the storm ended, and only 66 percent ever discharged. (Discharge was defined as flow of 0.01 inch per hour or greater in the discharge pipe. Outflow at less than 0.01 inch per hour continued for hours.)

For all four reservoir materials, peak discharge was typically 30 percent that of rainfall, whereas an impervious pavement would produce surface runoff at closer to 90 percent. The peak discharge was typically five to ten minutes later than the peak rainfall, whereas an impervious pavement on the same site would have a time of concentration of only two to three minutes.

Table 4.10 lists relationships of discharge volume to precipitation in two ways. The ratios describe annual total reservoir discharge; they range from 0.34 to 0.47. The equations describe discharge during individual rainfall events. They indicate that the pavement materials absorbed and evaporated the first 0.07 to 0.09 inch of rainfall; above that amount discharge increased at the rate of 0.68 to 0.81 inch of discharge for every additional inch of rain. Among individual storms, discharge could vary widely with antecedent conditions and rainfall intensity; nevertheless, the equations' R^2 values of 0.80 to 0.86 indicate favorable overall fit to the observed data.

Table 4.11 lists characteristics of the Nottingham base materials that explain their different discharges. Slag produced the lowest discharge because its honeycomb particles offered abundant storage for rainwater. Granite had a high discharge because of the low surface area of its particles and limited capacity for absorption.

TABLE 4.10
Relationships between Base-reservoir Discharge and Rainfall at the Nottingham Trent University Parking Lot

Base Reservoir Material	Annual Total Discharge ÷ Annual Total Rainfall, inch/inch	Event Precipitation (P) and Discharge, both in inches
Gravel	0.37	Base discharge = 0.69 P − 0.08
Blast furnace slag	0.34	Base discharge = 0.68 P − 0.09
Crushed granite	0.47	Base discharge = 0.76 P − 0.07
Crushed limestone	0.47	Base discharge = 0.81 P − 0.09

Pratt et al. (1995).

TABLE 4.11
Characteristics of Base Materials at the Nottingham Trent Parking Lot

Base Aggregate Material	Particle size (inches)	Void Space (percent)	Water Retention in Material (inch/inch)
Gravel	0.4	31	0.04
Blast furnace slag	1.5	48	0.06
Crushed granite	0.1 to 1.5	42	0.04
Crushed limestone	0.2 to 1.5	43	0.03

Pratt et al. (1989).

Gravel and limestone had intermediate discharge volumes, with gravel having a moderate surface area for wetting and limestone a moderate capacity for absorption.

The Nottingham observations were confirmed and supplemented in Wheatley, England, where Abbott and Comino-Mateos (2003) monitored reservoir discharge from a porous parking lot at a highway service station. The pavement structure was analogous to that at Nottingham, with Formpave open-jointed block. A geomembrane confined reservoir outflow to perforated drainage pipes, where discharge was monitored. The reservoir aggregate contained substantial fines, where abundant small pores were capable of retaining significant amounts of water.

Table 4.12 summarizes the observed discharge. Like the Nottingham discharge, it lagged behind rainfall and continued at low rates after rainfall stopped. The delay was more pronounced than at Nottingham due to the slow movement of water through the reservoir's small pores. The discharge rate was smaller than the rainfall; for a rainfall intensity of 0.5 inches per hour the peak discharge was only 0.01 inches per hour. During some events, however, the total volume of drainage water was greater than that of the rainfall due to antecedent moisture retained in the reservoir's small voids; in effect, the slow drainage from previous storms overlapped into later storm events.

INFILTRATION INTO SUBGRADE

In some porous pavements, infiltration into the subgrade is the reservoir's only discharge route. In other pavements, infiltration combines with lateral pipe discharge or surface overflow to contribute to total discharge. The subgrade's infiltration rate

TABLE 4.12
Discharge from the Bottom of an Open-jointed Block Parking Lot in Wheatley, England

	Range	Average
Time from start of rainfall to start of pipe drainage	< 5 minutes to > 2 hours	
Time from peak rainfall to peak outflow	5 minutes to > 9 hours	
Duration of outflow	5 to 31 times duration of rainfall	14 times duration of rainfall
Total water draining from each event	30 to 120 percent of rainfall	67 percent of rainfall
Peak rainfall intensity	0.19 inch/h to 1.32 inch/h	
Peak outflow rate	0.01 inch/h to 0.15 inch/h	

Data from Abbott and Comino-Mateos (2003).

TABLE 4.13
Terms with Particular Application to Subgrade Infiltration.

Term	Definition
Aquifer	Porous earth material that contains and transmits water
Base flow	Low flow of a stream during dry weather
Exfiltration	Discharge of water from a storage reservoir by infiltration into a soil surface
Ground water	Water in the earth filling the pores to saturation
Horizon	Distinctive layer in a soil profile
Recharge	Entry of water to an aquifer
Soil moisture	Water in the earth above the water table, filling the pores to less than saturation
Water table	Top of the saturated zone in a porous material

influences the reservoir's ponding time and depth, and with them the remaining storage capacity when the next storm occurs, the restoration of aerobic conditions for biodegradation of pollutants, the pavement's susceptibility to frost damage, and whether downstream conveyances will be required to carry frequent overflows. Table 4.13 lists some terms with particular application to subgrade infiltration.

SOIL INFILTRATION RATE

A subgrade soil's infiltration rate is a function of the soil's texture (combination of particle sizes), structure (aggregation), and compaction. It can be measured in infiltration tests, estimated from soil characteristics reported in surveys, or observed in borings. Multiple phases of soil identification and infiltration testing may be justified as a project progresses from general site layout to detailed hydrologic modeling and design.

Each naturally occurring soil has a series of horizons which may have different infiltration rates. In designating soil infiltration rates for porous pavement design, it is prudent to use the value for the least permeable layer within a few feet below a proposed pavement's floor. Shallow bedrock or groundwater may limit infiltration where it occurs within a few feet of the floor. A pavement's excavation depth can be planned to keep the floor well above a limiting layer or to expose deeper horizons with higher infiltration rates. Alternatively, a distinct "drainage well" or trench can pass water through slowly permeable layers and into more favorable layers.

Table 4.14 lists approximate infiltration rates associated with soil texture. The listed values are based on soils that have no aggregated structure, and that have been only lightly compacted. Thus, they may be representative of subgrade soils that have not been significantly compacted.

Subgrade compaction during construction reduces infiltration rate. For many pavements, structural requirements make subgrade compaction unavoidable. Pitt and Lantrip (2000) found the average infiltration rates shown in Figure 4.15 for 150

TABLE 4.14
Approximate Infiltration Rates in Unstructured Soils

Texture	Infiltration Rate (inch/hour)
Sand	8.27
Loamy sand	2.41
Sandy loam	1.02
Loam	0.52
Silt loam	0.27
Sandy clay loam	0.17
Clay loam	0.09
Silty clay loam	0.06
Sandy clay	0.05
Silty clay	0.04
Clay	0.02

Rawls et al. (1982), Table 2.

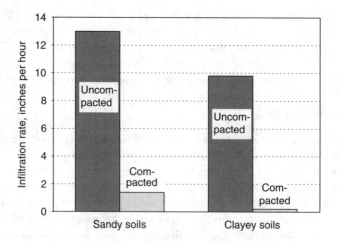

FIGURE 4.15 Average infiltration rates of compacted and uncompacted soils in wet conditions (data from Pitt and Lantrip, 2000).

compacted and uncompacted soils in Alabama. The "sandy" soils were sand, loamy sand, and sandy loam. The "clayey" soils were clay and clay loam. For sandy soils the infiltration rate of compacted soil was 0.108 that of uncompacted soil; for clayey soils it was 0.020.

For assigning an infiltration rate to a finished subgrade, neither on-site testing nor indirect indicators such as soil texture provide completely consistent or reliable results. Compaction during construction and sedimentation afterwards reduce infiltration rate in ultimately unpredictable ways (Rollings and Rollings, 1996, pp. 216–217). Consequently it is prudent to assume a subgrade infiltration rate lower than that found in preconstruction infiltration tests or inferred from soil texture, even after taking planned compaction into account. A common practice is, after finding a value from testing or by indication from texture, to multiply that value by a safety factor S_f with a value between 0 and 1. A safety factor of 0.5 has been derived from infiltration monitoring tests, and has precedents for use in practice (Bouwer, 1966; Ferguson, 1994, p. 96; Rawls et al., 1983; Paul Thiel Associates Limited, 1980, p. 179). With this safety factor value, the infiltration rate used in design is in effect half the rate otherwise indicated before construction.

A more precise procedure is to measure subgrade infiltration rate directly after compaction has been completed. In practice in Florida, subgrade infiltration rate is measured in the field after compaction to confirm the needed thickness of the base reservoir. Where the compacted infiltration rate turns out lower than that assumed in design calculations, additional reservoir thickness is added to increase storage capacity.

A soil's preconstruction infiltration rate can be preserved during construction by scrupulously preventing compaction. Necessary construction procedures must be specified strictly and construction in progress must be overseen to make sure no errors are made. Many of Cahill's (1993, pp. 15–16) porous asphalt installations have been completed without subgrade compaction even where the base course above it is to be compacted. Cahill's specifications, enforced by oversight during construction, explicitly require that the subgrade surface not be compacted or subjected to construction traffic before placement of the base material. If the excavated subgrade surface is subjected to rainfall before placement of the base, the resulting surface crust must be excavated (to perhaps 1 inch deep) or the surface must be raked to break up the crust.

INFILTRATION PONDING TIME AND RESERVOIR THICKNESS

The ponding time during infiltration of the rainwater from a given storm can be estimated by,

Ponding time = Water depth / (Subgrade infiltration rate $\times$ S_f)

where S_f is the safety factor with a value between 0 and 1.

If the ponding time found by the above equation is less than or equal to a designated maximum allowable time, then the pavement design works: the water infiltrates into the subgrade rapidly enough. On the other hand, if ponding time is greater than the allowable time, then the water does not infiltrate fast enough. Swales or pipes are necessary to convey overflow during the design storm, and, in some jurisdictions, a downstream reservoir might be required to manage it.

Where infiltrating the entire design-storm precipitation is not possible, infiltrating some smaller amount may still have environmental advantages, as described at the end of this chapter. The maximum depth of water that can be infiltrated within an allowable ponding time can be found by,

Water depth = Subgrade infiltration rate $\times S_f \times$ Ponding time.

Because water occupies only the void space in the reservoir material, the total required reservoir thickness is given by,

Total material thickness = Water depth / V_d.

Note that this simple calculation for reservoir thickness does not take into account infiltration during the storm event, which would reduce the peak depth of water stored during the storm. This is a prudently conservative approach to design. As an alternative it is possible to perform a reservoir routing, relating the reservoir's inflow, exfiltration, and change in storage over successive time increments during a storm event. The result may yield a thinner required reservoir.

LATERAL DISCHARGE FOLLOWING PARTIAL INFILTRATION

With or without detention in a reservoir, subgrade infiltration of at least some rainwater reduces and delays the discharge of the remaining overflow water (Erie, 1987; Ferguson, 1994, pp. 112–113; Ferguson, 1995a and 1995b; Ferguson, 1998, pp. 208–209; Ferguson and Deak, 1994; U.S. Soil Conservation Service 1986, p. F-1, equation for Figure 6.1). In some cases it reduces peak discharge rate. The reductions can be substantial during small, frequent storms. If, despite these transformations, a downstream detention basin is required to make the discharge comply fully with a local requirement, then the elimination of part of the flow volume reduces the basin's required size.

The quantitative effect of a proposed pavement's infiltration on downstream flow can be evaluated by applying a hydrologic model to the pavement or to the watershed of which the pavement is part. The resulting discharge would reflect both the volume reduction of the infiltration and the detention of the remaining water in the pavement reservoir.

Cahill (1993, p. 19) calculated that, for developments in the mid-Atlantic area, infiltrating a volume of rainwater equal to that of the two-year storm tends to reduce peak discharge rate during the 100-year storm to its predevelopment level.

Ferguson (1995a) modeled infiltration's downstream effects in complex watersheds where infiltration is practiced in the developed areas. His model routed the combined flows from developed and undeveloped areas through the stream system to the bottom of each watershed. Figure 4.16 shows the resulting peak flow rate. The horizontal axis is the amount of infiltration in the developed areas as a proportion of the runoff that would otherwise exist. The vertical axis is the peak flow at the bottom of a watershed as a proportion of peak flow before any development; at 1.0 a

Porous Pavement Hydrology

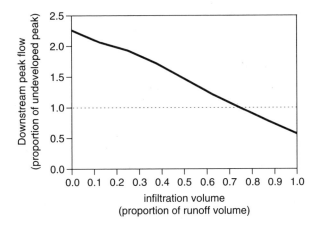

FIGURE 4.16 Peak storm flow in 30 artificial "watersheds," where developed areas infiltrate varying proportions of their runoff (data from Ferguson 1995a).

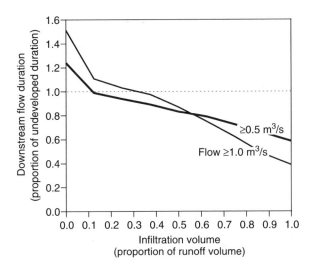

FIGURE 4.17 Storm flow duration in 30 artificial "watersheds," where developed areas infiltrate varying proportions of their runoff (data from Ferguson 1995a; m^3/s is cubic meters per second).

horizontal line indicates the predevelopment peak flow. The sloping curve indicates that increasing infiltration progressively reduced the downstream peak rate. It reduced peak rate below the predevelopment rate where infiltration equaled at least 75 percent of the developed areas' runoff.

Figure 4.17 shows the results for downstream flow duration. Long duration at moderate to high rates would increase stream erosion. Predevelopment duration is marked with a horizontal line. The chart's two curves reflect the results at two sample flow rates. The curves' downward slopes indicate that increasing infiltration volume progressively reduced downstream flow duration. At both rates the duration reached predevelopment levels with infiltration of only one third of the developed areas' runoff.

SUBSURFACE DISPOSITION OF INFILTRATED WATER

Researchers in Pennsylvania (Urban and Gburek, 1980; Gburek and Urban, 1980) observed directly how water, after infiltrating into a pavement subgrade, becomes part of subsurface soil moisture and groundwater.

For their observations they constructed a panel of porous asphalt at the Willow Grove Naval Air Station near Philadelphia. The panel was 150 feet × 150 feet in area and surrounded by grass. For comparison they set aside similar panels of grass and dense asphalt. Figure 4.18 shows the porous asphalt's setting and construction. Construction excavation slowly removed permeable clay subsoil and left only permeable broken stone between the pavement and the sandstone aquifer. Subsurface monitoring wells were arrayed in and around the experimental panels.

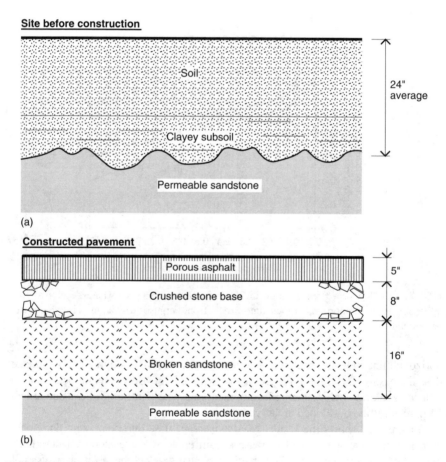

FIGURE 4.18 Setting and construction of porous asphalt panel in Willow Grove, Pennsylvania (after Urban and Gburek, 1980).

Porous Pavement Hydrology   **149**

Figure 4.19 illustrates the porous pavement's contribution to groundwater recharge. It shows how the water table under the porous asphalt responded to an

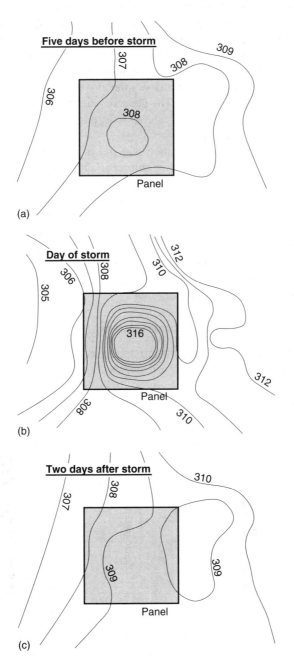

FIGURE 4.19 Water table elevations in response to the storm of August 28, 1978, under a 150 feet 3150 feet panel of porous asphalt; the panel's surface elevation is 326 (1 feet contours of water table elevation after Gburek and Urban, 1980).

intense 2-inch storm on August 28, 1978. Before the storm, the water table in the area was approximately level, from 17 to 20 feet below the pavement surface. On the day of the storm, as water infiltrated through the pavement into the subgrade, the water table under the pavement mounded up to within 10 feet of the pavement surface, with a peak in the center of the pavement 6 to 10 feet higher than the water table under the surrounding lawn. Although water infiltrated into the lawn, the water table there did not respond as quickly because the lawn retained the native soil's water retention capacity and slow permeability. Two days after the storm, the porous pavement's groundwater mound had dissipated. The general water table was now about 1 foot higher that it had been before the storm, reflecting both the pavement's rapid recharge and the lawn's more gradual recharge.

Figure 4.20 compares the disposition of water in all three experimental panels during a less intense storm the following summer (Gburek and Urban, 1980). Of the water falling on the porous asphalt, 70 percent recharged the groundwater; the other 30 percent remained in the unsaturated zone above the water table. The water table under the porous asphalt rose 2 to 5 inches for each inch of rainfall, then subsided after the storm. Under the grass panel, the fine-textured soil held essentially all water in soil moisture, where it was available for either evapotranspiration or gradual drainage to groundwater (perhaps porous pavement construction that left the fine-textured soil intact would have had a similar effect). The dense asphalt panel deflected almost all water into surface runoff.

WATERSHED DISCHARGE OF INFILTRATED WATER

Infiltrated water eventually discharges from a watershed by some combination of evapotranspiration (as plant roots withdraw soil moisture) and stream base flow (as groundwater drains out to low-lying stream channels). To illustrate this, Figure 4.21 shows the total annual discharge from a 67-acre school site in Georgia (Ferguson et al., 1991).

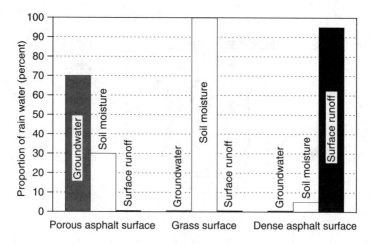

FIGURE 4.20 Disposition of rainwater through contrasting land covers during a one-inch summer storm in Willow Grove, Pennsylvania (data from Gburek and Urban, 1980).

Porous Pavement Hydrology 151

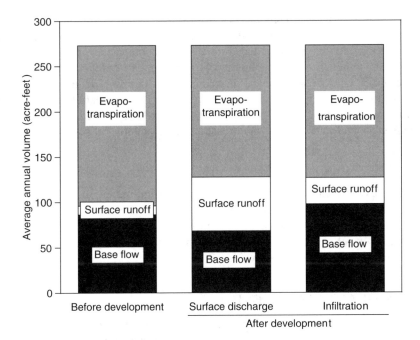

FIGURE 4.21 Annual discharge from a school site in Georgia under alternative development and stormwater management conditions (data from Ferguson et al., 1991).

Fortuitously, almost the entire site drained to a single watershed discharge point. Water-balance modeling compared the site's predevelopment hydrology with two prospective ways of managing stormwater on the developed site: one with surface drainage including impervious pavements and runoff detention basins, and the other with infiltration through porous pavements and infiltration basins. The average annual precipitation on this site amounted to 273 acre-feet in water. Each column in the figure shows one condition of development and stormwater management.

To the left is discharge before development. Before development the site was entirely vegetated and porous, so almost all rainwater infiltrated the surface and became soil moisture. Consequently evapotranspiration was large. Soil moisture that did not evapotranspire recharged the groundwater, whence it drained out in steady, continuous stream base flow. Surface runoff occurred only during large storms, and even then only in moderation.

In the middle of the figure is discharge from the developed site using surface stormwater management. The development's buildings and pavements reduced evapotranspiration and stream base flow by sealing over part of the once-vegetated surface. The same impervious roofs and pavements made annual surface runoff six times greater than it was before development. Although total annual stream flow was greater than that before development, only half of it was in steady base flow; the other half was in frequent flash storm flows.

To the right of the figure is discharge from the developed site using the same layout of buildings and pavements, but adding stormwater infiltration. Sufficient infiltration was provided to reduce peak flow in the 25-year storm to the same level achieved by the detention basins in the surface-discharge plan. The figure shows that stormwater infiltration greatly reduced surface runoff compared with the surface-discharge plan, although some residual runoff came from peripheral portions of the site that could not direct their runoff into infiltration basins. The infiltrated water recharged groundwater, after which it reappeared again as stream base flow. Base flow was even greater than it was before development, because the infiltration basins recharged groundwater without significant evapotranspirative losses.

Where Not to Infiltrate Stormwater

Despite the environmental advantages of stormwater infiltration, water should not be infiltrated into the subgrade indiscriminately. In some places infiltration could pollute groundwater, endanger structural stability, or sacrifice an opportunity to harvest and use the water on-site. Table 4.15 lists some types of settings where stormwater infiltration should ordinarily be avoided.

WATER-QUALITY TREATMENT

The large area that pavements cover, and the automobiles that use them, make pavements a locus of urban pollutant generation. Eliminating pavement stormwater

TABLE 4.15
Types of Places Where Infiltration Into Pavement Subgrade Could Pollute Ground Water, Endanger Structural Stability, or Forfeit a Resource

Place or Condition	Hazard
Steep subgrade slope (for example greater than 5%)	Water moving in the pavement base could erode the subgrade
Over a septic tank leaching field	Excess soil moisture could compromise the soil's treatment capacity
Swelling soil	Soil movement during changes in moisture can crack or displace concrete, asphalt and paving blocks (pavement surfaces of grass and aggregate can tolerate this movement)
Grossly permeable soil which would act as a conduit for untreated water into groundwater	Groundwater contamination from pavement pollutants
Steep unstable hillside	Excess moisture could further destabilize the slope
Over a toxic soil deposit in an old "brownfield" industrial area	Excess water could leach contaminants out of the soil and into the environment
Adjacent to a building with a basement (within about 20 feet)	Infiltrating water could flood the basement
Where water is harvested for direct use	Release of water to the environment would sacrifice an on-site resource
Fill soil that supports a heavily traveled road or other sensitive structure	Excess moisture could weaken the supporting fill

Porous Pavement Hydrology

TABLE 4.16
Terms with Particular Application to Porous Pavements' Treatment of Water Quality

Term	Definition
Adsorption	Adherence of chemical ions to the surfaces of solid particles
Biodegradation	Decomposition by biochemical activity into simpler chemical components
BOD	Biological oxygen demand
CEC	Cation exchange capacity
COD	Chemical oxygen demand
First flush	The first runoff from an impervious surface during a storm event, carrying constituents that had accumulated on the surface during preceding dry weather
Load	Total amount of a constituent discharged over time
mg/l	Milligrams of constituent per liter of water
Nonpoint pollution (nonpoint source pollution)	Any pollution, such as urban stormwater and its constituents, that originates over large, dispersed areas
NPDES	National Pollutant Discharge Elimination System, the U.S. federal regulation system that requires control of nonpoint pollution including urban stormwater
NPS	Nonpoint source
Organic compound	Chemical compound containing carbon
Particulates	Solid particles borne in water, as distinct from dissolved constituents
pH	A measure of hydrogen ion activity (the acidity or alkalinity of water)
SS	Suspended solids
TSS	Total suspended solids
VSS	Volatile suspended solids

contamination could reduce total urban contaminant loads by 75 percent (Bannerman et al., 1993). Table 4.16 lists terms with particular application to porous pavements' treatment of water quality.

The concentration of a water-quality constituent is the amount of constituent per amount of water, such as mg/l. In contrast a constituent's total load is the total amount discharged over time; it is equal to the concentration times the amount of water. A high concentration may be counterbalanced by low discharge of water, yielding a small total load. Where both concentration and water discharge are low, the total constituent load is small indeed.

On impervious pavements, surface runoff carries untreated constituents into the environment. When a pavement has accumulated constituents during a period of dry weather, they tend to be most concentrated in the next "first flush" of rainfall and runoff.

In contrast, porous pavements treat water quality during water's infiltration and storage in the pavement structure and subgrade. In general, porous pavements are effective at treating the particulates, oils, nutrients, and bacteria that occur in the course of pavements' normal use and maintenance. The treatment involves the removal of solid particles and their attached chemical ions from water, and bringing oil into contact with microorganisms for biochemical degradation.

Constituents in Pavement Stormwater

Table 4.17 lists substances that occur typically on urban pavements and the concerns about them when they occur in excessive amounts. They originate from everyday pavement use and maintenance. They tend to be widely distributed on pavements and, in most locations, low in concentration, but their concentrations can vary with traffic intensity and other factors.

Oils occur most prominently on busy streets and the busy portions of parking lots (Bannerman et al., 1993; Pitt, 1996, pp. 1–16; Steuer et al., 1997). In some cases they are modified by their interaction with asphalt. Relatively small amounts occur on quiet residential streets and in the lightly used portions of parking lots. Essentially no oils occur on sidewalks and bicycle paths. Excess oil blankets a water surface, inhibiting reaeration. Its decomposition deoxygenates water from within, weakening biotic communities in streams and lakes.

Excess nutrients come from dumpsters and trash-handling areas. A large proportion is carried on the surfaces of solid particles. Nutrients cause "blooms" of algae, which on decomposition deprive water of oxygen and reduce the diversity of aquatic communities. A potential oxygen loss can be expressed in terms of biological and chemical oxygen demand (BOD and COD).

Solid particles of soil and other materials are deposited from the atmosphere, blown in from surrounding land, dropped from overhanging vegetation, scattered from lawn maintenance and building construction, washed in by erosion, dropped from vehicles, ground from surface paving materials, and deliberately deposited for winter traction. All these particles are ground up and moved around by wind, rain, traffic, snowplowing, and pavement maintenance. In suspension, excess sediment makes water turbid, inhibiting aquatic plant growth and reducing species diversity. Small particles' surfaces carry a significant portion of a pavement's chemical pollutants (Tessier, 1992).

TABLE 4.17
Stormwater Constituents that Typically Occur on Pavements

Constituent	Source on Pavements	Effect in Excessive Amounts
Oil (organic hydrocarbon compounds)	Dropping from car engines	Deprive water of oxygen by decomposition
Nutrients	Organic debris; food waste	Algae blooms which deprive water of oxygen by decomposition
Particulates (solids)	Automobiles; organic debris; soil; litter	Turbid water; reduced plant growth; vectors for movement of attached metals
Trace metals	Corrosion and wear of car components and highway signs; deicing salts	Reduce organisms' disease resistance; reduce reproductive capacity; alter behavior
Chloride	Deicing salts	Reduce water intake and biotic growth
Bacteria	Animals; dumpsters; trash handling areas	Risk of disease

Trace metals such as lead, zinc, cadmium, and nickel come from the corrosion and wear of vehicle parts and highway signs, and from certain deicing salts (Pitt et al., 1996, pp. 1–16; Sansalone, 1999a). They occur mostly as ions attached to the surfaces of small solid particles; in areas with acid rain a significant amount can be dissolved. In excessive amounts they are poisonous to organisms. Because metals are chemical elements they cannot be biodegraded; they can only be captured and accumulated.

Chloride occurs where deicing salts such as sodium chloride are used in the winter (Keating, 2001; Pitt et al., 1996, pp. 1–16; Wegner and Yaggi, 2001). Salt dissolves readily in melting snow and ice; the resulting ions are not associated with particulates. Chloride ions inhibit plants' water absorption and reduce root growth. Sodium ions can injure tree roots and contaminate drinking wells. Although salt constituents appear in surface runoff only in the winter, they can remain in groundwater year-round. Chloride causes a distinctive problem for water quality; it is one of the few common pollutants that can pass through both pavement and subgrade without alteration. Porous pavements' inability to remove chloride from a solution is shared by all other passive, land- and water-based stormwater treatment approaches such as ponds, wetlands, and vegetated filters. Short of treatment plants using artificial chemical precipitation, the only way to reduce chloride pollution is to reduce it at the source in the application of deicing salts. A number of deicing alternatives to sodium chloride are available, all of which involve trade-offs of cost and effect.

In general, stormwater in residential areas tends to be relatively uncontaminated (except, in certain areas, by deicing salt). In contrast, stormwater in certain manufacturing areas may contain exotic pollutants and may have to be treated before being allowed to discharge or to infiltrate the soil (Pitt et al., 2003). In some locales special "hot spots" add locally high concentrations or exotic constituents; possible examples are gas stations, vehicle maintenance shops, and dumpster pads. On highways that carry industrial traffic, there is a concern about spills of toxic substances that could pollute groundwater. Where a toxic spill occurs, it is highly concentrated and localized at that spot; its cleanup is the responsibility of those who spilled it.

CAPTURE OF SOLIDS AND ATTACHED METALS

A porous pavement traps solid particles on the pavement surface along with the metal ions adsorbed to the particles (Balades et al., 1995; Brattebo and Booth, 2003; Hogland et al., 1990; James and Gerrits, 2003). Capture begins with the settling of sand grains and small gravel particles; then smaller particles lodge around the sand grains. Particle capture is one of the processes that can reduce the surface infiltration rate.

Particles that pass through the surface pores are likely to continue to the bottom of the pavement. They can settle on the pavement's floor or discharge through a drainage pipe if one is present. On the whole, solids accumulate most at the surface of a pavement or at the bottom; the smallest accumulations tend to be in the middle.

The following observations exemplify in detail the ability of porous pavements to filter out solids and attached metals. Sansalone (1999a and 1999b) suggested the use of oxide-coated sand as a water-quality filter because it can simultaneously adsorb dissolved metals out of solution, and filter out solid particles; this kind of material could be used as part of a porous pavement's base course, or added in a separate drainage trench.

Observations in Rezé, France

Researchers in France observed the capture of solids and metals by monitoring a porous asphalt street in Rezé (Colandini et al., 1995; Legret et al., 1996; Legret et al., 1999). The street carried 2000 vehicles per day. The pavement's aggregate base reservoir was separated from the subgrade by a geotextile and was drained at the bottom by a perforated pipe.

Figure 4.22 compares the reservoir's discharge with surface runoff from a nearby impervious street. The porous pavement discharged suspended solids at only 36 percent of their concentration in runoff; the particles discharging from the porous base were on average finer than those in the surface runoff. The adsorbed metals cadmium, zinc, and lead had concentrations only 21 to 31 percent of those in surface runoff.

The reductions in total loads were even greater, if it is assumed that the porous pavement discharged less water than the dense pavement. The Rezé researchers did not report discharge volume, but if the total discharge was half of that from the dense surface, then the total constituent loads (concentration × water volume) were only 11 to 16 percent of those from the dense pavement.

To find out where the pollutants had gone after four years of monitoring, the researchers gathered samples of particulate material from the pavement's surface pores with a road suction sweeper, and from the base pores and subgrade by digging a trench through the pavement. They analyzed the metal content in all particles smaller than 2 millimeters. Figure 4.23 shows the results for lead, which were representative of those for all the metals studied. Lead had primarily accumulated with the solid particles, particularly the smallest particles, to which it was attached at the pavement surface and secondarily at the bottom of the base reservoir. It did not

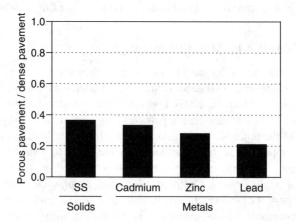

FIGURE 4.22 Constituent concentrations discharging from the bottom of a porous asphalt street in Rezé, France, as a proportion of those running off a nearby dense pavement (data from Legret et al., 1996).

Porous Pavement Hydrology

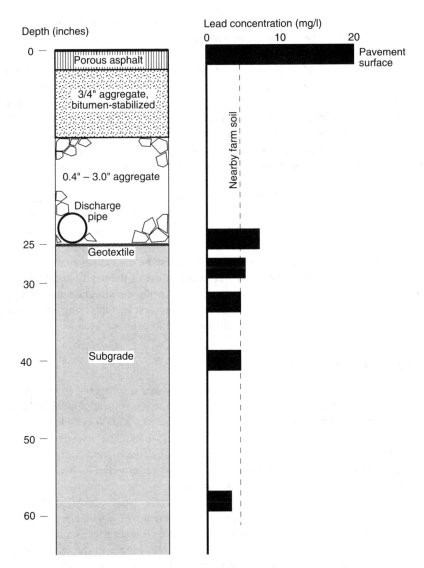

FIGURE 4.23 Lead concentrations in small particles accumulated in the Rezé, France porous asphalt street (data from Legret et al., 1996).

migrate below the structure into the soil; the subgrade's lead levels were similar to those in a nearby agricultural soil.

Observations in Nottingham, England

The Nottingham researchers (Pratt et al., 1995) observed a similar capture of solids and metals in the Nottingham Trent University parking lot that was described earlier in this chapter. The pavement's surface was made of open-celled concrete grids with aggregate fill. Partitioned portions of the base were made of four different types of aggregate. Water quality was monitored at pipes discharging from the bottom of the reservoir.

The concentration of suspended solids discharging from all four reservoir types was usually about 20 mg/l and always less than 50 mg/l, whereas that in surface runoff from nearby dense pavements was typically from 50 to 300 mg/l. The remaining solids and their attached metals were trapped mostly in the pavement's bedding aggregate. The pavement's combination of reduced discharge volume (described earlier in this chapter) and reduced concentration meant that the porous pavement's total particulate load was very low compared with that from dense pavements.

The various base materials discharged different amounts of metals because the materials' different levels of pH, hardness, and alkalinity mobilized metals into solution to different degrees. Slag discharged higher lead concentrations than the other base materials, but even those levels were only one tenth of those in the surface runoff from nearby dense pavements.

Concentrations of solids and metals were highest during the pavement's first six months while small particles and ions were washed out of the newly installed structure. Their concentrations declined as the source in the construction material was exhausted; after six months they stabilized at low levels. Only slag remained a notable source of lead and total dissolved constituents after six months.

Observation in Weinsberg, Germany

Stotz and Krauth (1994) observed the quality of water discharging from a porous asphalt overlay on a highway in Germany. The highway was located in Weinsberg in the province of Baden-Württemberg; its daily traffic was 35,000 vehicles, of which 25 percent were trucks. The overlay was 1.6 inches. thick. The investigation began after the overlay had been in service for two-and-a-half years, and continued for 12 months. The overlay's discharge was monitored at a gutter in the highway shoulder, sealed to collect discharge only from the overlay.

The total load of suspended solids from the overlay was approximately half of that in the runoff from a dense asphalt pavement in a nearby town on a per-acre basis. This indicates that even a thin porous pavement surface layer acts as a filter.

Concentrations of discharging solids were slightly higher in the winter than in the summer because higher winter rainfall and greater overlay permeability reduced the overlay's filtering efficiency. Concentrations of metals were higher in the winter due primarily to the seasonal application of deicing salt; analysis of the salt found that it was a source of metals such as zinc, lead, copper, chromium, and cadmium.

Observations in the Netherlands

Berbee et al. (1999) collected drainage from a porous asphalt highway overlay near Amsterdam similar to the one observed in Germany. The overlay carried 83,000 vehicles per day; it was three years old and was 2 inches thick. For comparison, the researchers collected runoff from another highway with an impervious asphalt surface, carrying 53,000 vehicles per day.

The researchers found that discharge from the overlay contained lower concentrations of metals, oils, and suspended solids than did runoff from the impervious asphalt, confirming the observations made in Germany.

They then experimented with treatment of the discharge water through methods such as settlement and sand filtration that could be used to further improve downstream water quality. The treatment removed the remaining metals from the overlay's discharge water less efficiently than from the dense pavement's runoff because the overlay's discharge had lower concentrations of solids and greater proportions of metals in solution rather than attached to particles. This suggests that not only is the treatment of porous pavement discharge less necessary than that of dense pavement runoff, but it may also be less feasible.

BIODEGRADATION OF OIL

A porous pavement destroys oil pollutants through the biochemical activity of microbiota which use the pavement as a substrate. Many naturally occurring types of bacteria and fungi utilize hydrocarbons as food. By consuming hydrocarbons they biodegrade them, which means that they break complex substances down into simpler chemical components. The components then disperse safely into the atmosphere as carbon dioxide and water, and cease to exist as pollutants (Atlas, 1981).

The combination of air, water, and food on a porous pavement's abundant internal surfaces supports the microbiotic community. The community's biotic composition shifts with the seasons. Biodegradation is faster at summer temperatures than at freezing or near-freezing temperatures. When the quantity of petroleum hydrocarbons increases, the biotic population expands to utilize them.

Biodegradation was observed, at least indirectly, in one of the first porous pavements. At an experimental 1975 porous asphalt parking lot at The Woodlands, Texas, percolating water was found to experience transformations of nitrogen and reductions of organic carbon and chemical oxygen demand. Researchers could attribute these improvements only to microbial activity within the pavement (Thelen and Howe, 1978, p. 12).

More recently and directly, the Nottingham Trent University researchers observed petroleum-product biodegradation in their open-celled grid pavement (Newman et al., 2001; Pratt et al., 1999). The microbial population was visible under an electron microscope as a highly diverse microbial "biofilm". In various studies, from 97.6 to over 99 percent of the applied motor oil was trapped in the pavement structure and biodegraded. A geotextile separating the grids' setting bed from the aggregate base course was a major site of oil retention, biofilm development, and gradual release of oil to further microbial populations deeper in the structure. The researchers witnessed microbial biodegradation in process through the production of microbial enzymes and respiratory carbon dioxide, the consumption of oxygen, the elevation of temperature, and the visible growth of bacterial populations. The microbial community responded quickly to additions of fresh oil. Under intense oil applications, indigenous microbial populations, when fertilized, were as efficient at biodegradation as commercially obtained oil-degrading microbial mixtures.

The Nottingham researchers speculate that adding organic material such as peat or carbon granules in the voids of base aggregate would increase the removal of organic pollutants (Pratt et al., 1989). This is an argument in favor of making a base course into a "structural-soil" rooting medium like those that will be described in Chapter 5. In the soil portions of such mixtures, clay, organic matter, and living organisms support the kind of biochemistry that protects water quality.

To stimulate and augment degrading microbes, nutrient and microbial supplements such as those listed in Table 4.18 are available. They come in powder and liquid forms. Some products add nitrogen and phosphorus nutrients that stimulate microbial activity by rounding out the nutrient supply. Others add large numbers of oil-degrading microorganisms. Products like these could be applied to entire pavements at regular intervals as preventive maintenance, or selectively to spots where oil has visibly accumulated.

Supplemental Treatment by Subgrade Infiltration

Beneath a porous pavement, almost any subgrade soil can further protect the quality of water before it percolates to groundwater (Pitt et al., 1996, pp. 1–16). Natural particles of clay and organic matter have electrochemically active surfaces that interact with the dissolved chemicals in percolating water. Naturally occurring soil microbiota degrade complex chemicals into simpler constituents. Like pavement structures, soils can treat infiltrating stormwater that contains the common constituents associated with ordinary pavement use and maintenance, but not exotic chemicals discharged from industry.

In the Central Valley of California, Nightingale (1987) found groundwater quality beneath stormwater infiltration basins indistinguishable from that elsewhere in the regional aquifer. Stormwater contaminants were still being trapped in the uppermost few inches of soil even where stormwater had been infiltrating for 20 years.

On Long Island, New York, Ku and Simmons (1986) found that the underlying aquifer was not measurably polluted by stormwater infiltration, either chemically or microbiologically. Where the aquifer had previously been polluted from non-stormwater sources such as septic tanks, recharging stormwater improved the groundwater by diluting excess nitrogen with freshwater.

In Denmark, pollutant concentrations reached natural background levels less than 5 feet from the surface even in soils that had been infiltrating urban stormwater for 45 years (Mikkelsen et al., 1996; Mikkelsen et al., 1997).

TABLE 4.18

Examples of Suppliers of Products Intended to Augment or Stimulate Microbial Activity

Company	Product	Contact Information
Bioremediate	Petro-Clear	www.bioremediate.com
Bioscience Environmental Products	Microcat	www.bioscienceinc.com
Kusuri	Bacta Pur	www.kusuri.co.uk/bactapur.htm
Stoney Creek Materials	Remediation material	www.stoneycreekmaterials.com

A soil's cation exchange capacity (CEC) indicates its ability to remove many kinds of pollutants from water. Natural mineral soils' CEC can range from 2.0 meq (milli-equivalents per 100 grams of soil) for sands to over 20 meq for some clays (Brady, 1974, p. 102).

A soil's CEC, multiplied by the thickness of the soil mantle, indicates the total renovation capability of a soil profile. It takes only a few inches of most kinds of soil to trap and transform oils, metals and nutrients. Great soil thickness can compensate for low CEC. Some agencies recommend that aerated soil be 4 feet thick over limestone aquifers where rapidly permeable solution channels and sinkholes would threaten the effective thickness of the soil mantle (Cahill, 1993, p. 26).

Where a natural subgrade's CEC is too low to assure treatment, a "treatment liner" of soil with higher CEC can be installed. In Washington State such a liner is to have CEC of at least 5 meq and to be at least 2 feet thick (Washington State Department of Ecology, 2001, pp. 48, 121, 128, 160). After it is decided that a liner of this type will be required, hydrologic modeling must be redone taking into account the new artificial layer's infiltration rate.

THE IMPORTANCE OF SMALL STORMS

A porous pavement that can store or control only a small amount of water can nevertheless have an important long-term hydrologic role. Pavements that control small amounts of water tend to be small in stature and therefore economically viable, so it is important to understand their potential effects.

FREQUENCY OF SMALL STORMS

Figure 4.24 shows the frequency of rainfall events of different sizes in four U.S. cities. All four curves show that by far the most common rainfall events are the smallest. Small events replenish soil moisture and groundwater many times per year, sustaining local ecosystems and stream base flow during the year. They control long-term water quality, because every time more than a minimal amount of rain falls on a pavement surface it potentially picks up pollutants and moves them through the environment.

Figure 4.25 shows that the rainfall in small events accumulates into large amounts of water over the course of an average year. The curves' steep rise on the left side of the graph indicates that small storms contribute greatly to total annual rainfall despite their small size, because they are so numerous. Storms smaller than 1 inch per day yield more than half of all the rain that falls in an average year. Storms larger than about 1.5 inches per day contribute relatively little to the annual total despite their size, because they occur so seldom. Most of the water that is annually available for maintenance of groundwater, stream base flow, and rooting-zone soil moisture comes from small, frequent storms.

CUMULATIVE RAINWATER INFILTRATION

A porous pavement's capacity to infiltrate water operates during every storm, large and small alike. Over time, only a pavement of small capacity can infiltrate a substantial amount of rainwater.

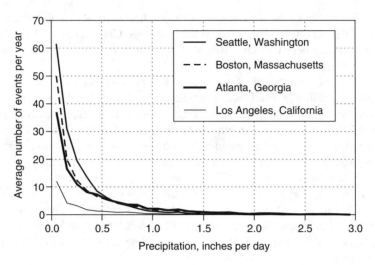

FIGURE 4.24 Average frequency of rainfall events ranked by size (calculated from daily precipitation data in increments of 0.1 inch, 1950–1989, from U.S. National Climatic Data Center, www.ncdc.noaa.gov).

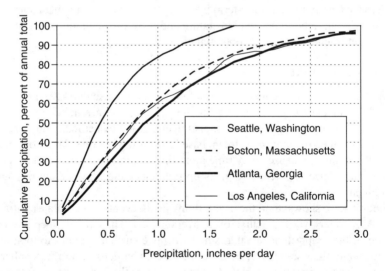

FIGURE 4.25 Cumulative amount of precipitation from events ranked by size (calculated from data in Figure 4.24).

Figure 4.26 illustrates how a porous pavement with a given infiltration capacity absorbs water from a sequence of storm events of diverse sizes. The columns show daily rainfall amounts during a three-month period in Atlanta. A horizontal line shows a hypothetical infiltration capacity at the low arbitrary level of 0.25 inch per day. The dark parts of the columns show the water that would infiltrate through a pavement with that capacity. They show that infiltration would include all the water from events

Porous Pavement Hydrology

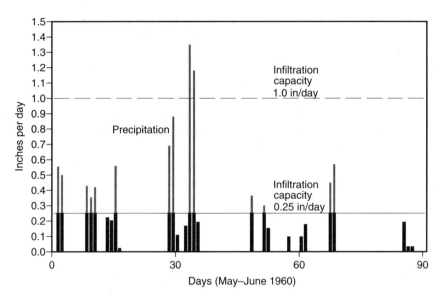

FIGURE 4.26 Precipitation in Atlanta in the spring of 1960, with two hypothetical infiltration capacities (precipitation data from National Climatic Data Center).

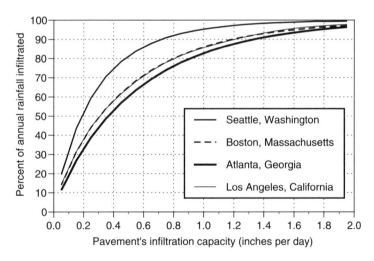

FIGURE 4.27 Portion of annual rainfall infiltrated as a function of pavement infiltration capacity (calculated from data in Figure 4.24).

smaller than 0.25 inch, and the first 0.25 inch of water from larger events. Every first flush is absorbed, and a substantial total amount of water is infiltrated over time. The light parts of the columns show the amount of water that would overflow and continue downstream. At the moderate infiltration capacity of 1.0 inch per day, indicated by another horizontal line, a pavement could absorb almost all water; little would overflow.

Figure 4.27 shows how much of the total annual rainfall porous pavements of different capacities can infiltrate. The horizontal scale is the amount of rainfall that a pavement can infiltrate within 24 hours. The vertical scale is the proportion of average annual rainfall that the pavement infiltrates. The curves for four cities were determined by analyzing rainfall data in the manner conceptualized in the previous figure. Rain amounts smaller than or equal to a pavement's capacity are infiltrated; the remainder overflows. The chart shows that even in the rainy city of Atlanta, half the rain that falls in an average year could be infiltrated by a pavement with the capacity to infiltrate only 0.37 inch of water per day.

TRADEMARKS

Table 4.19 lists the holders of registered trademarks mentioned in this chapter.

TABLE 4.19
Holders of Registered Trademarks Mentioned in This Chapter

Registered Trademark	Holder	Headquarters Location
Advanced Drainage Systems	Advanced Drainage Systems	Hilliard, Ohio
Aquaflow	Formpave	Coleford, UK
Bacta Pur	Kusuri	Newton Abbot, UK
Checkerblock	Nicolia Industries	Lindenhurst, New York
Contech	Contech Construction Products	Middletown, Ohio
Cultec	Cultec	Brookfield, Connecticut
Double Trap	StormTrap	Morris, Illinois
EnviroChamber	Hancor	Findlay, Ohio
Formpave	Formpave	Coleford, UK
Hancor	Hancor	Findlay, Ohio
Hydrologic Solutions	Hydrologic Solutions	Occoquan, Virginia
Invisible Structures	Invisible Structures	Golden, Colorado
Jensen Precast	Jensen Precast	Sparks, Nevada
LandMax	Hancor	Findlay, Ohio
Microcat	Bioscience Environmental Products	Bethlehem, Pennsylvania
Monoslab	E. P. Henry	Woodbury, New Jersey
Petro-Clear	Bioremediate	Cocoa, Florida
Rain Store	Invisible Structures	Golden, Colorado
Recharger	Cultec	Brookfield, Connecticut
Single Trap	StormTrap	Morris, Illinois
Stoney Creek Materials	Stoney Creek Materials	Austin, Texas
Storm Compressor	Advanced Drainage Systems	Hilliard, Ohio
StormChamber	Hydrologic Solutions	Occoquan, Virginia
Stormtech	Stormtech	Wethersfield, Connecticut
StormTrap	StormTrap	Morris, Illinois
StormVault	Jensen Precast	Sparks, Nevada
Uni-Group USA	Uni-Group USA	Palm Beach Gardens, Florida

ACKNOWLEDGMENTS

The following persons generously provided information used in this chapter: Derek Booth and Steve Burges of the University of Washington; Soenke Borgwardt of Borgwardt Wissenschlaftliche Beratung (Norderstedt, Germany); Thomas Cahill and Michele C. Adams of Cahill Associates (www.thcahill.com); Donna Dininno of Uni-Group USA (www.uni-groupusa.org); staff at Infiltrator Systems (www.infiltratorsystems.com); Margaret Kramer of The Woodlands (Texas) Community Service Corporation; Harald von Langsdorff of F. von Langsdorff Licensing (www.langsdorff.com); Per Mikkelsen of Technical University of Denmark; Helga Piro of SF Concrete Technology (www.sfconcrete.com); Christopher Pratt and his staff, Coventry University (UK); Wilton "Bud" Roberts of the Oregon Department of Transportation; and Kathleen Simmonds of the Town of Clifton Park (New York) Department of Planning.

The following persons constructively critiqued early drafts of this chapter: Derek Booth of the University of Washington; Stuart Echols of Pennsylvania State University; Robert E. Pitt of the University of Alabama; and David R. Smith of the Interlocking Concrete Pavement Institute (www.icpi.org).

REFERENCES

Abbott, C.L., and L. Comino-Mateos (2003). In-Situ Hydraulic Performance of a Permeable Pavement Sustainable Urban Drainage System, *Water and Environmental Management Journal* 17, 187–190.

Adams, Michelle C. (2003). Porous Asphalt Pavement with Recharge Beds: 20 years & Still Working, *Stormwater* 4, 24–32.

American Association of State Highway and Transportation Officials (1985). *Materials*, Washington, D.C.: American Association of State Highway and Transportation Officials.

American Association of State Highway and Transportation Officials (1986). *AASHTO Guide for Design of Pavement Structures,* Vol. 2, Washington, D.C.: American Association of State Highway and Transportation Officials.

American Association of State Highway and Transportation Officials (1993). *AASHTO Guide for Design of Pavement Structures 1993*, Washington: American Association of State Highway and Transportation Officials.

Andersen, C.T., I.D.L. Foster, and C.J. Pratt (1999). The Role of Urban Surfaces (Permeable Pavements) in Regulating Drainage and Evaporation: Development of a Laboratory Simulation Experiment, *Hydrological Processes* 13, 597–609.

Atlas, Ronald M. (1981). Microbial Degradation of Petroleum Hydrocarbons: An Environmental Perspective, *Microbiological Reviews* 45, 180–209.

Balades, J.-D., M. Legret, and H. Madiec (1995). Permeable Pavements: Pollution Management Tools, *Water Science and Technology* 32, 49–56.

Bannerman, R. T., D. W. Owens, R. B. Dodds, and N. J. Hornewer (1993). Sources of Pollutants in Wisconsin Stormwater, *Water Science and Technology*. 28, 241–259.

Berbee, R., G. Rijs, R. De Brouwer, and L. Van Velzen (1999). Characterization and Treatment of Runoff from Highways in the Netherlands Paved with Pervious Asphalt, *Water Environment Research* 71, 183–190.

Booth, Derek B., and Jennifer Leavitt, (1999). Field Evaluation of Permeable Pavement Systems for Improved Stormwater Management, *Journal of the American Planning Association* Summer 1999, pp. 314–325.

Booth, Derek B., Jennifer Leavitt, and Kim Peterson (1997). *The University of Washington Permeable Pavement Demonstration Project: Background and First-Year Field Results*, Seattle: University of Washington Center for Urban Water Resources Management.

Borgwardt, Soenke (1997a). *Leistungsfahigkeit und Einsatzmoglichkeit Versickerungsfahiger Pflastersysteme (Performance and Fields of Application for Permeable Paving Systems), Betonwerk + Fertigteil-Technik*, 1997 No. 2.

Borgwardt, Soenke (1997b). *Versickerungsfahige Pflastersysteme aus Beton: Voraussetzungen, Anforderungen, Einsatz (Infiltratable Concrete Block Pavement Systems: Prerequisites, Requirements, Applications)*, Bonn: Bundesverband Deutsche Beton- und Fertigteilindustrie.

Borgwardt, Soenke (1999). *Survey and Expert Opinion on the Distribution, Performance, and Possible Application of Porous and Permeable Paving Systems*, West Yorkshire, England: Marshalls Mono Ltd.

Bouwer, H. (1966). Rapid Field Measurement of Air Entry Value and Hydraulic Conductivity of a Soil as Significant Parameters in Flow System Analysis, *Water Resources Research* 2, 729–738.

Brady, Nyle C. (1974). *The Nature and Properties of Soils*, 8th ed., New York: Macmillan.

Brattebo, Benjamin O., and Derek B. Booth (2003). Long-term Stormwater Quantity and Quality Performance of Permeable Pavement Systems, *Water Research* 37, 4369–4376.

Cahill, Thomas (1993). *Porous Pavement with Underground Recharge Beds*, West Chester, Pennsylvania: Cahill Associates (www.thcahill.com).

Cahill, Thomas (1994). A Second Look at Porous Pavement/Underground Recharge, *Watershed Protection Techniques* 1, 76–78.

Cao, Su Ling, Daryl Pocuska and Dan G. Zollinger (1998). *Drainage Design and Performance Guidelines for UNI Eco-Stone Permeable Pavement*, Palm Beach Gardens, Florida: UNI-Group U.S.A.

Chow, Ven Te, David R. Maidment and Larry W. Mays (1988). *Applied Hydrology*, New York: McGraw-Hill.

Colandini, Valérie, Michel Legret, Yves Brosseaud and Jean-Daniel Balades (1995). Metallic Pollution in Clogging Materials of Urban Porous Pavements, *Water Science and Technology* 32, 57–62.

Day, Gary E., David R. Smith, and John Bowers (1981). *Runoff and Pollution Abatement Characteristics of Concrete Grid Pavements*, Bulletin 135, Blacksburg: Virginia Polytechnic Institute and State University.

Debo, Thomas N. (1994). Computer Model for Infiltration System Design, *Journal of Computing in Civil Engineering* 1994, 241–248.

Debo, Thomas N. (1994). *Stormwater Infiltration Structure Design*, Washington: National Stone Association.

Debo, Thomas N., and Andrew J. Reese (2002). *Municipal Storm Water Management*, 2nd ed., Boca Raton: Lewis Publishers.

Diniz, Elvidio V., (1976). Quantifying the Effects of Porous Pavements on Urban Runoff, in *Proceedings, National Symposium on Urban Hydrology, Hydraulics, and Sediment Control*, pp. 63–70, Lexington: University of Kentucky.

Diniz, Elvidio V. (1980). *Porous Pavement, Phase I, Design and Operational Criteria*, EPA-600/2-80-135, Cincinnati: U.S. Environmental Protection Agency, Municipal Environmental Research Laboratory.

Diniz, Elvidio V., and William H. Espey, Jr. (1979). *Maximum Utilization of Water Resources in a Planned Community: Application of the Storm Water Management Model*, Vol. 1, Cincinnati: U.S. Environmental Protection Agency, Municipal Environmental Research Laboratory.

Erie, Len, 1987, Stormwater Retention Criterion for Urban Drainage Basin Management, in *Computational Hydrology 87, Proceedings of the 1st International Conference*, pp.44–48,Theodore N. Hromadka II and Richard H. McCuen, Eds., Mission Viejo, California: Lighthouse Publications.

Ferguson, Bruce K. (1994). *Stormwater Infiltration*, Boca Raton: Lewis Publishers.

Ferguson, Bruce K. (1995a). Downstream Hydrographic Effects of Urban Stormwater Detention and Infiltration, in *Proceedings of 1995 Georgia Water Resources Conference*, pp.128–131, Kathryn J. Hatcher, Ed., Athens: University of Georgia Carl Vinson Institute of Government.

Ferguson, Bruce K. (1995b). Storm-Water Infiltration for Peak-Flow Control, *Journal of Irrigation and Drainage Engineering* 121, 463–466.

Ferguson, Bruce K. (1998). *Introduction to Stormwater: Concept, Purpose, Design*, New York: John Wiley and Sons.

Ferguson, Bruce K. (2002). Stormwater Management and Stormwater Restoration, in *Handbook of Water-Sensitive Planning and Design*, pp.11–28, Robert France, Ed., Boca Raton: Lewis Publishers.

Ferguson, Bruce K., and Tamas Deak (1994). Role of Urban Storm-Flow Volume in Local Drainage Problems, *Journal of Water Resources Planning and Management* 120, 53–530.

Ferguson, Bruce K., M. Morgan Ellington and P. Rexford Gonnsen (1991). Evaluation and Control of the Long-term Water Balance on an Urban Development Site, in *Proceedings of 1991 Georgia Water Resources Conference*, pp.217–220, Kathryn J. Hatcher, Ed., Athens: University of Georgia Institute of Natural Resources.

Gburek, William J., and James B. Urban (1980). Storm Water Detention and Groundwater Recharge Using Porous Asphalt — Initial Results, in *International Symposium on Urban Storm Runoff*, pp.89–97, Lexington: University of Kentucky.

Goforth, Gary F., Elvidio V. Diniz, and J. Brent Rauhut (1983). *Stormwater Hydrological Characteristics of Porous and Conventional Paving Systems*, EPA-600/2-83-106, Cincinnati: U.S.E.P.A. Municipal Environmental Research Laboratory, National Technical Information Service No. PB-84-123728.

Hamilton, George W. (1990). *Infiltration Rates on Experimental and Residential Lawns*, M.S. thesis, State College, PA.: Pennsylvania State University.

Hogland, W., M. Larson, and R. Berndisson (1990). The Pollutant Build-Up in Pervious Road Construction, *5th International Conference on Urban Storm Drainage*, pp. 845–852, Osaka.

Huber, Wayne C., and Robert E. Dickinson (1988). *Storm Water Management Model, Version 4: User's Manual*, Athens, Georgia: U.S. Environmental Protection Agency Environmental Research Laboratory.

Jackson, Thomas J., and Robert M. Ragan (1974). Hydrology of Porous Pavement Parking Lots, *Journal of the Hydraulics Division* (American Society of Civil Engineers) 100, 1739–1752.

James, W. Robert, William James and Harald von Langsdorff (2001). Stormwater Management Model for Environmental Design of Permeable Pavement, in *Models and Applications to Urban Water Systems*, Monograph 9, pp. 423–444,William James, Ed., Guelph, Ontario: Computational Hydraulics International.

James, William, and Christopher Gerrits (2003). Maintenance of Infiltration in Modular Interlocking Concrete Pavers with External Drainage Cells, in *Practical Modeling of Urban Water Systems*, Monograph 11, pp. 417–435, William James, Ed., Guelph, Ontario: Computational Hydraulics International.

James, William, Wayne C. Huber, and W. Robert James (1998). *Water Systems Models Hydrology: Student's Guide to SWMM4 RUNOFF and Related Modules*, Guelph, Ontario: CHI Publications.

Keating, Janis (2001). Deicing Salt: Still on the Table, *Stormwater* May/June 28–35.

Kipkie, Craig W., and William James (2000). Feasibility of a Permeable Pavement Option in SWMM for Long-Term Continuous Modeling, in *Applied Modeling of Urban Water Systems*, Monograph 8, pp. 303–322, William James, Ed., Guelph, Ontario: Computational Hydraulics International.

Kosmatka, Steven H., and William C. Paranese (1988). *Design and Control of Concrete Mixtures*, 13th ed., Skokie, Illinois: Portland Cement Association.

Ku, Henry F.H. and Dale L. Simmons (1986). *Effect of Urban Stormwater Runoff on Ground Water Beneath Recharge Basins on Long Island, New York*, Water-Resources Investigations Report 85-4088, Syosset, New York: U.S. Geological Survey.

Legret, M., and V. Colandini (1999). Effects of a Porous Pavement with Reservoir Structure on Runoff Water: Water Quality and Fate of Heavy Metals, *Water Science and Technology* 39, 111–117.

Legret, M., M. Nicollet, P. Miloda, V. Colandini, and G. Raimbault (1999). Simulation of Heavy Metal Pollution from Stormwater Infiltration through a Porous Pavement with Reservoir Structure, *Water Science and Technology* 39, 119–125.

Legret, M., V. Colandini, and C. Le Marc (1996). Effects of a Porous Pavement with Reservoir Structure on the Quality of Runoff Water and Soil, *The Science of the Total Environment* 190, 335–340.

Lull, Howard W., and Kenneth G. Reinhart (1972). *Forests and Floods in the Eastern United States*, Research Paper NE-226, Upper Darby, PA.: U.S. Forest Service Northeastern Forest Experiment Station.

Mikkelsen, P.S., M. Hafliger, M. Ochs, J.C. Tjell, P. Jacobsen and M. Boller (1996). Experimental Assessment of Soil and Groundwater Contamination from Two Old Infiltration Systems for Road Run-Off in Switzerland, *Science of the Total Environment* 189/190, 341–347.

Mikkelsen, P.S., M. Hafliger, M. Ochs, P. Jacobsen, J.C. Tiell and M. Boller (1997). Pollution of Soil and Groundwater from Infiltration of Highly Contaminated Stormwater — A Case Study, *Water Science and Technology* 36, 325–330.

Newman, A.P., C.J. Pratt, S.J. Coupe and N. Cresswell (2001). Oil Bio-Degradation in Permeable Pavements by Inoculated and Indigenous Microbial Communities, *NOVATECH 2001*, School of Science and the Environment, Coventry University, Coventry, UK.

Nightingale, Harry I. (1987). Water Quality Beneath Urban Runoff Management Basins and Accumulation of As, Ni, Cu, and Pb in Retention and Recharge Basins Soils from Urban Runoff, *Water Resources Bulletin* 23, 197–205; and 23, 663–672.

Overton, D.E., and M.E. Meadows (1976). *Storm Water Modeling*, New York: Academic Press.

Paine, John E. (1990). *Stormwater Design Guide, Portland Cement Pervious Pavement*, Orlando: Florida Concrete and Products Association.

Partsch, C.M., A.R. Jarrett, and T.L. Watschke (1993). Infiltration Characteristics of Residential Lawns, *Transactions of the American Society of Agricultural Engineers* 36, 1695–1701.

Paul, Thiel, Associates Limited (1980). Subsurface Disposal of Storm Water, in *Modern Sewer Design*, pp.175–193, Washington: American Iron and Steel Institute.

Pitt, Robert (1994). The Risk of Groundwater Contamination from Infiltration of Stormwater Runoff, *Watershed Protection Techniques* 1, 126–128

Pitt, Robert, and Janice Lantrip (2000). Infiltration through Disturbed Urban Soils, in *Applied Modeling of Urban Water Systems*, Monograph 8, pp. 1–22, William James, Ed., Guelph, Ontario: Computational Hydraulics International.

Pitt, Robert, Shen-En Chen, Shirley, Clark, Janice Lantrip, Choo Keong Ong, and John Voorhees, (2003). Infiltration through Compacted Urban Soils and Effects on Biofiltration Design, in *Practical Modeling of Urban Water Systems*, Monograph 11, pp. 217–252, William James, Ed., Guelph, Ontario: Computational Hydraulics International.

Pitt, Robert, Shirley Clark, Keith Parmer and Richard Field (1996). *Groundwater Contamination from Stormwater Infiltration*, Chelsea, Michigan: Ann Arbor Press.

Pratt, C.J., A.P. Newman, and P.C. Bond, (1999). Mineral Oil Bio-Degradation within a Permeable Pavement: Long Term Observations, *Water Science and Technology* 39, 103–109.

Pratt, C.J., J.D.G. Mantle, and P.A. Schofield (1989). Urban Stormwater Reduction and Quality Improvement through the Use of Permeable Pavements, *Water Science and Technology* 21, 769–778.

Pratt, C.J., J.D.G. Mantle, and P.A. Schofield (1995). UK Research into the Performance of Permeable Pavement Reservoir Structures in Controlling Stormwater Discharge Quantity and Quality, *Water Science and Technology* 32, 63–69.

Rawls, W.J., D.L. Brakensiek, and K.E. Saxton (1982). Estimation of Soil Water Properties, *Transactions of the American Society of Agricultural Engineers*, 25, 1316–1320 and 1328.

Rawls, W.J., D.L. Brakensiek, and N. Miller (1983). Green-Ampt Infiltration Parameters from Soils Data, *Journal of Hydraulic Engineering* 109, 62–70.

Rollings, Marion P., and Raymond S. Rollings, Jr. (1996). *Geotechnical Materials in Construction*, New York: McGraw-Hill.

Rollings, Raymond S., and Marian P. Rollings (1999). *SF-Rima, A Permeable Paving Stone System*, Mississauga, Ontario: SF Concrete Technology.

Sansalone, John J. (1999a). Adsorptive Infiltration of Metals in Urban Drainage: Media Characteristics, *The Science of the Total Environment* 235, 179–188.

Sansalone, John J. (1999b). In-Situ Performance of a Passive Treatment System for Metal Source Control, *Water Science and Technology* 39, 193–200.

Sansalone, John J., and S. G. Buchberger (1995). An Infiltration Device as a Best Management Practice for Immobilizing Heavy Metals in Urban Highway Runoff, *Water Science and Technology* 32, 119–125.

Shergold, F.A. (1953). The Percentage Voids in Compacted Gravel as a Measure of Its Angularity, *Magazine of Road Research* 5, 3–10.

Smith, David R. (2001). *Permeable Interlocking Concrete Pavements: Selection, Design Construction, Maintenance,* 2nd., Washington: Interlocking Concrete Pavements Institute.

Smith, Hillard L., and Luna B. Leopold (1942). Infiltration Studies in the Pecos River Watershed, New Mexico and Texas, *Soil Science* 53, 195–204.

St. John, Matthias S., and Richard R. Horner (1997). *Effect of Road Shoulder Treatments on Highway Runoff Quality and Quantity*, WA-RD-4291, Olympia: Washington State Department of Transportation.

Steuer, J., W. Selbig, N. Hornewer and J. Prey (1997). *Sources of Contamination in an Urban Basin in Marquette, Michigan and an Analysis of Concentrations, Loads, and Data*

Quality, Water Resources Investigations Report 97-4242, Washington: U.S. Geological Survey.

Stotz, G., and K. Krauth (1994). The Pollution of Effluents from Pervious Pavements of an Experimental Highway Section: First Results, *The Science of the Total Environment* 146/147, 465–470.

Tessier, A. (1992). Sorption of Trace Elements on Natural Particles in Oxic Environments, in *Environmental Particles*, Jacques Buffle and Herman P. van Leeuwen, Ed., Boca Raton: Lewis Publishers.

Thelen, Edmund, and L. Fielding Howe (1978). *Porous Pavement*, Philadelphia: Franklin Institute Press.

Thelen, Edmund, Wilford C. Grover, Arnold J. Holberg, and Thomas I. Haigh (1972). *Investigation of Porous Pavements for Urban Runoff Control*, 11034 DUY, Washington: U.S. Environmental Protection Agency.

U.S. Federal Aviation Agency (USFAA) (1965). *Airport Drainage*, Advisory Circular AC150/5320-5A, Washington: U.S. Federal Aviation Agency.

U.S. Soil Conservation Service (1986). *Urban Hydrology for Small Watersheds*, Technical Release 55, Washington: U.S. Soil Conservation Service.

Urban, James B., and William J. Gburek (1980). Storm Water Detention and Groundwater Recharge Using Porous Asphalt — Experimental Site, in *International Symposium on Urban Storm Runoff*, pp. 81–87, Lexington, Kentucky, July 28–31.

Van der Leeden, Frits, Fred L. Troise and David Keith Todd (1990). *The Water Encyclopedia*, 2nd ed., Chelsea, Michigan: Lewis Publishers.

Washington State Department of Ecology, Water Quality Program (2001). *Stormwater Management Manual for Western Washington, Volume V— Runoff Treatment BMPs*, Olympia: Washington State Department of Ecology.

Watanabe, Satoshi (1995). Study on Storm Water Control by Permeable Pavement and Infiltration Pipes, *Water Science and Technology* 32, 25–32.

Wegner, William, and Marc Yaggi (2001). Environmental Impacts of Road Salt and Alternatives in the New York City Watershed, *Stormwater* July/August 24–31.

Wei, Irvine W. (1986). *Installation and Evaluation of Permeable Pavement at Walden Pond State Reservation*, Report to the Commonwealth of Massachusetts Division of Water Pollution Control, Boston: Northeastern University Department of Civil Engineering.

Wingerter, Roger, and John E. Paine (1989). *Field Performance Investigation, Portland Cement Pervious Pavement*, Orlando: Florida Concrete and Products Association.

Wyant, David C. (1992). *Final Report, Field Performance of a Porous Asphaltic Pavement*, VTRC 92-R10, Charlottesville: Virginia Transporation Research Council.

5 Porous Pavement Tree Rooting Media

CONTENTS

Viable Rooting Media 174
Rooting Volume 175
The Role of a Pavement's Surface Course 177
Base Course Rooting Media 180
    Stone-Soil Mixture (Cornell Structural Soil) 180
    ESCS Mixtures (Arnold Air-Entrained Soil) 184
    Sand Mixture (Amsterdam Tree Soil) 187
    Provisions for All Base-Course Rooting Mixtures 189
Rooting Space in Subgrade 192
Special Treatment at the Tree Base 192
Trademarks 194
Acknowledgments 195
References 195

The benefits for which trees are planted, such as shading and cooling, come into being only where the trees grow long enough to realize some size (Geiger, 2003). For this reason, where trees are surrounded by pavements, the pavement structures must contribute to the habitat in which trees root and grow. The challenge is to maintain viable soil for abundant tree roots while supporting traffic load through the same structure.

Tree root systems are typically broad and shallow (Watson and Himelick, 1997, pp. 137–139). The rooting zone is typically within 24 or 36 inches of the surface. Even the few species with vertical taproots can have substantial lateral roots as do other trees. Growing roots meander and fork around stones and other obstructions to explore for aerated, moist, fertile soil; individual roots take on unusual shapes as they press through given pore spaces (Craul, 1992, pp. 127, 154). They branch repeatedly, terminating in numerous fine absorbing roots. They find their way to all the favorable soil zones around a tree, concentrating particularly in zones with high levels of air, water, and nutrients (Watson and Himelick, 1997, p. 145).

Figure 5.1 illustrates a number of rooting-zone provisions that will be described in this chapter. It shows a pavement constructed in Sydney, Australia's Homebush

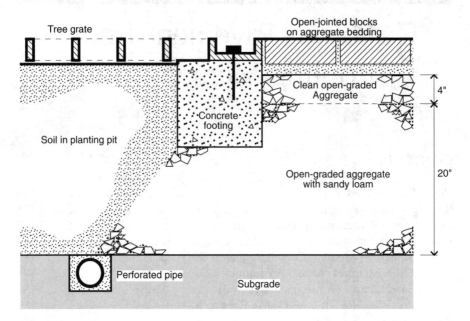

FIGURE 5.1 Installation of a rooting zone at Olympic Park, Sydney, Australia (after Bennett, 1999).

Olympic Park for the 2000 Olympic Games. Large fig trees were planted to shade seating areas at the edges of the park's central plaza. The trees needed a rooting medium under a pavement surface that would bear very heavy pedestrian traffic and occasional machinery. The base course is a mixture of open-graded aggregate and sandy loam in which tree roots can grow while the aggregate supports the pavement's traffic load and protects the loam from compaction. The aggregate-soil mixture is in contact with the soil of the tree planting pit so roots can extend directly from one into the other. The subgrade soil further expands the rooting zone if it is not so compacted as to preclude penetration by roots. The surface course of open-jointed blocks admits fresh air and water into the rooting zone. Between the surface course and the rooting zone is a layer of open-graded aggregate free of soil which protects the surface from root heaving by not attracting roots near the surface. A perforated drainage pipe prevents the accumulation of ponded water that would inhibit aeration.

A tree rooting zone should not be constructed where growing tree roots could damage utility lines, structures, or pavements. Table 5.1 lists some specific places where construction of rooting zones should be avoided.

Table 5.2 lists some terms with distinctive application to tree planting and rooting. Further information on tree rooting habitat can be obtained from the International Society of Arboriculture (www.isa-arbor.com), its *Journal of Arboriculture*, its manual *Principles and Practice of Planting Trees and Shrubs* (Watson and Himelick, 1997), and the useful books by Craul (1992, 1999). Concise guidelines for urban tree planting are in the tenth edition of *Architectural Graphic Standards* (Urban, 2000). Detailed information on soil and aggregate materials is given in Chapter 6 of this book.

TABLE 5.1
Examples of Places Where Construction of Tree Rooting Zones Should Be Avoided

Place	Hazard
In contact with underground sewer lines	Tree roots can intrude into and clog sewer lines
In contact with pavements that could be damaged by growing tree roots	Tree roots may extend under and heave pavements that were not built to accommodate them
In contact with old building basements made of drylaid masonry or masonry with soft mortar	Tree roots can intrude into joints and damage basement walls

Wayde Brown and Jason Grabosky, personal communications 2002.

TABLE 5.2
Some Terms with Distinctive Application to Tree Planting and Rooting

Term	Definition
Caliper	Trunk diameter at breast height (4.5 feet above the ground)
Canopy (crown)	A tree's overhead structure of branches and leaves
Cation exchange capacity (CEC)	Quantity of exchangeable cations that a rooting medium can adsorb and hold available for absorption by roots
Capillary	Micropore
CBR	California Bearing Ratio, a test of strength for granular materials
dbh	Trunk diameter at breast height (4.5 feet above the ground)
Dripline	The roughly circular outline of the area under a tree's canopy, or the distance of this outline from the tree's trunk
ESCS	Expanded shale, clay or slate
Gap-graded	Of two distinct size ranges, with intermediate sizes substantially absent
Hydrogel	Polymer capable of absorbing and releasing water, commonly added to rooting media to increase water-holding capacity
Lift	Layer of granular material placed at one time
Macropore	Pore wide enough that water flows through it freely without capillary tension
meq	Milliequivalents per 100 grams of rooting medium, a measure of CEC
Micropore	Pore narrow enough that water clings inside it
Open-graded (single-sized)	Of a narrow range of sizes
Planting medium (rooting medium)	Soil or soil-like material penetrable by roots and capable of supporting root growth by storing, conveying and exchanging air, water and nutrients
Planting pit (tree pit)	The immediate area where a tree is planted, consisting of the tree's root ball at the time of planting and the planting medium used for fill immediately around it
Root base (tree base, trunk flare, root flare, root buttress)	The spreading part of the trunk and the rising buttresses of the roots, where the trunk meets the ground
Rooting zone	Zone of planting medium providing conditions for potential root penetration and growth
Water-holding capacity	The quantity of water that can a rooting medium can hold, after free drainage, available for absorption by tree roots

VIABLE ROOTING MEDIA

Media that provide a viable root habitat are those that permit a combination of natural processes for the storage and transfer of air, water, and nutrients. Tree roots require from a medium the capacity to perform all the functions listed in Table 5.3.

Penetrability by roots must be unobstructed. Roots penetrate through large pores (macropores) like those between single-sized aggregate particles and through loose bodies of soil (Watson and Himelick, 1997, p. 145). A soil that is compacted is dense and its macropores are destroyed, so root penetration into it is impeded. Penetrability is indirectly indicated by permeability, because high permeability comes from connected networks of macropores or bodies of loose granular soils. Where macropores are abundant they are readily visible.

Aeration must be unobstructed. Living roots consume oxygen and produce carbon dioxide. Roots live and grow only where oxygen is present (Craul, 1992, p. 142; Evans et al., 1990). In a rooting medium excess carbon dioxide must be replaced by fresh oxygen from the atmosphere. The exchange happens through networks of macropores or the pores in loose granular soil. A rooting medium that is compacted is dense and its macropores are destroyed, so aeration is inhibited. A soil's capacity for aeration is indicated by its permeability and its contact with the atmosphere.

Drainage of excess water must be unobstructed (Bassuk et al., 1998; Watson and Himelick, 1997, p. 144). Excess water would displace the air in a rooting zone's voids, inhibiting gaseous exchange; carbon dioxide would build up without replacement of oxygen. Excess water drains out through networks of macropores or the pores in granular soil, so long as the pores are in contact with a drainage outlet at the bottom of the rooting zone. In large pores water is not held under the influence of capillary tension (Beven and Germann, 1982). In a soil that is compacted, macropores are destroyed and drainage is inhibited. The capacity for drainage is indicated by a soil's permeability and the presence of a positive drainage outlet. In urban areas many tree rooting zones are surrounded by compacted, impervious soil; in these cases a positive drainage outlet at the bottom of the rooting zone is especially vital

TABLE 5.3
Functions of Viable Rooting Media

Function	Critical Functional Feature	Indication of Function
Penetrability by roots	Networks of macropores or bodies of loose soil	Permeability
Aeration	Networks of macropores in contact with the atmosphere	Permeability
Drainage of excess water	Networks of macropores in contact with a drainage outlet	Permeability
Water-holding capacity	Numerous micropores	Water retention
Nutrient-holding capacity	Electrochemically active surfaces of clay, organic matter, or certain ESCS aggregates	Cation exchange capacity

Evans, Bassuk and Trowbridge (1990); Trowbridge and Bassuk (2004, pp. 4–6); Watson and Himelick (1997, pp. 3–10).

(Arnold, 2001). However, it is theoretically possible for excess water to drain through macropores too rapidly, failing to soak into fine pores for storage and later availability to tree roots.

Water-holding capacity must be abundant. Roots absorb water daily to supply the growth and metabolism of a tree's superstructure. The source of water might be either natural rainfall or artificial irrigation. After excess water has drained out, the remaining water is held available for roots in small pores (micropores) like those in fine-textured soils and certain ESCS aggregates. The amount of water available for roots is indicated in a soil's water retention (or "water release") at a range of capillary tensions, expressed as a percentage of the soil's dry weight. The American Society of Testing and Materials (ASTM) D 2325 specifies test methods for water retention. Total water-holding capacity of a rooting zone is equal to the capacity of a unit of rooting medium times the total volume of medium in the rooting zone.

Nutrient exchange capacity must be abundant. The electrochemically active surfaces of clay, organic matter, and certain ESCS aggregates adsorb nutrient ions and release them to tree roots. The source of the nutrients might be artificial fertilizers. Nutrient exchange capacity is indicated quantitatively by cation exchange capacity (CEC), which is expressed in milliequivalents per 100 grams of material (meq). Total CEC in a rooting zone is equal to the CEC in a unit of rooting medium times the total volume of medium in the rooting zone.

The combination of requirements for rooting must be provided together to work as a system (Arnold Associates, 1982). A network of macropores is necessary for root penetration, aeration, and drainage. Numerous micropores are necessary to hold available water after free drainage. The electrochemically active surfaces of soil or ESCS particles are necessary for cation exchange. Contact with the surface above is necessary for infiltration of air and water. Contact with a positive outlet below is necessary for drainage of excess water.

ROOTING VOLUME

It is vital that a viable rooting medium be ample in volume, because the rooting zone is the reservoir from which a tree draws moisture. A tree can transpire more than twice its weight in water each day (Craul, 1992, p. 127). A rooting zone must supply stored water in the periods between rain or irrigation events.

If the roots of a growing tree encounter a restricted rooting volume and a correspondingly restricted storage of available water, the tree's leaf size and rate of shoot growth decline (Craul, 1992, p. 153). The stressed and dwarfed tree is vulnerable to pests, diseases, and injury. It is likely to die prematurely, failing to reach the size at which it could have provided the shade and other benefits for which it was planted. Consequently, the need for tree maintenance becomes excessive, the need for tree replacement is common, and the success of the place of which the tree was to have been part is heavily compromised.

The required volume of viable rooting media is much larger than many people realize. It is roughly proportional to the size of the canopy of leaves and branches where the water will be used, and depends further on the soil's water-holding capacity, the supply of water to the zone, the species of tree, and the evapotranspirative water demand of the local climate (Craul, 1999, p. 216; Lindsey and Bassuk, 1991). Because of those complexities, the judgments of how much rooting volume trees need have been diverse. The judgments have been expressed variously as ratios of soil volume to crown area, to trunk caliper, and to number of trees. Even researchers using the same type of ratio have come up with different required volumes.

The viable thickness of a rooting zone is commonly limited to 2 or 3 feet (Craul, 1992, p. 135, and 1999, p. 274; Watson and Himelick, 1997, p. 12), because below that depth in most soils tree roots tend not to find adequate oxygen. The limited thickness of the rooting zone makes it imperative for the zone's horizontal area to produce the required rooting volume.

Figure 5.2 shows minimum rooting volume requirements in natural soil according to one of the expert researchers in this field (Perry, 1994). On the horizontal axis the tree diameter goes up to 25 inches, which is the diameter of a relatively mature full-sized tree. According to the figure, to grow to full size, an individual tree requires a rooting volume as big as the "footprint" of a small house. Even volumes as large as those shown in the figure may not permit a given tree to grow to the size of others of its species which grew up with larger volumes. Where a volume smaller than that indicated in the figure is the best that can be provided, the likelihood of early decline, high maintenance requirement, and short life increase. Greater volumes than those shown in the figure would reduce stress and make trees more likely to be long-lived, full-sized, and maintenance-free.

Urban (2000, and cited in Craul, 1999, pp. 215–216) independently arrived at values more or less similar to Perry's by estimating the rooting volumes of trees in cities of the northeastern U.S. and evaluating the trees' relative vigor and survival. He found that among small trees those that were healthy and vigorous had rooting

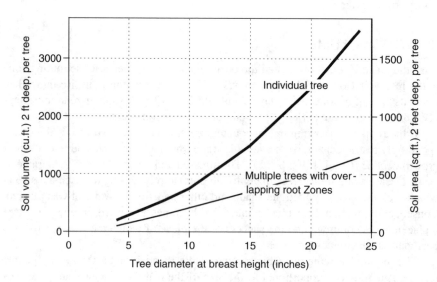

FIGURE 5.2 Minimum volume of typical natural soil required for tree rooting (after Perry, 1994).

volumes in excess of 600 cubic feet. All trees greater than 25-inch caliper had up to 1200 cubic feet.

Perry's values are roughly consistent in effect with an old rule of thumb which says that the area given to a tree in a 24-inch thick rooting zone should be at least as large as the area expected under the tree's canopy at full maturity (Grabosky et al., 2002; Thompson and Sorvig, 2000, p. 119; Watson and Himelick, 1997, pp. 12, 44). For example, a large tree with canopy diameter of 40 feet has a canopy area of about 1250 square feet, which is similar to the values near the right side of the chart.

A rooting zone may take on a very irregular shape, so long as the volume requirement is not compromised (Thompson and Sorvig, 2000, p. 119; Watson and Himelick, 1997, p. 12). The zone need not conform to the area directly under a tree's expected canopy or be symmetrical around a tree's root base. Many species can extend individual roots great distances in favorable directions while leaving unfavorable directions alone. For example, an adequate rooting area could be narrowly linear to fit available space along an urban street, or it could link patches of available soil in a complex setting.

The requirement for multiple trees planted together in a single connected rooting zone is smaller per tree than that for an individual tree, because the roots of multiple trees overlap in the soil, using a given volume of soil and its stored moisture more efficiently, and the canopies of multiple trees shield each other from drying sun and wind (Watson and Himelick, 1997, pp. 43–44). Connecting a series of rooting zones for individual trees into a single continuous trench enlarges the rooting volume for all the trees together (Evans et al., 1990; Thompson and Sorvig, 2000, p. 120; Watson and Himelick, 1997, p. 44). Even connecting zones with narrow rooting "channels" allows individual roots to explore large soil volumes.

Where a preexisting tree is to be preserved in the midst of pavement construction, its dripline is a handy and perhaps necessary aid to estimating the approximate location and extent of its rooting zone. The dripline area may encompass the majority of the tree's roots. However, far beyond the dripline long individual roots may occupy an additional, extensive, irregularly shaped soil area. Consequently, 1.5 times the dripline has been suggested as a more reliably encompassing guideline (Nina Bassuk, personal communication 2003).

THE ROLE OF A PAVEMENT'S SURFACE COURSE

A rooting medium must be able to exchange air and water with the atmosphere (Arnold, 1993, p. 128; Thompson and Sorvig, 2000, p. 118; Trowbridge and Bassuk, 1999; Evans et al., 1990). Under a pavement, moisture and fresh air attract roots from adjacent planting pits, making the medium an effective part of the rooting zone.

Figure 5.3 shows three ways to connect a rooting medium through a pavement surface course to sources of air and water above. Perforated pipes allow some aeration and irrigation beneath a dense (impervious) surface; the pipes must be closely spaced, and thoroughly and continuously vented. An example of a product specifically intended to work with a pipe system is the Wane Tree Feeder System (www.wane3000.com). Intermittent porous surfaces are produced by tree pits along a continuous rooting trench; they can aerate the base to a degree as long as the tree

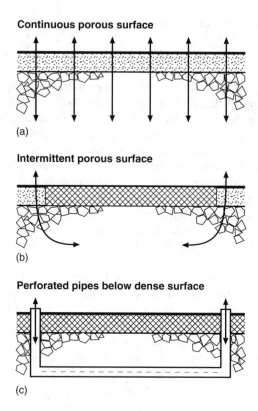

FIGURE 5.3 Alternative provisions to infiltrate air and water to a rooting medium through a pavement surface course.

pits are wide and are surfaced with permeable material in contact with the porous rooting zone. A layer of open-graded aggregate over the rooting zone can distribute air under the surface course and across the rooting zone (Kristoffersen, 1998). A continuous porous surface is the most thorough and reliable way to aerate the rooting zone, so long as the surface is truly porous and permeable and no dense material is introduced between the surface course and the rooting zone. A porous surface overlies the entire rooting zone, providing uniform aeration through myriad pathways.

A surface course may have to deal with the possibility of heaving roots growing in the rooting zone. The roots that may have heaved sensitive pavements in the past were those that grew very near the surface course (Grabosky and Bassuk, 1995; Watson and Himelick, 1997, p. 45; Kopinga, 1994; Kristoffersen, 1999). Because of their location, these roots could concentrate expansive pressure on the surface layer, potentially to lift the surface from below. Most of the damage has been located within a few feet of tree trunks, in effect within the tree root base, where roots grow to the greatest diameter.

However, the causal relationship between roots and pavement heaving is not conclusively known, and is for the moment controversial. The upward pressure exerted by an expanding root (in units such as pounds per square inch) is not known, so whether that pressure is adequate to lift a pavement surface course against the pavement's weight and interlocking resistance to movement is also not known (Jason Grabosky, personal communication 2002).

A persuasively skeptical inquiry into the possibility of root heaving was a series of studies in Cincinnati by Sydnor and his colleagues (Sydnor et al., 2000; D'Amato et al., 2002). The researchers found that deformations of concrete sidewalk pavements were no more common near trees than where trees were not present (Sydnor et al., 2000). Instead, a disproportionately high number of deformations occurred where pavement subgrades were of unusually low strength and susceptible to frost action. Of the sidewalk cracks located near trees, half were not occupied by roots (derived from data in Table 2 of D'Amato et al., 2002). The causes of the unoccupied cracks came, by default, not from roots but from the ordinary factors of pavement structural design such as thickness in its relation to subgrade strength, frost heave, and traffic load. For the cracks occupied by roots, it is not known whether the roots caused the cracks, or were attracted into the aerated soil under previously formed cracks.

Until the question of causality is resolved, it is prudent in pavement design to assume a possibility of heaving pressure from roots below. It is necessary either to select a type of surfacing that is unharmed by root heaving or to protect the surface layer from the pressure below.

Pavement surfaces that are tolerant of heaving are unbound aggregate, turf, geocells filled with aggregate or turf, and the "soft" paving materials. These surfaces are flexible and adaptable, and not disrupted by slight irregularities that could result from root heaving. They can be placed over tree rooting zones with little provision to protect them from heaving. Decks are even less sensitive to heaving because their surfaces are isolated from rooting zones.

Pavement materials that are sensitive to displacement are concrete, asphalt, blocks, and grids. These materials become difficult to travel on and look unsightly when they are displaced or cracked.

For a surface that is sensitive to heaving, a nonrooting layer intervening between the surface course and the rooting zone can limit and counteract heaving pressure (Craul, 1999, pp. 291–297; Kopinga, 1994). A nonrooting layer is made of clean open-graded mineral aggregate; roots tend not to grow into empty void spaces between inert aggregate particles where water and nutrients are not stored. An example was shown in Figure 5.1. The thickness of a nonrooting layer dissipates heaving pressure. The combined weight of the layer and the surface course, and their interlocking resistance to displacement, counteract the upward pressure of root expansion. Layers of this kind that have been built in the past have been 4 or more inches in thickness; the thickness actually needed has not been established by research.

Porous, penetrable, well-aerated base materials like those described below could further dissipate heaving forces by inviting roots to grow deep in the base layer (Grabosky et al., 2002). "Structural soils" tend to be better aerated than most native soils, so at a depth they provide a combination of reliable moisture supply, stable temperatures, and adequate aeration which does not exist near the surface. At a depth, the weight of overlying pavement materials counteracts and disperses the heaving pressure of expanding tree roots.

BASE COURSE ROOTING MEDIA

Experience since the early 1980s indicates that it is possible to grow and sustain healthy trees with pavement base courses made of specially designed rooting media. Figure 5.1 showed an example of placement of this type of base. The principle of the material is a combination of open-graded aggregate that forms the load-bearing structure, and a provision for water and nutrient storage either within the aggregate particles or in soil occupying the voids between aggregate particles. Roots can grow in the soil while the aggregate skeleton supports the traffic load and protects the soil from compaction.

A correct proportion of soil is vital to the mixture's success (Grabosky and Bassuk, 1996; Spomer, 1983). The quantity of soil must be less than the void space between aggregate particles in order to allow the aggregate particles to interlock structurally with each other and prevent compaction of the soil. The remaining void space is left empty for the movement of water and air. This limitation is particularly relevant for fine-textured soil, because fine-textured soil has little aeration or drainage through its own pores, and bodies of fine-textured material can be susceptible to frost heave. Craul (1999, pp. 291–293) has suggested as a rule of thumb a maximum soil volume of 20 or 25 percent of the volume of the aggregate, assuming aggregate with void space of 30 to 40 percent.

This chapter describes three types of load-bearing mixtures that have been developed.

STONE-SOIL MIXTURE (CORNELL STRUCTURAL SOIL)

Researchers at Cornell University (Grabosky and Bassuk, 1995, 1996, 1998; Grabosky et al., 2002; Bassuk et al., 1998) developed a load-bearing mixture of open-graded stone aggregate and fine-textured soil. This type of mixture has been variously called structural soil or skeletal soil. With it they have performed a great amount of technical research and have contributed greatly to the understanding of materials that both bear loads and support tree rooting. Updates on the continuing research are posted on the web site of Cornell's Urban Horticulture Institute (www.hort.cornell.edu/uhi).

The Cornell researchers' mixture and their method of adapting it to specific locally available aggregate and soil materials have been patented (Grabosky and Bassuk, 1998) and are now being distributed exclusively by Amereq and its licensees under the name CU Soil (www.amereq.com). The ability to purchase the proprietary product from a licensed distributor brings with it the reliability of a legitimately controlled product with given specifications and documented performance (Grabosky et al., 2002). The following discussion attempts to describe the patented mixture and its documented effects without purporting to advise on where the limits of the patent's protection lie or what would or would not constitute an infringement. More information about the specific claims of the patent is available in the original patent document (Grabosky and Bassuk, 1998).

Table 5.4 lists the components of the patented mixture. The researchers have specified quantities on a dry-weight basis; the volumetric proportions are roughly equal to those given by weight.

TABLE 5.4
Components of Stone-Soil Mixture Patented by the Cornell Researchers

Component	Description	Amount
Stone aggregate	ASTM No. 4	Approximately 80 percent of mixture's total weight
Soil	Fine-textured	12 to 33 percent of aggregate by dry weight (approximately 20 percent of mixture's total weight)
Hydrogel	Acrylamide	25 to 40 grams per 100 kg of aggregate (0.40 to 0.64 ounces per 100 pounds of aggregate) by dry weight
Compaction of mixture	Proctor standard	90 to 95 percent minimum
Strength criterion	CBR at least 40	Applied by testing compacted mixtures with alternative quantities of soil

Grabosky and Bassuk (1998).

The mixture's aggregate must be durable, coarse, open-graded, and angular. Cornell's research assumed that the aggregate is inert rock with no internal micropores, water-holding capacity, or cation exchange capacity. In practice, the aggregate is most typically one inch in size; aggregate up to 2.5 inches has been used (Jason Grabosky, personal communication 2002). (The Cornell research and patent documents cite gradations from ASTM C 33 which are identical to those in ASTM D 448.)

The mixture's soil must be fine-textured; it is usually loam, silt loam, or clay loam, with little organic matter. Fine-textured soil has water- and nutrient-holding capacities which dense, inert stone aggregate does not, so its addition to the mixture is necessary to make the mixture into a viable rooting medium. Fine-grained soil also has cohesiveness which, together with that of hydrogel, holds the soil uniformly to the aggregate particles during mixing and placement, and may resist downward erosion through the pores after placement.

A small amount of commercially available hydrogel is added to hold the soil to the aggregate particles and assure uniformity of the mixture. The quantity of hydrogel must be very limited to prevent interference with aggregate interlock and potential frost action (Grabosky and Bassuk, 1998). At the specified application rate, the hydrated hydrogel occupies less than 1 percent of the aggregate's void space. Amereq, the holder of the distribution license for CU Soil, supplies a hydrogel called Gelscape (www.amereq.com). Other hydrogels probably exist which are capable of being used for this purpose.

The Cornell researchers found that as long as the aggregate is durable and angular, high strength is produced by mixtures in which the quantity of soil is less than 25 percent the weight of the mixture. Amounts of soil greater than that tend to interfere with the bearing capacity of the aggregate interlock. Good tree rooting occurs only in mixtures in which the quantity of soil is between 10 and 35 percent. Therefore, in the overlapping range between 10 and 25 percent both strength and rooting objectives could be met (Grabosky and Bassuk, 1998).

However, for a given project, the exact amount of soil that can be blended into the aggregate without diminishing strength depends on the particular locally available stone and soil materials. Therefore, the quantity of components is adjusted to local materials using the governing criterion of strength (Grabosky and Bassuk, 1998). For each project a strength requirement is designated using the California Bearing Ratio (CBR, ASTM D 1883); a minimum CBR of 40 or 50 is sufficient for many pavements with pedestrian or light vehicular traffic. Test batches of the mixture containing different quantities of soil are tested for CBR. Among the test mixtures with acceptable CBR, that containing the greatest quantity of soil is selected for the project mixture.

During mixing, just enough water is added to the hydrogel to make a sticky slurry which coats the aggregate particles. When the soil is blended into the aggregate, the slurry holds the soil on the surfaces of the aggregate particles. Upon placement, the mixture's moisture is brought up to the Proctor optimum for compaction and the material is compacted to 95 percent Proctor density in 6-inch lifts.

Figure 5.4 shows trees growing in Cornell structural soil on State Street in Ithaca, New York. Beginning in 2000 five blocks of this street were rehabilitated

FIGURE 5.4 Trees rooted in Cornell structural soil on State Street in Ithaca, New York.

with reconstructed and enlarged sidewalks and new trees to supplement large preexisting trees. Soil was excavated from the entire sidewalk area except for a 12 feet × 7 feet area at each preexisting tree base. Structural soil was installed 35 inches deep between the tree bases and from them to the building front, replacing dense compacted soil with rooting zone. A dense concrete sidewalk was placed over the structural soil, and block pavements were added between tree bases for pedestrian access across the formerly compacted soil tree strip. After three years the new linden and koelreuteria trees were growing well. In addition the large preexisting ash and honey locust trees were doing surprisingly well following the stress of root cutting during excavation (Nina Bassuk, personal communication 2003). They benefited from the very penetrable, well-aerated environment of the structural soil. The viable rooting volume below the sidewalk is much larger than it used to be, and is in contact with adjacent lawns for still further rooting area.

Other installations in all parts of North America are listed on the Amereq web site. Table 5.5 lists some of those for which information about the site or project is easily available.

Cornell's research showed that the compacted mixture can adequately bear loads through the stone skeleton. In the Unified classification, the blended mixture might be classified as highly stable GM, GC, or GP. Because the stone-soil mixture contains little silt and organic matter, it meets Corps of Engineers criteria for low frost susceptibility. Compared with the sand-based Amsterdam tree soil the Cornell stone-soil mixture has greater load-carrying capacity and less susceptibility to frost damage.

The research showed that in the compacted mixture the young roots of oaks and lindens deformed to grow around aggregate particles while absorbing water and nutrients from the fine-textured soil. The roots were deeper and more abundant than those in compacted uniform soil because they were able to penetrate through the

TABLE 5.5
Examples of Installations of CU Soil

Name	Location	Source of Further Information
Statehouse Convention Center	Little Rock, Arkansas	www.littlerock.com
Tressider Hall Parking, Stanford University	Palo Alto, California	www.stanford.edu
Hewlett Foundation	Menlo Park, California	www.hewlett.org
Ritz Carlton	Washington, DC	www.ritzcarlton.com
Xcel Energy Center	St. Paul, Minnesota	www.xcelenergycenter.com
Landmark Plaza	St. Paul, Minnesota	www.landmarkcenter.org
Camden Water Front	Camden, New Jersey	www.camdenwaterfront.com
Nationwide Arena	Columbus, Ohio	www.nationwidearena.com
Heinz Field	Pittsburgh, Pennsylvania	www.steelers.com
Watson Institute	Providence, Rhode Island	www.watsoninstitute.org
Oyster Point Town Center	Newport News, Virginia	www.oysterpointonline.com
West Seattle High School	Seattle, Washington	http://wshs.seattleschools.org

www.amereq.com.

mixture's macropores and they found air and water in balance deep in the well-aerated, well-drained mixture.

Tree shoot growth in paved areas of Cornell structural soil is equal to or greater than that in nearby areas of lawn in the first three years after planting (Grabosky et al., 2002). The trees where this was observed were of several species and had been planted at 2- to 2.5-inches caliper. The relatively low rate of growth in the lawns could be attributable to the greater compaction of the lawn soil than of the soil in the stone-soil mixture (or other factors of lawn soil quality), root competition from the turf, or mower damage. The researchers speculate that as the trees in structural soil enlarge they may become stressed by the small size of the pavement openings around their bases. Additional potential limiting factors for large trees could be the total volume of available rooting zone, and whether air exchange occurs through the pavement surface. Continuing research is planned.

ESCS Mixtures (Arnold Air-Entrained Soil)

Landscape architect Henry Arnold (1993, pp. 129–130) developed a base-course rooting mixture based on ESCS aggregate in the 1980s. He calls the mixture "air-entrained soil," referring to the presence of open pores in and between ESCS aggregate particles despite compaction of the mixture during construction. Air-entrained soil predates the Cornell mixture and is physically different. The defining difference is the mixture's use of porous ESCS aggregate rather than dense mineral aggregate.

Arnold built the idea on the work of federal agronomist James Patterson (1976; Patterson and Bates, 1994), who had mixed ESCS aggregate into the soil of high-traffic lawns in Washington, DC. Arnold was inspired to apply the idea to paved sites by the successful rooting of trees in or under aggregate walking surfaces in old European parks (Arnold Associates, 1982).

There is no patent on the mixture (Arnold 2001) nor has there been systematic research as there was for Cornell structural soil. Instead, over the years, Arnold and other designers have used regionally available ESCS products and adjusted the mixture from project to project with experience (Arnold 1993, p. 130). Table 5.6 lists examples of installations of the mixture by both Arnold and other designers. The ESCS suppliers' web sites list many more recent projects.

ESCS is a common component of horticultural planting media because of its aerating and water-holding capabilities (ASTM D 5883). The particles' small pores moderate moisture fluctuations in planting media by absorbing excess water when it is present and releasing it gradually as the surrounding soil dries out. Retention of water available to tree roots can be 12 to 35 percent of the aggregate's oven-dry weight.

Open-graded sizes of ESCS aggregates can be selected from among those defined in ASTM C 330 or 331. ESCS products that meet the standards of ASTM C 330 or C 331 are strong, durable, and uniform. Most ESCS aggregates can be compacted to 95 percent Proctor density.

Some regionally available ESCS aggregates have high cation exchange capacity (CEC). The CEC of the Stalite product is reportedly 26 meq (Chuck Friedrich, personal communication 2003), which is comparable to clay and higher than many natural soils. Other ESCS aggregates can have higher or much lower CEC due to their

TABLE 5.6
Examples of Projects using ESCS Tree Rooting Mixtures

Name	Date	Location	Reference
Community Park School	1985	Princeton, New Jersey	Lederach
Peachtree Plaza Hotel	1988	Atlanta, Georgia	Lederach
Charlotte-Mecklenburg Government Center	1988	Charlotte, North Carolina	Lederach
Newport Plaza	1991	Jersey City, New Jersey	www.nesolite.com; Lederach
Metrotech	1992	Brooklyn, New York	Lederach
One Peachtree Plaza	1995	Atlanta, Georgia	Lederach
Marietta Street median	1996	Atlanta, Georgia	Friedrich
Downtown street trees	1997	Cary, North Carolina	Friedrich
Korean War Memorial Plaza	1998	Washington, DC	www.permatill.com; Friedrich
Robert F. Wagner Jr. Park	1999	Battery Park City, New York	www.nesolite.com
East Chase Mall	2002	Montgomery, Alabama	Friedrich
Atlanta College of Art	2003	Atlanta, Georgia	Friedrich

"Friedrich" is Chuck Friedrich, personal communication (2003); "Lederach" is Stephen Lederach, personal communication (2002).

geologic sources or details of processing. One can select specific combinations of ESCS and soil materials to produce composite cation-exchange and water-holding capacities in a project-specific rooting medium. Some ESCS suppliers provide guideline specifications for mixtures using their specific products.

Some of Arnold's projects have called for porous ESCS aggregate 3/8 inch in size, into which is mixed 5 to 10 percent (of volume of aggregate) peat moss and loam topsoil and 15 percent clay. Because ESCS aggregate is light in weight, those volumetric proportions give the soil-organic matter content up to 50 percent of the aggregate's bulk weight (Arnold, 2001).

Figure 5.5 shows one of Arnold's largest projects, the Commons in the Metrotech Business Improvement District in Brooklyn, New York. The district is an urban revitalization area occupied by colleges, schools, and government and corporate offices (www.metrotechbid.org). At the center of the district is a pedestrian open space called The Commons, set with trees and benches; pedestrian traffic under the trees is continuous and heavy. The base course is of low-CEC expanded clay aggregate, with soil blended into the aggregate's voids. The rooting mixture is 2 feet deep. The pavement surface is a 1.5-inch layer of clean, angular No. 10 aggregate which bears very heavy pedestrian traffic while admitting air and water to the rooting medium. The numerous honeylocusts and oriental cherries rooted in the ESCS mixture were large and healthy 12 years after installation. The same rooting medium was used for street trees in eight surrounding city blocks (Stephen Lederach, personal communication 2002).

A more recent project is Pier A Park, located on a reclaimed cargo pier in Hoboken, New Jersey (Richardson et al., 2000; Thompson, 2001). The park's porous

FIGURE 5.5 Cherry trees rooted in an ESCS-soil mixture at Metrotech Commons, Brooklyn, New York, 12 years after installation.

surfaces were introduced in Chapter 2. Table 5.7 lists the components in the park's air-entrained soil base-course mixture. The ESCS is open-graded aggregate of C 331 1/2 inch gradation. The quantity of topsoil and peat was sufficient almost to fill the voids between aggregate particles; some experts speculate that the amount of soil was high enough to be susceptible to compaction. The mixture was placed in 8-inch lifts that were compacted to 90 to 95 percent Proctor density. Compared with dense mineral aggregate, the lightweight aggregate limited loads on the pier deck, avoiding the considerable expense of pier reinforcement or replacement. The rooting zone is drained by drainage mats beneath the air-entrained soil layer. Where the surfacing is aggregate, a 1.5-inch thick surface layer of open-graded No. 10 traprock was placed directly on top of the air-entrained soil (Henry Arnold, personal communication 2001).

TABLE 5.7
Components of Air-Entrained Soil at Pier A Park, Hoboken, New Jersey

Component	Quantity
ESCS aggregate	1/2" gradation
Topsoil	30 percent of aggregate's volume
Peat	5 percent of aggregate's volume
5-10-10 fertilizer	As required according to agronomic soil tests

Calculated from data in Thompson (2001).

In other places, a layer of the same No. 10 aggregate is the setting bed for open-jointed paving blocks.

ESCS products with high CEC theoretically need little or no blended-in soil for nutrient-holding. Plants have been rooted in this type of ESCS aggregate without added soil, under carefully operated irrigation. However, in a setting without such careful irrigation the very high permeability of the open-graded material would allow a large part of natural rainwater or casually applied irrigation water to drain rapidly through the material, leaving insufficient water stored in the particles' small pores. Therefore, soil is blended even into high-CEC ESCS to increase water-holding capacity. The added soil is usually coarse-textured, inorganic materials such as sand, sandy loam, sandy clay loam, or fine particles of ESCS, which are well-drained and limit the mixture's susceptibility to frost damage. The amount of soil typically has been 20 percent of the bulk volume of the aggregate. In some projects using high-CEC ESCS, hydrogel was not necessary while mixing in sand; instead, moistening the ESCS aggregate with water was sufficient to cause sand particles to coat the aggregate during mixing, transportation, and placement (Chuck Friedrich, personal communications 2002 and 2004).

Two decades of field experience with ESCS mixtures is valuable, but it is only a partial substitute for systematic research like that which upholds CU Soil. ESCS air-entrained soil requires research and documentation equal to that given to the Cornell stone-soil mixture. ESCS producers need to publish uniform information about specific aggregate products and aggregate-soil mixtures relevant to pavement design and tree rooting, including CBR freeze-thaw durability (the magnesium sulfate test), water-holding capacity, and CEC.

SAND MIXTURE (AMSTERDAM TREE SOIL)

An early type of load-bearing rooting medium was developed by the Amsterdam city government; it has been referred to as "Amsterdam tree soil" or "Amsterdam sand" (Couenberg, 1993; Watson and Himelick, 1997, p. 44). The city's personnel hoped to prolong the life of street trees by providing a better rooting medium than the heavy clay subgrade that characterizes much of the city. Amsterdam sand does not have the same advantages as the other mixtures described above, either for rooting or structural stability. However, it represents a resourceful use of materials in a locale with undemanding structural standards and limited options. It continues to be used in some areas.

The mixture's aggregate skeleton is open-graded coarse sand; larger aggregate is difficult to obtain in the Amsterdam area. A fine-textured soil of clay and organic matter is mixed into the sand's voids (Couenberg, 1993). The mixture could be classified as a loamy sand (Craul, 1999, pp. 291–292). The amount of fine-textured soil is limited to prevent excessive settling after the mixture is installed, although this also limits the mixture's water- and nutrient-holding capacities. Fertilizer is added according to agronomic soil tests. Upon placement, the mixture is compacted to 70 to 80 percent Proctor density, which is low compaction by U.S. standards. The mixture is surfaced with a flexible pavement such as blocks with sand-filled joints. Table 5.8 lists the components. Figure 5.6 shows typical construction.

At a 1984 installation on the Transvaalkade (a street alongside the Ringvaart canal in Amsterdam), the tree soil supports a series of elm trees (Couenberg, 1993).

TABLE 5.8
Components of Amsterdam Tree Soil

Component	Description	Amount (Proportion of Mixture's Total Weight)
Coarse sand	50 percent (by weight) at least 220 μm in size, and no more than 2 percent smaller than 2 μm	91 to 94 percent
Organic matter	Compost or sphagnum	4 to 5 percent
Clay		2 to 4 percent

Couenberg (1993).

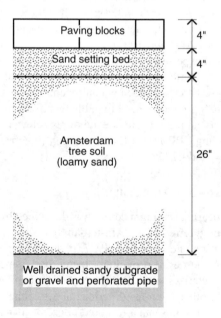

FIGURE 5.6 Typical installation of sand rooting mixture in Amsterdam (data from Couenberg, 1993).

The mixture fills a trench 3 feet wide, connecting the tree planting pits and providing about 140 cubic feet per tree. The tree pits are located 50 feet apart; between them the trench is surfaced with paving blocks for parking spaces. A setting bed of open-graded sand distributes air from the tree pits across the rooting zone. After four years the trees were doing well and had grown considerably; after nine years they "were still doing reasonably well."

A similar installation on the Plantage Middenlaan (a street in a historic district of the city) supports trees planted in 1986. A trench filled with tree soil 6-feet wide and 3-feet deep was surfaced with a sand-setting bed and paving blocks, tight-jointed for stability in parking spaces and open-jointed for air exchange elsewhere. At seven to ten years after planting the trees had "grown very well" and were "still growing."

The material has been used in other European countries such as Denmark. At Christians Havns Square in Copenhagen seven linden trees were planted in a trench of tree soil 10 feet wide and 180 feet long (Kristoffersen, 1998). Similar material is being used in Britain under the name "tree sand" (www.treesand.co.uk).

Amsterdam's experience has been that the tree soil can bear light loads such as those of sidewalks while supporting the rooting of trees. In sidewalks and parking stalls the city tolerates one inch of surface settlement in three years. However, the mixture can be compacted by heavy construction equipment or heavy traffic. As described in Chapter 3, loamy sand has only a moderately stable ML category in the Unified classification of soil and aggregate, and at best a "medium" susceptibility to frost heave. The uniformly small size of the pores between sand particles in Amsterdam tree soil might inhibit root growth and movement of air (Craul, 1999, pp. 291–292), especially where the material has been compacted.

Provisions for All Base-Course Rooting Mixtures

For all rooting-zone mixtures a functioning rooting-zone layer must be at least 24 inches, and preferably 36 inches, thick (Bassuk et al., 1998). The total volume of a rooting zone must be adequate for the trees planted in it. Because the layer is a structural component of a pavement, its thickness must, in addition, meet any applicable structural requirement for the pavement as described in Chapter 3.

A pavement layer used for rooting should not also be used as a reservoir for prolonged storage of water. The displacement of air from the voids would lead to anoxic conditions. If the storage of water like that described in Chapter 4 is desired, a distinct reservoir layer can be added beneath a rooting-zone layer.

All rooting-zone materials require regional adaptations for available aggregate materials, tree species, climate, and contractor qualifications. Aggregate used in pavement construction must meet local requirements for durability and stability. Like any planting medium, a mixture should be given agronomic tests, and where indicated, fertilizer should be added. Like other pavement aggregate materials, rooting mixtures should be installed in lifts and compacted lift by lift.

Given a correct proportion of soil, the soil must be uniformly blended into the stone-soil mixture to avoid dearth in some places and over-concentration elsewhere. Kristoffersen (1998) compared three methods of blending and installing aggregate-soil mixtures. (1) In the first method, the aggregate and soil are mixed before installation and then placed and compacted in lifts. In this method, hydrogel and water are important in holding soil and aggregate particles together to obtain uniform mixing. (2) In the second, a lift of aggregate is placed and compacted, then an appropriate amount of soil is spread over the aggregate and washed into the voids with water. This method is most suitable with large aggregate sizes. It ensures structural stone-to-stone contact and that the soil is not compacted. It can be done in either dry or wet weather. It can require large volumes of water. (3) In the third method, the aggregate is placed in lifts and dry soil is mixed into each lift's voids by sweeping and vibration. This method is suitable only with very large aggregate (3 to 6 inches in size) and in dry weather. It ensures structural stone-to-stone contact without soil compaction.

The selection of tree species must be matched to the mineral composition of the aggregate that will be used in the rooting zone. Soft limestone and crushed concrete rubble leach out alkalinity to which only certain species are tolerant. Some locally available aggregates leach out undesirable metals or other constituents. Trowbridge and Bassuk (2004, pp. 61–73) listed examples of trees appropriate for use with structural soils; they include trees that are tolerant of both alkalinity (from concrete or limestone aggregate) and drought. For an aggregate material that may leach unwanted constituents, potential leaching can be limited by using only strictly coarse, open-graded (single-size) aggregate, and washing out clinging fine particles before mixing in any soil. The remaining large aggregate particles have relatively small surface area from which minerals could be leached.

Drainage and water-holding capacity are mutually offsetting functions. To have both of them, a rooting medium must have a balanced combination of macropores and micropores. Figure 5.7 compares the quantities of macropores and micropores in typical mixtures. Natural soil typically has high total porosity in a balanced combination of macropores and micropores, with the potential for equally balanced drainage and water-holding capacity. Open-graded stone and ESCS aggregates have high volumes of macropores between aggregate particles; ESCS also has some microporosity. In both types of aggregate, the addition of soil to make aggregate-soil mixtures reduces the macroporosity but also adds micropores which hold available water. The ESCS-based mixtures end up with relatively balanced macroporosity and microporosity, more or less emulating natural soil. The portion of ESCS's water-holding capacity in the aggregate's micropores is permanently positioned in a rooting zone and not susceptible to settling as the soil portion of stone-soil mixtures may be.

For all types of load-bearing rooting media, research is needed into the mixtures' quantitative water availability and its implications for rooting-zone design. Trees rooted in mixtures with low water-holding capacity may be more dependent on artificial irrigation than those rooted in natural soils. A need for supplemental irrigation may be most pronounced in the first one or two years after planting, before the roots have elongated sufficiently to activate much of the available rooting zone (Grabosky et al., 2002). Mixtures with low water-holding capacity may need to be placed in larger rooting zones than those suggested in Figure 5.2 in order to build up the total water-holding capacity required by a full-sized tree (Kim Coder and James Urban, personal communications 2001).

Porous Pavement Tree Rooting Media

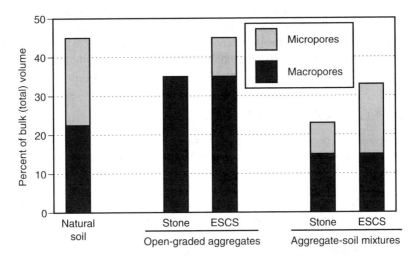

FIGURE 5.7 Approximate void space in typical rooting media (natural soil from Brady, 1974, pp. 54–55; aggregates and mixtures calculated assuming aggregates have 35 percent void space between particles, amount of soil in mixture is 20 percent of aggregate's bulk volume, microporosity in soil is 40 percent of soil volume, and microporosity in ESCS particles is 10 percent of aggregate's bulk volume).

On the other hand, Arnold has speculated that inert porous pavement surfacing may inhibit evaporation from a rooting zone as would a mulch (Arnold Associates, 1982). If so, then trees rooted under porous pavements have lower replacement water requirements than those growing in vegetated lawns or planting beds. This possibility needs to be confirmed by research.

Among other subjects requiring further research are (1) whether the soil washes downward through the pores over a period of years, reducing the uniformity and effectiveness of the rooting medium (factors that could inhibit erosion are the cohesiveness of clay and hydrogel, the filling of half or more of the void space by properly blended-in soil, and networks of fine roots when they develop); (2) whether roots can grow deeper than 36 inches in very well-aerated rooting media based on coarse open-graded aggregate; (3) whether the roots growing in a base-course rooting zone can heave a pavement surface (factors that might inhibit heaving are deep rooting habits in well-aerated material, the ability of roots to conform to the shapes of available void spaces that they encounter, a root barrier of clean open-graded aggregate just below the surface, and the resistance of overlying pavement materials); and (4) whether, on the contrary, roots are strangled between aggregate particles that they cannot push aside.

A viable planting medium is only one part of a system to maintain root growth and health. The other components of the system are a means for air and water to enter the soil, and a means for draining excess water out.

Where unpaved soil such as turf areas or planting beds exists near tree pits, the expense of a subpavement rooting zone need not necessarily be borne for an entire paved area around the trees. Instead, narrow zones of subpavement rooting zone could form "root channels" connecting tree pits to each other and to the unpaved areas. Where a narrow pavement such as that of a sidewalk has trees on one side and unpaved soil on the other, "root channels" located at intervals along the pavement could connect one side to the other, enlarging the soil volume available to the trees.

ROOTING SPACE IN SUBGRADE

The subgrade soil beneath a pavement structure is a potential rooting zone. Where a rooting medium is present in a base course, additional rooting in the subgrade supplements the rooting-zone volume and the moisture reservoir from which a tree can draw. Where a rooting medium is not present in the base course, clean open-graded base material can distribute air and water to a subgrade rooting zone while counteracting the pressure of root heaving.

A vital requirement is that the subgrade not be compacted to any degree: roots cannot penetrate compacted subgrade (Grabosky and Bassuk, 1998). Specifications must clearly require that the subgrade is not to be compacted and that construction traffic on the subgrade surface is forbidden even though the base material to be placed on it will be compacted. Some rooting may be possible following carefully limited compaction (Gray, 2002).

Some pavements are installed over the rooting zones of pre-existing trees. To protect the trees it is vital to avoid compaction of the subgrade where the roots are located. Where the soil has been compacted by traffic that previously occurred, it may be necessary to loosen the soil before placing new pavement material. The most severe compaction is usually within the top one inch of soil, so only surface loosening may be necessary. Leveling of the rooting area should be done by building up low spots with extra thickness of open-graded aggregate, not by excavating into high spots.

SPECIAL TREATMENT AT THE TREE BASE

The "root flare" or tree base requires special treatment. This is the area in the first few feet around a trunk where thick roots radiate and divide outward while developing into buttresses to resist the tensile stress of the swaying superstructure. The tree base is the area where most or all true lifting of pavements by roots seems to happen. A tree base can grow outward twice as fast as the trunk at breast height. Figure 5.8 shows a base that has expanded far beyond the expectations of those who, years ago, placed a tree and a sidewalk side by side. Outside the flare zone most roots are less than 2 inches in diameter, long and rope-like (Urban, 2000).

A surprisingly large area must be set aside for the development of a mature tree base. Flexible pavement materials like unbound aggregate and open-jointed blocks can be placed in the zone and releveled or removed incrementally as the base expands. If space is available for setting aside an area exclusively for the tree base, then traffic can be excluded from it by fences, bollards, or fence-like arrangements of benches or bike racks. However, if a fixed pavement is placed in a zone within about 6 feet of the trunk, future structural conflicts between the pavement and the growing tree are likely.

Tree grates are bridge-like structures that carry traffic into a tree-base area in place of a pavement. However, many models of tree grates present fixed obstructions that can girdle and kill a tree as the base expands outward. Some other models allow

FIGURE 5.8 Base of a mature oak tree.

TABLE 5.9
Examples of Suppliers of Tree Grates, Tree Fences and Related Products

Supplier	Contact Information
Alpha Precasts	www.alphaprecasts.com
Balco	www.balcousa.com
Canterbury International	www.canterburyintl.com
Crescent Foundry	www.crescentfoundry.com
DuMor	www.dumor.com
Dura Art Stone	www.duraartstone.com
East Jordan Iron Works	www.ejiw.com
Fairweather	www.fairweathersf.com
Ironsmith	www.ironsmith.cc
Litchfield Landscape Elements	www.landscapeelements.com
NDS	www.ndspro.com
Neenah Foundry	www.nfco.com
Petersen Manufacturing	www.petersenmfg.com
Quick Crete Products	www.quickcrete.com
Structural Plastics	www.plastictreegrates.com
Syracuse Castings	www.syrcast.com
Toronto Fabricating	www.tfmc.com
Urban Accessories	www.urbanaccessories.net
Victor Stanley Inc.	www.victorstanley.com
Wabash Valley Manufacturing	www.wabashvalley.com
Wausau Tile	www.wausautile.com

the central opening to be enlarged by removal in sections. Metal grates have high thermal conductivity; a hot grate might overheat a tree if the base grows into contact with it. Table 5.9 lists examples of suppliers of tree grates, tree fences and related products.

To permit roots to extend from a tree-base area into a rooting zone under the surrounding pavement, the pavement's edge should be free of continuous structural footings that could obstruct root growth. A curb at the edge could be broad rather than deep for stability. If a curb must be deeply seated, then its footing can be shaped with gaps or root channels to allow roots to penetrate outward. Where a footing is required for a tree grate, it could be made in the form of a series of piers rather than as a continuous wall.

TRADEMARKS

Table 5.10 lists the holders of registered trademarks mentioned in this chapter.

TABLE 5.10
Holders of Registered Trademarks Mentioned in This Chapter

Registered Trademark	Holder	Headquarters Location
Alpha Precasts	Knecht & Berchtold	Brampton, Ontario
Amereq	Amereq	New City, New York
Balco	Balco Inc.	Wichita, Kansas
Canterbury International	Canterbury International	Los Angeles, California
Crescent Foundry	Crescent Foundry Company	Kolkata, Indiana
CU Soil	Amereq	New City, New York
DuMor	DuMor Site Furnishings	Mifflintown, Pennsylvania
Dura Art Stone	Dura Art Stone	Hayward, California
East Jordan Iron Works	East Jordan Iron Works	East Jordan, Michigan
Fairweather	Fairweather Site Furnishings	Port Orchard, Washington
Ironsmith	Ironsmith	Palm Desert, California
Litchfield Landscape Elements	Litchfield Industries	Litchfield, Michigan
NDS	NDS, Inc.	Lindsay, California
Neenah Foundry	Neenah Foundry	Neenah, Wisconsin
Petersen Manufacturing	Petersen Manufacturing Co.	Denison, Iowa
Quick Crete Products	Quick Crete Products	Norco, California
Structural Plastics	Structural Plastics Corp.	Holly, Michigan
Syracuse Castings	Syracuse Castings	Cicero, New York
Toronto Fabricating	Toronto Fabricating & Mfg. Co.	Mississauga, Ontario
Urban Accessories	Urban Accessories	Woodinville, Washington
Victor Stanley Inc.	Victor Stanley Inc.	Dunkirk, Maryland
Wabash Valley Manufacturing	Wabash Valley Manufacturing	Silver Lake, Indiana
Wausau Tile	Wausau Tile	Wausau, Wisconsin

ACKNOWLEDGMENTS

The following persons generously provided information used in this chapter: Henry Arnold and Stephen Lederach of Arnold Associates (www.arnoldassociates.org); Nina Bassuk of Cornell University's Urban Horticulture Institute (www.hort.cornell.edu/uhi); Wayde Brown and Mark Reinberger, University of Georgia School of Environmental Design; Kim Coder of University of Georgia Warnell School of Forest Resources; Patrick Flynn, Eagle Picher Minerals (www.epcorp.com); Chuck Friedrich of Carolina Stalite Company (www.permatill.com); Jason Grabosky of Rutgers University (formerly of Cornell University's Urban Horticulture Institute); George Hargreaves, Hargreaves Associates (www.hargreaves.com); Dudley Hartel, consulting urban forester of Comer, Georgia; Scott King of ERTH Products (www.erthproducts.com); Chet Thomas, Jaeger Company (www.jaegerco.com); Brian Trimble, Brick Institute of America (www.bia.org); James Urban, landscape architect of Annapolis, Maryland.

The following persons constructively critiqued early drafts of this chapter: Nina L. Bassuk, Cornell University Urban Horticulture Institute (www.hort.cornell.edu/uhi); Chuck Friedrich of Carolina Stalite Company (www.permatill.com); and James Urban, landscape architect of Annapolis, Maryland.

REFERENCES

Arnold Associates (1982). *Urban Trees and Paving*, Report for The Gateway Project, Princeton, New Jersey: Arnold Associates.
Arnold, Henry F. (1993). *Trees in Urban Design*, 2nd ed., New York: Van Nostrand Reinhold.
Arnold, Henry F. (2001). The Down and Dirty on Structural Soil [letter to the editor], *Landscape Architecture* 91, 9–11.
Bassuk, Nina, Jason Grabosky, Peter Trowbridge, and James Urban (1998). Structural Soil: An Innovative Medium Under Pavement that Improves Street Tree Vigor, in *1998 ASLA Annual Meeting Proceedings*, pp. 183–185 Diane L. Scheu, Ed., Washington: American Society of Landscape Architects.
Bennett, Paul (1999). Big Ideas Down Under, Innovating in Sydney, *Landscape Architecture* 89, 20, 60–67, and 88–95.
Beven, Keith, and Peter Germann (1982). Macropores and Water Flow in Soils, *Water Resources Research* 18, 1311–1325.
Brady, Nyle C. (1974). *The Nature and Properties of Soils*, 8th ed., New York: Macmillan.
Couenberg, E.A.M. (1993). Amsterdam Tree Soil, in *The Landscape Below Ground*, pp. 24–33, G. Watson and D. Neely, Eds., Savoy, Illinois: International Society of Arboriculture.
Craul, Phillip J. (1992). *Urban Soil in Landscape Design*, New York: John Wiley and Sons.
Craul, Phillip J. (1999). *Urban Soils, Applications and Practices*, New York: John Wiley and Sons.
D'Amato, Nicholas E., T. Davis Sydnor, Michael Knee, Robin Hunt, and Bert Bishop (2002). Which Comes First, the Root or the Crack?, *Journal of Arboriculture* 28, 277–282.

Day, Susan D, and Nina L. Bassuk (1994). A Review of the Effects of Soil Compaction and Amelioration Treatments on Landscape Trees, *Journal of Arboriculture* 20, 9–17.

Evans, Matthew, Nina Bassuk, and Peter Trowbridge (1990). Sidewalk Design for Tree Survival, *Landscape Architecture* 80, 102–103.

Friedrich, Charles R. (2001). More on Structural Soils [letter to the editor], *Landscape Architecture* 91, 36.

Geiger, James R. (2003). The Case for Large Trees vs. Small Trees, in *Urban Forest Research* Fall 2003 (http://cufr.ucdavis.edu/newsletter.asp).

Goldsmith, Wendi, Marvin Silva, and Craig Fischenich (2001). *Determining Optimal Degree of Soil Compaction for Balancing Mechanical Stability and Plant Growth Capacity*, ERDC TN-EMRRP-SR-26, Vicksburg: U.S. Army Engineer Research and Development Center.

Grabosky, Jason, and Nina Bassuk (1995). A New Urban Tree Soil to Safely Increase Rooting Volumes Under Sidewalks, *Journal of Arboriculture* 21, 187–201.

Grabosky, Jason, and Nina Bassuk (1996). Testing of Structural Urban Tree Soil Materials for Use Under Pavement to Increase Street Tree Rooting Volumes, *Journal of Arboriculture* 22, 255–263.

Grabosky, Jason, and Nina Bassuk (1998). *Urban Tree Soil to Safely Increase Rooting Volumes*, patent no. 5,849,069, Washington: U.S. Patent and Trademark Office (www.uspto.gov).

Grabosky, Jason, Nina Bassuk, and B.Z. Marranca (2002). Preliminary Findings from Measuring Street Tree Shoot Growth in Two Skeletal Soil Installations Compared to Tree Lawn Plantings, *Journal of Arboriculture* 28, 106–108.

Grabosky, Jason, Nina Bassuk, and Peter Trowbridge (2002). *Structural Soils: A New Medium to Allow Urban Trees to Grow in Pavement*, revised ed., Landscape Architecture Technical Information Series, Washington: American Society of Landscape Architects (www.asla.org).

Gray, Donald H. (2002). Optimizing Soil Compaction, *Erosion Control* 9, 34–41.

Hammatt, Heather (2001). On the Move, *Landscape Architecture* 91, 22–24 and 126.

Kopinga, Jitze (1991). The Effects of Restricted Volumes of Soil on the Growth and Development of Street Trees, *Journal of Arboriculture* 17, 57–63.

Kopinga, Jitze (1994). Aspects of the Damage to Asphalt Road Pavings Caused by Tree Roots, in *The Landscape Below Ground, Proceedings of an International Workshop on Tree Root Development in Urban Soils*, pp. 165–178, Gary W. Watson and Dave Neely, Eds., Savoy, Illinois: International Society of Arboriculture.

Kristoffersen, Palle (1998). Designing Urban Pavement Sub-bases to Support Trees, *Journal of Arboriculture* 24, 121–126.

Kristoffersen, Palle (1999). Growing Trees in Road Foundation Materials, *Agricultural Journal* 23, 57–75.

Lichter, John M., and Patricia A. Lindsey (1994). Soil Compaction and Site Construction: Assessment and Case Studies, in *The Landscape Below Ground: Proceedings of an International Workshop on Tree Root Development in Urban Soils*, pp. 126–130, Gary W. Watson and D. Neely, Eds., Savoy, Illinois: International Society of Arboriculture.

Lindsey, Patricia, and Nina L. Bassuk (1991). Specifying Soil Volumes to Meet the Water Needs of Mature Urban Street Trees and Trees in Containers, *Journal of Arboriculture* 17, 141–149.

Patterson, James C. (1976). Soil Compaction and Its Effects upon Urban Vegetation, *Better Trees for Metropolitan Landscapes*, pp. 91–102, U.S. Forest Service General Technical Report NE-22, Upper Darby, Pennsylvania.

Patterson, James C, and Christine J. Bates (1994). Long Term, Light-Weight Aggregate Performance as Soil Amendments, in *The Landscape Below Ground: Proceedings of an International Workshop on Tree Root Development in Urban Soils*, pp. 149–156, Gary W. Watson and Dan Neely, Eds., Savoy, Illinois: International Society of Arboriculture.

Perry, Thomas O. (1994). Size, Design, and Management of Tree Planting Sites, in *The Landscape Below Ground: Proceedings of an International Workshop on Tree Root Development in Urban Soils*, pp. 3–14, Gary W. Watson and Dan Neely, Ed., Savoy, Illinois: International Society of Arboriculture.

Pregitzer, Kurt, John S. King, Andrew J. Burton and Shannone E. Brown (2000). Responses of Tree Fine Roots to Temperature, *New Phytologist* 147, 105–115.

Richardson, Colin, John Lizzo, Phillip Dinh, and Dwight Woodson (2000). On the New Waterfront, *Civil Engineering* 70, 60–63.

Sorvig, Kim (2001). Soil Under Pressure: Why the Debate over the Merits of Structural Soil?, *Landscape Architecture* 91, 36–43.

Spomer, L. Art (1983). Physical Amendment of Landscape Soils, *Journal of Environmental Horticulture* 1, 77–80.

Sydnor, T. Davis, D. Gamstetter, J. Nichols, B. Bishop, J. Favorite, C. Blazer, and L. Turpin (2000). Trees Are Not the Root of Sidewalk Problems, *Journal of Arboriculture* 26, 20–29.

Thompson, J. William (2001). Roots Over the River, *Landscape Architecture* 91, 62–69.

Thompson. J. William, and Kim Sorvig (2000). *Sustainable Landscape Construction, A Guide to Green Building Outdoors*, Washington: Island Press.

Trowbridge, Peter J., and Nina L. Bassuk (1999). Redesigning Paving Profiles for a More Viable Urban Forest, in *1999 Annual Meeting Proceedings*, pp. 350–351, Washington: American Society of Landscape Architects.

Trowbridge, Peter J., and Nina L. Bassuk (2004). *Trees in the Urban Landscape: Site Assessment, Design, and Installation*, New York: John Wiley and Sons.

Urban, James R. (1996). Room to Grow, *Landscape Architecture* 86, 74–79 and 96–97.

Urban, James R. (1999). New Approaches to Planting Trees in Urban Areas, in *1999 Annual Meeting Proceedings*, pp. 341–344, Washington: American Society of Landscape Architects.

Urban, James (2000). Planting Details, Tree Planting and Protection, Tree Planting in Urban Areas, in *Architectural Graphic Standards*, pp. 178–182, 10th ed., John Ray Hoke, editor in chief, New York: John Wiley and Sons.

Watson, Gary W., and E.B. Himelick (1997). *Principles and Practice of Planting Trees and Shrubs*, Savoy, Illinois: International Society of Arboriculture.

6 Porous Aggregate

CONTENTS

General Characteristics of Aggregate Materials 200
    Particle Shape 201
    Particle Size 202
    Gradation 202
    Porosity and Permeability 206
    Bearing Strength 206
Mineral Aggregate Materials 207
    Standard Gradations of Mineral Aggregate 207
    Durability of Mineral Aggregate 209
    Specifications for Mineral Aggregate 210
ESCS Aggregate Materials 211
    Standard Gradations of ESCS Aggregates 213
    Durability of ESCS Aggregates 213
Recycled Aggregate Materials 214
Porous Aggregate Surfaces 215
    Appropriate Settings 215
    Appropriate Surface Materials 216
    Placement of Aggregate Material 217
    Stabilization of Aggregate Surfaces 217
    Maintenance of Aggregate Surfaces 219
Porous Aggregate Surfaces in Pedestrian Plazas 220
    Wells-Lenny Office, Cherry Hill, New Jersey 220
    Herty Mall, Athens, Georgia 220
    Steelhead Park, Los Angeles, California 222
Porous Aggregate Surfaces in Residential Driveways 222
    Medford, New Jersey 223
    Garden City, New York 224
Porous Aggregate Surfaces in Parking Lots 225
    Quiet Waters Park, Annapolis, Maryland 225
    Lincoln Street Studio, Columbus, Ohio 227
    Riverbend East, Athens, Georgia 228

Porous Aggregate Surfaces in Work and Equipment Platforms 229
    Electric Substation 229
    Plant Nurseries 229
Porous Aggregate Surfaces in Playgrounds 229
    Vandergrift Park, Arlington, Texas 231
    Barnes Hospital, St. Louis, Missouri 232
Porous Aggregate Base Courses and Reservoirs 233
Porous Aggregate Filter Layers and Use of Geotextiles 235
    Aggregate Layer 236
    Separation Geotextile 237
Trademarks 238
Acknowledgments 239
References 239

"Aggregate" is a generic term for any granular substance used as a construction material. Together, numerous granular particles make an agglomerated mass. In this book, and especially in this chapter, the term "aggregate" is used to refer to strong, durable, granular materials such as crushed stone, natural gravel, crushed concrete, crushed brick, and expanded-clay products. Other "soft" kinds of granular materials, such as biodegradable mulches, will be covered in Chapter 13.

The use of aggregate in pavements is almost ubiquitous. Aggregate forms the setting beds of paving blocks and grids, and fills their cells and joints. It forms the base course in almost any pavement. It is the largest component of asphalt and concrete materials. It is the vital structural skeleton of "structural soils." In some pavements, it is used alone to form a pavement structure entirely of unbound aggregate material.

Aggregate materials must be durable under the conditions to which they will be exposed in service. The best source of information about the suitability of an aggregate material for a particular application is the material's carefully observed and documented service record in the local conditions to which it will be exposed. In the absence of proven local experience, laboratory tests provide predictive information that can be compared with standard benchmarks. The American Society for Testing and Materials (ASTM) maintains definitions and standards for aggregate materials and their evaluative tests. Many state and local agencies, such as highway departments, adopt ASTM's conventions or replace them with locally applicable variations. It is very valuable to find out what the local standards are for a given project, and to refer to them to the greatest appropriate extent in specifying a desired material.

This chapter introduces the important characteristics of aggregate materials and their applications in surface and base courses.

GENERAL CHARACTERISTICS OF AGGREGATE MATERIALS

All granular materials have characteristics that are fundamental to the materials' structural and hydrologic capabilities. Table 6.1 defines some terms that have particular application to aggregate materials.

Porous Aggregate

TABLE 6.1
Some Terms with Particular Application to Aggregate Materials

Term	Definition
Aggregate	Any mass of particulate material; in this book the term is used to refer specifically to hard, durable, granular material
Bulk density	Density of an aggregate mass including the empty void space between and within aggregate particles
Bulk volume	Volume of an aggregate mass including empty void space between and within aggregate particles
Calcined	Heated to drive off impurities and to alter physical and chemical properties
Crushed gravel	Rounded alluvial stone that has been crushed to produce partly angular particles
Crushed stone	Mineral aggregate that has been crushed to produce angular particles
Decomposed granite	Naturally occurring granular product of the weathering of granitic rocks; or, generically, almost any naturally occurring granular material
Density	Weight per unit volume
Durability	Ability of a material to retain its original qualities or abilities over time, withstanding conditions of service including load and weather
ESCS	Expanded (calcined) shale, clay or slate
Fines	Fine (small) particles
Gravel	Rounded river stone, or generically, any aggregate containing particles larger than sand
Limerock	Limestone
Screening (sieving)	Sorting by size using a series of sieves
Screenings	Aggregate material that has been separated out of an aggregate mass by screening (sieving); the term is variously used to refer to either the larger or the smaller particles so separated
Traprock	Dark, hard, fine-grained igneous rock such as basalt

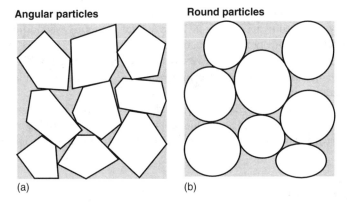

FIGURE 6.1 Round and angular particle shapes.

PARTICLE SHAPE

For an aggregate mass to be structurally stable it is mandatory that the particles be angular in shape. Figure 6.1 contrasts rounded and angular particles. Particles that have at least some planar faces interlock with each other to resist rotating and shifting (Rollings and Rollings, 1996, pp. 50–51). In contrast, round particles like those of river gravel rotate and slip past each other, so the material "gives" under a load.

Most commonly, the source material must be artificially crushed to produce angular faces. The terms "crushed stone" and "crushed gravel" refer to mineral aggregates that have been crushed at a processing plant. Crushed concrete and brick can have similarly suitable angular shape. Surface roughness and cleanness (absence of clays and other adhering matter) also help friction and binding among particles.

Particle Size

The size of the particles in an aggregate influences the material's porosity, permeability, stability under traffic load, and accessibility to pedestrians. Table 6.2 lists some units used to describe particle size.

Aggregate particle sizes are measured by passing the material through a series of sieves. Particles smaller than a sieve's apertures pass through; larger particles are retained. A standard ASTM sieve has square apertures formed by a grid of wires. Table 6.3 lists sieve opening sizes defined by ASTM. According to ASTM's convention, "coarse" aggregate is retained on openings of 1/4 inch and larger and its size is designated in inches; "fine" aggregate passes through openings smaller than 1/4 inch and its size is designated by a sieve number or "gage." Sieve number increases with smaller opening size, because greater numbers of wires in a sieve's grid produce smaller apertures.

Gradation

At processing plants, aggregate particles of different sizes are separated and sometimes selectively recombined to give an aggregate mass a desired range of sizes. An aggregate's combination of sizes, or gradation, is one of the most important determinants of the material's porosity, permeability, and stability under a load so it must be explicitly specified in every project. Table 6.4 defines terms that portray the general character of various types of gradation.

Quantitatively, gradation is expressed in terms of the percentage of the material's total mass that passes a series of sieve sizes. ASTM C 136 specifies test methods for aggregate gradation.

Figure 6.2 contrasts four types of gradations in a chart. The horizontal axis is sieve aperture size; the vertical axis is the proportion of the weight of the material that passes each size. The values on the vertical axis are cumulative. On encountering

TABLE 6.2
Some Units Used to Describe Particle Sizes in Aggregate

Unit	Equivalent units	Equivalent units
1 µm	meter × 10^{-6}	0.000039 inch
1 mm	meter × 10^{-3}	0.039 inch
1 inch	0.0254 m	25,400 µm; 25.4 mm

TABLE 6.3
Opening Sizes of ASTM Standard Sieves (ASTM E 11)

ASTM Designation	Sieve Opening Size (inch)	Equivalent Size (mm)
Coarse openings		
4 inch	4.0	100
31/2 inch	3.5	90
3 inch	3.0	75
21/2 inch	2.5	63
2 inch	2.0	50
11/2 inch	1.5	37.5
1 inch	1.0	25.0
3/4 inch	0.75	19.0
1/2 inch	0.5	12.5
3/8 inch	0.375	9.5
1/4 inch	0.25	6.3
Fine openings		
No. 4 sieve	0.187	4.75
No. 5 sieve	0.157	4.00
No. 6 sieve	0.132	3.35
No. 8 sieve	0.0937	2.36
No. 10 sieve	0.0787	2.00
No. 12 sieve	0.0661	1.70
No. 16 sieve	0.0469	1.18
No. 20 sieve	0.0331	0.850 (850 μm)
No. 30 sieve	0.0234	0.600 (600 μm)
No. 40 sieve	0.0165	0.425 (425 μm)
No. 50 sieve	0.0117	0.300 (300 μm)
No. 60 sieve	0.0098	0.250 (250 μm)
No. 80 sieve	0.0070	0.180 (180 μm)
No. 100 sieve	0.0059	0.150 (150 μm)
No. 200 sieve	0.0029	0.075 (75 μm)
No. 400 sieve	0.0015	0.038 (38 μm)
No. 500 sieve	0.0010	0.025 (25 μm)

similar graphs in practice, study them carefully: on some of them the horizontal and vertical axes are the reverse of those in Figure 6.2, and some use logarithmic scales.

Any one point on one of the figure's curves can be designated by the symbol D_x, in which D is the sieve size and x is the percentage of the weight that passes that size. For example, on the open-graded curve D_{60} is 0.36 inch. A combination of sizes on a curve can be roughly characterized by a coefficient of uniformity, which is usually the ratio D_{60}/D_{10}.

The open-graded curve in the figure is almost vertical, indicating a narrow range of particle sizes. Most aggregates with a coefficient of uniformity less than about 2.0 or 2.5 could be considered relatively open-graded. The stability of an open-graded aggregate comes from the interlock of the single-sized particles with each other, especially along the flat faces of angular particles. Between the particles are open

TABLE 6.4
Some Terms Describing General Character of Aggregate Gradations

Term	Definition
Coarse-graded	Dense-graded, with predominance of coarse (large) particles
Crusher run	Crushed material that has not been screened (sorted) into selected sizes following crushing; this is ordinarily a dense-graded gradation
C_u	Coefficient of uniformity, usually the ratio D_{60}/D_{10}
Dense-graded	Having a combination of particles of different sizes, such that the relatively small particles fill the voids between the large ones, giving the aggregate mass low porosity and permeability
D_x	The sieve size which x percent (by weight) of an aggregate mass passes
Fine-graded	Dense-graded, with predominance of fine (small) particles
Fraction	A separately collected portion of an aggregate mass, for example, after separation by sieving for size
Gap-graded	Of two distinct size ranges, with intermediate sizes substantially absent
Gradation	Combination of particle sizes in an aggregate mass
Graded	Dense-graded; well graded
Open-graded	Single-sized or of a narrow range of sizes, such that the voids between particles tend not to be filled by relatively small particles
Pit run	Naturally granular material that is not screened (sorted) into selected sizes; this is ordinarily a dense-graded gradation
Screenings	Particles remaining after screening (sieving) for size; this term is sometimes used to refer to the relatively small particles that pass the screen, and sometimes oppsitely to the coarse particles retained on the screen
Single-sized	Of a single narrow range of size; open-graded
Stone dust	Fine screenings
Top size	Maximum particle size in a gradation; sometimes used to refer to a nominal maxmum size such as the sieve size through which 90 percent of the material passes
Well graded	Dense-graded; of a relatively balanced mixture of all particles sizes

Adapted from Webster (1997, pp. 13–14).

voids; the aggregate mass is highly porous, highly permeable, and well-drained. An open-graded aggregate can be nonplastic and nonsusceptible to frost damage. For these reasons decisively open-graded aggregate is a vital component in porous pavements. In many applications it is advisable even to wash open-graded stone clean of clinging fine particles which would clog some of the pores.

The dense-graded curve in the figure has a relatively sloping line, indicating a continuous and more or less even range of sizes from small to large. There may be sufficient large particles to interlock with each other, while fine particles pack the space between them. Dense-graded aggregate can be very dense, slowly permeable, and highly stable, although it may be frost-susceptible because of the retention of moisture in its numerous small pores. Figure 6.3 contrasts how the particles in dense-graded and open-graded gradations pack together into a mass.

The fine-graded and coarse-graded curves in Figure 6.2 represent dense gradations with a predominance of, respectively, relatively fine and relatively coarse particles. Fine-graded aggregates tend to have comparatively little stability, because the great amount of fines allows little direct interlock between the large particles.

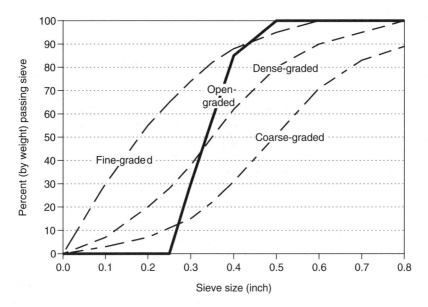

FIGURE 6.2 A graph illustrating aggregate gradation, with four hypothetical curves.

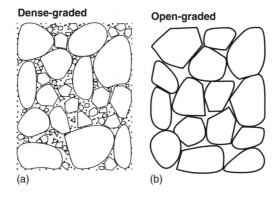

FIGURE 6.3 Packing of particles in dense-graded and open-graded gradations.

Fine-graded aggregates can be susceptible to frost action because the numerous small pores retain moisture. They tend to be poorly aerated, poorly drained, and slowly permeable. In contrast, coarse-graded aggregate can be stable.

A gradation can be custom-mixed to meet the requirements of any specific project. However, it is most economical and convenient to select from the standard gradations that local processing plants keep on inventory. Where gradation standards are adopted by local agencies such as departments of transportation, aggregate suppliers keep them in stock and make them readily available. Different standard gradations are used for different types of materials. It can greatly expedite a project to find out what the local standards are and to use them to specify gradations that will meet project objectives. Applicable national standards will be described in later sections of this chapter.

Porosity and Permeability

In open-graded aggregates, as particle size increases, the voids between them enlarge proportionately. According to one rule of thumb, the diameter of the voids is up to onefifth the diameter of the particles. In dense-graded aggregate with numerous fine particles, all void spaces are small.

In open-graded aggregates in which the particles have no internal pores, the total porosity (ratio of void space to total volume of the material) does not vary measurably with particle size; it is in a relatively constant range of 30 to 40 percent. The exact value depends on the exact gradation of the material, particle shape, and degree of compaction. Aggregates of extraordinarily uniform particle size have high porosities of 33 to 45 percent in which the highest values are associated with rigorously angular particles and the lowest ones with rounded gravels (Shergold, 1953). If in addition the aggregate particles have internal pores, the total porosity can be even higher. At the other extreme, dense-graded aggregates have low porosities because small particles fill the space between large particles. ASTM C 29 specifies standard tests for void space.

The permeability of an aggregate comes partly from its total porosity, and partly from the size of the individual voids. Table 6.5 lists approximate permeabilities of a few general gradations. Open-graded aggregates have high permeabilities because they have both high porosity and large individual voids. The permeability of a clean, coarse, open-graded aggregate is surely the highest of any of the pavement materials described in this book. ASTM D 5084 specifies standard tests for aggregate permeability.

The runoff coefficients of open-graded aggregate surfaces have not been directly measured. The very high permeability of open-graded aggregate materials suggests that the values need not be higher than those for turf.

Bearing Strength

The bearing strength of an aggregate mass comes from a combination of the resistance of particles to crushing, and the interlock of particles with each other. The California Bearing Ratio (ASTM D 1883), which was described in Chapter 3, compares the strength of an aggregate material to that of a standard (very strong) aggregate. Values above 40 or 50 are adequate for most base-course materials, and are commonly attained by durable, angular, open-graded material. Open-graded materials get their

TABLE 6.5
Approximate Permeabilities of Aggregate Materials

Gradation	Permeability, inches per hour
1 inch aggregate (uniform size)	25,000
1/2 inch aggregate (uniform size)	7500
1/4 inch aggregate (uniform size)	1250
Coarse sand	50
Dense-graded sand and gravel	0.25

AASHTO (1985, p. AA-18).

strength from the interlock of particles with each other, especially along the planar facets of angular particles. Dense-graded materials get their strength from some combination of particle interlock and the packing of particles around each other.

MINERAL AGGREGATE MATERIALS

Most aggregate materials are natural mineral matter. The mineral aggregate-producing industry is huge: the U. S. produces about two billion tons of crushed stone each year, and another 1.2 billion tons of sand and gravel (Bolen, 2002; Tepordei, 2002). There are mineral aggregate suppliers in every region where there are construction activities of any kind. Mineral aggregates are not expensive to produce, so by default in some locales transportation is a large part of the material's delivered cost (Huhta, 1991, p. 2-2). Local suppliers can be identified from local business and telephone directories, or from the member lists of state aggregate industry associations. Links to many state industry associations are maintained by NSSGA (www.nssga.org).

The characteristics of locally available mineral aggregates vary with geology. Potential natural sources are rock quarries, alluvial gravel deposits, and surficial formations such as caliche and decomposed granite. Of the crushed stone produced in the U.S., 71 percent is limestone and dolomite; 15 percent is granitic rock; the remainder is mostly traprock, sandstone, marble, and quartzite (Huhta, 1991, pp. 1–4; Rollings and Rollings, 1996, p. 45; Tepordei, 2002). Porous aggregate particles can be obtained from naturally porous rocks such as pumice and tuff, but commercial availability of such materials is rare in North America. Most of the pumice is produced in the western states; an example of a producer is Sierra Cascade of Oregon (www.sierracascadellc.com).

Each potential source material must be evaluated individually because, even within a given classification of rock, the quality varies from region to region and from quarry to quarry. Granite, hard sandstone, hard limestone, and traprock can make some of the strongest and most durable aggregate. On the other hand, some soft varieties of limestone make poor aggregate because they crush easily and deteriorate in the presence of water.

Those seeking further information on mineral aggregates are referred to the National Stone, Sand and Gravel Association (www.nssga.org), the association's *Aggregate Handbook* (Barksdale, 1991), and Rollings and Rollings' (1996) overview.

STANDARD GRADATIONS OF MINERAL AGGREGATE

The American Society of Testing and Materials (ASTM) quantitatively defines standard gradations for typical pavement construction in D 448, *Standard Classification for Sizes of Aggregate for Road and Bridge Construction* (the gradations listed in ASTM C 33, *Standard Specification for Concrete Aggregates*, are identical). These gradations are adopted in the standard specifications of many state departments of transportation.

Table 6.6 lists several of ASTM's open-graded gradations. ASTM D 448 designates each gradation by a number (gradation numbers should not be confused with

TABLE 6.6
Some ASTM Standard Gradations for Mineral Aggregate

Sieve Designation	Size (inches)	Percent (By Weight) Passing Sieve				
		ASTM No. 4	ASTM No. 5	ASTM No. 57	ASTM No. 6	ASTM No. 67
2 inch	2.0	100				
1.5 inch	1.5	90 to 100	100	100		
1 inch	1.0	20 to 55	90 to 100	95 to 100	100	100
3/4 inch	0.75	0 to 15	20 to 55	Not specified	90 to 100	90 to 100
1/2 inch	0.5	Not specified	0 to 10	25 to 60	20 to 55	Not specified
3/8 inch	0.375	0 to 5	0 to 5	Not specified	0 to 15	20 to 55
No. 4 sieve	0.187			0 to 10	0 to 5	0 to 10
No. 8 sieve	0.0937			0 to 5		0 to 5
No. 16 sieve	0.0469					
No. 50 sieve	0.0118					
No. 100 sieve	0.0059					
		ASTM No. 7	ASTM No. 78	ASTM No. 8	ASTM No. 89	ASTM No. 9	ASTM No. 10
2 inch	2.0						
1.5 inch	1.5						
1 inch	1.0						
3/4 inch	0.75	100	100				
1/2 inch	0.5	90 to 100	90 to 100	100	100		
3/8 inch	0.375	40 to 70	40 to 75	85 to 100	90 to 100	100	100
No. 4 sieve	0.187	0 to 15	5 to 25	10 to 30	20 to 55	85 to 100	85 to 100
No. 8 sieve	0.0937	0 to 5	0 to 10	0 to 10	5 to 30	10 to 40	Not specified
No. 16 sieve	0.0469		0 to 5	0 to 5	0 to 10	0 to 10	Not specified
No. 50 sieve	0.0118				0 to 5	0 to 5	Not specified
No. 100 sieve	0.0059						10 to 30

(ASTM D 448).

ASTM's gage numbers for individual particle size.) The first digit represents the general particle size; a one-digit number represents a very open-graded gradation with a narrow range of sizes. Where second and third digits are present, the composite numbers represent blends of, or gradations between, the single-size gradations.

Figure 6.4 illustrates three D 448 gradations. It shows that the range of contents allowable within each of ASTM's gradations is defined by two bounding lines. The steep bounding lines of No. 6 and No. 7 indicate open-graded gradations. The intermediate No. 67 gradation has proportionally more large particles than No. 7, more small particles than No. 6, and a broader range of allowable quantities for all particle sizes. Although No. 67 is not as open-graded as the narrow No. 6 and No. 7, it is still considered relatively open-graded. Some aggregate plants make two-digit gradations more available than the strict single-digit gradations because it is easier to produce material with the two-digit range of tolerance.

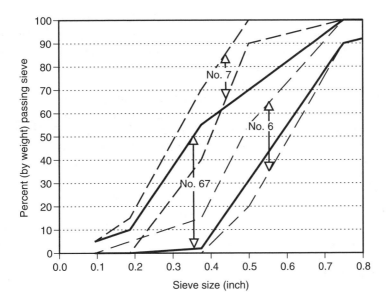

FIGURE 6.4   Combination of ASTM gradation numbers 6 and 7 to make number 67 (data from ASTM D 448).

DURABILITY OF MINERAL AGGREGATE

Aggregate must resist abrasion in order not to become rounded and lose its stabilizing interlock of particles, and in order not to generate fine particles that could clog pores or generate dust. It must also resist deterioration in freezing weather: when water freezes it expands; repeated freezing in the micropores of aggregate particles could weaken or crack the particles.

The "Los Angeles" test evaluates resistance to abrasion. It simulates impacts that an aggregate might experience during processing and placement or, to a lesser extent, under traffic (ASTM C 131 and C 535; Rollings and Rollings, 1996, pp. 159–160). It involves rotating a sample of aggregate particles with steel balls in a revolving drum (the test is sometimes referred to as the "Los Angeles rattler"). The result is expressed as the percentage of material that was chipped away as fine particles smaller than the No. 12 sieve. It is a rigorous test; some aggregates degrade more in this test than they do under actual construction and traffic conditions. A test result of up to 30 or 40 percent is considered satisfactory for many pavement applications. Aggregates that will be used in a base course and out of contact with traffic are sometimes allowed to have losses as high as 50 percent. The results are most useful when they are correlated with local construction experience.

The "magnesium sulfate" test evaluates soundness in severe weather. It simulates expansive forces that an aggregate might experience during freezing of water (ASTM C 88; Rollings and Rollings, 1996, p. 160). Before the test, the gradation of an aggregate sample is precisely analyzed. Then the sample is subjected to repeated cycles of immersion in a saturated solution of magnesium sulfate (or, alternatively, sodium sulfate) during which salt crystals form in the aggregate particles' internal pores. After each immersion the sample is oven-dried. Upon each re-immersion the growth of the salt crystals generates expansive force like that of the expansion of water on freezing. After this artificial "weathering" the particles are again analyzed for size, and their gradation is compared with that form before the weathering cycles. The difference indicates the degree to which the aggregate has fractured into smaller particles. The loss of weight to smaller size categories is reported as a percentage of the aggregate's total dry weight. However, the result is considered only a "preliminary estimate" of soundness during freezing and is not considered superior to the service record of a material exposed to actual weathering conditions.

The American Association of State Highway and Transportation Officials (AASHTO) has a test that evaluates an aggregate's durability under freezing and thawing more directly (Rollings and Rollings, 1996, p. 160). AASHTO's test T103 is like the magnesium sulfate test, except that cycles of freezing are used rather than cycles of salt-crystal growth. Degradation is reported as the percentage (by weight) of aggregate material that degraded to smaller sieve sizes as a result of chipping and breaking during freeze-thaw cycles.

SPECIFICATIONS FOR MINERAL AGGREGATE

Specifying a mineral aggregate material can be simple, but it must be done knowledgeably and completely in order for the resulting pavement structure to have the intended porosity, permeability, and durability.

On the national level, the only ASTM standard that specifies any desired characteristics for unbound mineral aggregate is D 2940, *Standard Specification for Bases or Subbases for Highways or Airports*. Unfortunately, it anticipates the use of unbound aggregate only in dense, impervious gradations, which are useless in most porous pavements. It requires that coarse aggregate should consist of durable angular particles of crushed stone, gravel, or slag capable of withstanding the effects of handling, spreading, and compacting, but does not cite the Los Angeles test or any other definite measure of resistance to abrasion, nor any test for durability in freeze-thaw conditions.

In almost any region, aggregate suppliers are organized to supply materials meeting local department of transportation (DOT) standards, so it is very useful to get to know those standards. The Federal Highway Administration maintains links to state specifications at its National Highway Specifications web site (accessible through www.fhwa.dot.gov). One should get to know which of the DOT-specified materials are open-graded and have the durability and other characteristics necessary for specific types of applications.

In many regions, material of measured durability is marketed in reliably open-graded gradations. In those regions, specifications need do little more than refer to the applicable DOT or other standards for gradation and durability. A designer who specifies one of those categories for a local project works within a known and established system to meet project objectives.

But in equally as many regions commonly marketed aggregate products include densely graded or nondurable products that are hostile to the purposes of porous pavement. Customary references to these locally well-known materials are insufficient; project specifications must overcome local conventions to obtain material that will satisfy project objectives. In the absence of appropriate local standards, one must write one's own requirements for gradation and durability. The following paragraphs give examples of regionally available materials that could cause confusion or failure in porous pavement specification.

In California and other southwestern states "decomposed granite" tends to be sold, when not specified otherwise, as a naturally dense-graded mixture of particle sizes. Decomposed granite is a naturally occurring granular product of the weathering of granitic rocks; geologists refer to the material as "grus" (Wagner, 1991). The same term is also used generically to refer to an uncontrolled variety of aggregate materials, without pinning down the durability and angularity that a load-bearing aggregate ought to have. A project in California must explicitly and quantitatively specify for its aggregate an open-graded gradation and measurable durability in order not to obtain dense-graded material with low porosity, low permeability, and of unknown stability.

In Florida "limerock" tends to be sold, when not specified otherwise, as a naturally soft material with very low durability. A project in Florida must explicitly specify measurable durability in order not to obtain material that can turn in a few years to dust and mud.

In the southeastern Piedmont region "crusher run" is widely marketed in its naturally occurring dense gradation. "Crusher run" and "pit run" refer to any mixtures of particles that emerge from quarries or processing plants without any sorting for size. The quantitative content of particle sizes is ultimately uncontrolled because the terms have no fixed quantitative meaning. A project in the Piedmont must explicitly specify a quantitatively defined open-graded gradation in order not to obtain material that will be effectively impervious as soon as it is placed and compacted.

In the midwestern states "graded aggregate base" is widely installed in the form of dense-graded aggregate material. A project in the Midwest must explicitly and quantitatively specify an open-graded gradation in order not to obtain material that will be essentially nonporous and impermeable.

In the northeastern states "pea gravel" is widely marketed in the form of naturally rounded river gravel. In the Northeast, a project must explicitly specify angular particles in order not to receive rounded material that is unstable under traffic.

Most narrowly, the term "gravel" refers to rounded river stone; the term is also used informally to refer to almost any aggregate. Unprocessed river gravel can be low in cost. But crushing at a processing plant is almost always necessary to turn river gravel into particles angular enough to form a stable structure. A definite specification must call for crushed gravel, crushed stone, or other specifically angular material in order not to obtain material that rolls and shifts under a load.

ESCS AGGREGATE MATERIALS

Aggregates of expanded shale, clay and slate (ESCS) are made by firing lumps of shale, clay, or slate at high temperature in a kiln. Hot gases create air pockets in the material; the material "bloats" to 1.5 to 2 times its original size. The fired (calcined) product is a stable, strong ceramic material. It may be crushed to make angular particles, and sorted for size. The particles are hard and durable; their surfaces are rough.

ESCS differs importantly from most mineral aggregates in that it is highly porous. Like other aggregates the void spaces between particles amount to 30 to 40 percent of the aggregate's bulk volume. In addition, the particles' internal pores add approximately another 10 percent to the material's total porosity. The combination can make the material's total porosity approach 50 percent.

ESCS is well-known as an aggregate in concrete (ASTM C 330 and C 331) where its pores make it light in weight for bridges and tall structures. It is also well known among horticultural planting media (ASTM 5883), where its pores give it aerating and water-holding capabilities which most mineral aggregates do not have. Some ESCS products even have nutrient-holding capacities akin to those in natural soil. ESCS is a distinctive component of some of the tree-rooting mixtures described in Chapter 5.

The availability and cost of ESCS aggregate vary from region to region with the proximity of producing plants. However, the light weight of the material reduces the cost of transportation to some project sites, so cost relative to other types of aggregate is always project-specific.

The variety of ESCS products have different geologic origins, varying production processes, and different physical and chemical properties in the finished product. Each must be evaluated individually for its ability to meet the needs of a given project. ASTM C 330 and C 331 specify standards for ESCS aggregate. Although those documents anticipate the use of the aggregate in lightweight concrete, their provisions for sizing and durability are useful for unbound applications in porous pavements as well.

Table 6.7 lists examples of suppliers of ESCS aggregate. Further companies can be identified through the membership list of the Expanded Shale, Clay and Slate Institute (www.escsi.org). The institute also has further information on ESCS aggregates in general.

TABLE 6.7
Examples of Suppliers of ESCS Aggregate

Company	Products	Contact Information
Arkalite	Lightweight aggregate	www.generalshale.com
Big River Industries	Gravelite	www.bigriverind.com
Buildex	Lightweight aggregate	www.buildex.com
Carolina Stalite	PermaTill	www.permatill.com
Carolina Stalite	Stalite	www.stalite.com
Garick	HydRocks	www.hydrocks.com
Lehigh Cement Company	Lelite	www.lehighcement.com
Norlite	Norlite	www.norliteagg.com
Northeast Solite	Northeast Solite	www.nesolite.com
Profile Products	Profile	www.profileproducts.com
Solite	Solite	www.solitecorp.com
Tri-Texas Industries	Expanded shale and clay	www.txi.com

Standard Gradations of ESCS Aggregates

ASTM C 330 specifies the standard gradations for ESCS aggregates listed in Table 6.8. Their steep curves in Figure 6.5 show that they are open-graded. They are not precisely the same as any of the open-graded gradations for mineral aggregate specified in ASTM D 448. ESCS suppliers tend to keep the C 330 gradations in stock, and to refer to them as "1 inch," "3/4 inch," and "1/2 inch."

Durability of ESCS Aggregates

ESCS aggregates should stand up to abrasion and weathering as should mineral aggregates. Most of the provisions in ASTM C 330 apply to the lightweight concrete

TABLE 6.8
Coarse Open-Graded Gradations Specified for ESCS Aggregate in ASTM C 330

Sieve Designation	Size (inches)	Percent (By Weight) Passing Sieve		
		1 inch to No. 4	3/4 inch to No. 4	1/2 inch to No. 4
1 inch	1.0	95 to 100	100	
3/4 inch	0.75	Not specified	90 to 100	100
1/2 inch	0.5	25 to 60	Not specified	90 to 100
3/8 inch	0.375	Not specified	10 to 50	40 to 80
No. 4 sieve	0.187	0 to 10	0 to 15	0 to 20
No. 8 sieve	0.0937			0 to 10

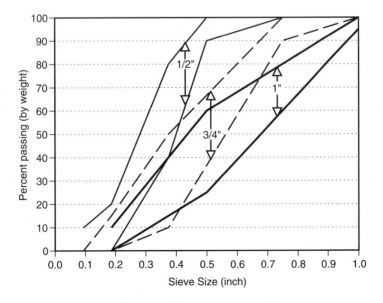

FIGURE 6.5 Gradations specified for ESCS aggregate in ASTM C 330.

in which ESCS aggregate might be used, and not to the aggregate itself. For use as unbound material in pavements, one can also evaluate an ESCS aggregate for resistance to abrasion using the Los Angeles test, and to soundness using the magnesium sulfate test. Some ESCS suppliers supply test data for their products using these tests, and their products show excellent durability.

The porosity of ESCS particles invites a logical question about their durability in freezing and thawing. The pores harbor internal moisture; when the water freezes it might concentrate destructively expansive force on the surrounding material. However, the high performance of many ESCS materials in the magnesium sulfate test indicates that this is not what happens. Apparently the same pores that hold moisture also provide relief for pressure when the water freezes. Ice crystals expand in the direction of least resistance, through the pores and out to the particle surface. Where pressure develops in more confined pockets, the great strength of the ESCS solid material resists fracturing.

RECYCLED AGGREGATE MATERIALS

Recycled aggregate materials have the environmental benefit of low consumption of natural materials. However, like any other aggregates they must be specified and evaluated for their gradation and durability as necessary to meet a project's objectives. Some salvaged materials are available through recycled-materials exchanges like those listed in Table 6.9.

Crushed brick and crushed vitrified clay come from the plants that make bricks and vitrified-clay products. The plants discard chipped, deformed, or otherwise unwanted products, and crush them for use as aggregate. Their particles can be stable, inert, angular, and strong. They are low in cost within a limited distance of the plants where they are produced. However, the plants where they originate may not keep standard open-graded gradations on stock; instead they keep only "crusher-run" material which is likely to be dense-graded; in these cases the material may require special sorting for size.

In many locales, concrete rubble becomes available from time to time from demolition activities. Its particles can be very strong and angular. However, concrete aggregate must be sorted for size to produce porosity and permeability, like other aggregate; without sorting it may have a great amount of very fine material. It is

TABLE 6.9

Examples of Sources of Information on Availability of Demolition Debris and Other Recycled Materials

Source	Contact Information
California Integrated Waste Management Board	www.ciwmb.ca.gov
King's Material	www.kingsmaterial.com
Recycle.net	http://build.recycle.net
Recycler's World	www.recycle.net
Used Building Materials Association	www.ubma.org

capable of leaching out alkalinity which could disrupt the health of nearby plants and trees. Concrete aggregate has the capacity to recement itself into a solid mass. It may be susceptible to abrasion. Both leaching and recementing may be minimized, but not eliminated, by using strictly only coarse, single-sized material, the particles of which have low surface area exposed to weather.

Regional heavy industries such as those of iron and steel generate scoria or slag as an industrial by-product. The particles can be distinctively porous. Many slag suppliers stockpile materials in ASTM D 448 standard gradations. The use of slag as an aggregate in pavement construction is anticipated in ASTM D 2940, *Standard Specification for Graded Aggregate Material for Bases or Subbases for Highways or Airports*. However, the material may leach chemicals that are undesirable in soils or aquifers. Some slag particles are subject to expansion with changes in moisture. For projects where expansion would disrupt an overlying surface course, slag aggregate must be obtained from a source known to have a satisfactory service record, or should have low expansion when tested with an expansion test such as that of ASTM D 4792. Further information is available from the National Slag Association (www.nationalslagassoc.org).

POROUS AGGREGATE SURFACES

Fortuitously, open-graded aggregate tends to combine in a single material the highest permeability and the lowest cost of almost any pavement surfacing material. Aggregate is extremely simple to install and to maintain. An aggregate surface directly bears a traffic load and the traffic's forces of abrasion and displacement. The surface must be made of suitable material, placed in an appropriate setting, and correctly installed.

Appropriate Settings

Because unbound aggregate can be displaced, it is appropriate only in very low-traffic settings such as pedestrian walkways and plazas, residential driveways, lightly used parking stalls, long-term parking such as that of seasonally used recreational vehicles, and platforms for fixed equipment such as electric substations. It should not be used where vehicles move or turn frequently or rapidly. Heavy traffic would push unbound particles out of tire tracks and toward the periphery of the surface, building up into "waves" on the pavement surface.

Unbound aggregate is distinctively appropriate in environmental conditions that could deform a pavement structurally, including swelling subgrade soil, deep winter freezing, growing tree roots, and uncompacted subgrade. Unbound aggregate is loose and adaptable to gentle movement, and when necessary it is easy to relevel. Consequently, these environmental conditions can usually be disregarded in the design of unbound aggregate surfaces (AASHTO, 1993, p. II-11).

Aggregate is not suited to steep pavement surfaces; a steep slope aggravates displacement under any surface force. The exact gradient at which displacement could begin depends on particle density, shape, and gradation, and the magnitude of the displacing force. Thompson (1991, pp. 11–64) suggested 3 percent as an upper limit of tolerance.

Aggregate can be used in the dense shade of trees, unlike living turf which can be suppressed by dense shade.

In settings that will be snow-plowed in the winter, unbound aggregate should be used only where plowing will be done carefully; careless plowing can gouge and abrade an aggregate surface. Attachment of skids or rollers to keep the blade an inch above the surface can prevent displacement. When necessary, displaced material is easily returned to its place.

Although the places where unbound aggregate surfacing can be appropriately used are small in scale and geographically fragmented, they are numerous in urban areas. They add up to a substantial total area that is available for this type of environmental restoration.

Appropriate Surface Materials

For an aggregate surface to be porous and permeable, a strictly open-graded (single-size) gradation is mandatory. For the surface to be stable under traffic it is vital that the particles be of durable material and angular in shape.

The open-graded ASTM numbers 57, 67, 78, 89 and 10 encompass a range of sizes that could reasonably be selected for use in aggregate surfaces. They typically have 30 to 40 percent porosity and good permeability.

Larger sizes such as 57, 67 and 78 have particles up to about 1 inch in size. Their large open pores produce very rapid permeability and very little susceptibility to clogging.

On the other hand, smaller sizes such as 89 and 10 make rather smooth surfaces that are easy to walk on, while not lacking in permeability. For this type of advantage, Thompson (1991) suggested the nonstandard gradation listed in Table 6.10, which is just slightly larger than ASTM No. 10. Only small gradations should be used where universal accessibility is required.

The effects of durable open-graded aggregate surfaces are completely unlike those of historical or contemporary dense-graded surfaces. In the early part of the twentieth century, dense-graded aggregate was used for auto roads. Clay and other fine particles were deliberately added to the mixture for cohesion and to fill voids for dense compaction. Dense-graded aggregate is still used today in low-cost rural roads. Dense-graded material produced the notorious dust of early rural roads

TABLE 6.10
Gradation for Aggregate Walking and Parking Surfaces Suggested by Thompson

Sieve Designation	Particle size, inch	Percent Passing Sieve
1/2 inch	0.5	100
3/8 inch	0.375	90 to 100
No. 4	0.187	75 to 90
No. 16	0.0469	20 to 40
No. 100	0.0059	10 to 20
No. 200	0.0029	5 to 15

1991, pp. 11-63–11-64.

(Hewes, 1942, p. 317) and is subject to impervious surface crusting (Ferguson, 1994, pp. 188–189). The addition of fine particles was arguably an attempt to "push" low-cost aggregate surfacing into applications beyond its natural traffic-carrying capacity. For open-graded angular material under very low traffic loads such as those in driveways, walkways, and parking stalls, the old clay binder is structurally unnecessary. The omission of the binder eliminates a source of dust, leaves void spaces open for high permeability, and makes the material nonsusceptible to frost heave.

AASHTO M 147, *Standard Specification for Materials for Aggregate and Soil-Aggregate Subbase, Base and Surface Courses*, specifies several possible gradations for use in surface courses. Unfortunately, they are all relatively dense-graded; porosity and permeability are evidently not among their purposes. For all of them it specifies durability and stability in general but useful terms: composition, durability in freezing and thawing, and wetting and drying, durability in wearing, and liquid limit.

PLACEMENT OF AGGREGATE MATERIAL

The subgrade under an aggregate pavement seldom needs to be compacted. Foregoing subgrade compaction preserves the soil's permeability and root habitat. Unbound aggregate is adaptable to movement of soft, uncompacted soil. On certain subgrade soils a geotextile is necessary to separate the soil from the aggregate, as described later in this chapter.

Whether or not the subgrade is compacted, compaction of the aggregate material as it is placed in "lifts" is useful. Compaction settles the particles into place so they resist rutting and displacement under traffic.

At the edges, many aggregate installations require firm edge restraints or clear edge markers to prevent traffic from deforming the edge. (Alternative edge treatments were described in Chapter 3.)

STABILIZATION OF AGGREGATE SURFACES

Small quantities of stabilizing binders are sometimes added to aggregate surfaces to inhibit particle displacement. Small amounts of Portland cement bind particles together, making in effect a thin form of porous concrete.

Table 6.11 lists some additional aggregate binders. Some are organic psyllium which when wetted semiharden to a consistency like that of asphalt. Some are emulsions of pine rosin which harden with evaporation of the water carrier. Some are polymer emulsions which cure into water-insoluble binders. All these products are useful in certain circumstances because their transparency leaves a naturalistic appearance, and because they do not require the same equipment that is needed for installation of asphalt or concrete (Gourley, 2001; Keating, 2003). However, there are no known installations of these products with clearly open-graded aggregate, and the manufacturers of some of them have stated that their products are in fact intended for use with dense-graded aggregate. Therefore, the resulting pavement material cannot be considered significantly porous and permeable.

Battle Road Trail in Minuteman National Historic Park, Massachusetts, illustrates a combination of effects of one of the binders (Hammatt, 2002). Figure 6.6

TABLE 6.11
Examples of Aggregate-Stabilizing Products

Product	Manufacturer	Contact Information
Envirotac	Environmental Products and Applications	http://envirotac.com
PaveCryl	Rohm and Haas	www.rohmhaas.com
Penzsuppress	American Refining Group	www.pennzsuppress.com
PolyPavement	PolyPavement	www.polypavement.com
Road Oyl (Resin Pavement)	Soil Stabilization Products	www.sspco.org
Soil-Sement	Midwest Industrial Supply	www.midwestind.com
Stabilizer	Stabilizer Solutions	www.stabilizersolutions.com
T-NAPS	George L. Throop Co.	www.t-naps.com

FIGURE 6.6 Stabilized aggregate trail in Minuteman National Park, Concord, Massachusetts.

shows the trail a year or two after installation. The surface is a dense-graded mixture of crushed stone and sand bound with Stabilizer. There is no visible structural edging. The trail carries more than a million walkers and bikers per year. It is not plowed for snow removal. The surface's structural condition is excellent: it is smooth, with no significant erosion, no spalling of stabilized patches, and no cracking, heaving, or displacement. It is firm and regular for walking on, and is considered universally accessible. The appearance of the trail's local aggregate materials is compatible with the park's historic and natural setting. The permeability is low but positive. Polished-looking, fine-textured patches result from the imperfect mixing of material, and subsequent traffic compaction and surface washing.

MAINTENANCE OF AGGREGATE SURFACES

An aggregate surface has low maintenance requirements, when located in an appropriately low-traffic setting and made of appropriately open-graded, angular, durable material. When maintenance is required, the causes for it should be understood before any action is taken in order to avoid unnecessary difficulties and recurring costs.

Particle displacement occurs where vehicles make sharp turns at high speed (AASHTO, 1993 p. II-12) or where the surface is carelessly snowplowed. The resulting uneven surfaces can be releveled by raking. Where a significant amount of aggregate is lost beyond the edges of a pavement, the material will need to be replaced. On some sites the control or redirection of vehicles may reduce displacement by traffic. Snowplowing displacement can be reduced by attaching rollers or skids to a plow blade to keep it 1/2 inch or more above the surface (Keating, 2001).

Over time slow-moving traffic can realign surface particles, without displacing them, into visibly flattened and compacted tracks. Compacted tracks are more stable but less permeable than adjacent uncompacted areas. As long as the compacted areas are limited to narrow tracks they do not detract from a surface's overall hydrology, because any runoff migrates quickly to adjacent uncompacted areas and infiltrates into open voids, whence it spreads throughout the aggregate mass (Ferguson, 1994, p. 189). However a compacted track may be a cosmetic concern, as the surface no longer presents a uniformly textured plane. Appearance and permeability can be restored by loosening the surface with a rake.

Where traffic has produced fine particles by grinding aggregate, surface runoff can wash the fine particles into polished-looking, slowly permeable crusts in low spots. The crusting is typically within the top inch of material. Appearance and permeability can be restored by excavating the surface layer of material and replacing it with fresh open-graded material. Where this kind of maintenance is required, it may be a sign that the original aggregate material was too dense-graded or not durable enough to withstand the traffic, and stronger or more decisively open-graded material should be considered for replacement.

Weed seeds can lodge in the interstices of an aggregate surface, where they can sprout if there is enough moisture, as there may be in shady areas or areas of irrigation overspray. Spreading lawn grasses can encroach onto an aggregate surface from the side. In little-used peripheries of an aggregate surface there is little abrasion by traffic to set back incipient vegetation. Weeds can be suppressed with herbicide; but herbicides ought to be applied only in the zones where weeds are actually seen to occur. In settings that are sensitive to the cosmetic effect of incipient weeds, weeding along the edges might be necessary two or more times per year. In open areas the naturally droughty character of open-graded aggregate suppresses most vegetation.

POROUS AGGREGATE SURFACES IN PEDESTRIAN PLAZAS

Unbound aggregate is useful in plazas with a wide range of pedestrian traffic. Pedestrian settings can use a great variety of open-graded aggregate materials. Many pedestrian plazas include trees rooted in or below the aggregate layer.

WELLS-LENNY OFFICE, CHERRY HILL, NEW JERSEY

The Wells-Lenny Office in Cherry Hill, New Jersey exemplifies the use of porous aggregate surfacing in immediate proximity to a small building. Figure 6.7 shows the aggregate courtyard adjacent to an "underground" building with its floor level 7 feet below the ground. Architect Malcolm Wells installed the office and its courtyard in 1974 (Wells, 1974); as of 1996 it was occupied by real estate appraiser Michael J. Lenny.

The aggregate is crushed limestone about 1 inch in size. A visit 22 years after it was installed showed that, the aggregate surface had settled slightly relative to the building and pedestrian traffic had compacted a relatively flat-looking central track. The entire surface was very easy to walk on. The addition of an inch of new aggregate would be sufficient to restore the original surface level.

The porous surface infiltrates the rain falling directly into the courtyard, and additional water dripping slowly from the building's earth-covered "green roof" long after storms are past. The subgrade is sandy. Drip and shading make this a slightly moist environment, where some weeds sprout in the aggregate's voids and some moss and algae occupy aggregate particles in the particularly moist area beneath the roof's drip line. A surface drainage inlet is present, but has evidently never been called into use; the courtyard shows no evidence of standing surface water. Despite the integration of the courtyard with the structure, no complaints concerning foundation problems or interior moisture have been reported.

HERTY MALL, ATHENS, GEORGIA

Walkways on Herty Mall in Athens, Georgia exemplify the use of crushed-brick aggregate. The mall was constructed on the University of Georgia campus in 1999. Its central location draws heavy pedestrian traffic and occasional events. Figure 6.8 shows the aggregate walks separated from adjacent lawns and planting beds by raised granite curbs.

The red crushed brick came from a regional brick manufacturer who crushed scraps from the manufacturing process into aggregate (Scott Beveridge, personal communication 2002). The cost was comparable to that of other decorative aggregates; transportation from the source in North Carolina was a substantial part of the cost.

FIGURE 6.7 Aggregate courtyard at the Wells-Lenny Office in Cherry Hill, New Jersey.

The material was not sorted by size. It consists of angular particles up to about 3/8 inch in size. Despite the dense gradation the surface is permeable; most rainwater infiltrates into it. Compacted clay subgrade is the limiting layer for infiltration. Small surface drainage inlets along the paths receive runoff during prolonged rains that fill the material's void space to overflowing.

The loose aggregate surface does not invite rapid walking, so pedestrians tend to stay on dense concrete paths paralleling the plaza. Nevertheless, the traffic on the aggregate walkways is heavy enough to compact the center of each path and displace loose particles to the sides after a few days' heavy use.

The ends of the paths meet flush with dense concrete walkways. Particles of aggregate are commonly scattered onto the concrete, where they become an annoyance for

FIGURE 6.8 Crushed brick walkway on Herty Mall, Athens, Georgia.

walkers. Adding a gently sloping ramp at each intersection might confine the particles to their proper place in the plaza.

STEELHEAD PARK, LOS ANGELES, CALIFORNIA

Steelhead Park on Riverside Drive in Los Angeles, California exemplifies the use of decomposed granite. Steelhead is one of a string of "pocket parks" along the Los Angeles River Greenway sponsored by the Santa Monica Mountains Conservancy and other agencies. Other segments of the greenway built with similar materials are Zanja Madre Park, Rattlesnake Park, and the Anza Picnic Area (Gustaitis, 2001).

Figure 6.9 shows the park. The decomposed granite has a small top size and is rather dense-graded. The particles are angular. The material directly under the trees, where traffic is not expected, is unbound; in these areas the permeability is moderate.

The decomposed granite in the traffic areas, away from the trees, is bound by Stabilizer, a clear, organic binder. The permeability is lower than that in the unbound areas, but still positive. Visually, the stabilized and unstabilized areas are indistinguishable except that the surface of the stabilized material appears more uniformly level and packed-down than the unstabilized, untrafficked material.

POROUS AGGREGATE SURFACES IN RESIDENTIAL DRIVEWAYS

Residential driveways bear very light, slow-moving vehicular traffic. Unbound aggregate is very suited to this type of traffic. The low cost of aggregate keeps down

Porous Aggregate

FIGURE 6.9 Pedestrian area of decomposed granite partly stabilized with Stabilizer, Steelhead Park, Los Angeles, about one year after installation.

the price of housing. The following two examples were installed at very different times, and are of very different ages.

Medford, New Jersey

The driveway shown in Figure 6.10 was built in the mid-1990s in Medford Township, New Jersey of open-graded crushed limestone in order to comply with local ordinances limiting impervious cover and regulating stormwater runoff. The subgrade is sandy. The edging is constructed of rustic timber, in character with the well-preserved native woods around the residence. The driveway is easily able to infiltrate all of the annual precipitation of 40 inches.

FIGURE 6.10 Aggregate driveway at a residence in Medford Township, New Jersey, about four years after installation.

In the foreground is the open lawn-like environment of the cleared street right-of-way. The edge of the nearby woods makes an "ecotone" which harbors diverse species and invites weeds to sprout in the driveway aggregate. Automobile traffic suppresses the growth in the two wheel tracks. The woodland shade on the interior of the private lot suppresses weeds. In both settings the porous aggregate driveway infiltrates rainwater and aerates the soil as do the surrounding grass and forest floor.

GARDEN CITY, NEW YORK

The driveway shown in Figure 6.11 is in an older and more formal setting. It is located in Garden City, Long Island, New York. The aggregate is open-graded crushed stone about 1 inch in size. Light traffic has settled the material into a relatively even, walkable surface. Raised concrete curbs and a concrete apron crisply delineate the driveway and stabilize the edge.

Judging from the architecture of the house in the background, the driveway was installed about 100 years ago. During the years since then the surface has been occasionally raked for smoothing and leveling. From time to time lost aggregate has been replenished. The mostly sunny, dry setting has inhibited the development of weeds.

Beneath this driveway is the sandy soil of the aquifer from which most of Long Island draws its water supply. Although this driveway by itself is small, it occupies an area that could have been impervious, but is not. For 100 years the driveway has been infiltrating rainwater into the aquifer at the average rate of 40 inches per year.

Porous Aggregate   225

FIGURE 6.11   A residential driveway in Garden City, New York.

POROUS AGGREGATE SURFACES IN PARKING LOTS

Parking lots, bear widely varying amounts of vehicular traffic. Within a given parking lot, the traffic in dead-end parking stalls is always much less than that in the nearby driving lanes. The following examples of parking lots are within, or almost within, the low capacity traffic to which unbound aggregate surfacing is suited.

QUIET WATERS PARK, ANNAPOLIS, MARYLAND

An aggregate parking lot in Quiet Waters Park in Annapolis, Maryland exemplifies the use of aggregate under public vehicular traffic. The parking lot is located near the park's picnic grounds and performance area, so its use is limited mostly to weekends and events in the summer season. In the winter it is not plowed for snow removal. The same park has porous asphalt parking lots that bear greater amounts of traffic. The motivation for selecting porous surfaces was to protect water quality in the adjacent Chesapeake Bay. The park was built in 1990. Figure 6.12 shows the parking lot ten years later.

The surface slopes at a few percent. A central turning lane follows the topographic contour. Wheel stops define parking spaces on the lane's uphill and downhill sides. The size of the parking lot invites periodically abundant, fast-moving traffic.

The aggregate particles are hard, angular crushed limestone. The gradation is coarse and permeable. Most surface particles are 1/2 inch in size; a small portion are up to 1 inch and even 2 inches in size. The permeability is high.

The parking lot is in a mostly grassy setting with abundant sources of weed seeds at nearby hedgerows. Weeds germinate here and there in the aggregate.

FIGURE 6.12 Aggregate parking lot in Quiet Waters Park, Annapolis, Maryland, ten years after installation.

In the driving lane the particles are packed together into a relatively continuous surface. Shallow ruts formed by fast-moving, turning vehicles are visible. In the parking stalls the surface is comparatively coarse and open.

Over ten years the surface particles have become worn with their movement against one another: the angles are slightly rounded, and the surfaces are whitish as if repeatedly scratched and abraded. Particles below the surface, where there is less movement, are less worn.

As a whole the aggregate surface has "flowed" slightly from the central turning lane to the sides and downhill, forming banks of aggregate a few inches high at the downhill wheel stops, and lower banks at the uphill stops. The movement of gravel toward the edges could be counteracted with a little manual labor with shovels and rakes, if park personnel feel such an investment necessary.

LINCOLN STREET STUDIO, COLUMBUS, OHIO

Crushed vitrified clay forms an aggregate parking area at Lincoln Street Studio, an architectural office in Columbus, Ohio (www.lincolnstreet.com; Frank Elmer, personal communication 2001). The aggregate is recycled scrap material from a regional manufacturing company. The material is very inexpensive. The parking area was installed in 1995 or 1996. Figure 6.13 shows the place four or five years after installation.

The parking lot is located in a narrow alley-like setting in an old part of Columbus. Eight diagonal parking stalls line a driving lane of dense asphalt. A planting bed's edge configuration defines individual stalls. The aggregate material in the parking stalls is separated from the planting bed by a metal strip and from the asphalt lane by a 36-inch-wide concrete band with jointing that parallels the diagonal parking arrangement.

The aggregate material contains 1-inch particles and a substantial quantity of particles 1/16 inch and smaller. The material is permeable to a degree despite its relatively dense gradation. Below the surface it is dark from the moisture held in its small pores. The aggregate surface is loose and open except in the compacted wheel tracks, where aligned particles form a more or less smooth, tight surface. The surface is sometimes raked after displacement by trucks that visit the construction company that shares the building.

The original installation of aggregate was only 3 inches thick. After a couple of years, another inch of identical material was added to compensate for the material's settlement into the frost-susceptible clay subgrade (the nominal frost depth here is 30 inches).

FIGURE 6.13 Parking stalls of crushed vitrified clay at Lincoln Street Studio, Columbus, Ohio.

There has been no standing water in the aggregate area, even during the very rainy May and June of 2001 and despite the absence of storm drains in this old part of the city.

RIVERBEND EAST, ATHENS, GEORGIA

Figure 6.14 shows unbound aggregate in a residential parking lot. This is a small parking expansion at the Riverbend East condominium in Athens, Georgia. The setting is woody and informal despite the moderate density of housing. A dense asphalt driving lane had been built with the rest of the development in the mid-1970s. Unplanned overflow parking had been occurring at this unpaved spot for many years, compacting the soil and forming muddy ruts. The addition in around

FIGURE 6.14 A retrofit parking expansion of unbound aggregate in the Riverbend East residential condominium in Athens, Georgia.

1990 of open-graded aggregate over the soil stabilized and drained the parking surface. The aggregate is No. 57 crushed granitic rock (a slightly smaller particle size would arguably have been adequate). The edging is of timbers. Fallen pine needles line the periphery of the aggregate with quickly decomposing natural mulch. After 10 years of continued light vehicular use of the spot, no displacement or compaction of the aggregate was visible, nor was there any raveling at the adjoining asphalt edge.

POROUS AGGREGATE SURFACES IN WORK AND EQUIPMENT PLATFORMS

Commercial and industrial enterprises require well-drained, stable, economical surfaces for outdoor work and placement of equipment. Where the traffic is light, unbound porous aggregate is highly suitable.

ELECTRIC SUBSTATION

Figure 6.15 shows a typical electric substation of the Georgia Power Company. A thick layer of porous aggregate gives workmen and their trucks access to large electrical equipment about once per week. The aggregate is crushed granitic rock. The surface is very well drained, stable under very heavy occasional loads, and nearly maintenance-free. The electrical equipment is placed on concrete footings surrounded by the aggregate. Hundreds of substations in the company's network have the same type of flooring.

PLANT NURSERIES

Figure 6.16 shows an outdoor work area at Strader's Garden Center in Grove City, Ohio. The surface is No. 57 crushed limestone. The coarse-textured, well-drained surface is used as a work platform by company employees. (A smaller particle size would arguably have made walking more comfortable.)

Figure 6.17 shows a platform for outdoor retail display at Cofer's nursery in Athens, Georgia. The nursery uses the large platform for seasonally changing displays of plants. The surface is No. 89 crushed granitic rock. It is extremely well drained despite heavy natural rains and abundant irrigation. There is essentially never any surface ponding or runoff. It is level and firm enough for rolling the small wheels of shopping carts.

POROUS AGGREGATE SURFACES IN PLAYGROUNDS

Playgrounds require soft surfaces of rounded aggregate particles to absorb falls. This application is one of the very few exceptions to the requirement that aggregate particles be angular for stability under traffic. In a playground, nearly spherical particles are desirable; they roll against each other without interlock, yielding under pressure to cushion a child's fall.

ASTM F 1292 defines a test for evaluating the shock-absorbing properties of playground surfacing materials. For a given material it estimates a "critical height,"

FIGURE 6.15 Aggregate surfacing in an electric substation.

representing the maximum height from which a child could fall onto the surface without the likelihood of a life-threatening head injury (Consumer Products Safety Commission, no date). The surfacing material used around a given piece of playground equipment should have a critical height at least as great as the height of the highest accessible part of the equipment.

The shock-absorbing capacity of rounded unbound aggregate depends on its loose particulate character. It should not be installed over a hard base such as asphalt or concrete. It may require periodic raking to maintain appropriate thickness, and sifting to remove foreign matter. Figure 6.18 shows the gradations of two materials tested for critical height by the Consumer Products Safety Commission. Both were somewhat open-graded. Table 6.12 lists the critical heights of the materials. According to the commission's results, critical height increases with depth of the

Porous Aggregate 231

FIGURE 6.16 Aggregate work area at a plant nursery in Grove City, Ohio.

aggregate layer, but it can decrease with compaction resulting from construction or prolonged use.

VANDERGRIFT PARK, ARLINGTON, TEXAS

Figure 6.19 shows rounded aggregate at Vandergrift Park in Arlington, Texas. The particles are small river gravel. That the aggregate has yielded under pressure is visibly evidenced by the depression at the impact point at the bottom of the slide. Park personnel rake the aggregate from time to time to relevel the surface. The aggregate surface is edged by a concrete curb.

The subgrade soil is the notoriously swelling Houston black clay. Although the highly permeable playground surface invites cycles of drying and moistening in the

FIGURE 6.17 Aggregate display area at a plant nursery in Athens, Georgia.

soil, the resulting heaving is never noticed in this naturally uneven and dynamic surface. Scheduled raking relevels the surface whether or not heaving has occurred.

BARNES HOSPITAL, ST. LOUIS, MISSOURI

Figure 6.20 shows an analogous installation at Barnes Hospital in St. Louis, Missouri. The playground is located on a roof garden over a parking garage. Slightly rounded pebbles about 1/4 inch to 3/8 inch in size make up the playground surface. Play equipment rests on footings below the gravel surface. The gravel is so yielding under pressure that it is hard to walk on. The gravel surface is edged in many places with a concrete sidewalk which forms a raised curb, and elsewhere by walls of timber and concrete. The edging essentially eliminates loss of gravel by displacement.

Porous Aggregate

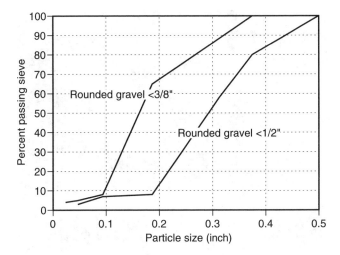

FIGURE 6.18 Gradations of two aggregates tested for critical height (data from Consumer Products Safety Commission, no date).

TABLE 6.12
Critical Heights (in feet) of Two Aggregate Materials

Material	Uncompacted Layer 6 inch Thick	Uncompacted Layer 9 inch Thick	Uncompacted Layer 12 inch Thick	Compacted Layer 9 inch Thick
Rounded gravel <3/8 inch	6	7	10	6
Rounded gravel <1/2 inch	5	5	6	5

Consumer Products Safety Commission, (no date).

Rainwater that percolates through the gravel to the garage roof is drained by the structure's system of inlets and pipes.

POROUS AGGREGATE BASE COURSES AND RESERVOIRS

A porous base reservoir simultaneously bears a structural load and conveys and stores water while it discharges (Nichols, 1991, pp. 15-20–15-23). Aggregate is the almost universal choice of material for base courses, because of its low cost and, when properly selected, its strength and durability.

ASTM D 2940 specifies aggregate for use in base courses, but is limited to dense-graded material which is useless in most porous pavements.

In porous pavements a base reservoir must be of clean, open-graded aggregate to produce porosity and permeability. A common open-graded gradation for base reservoirs is ASTM No. 57, which ordinarily has a porosity of 30 to 40 percent and very high permeability. Specifying that the material be washed eliminates clinging fine particles that would reduce the effective void space or clog the subgrade surface.

FIGURE 6.19 Playground surface of rounded aggregate in Vandergrift Park, Arlington, Texas.

Acceptable minimum CBR for base materials is ordinarily 40 or 50 for pedestrian and light vehicular loads, increasing to 80 for heavy vehicular traffic.

Aggregate base courses should be compacted during placement, even where the underlying subgrade is not compacted. Compaction of base material prevents settling or shifting just below the surface layer. Compaction in a layer 6 inches deep or greater tends not to influence the subgrade; the layer spreads the compactive load over a large area of subgrade. Cahill (www.thcahill.com) commonly specifies only light compaction of open-graded base material.

There has been speculation about whether it is necessary to "choke" an open-graded base course at intervals throughout its thickness with layers of smaller particles in order to prevent shifting and deformation. For example, this idea was the rationale for a layer of choking stone in the middle of the base course at the

Porous Aggregate 235

FIGURE 6.20 Playground at Barnes Hospital, St. Louis, Missouri.

University of Delaware's porous asphalt parking lot, which will be described in Chapter 12. This question was resolved at the Walden Pond porous asphalt parking lot (also described in Chapter 12), where several experimental base courses were installed with intermediate choking layers, and several without them. In-place structural tests found that the installations with intermediate choking layers were no stronger than those without them (Eaton and Marzbanian, 1980). Uniformly sized, open-graded aggregate buried in a base course is stable by virtue of the interlock of similar particles on all sides.

Blending sand into open-graded aggregate has sometimes been recommended for porous pavement base courses, for a variety of reasons (Bohnhoff, 1999; Chere Peterson, Petrus UTR, personal communication 2002). The recommended amount of sand is sufficient to fill or nearly fill the voids in open-graded material. The addition of sand or soil can assist rooting of grass or trees as described in Chapters 5 and 7. It probably enhances filtration of percolating water for water quality improvement, because it reduces pore size and increases internal surface area. Whether it increases bearing capacity is debatable; while blending large amounts of sand, uniform blending is difficult, and pockets of pure sand are difficult to avoid. The addition certainly reduces porosity and permeability.

POROUS AGGREGATE FILTER LAYERS AND USE OF GEOTEXTILES

Soil or small aggregate can move into the void space of open-graded aggregate if the encroaching particles are significantly smaller than those of the aggregate (Moulton,

1980, pp. 98–101). Encroachment reduces the porosity and permeability of the zone where the small particles accumulate, and may interfere with structural interlock between the large particles. In porous pavements this kind of encroachment is a hazard where a setting bed or surface layer of small aggregate overlies a base course of larger open-graded materials, and where fine-grained subgrade soil underlies an open-graded base course. To prevent internal erosion and encroachment, the materials placed adjacent to each other must be compatible.

Aggregate Layer

Before installing a surface course on top of an open-graded base course, it is sometimes necessary to smooth and stabilize the top of the base course by placing a setting bed or "choking" layer of finer open-graded material. The choking particles must be smaller than those of the base course, but not so small that they would fall through the voids. The choking material is spread over the surface in a layer a few inches thick and vibrated or rolled into the surface of the base material (Rollings and Rollings 1999, pp. 10–11).

Quantitative filter criteria have been established that evaluate the compatibility of adjacent pairs of material. A combination of criteria preserves permeability through the material while making voids small enough to prevent small particles from penetrating.

Rollings and Rollings (1999, pp. 10–13, 23–25) have reviewed aggregate criteria adopted by the U.S. Federal Highway Administration (Moulton, 1980) and other agencies. Figure 6.21 summarizes the criteria for a choke layer or setting bed overlying a

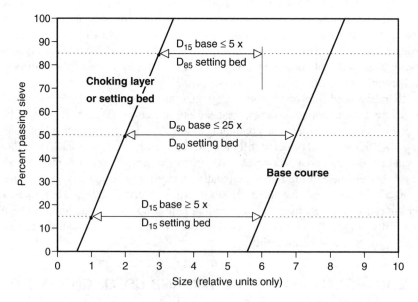

FIGURE 6.21 Filter criteria for choking layer or setting bed overlying a base course of larger aggregate (data from Rollings and Rollings 1999, p. 13).

base coarse of large aggregate. In the figure, D_{15} is the particle size of which 15 percent of the material is finer, D_{50} the diameter of which 50 percent is finer, and D_{85} the diameter of which 85 percent is finer. The horizontal arrows connecting the choking later gradation to that of the base course represent the three required filter criteria.

The D_{15}/D_{15} criterion (D_{15} of base $\geq 5 \times D_{15}$ of setting bed) ensures that the permeability of the base course is much higher than that of the overlying layer, so that no backup occurs in the upper layer. The other two criteria ensure that particles from the upper layer cannot migrate into the voids of the base course. In a complex pavement structure multiple layers of aggregate may be required, each layer satisfying filter criteria with the layers around it.

For example, ASTM No. 8 aggregate meets the filter criteria as choke stone over No. 57 base course (Rollings and Rollings, 1999, pp. 10–11). If No. 10 is to be used for a setting bed, it is too small to be placed directly on No. 57, so an intermediate filter layer of No. 8 or 89 would first have to be placed on the No. 57 base.

Table 6.13 lists maximum and minimum D_{15}, D_{50} and D_{85} of open-graded ASTM gradations. These figures can be used to evaluate whether one of these gradations might meet the filter criteria when placed adjacent to any other aggregate.

Where an aggregate filter is to prevent movement of particles out through a perforated drainage pipe, analogous criteria apply. For a pipe with circular holes, D_{85} of the adjoining aggregate must be greater than the hole diameter. For a slotted pipe, D_{85} of the aggregate must be greater than 1.2 times the slot width.

Separation Geotextile

A geotextile placed between aggregate and the surrounding soil can prevent the soil from migrating into the aggregate voids and reducing storage capacity and the stone-to-stone interlock that gives aggregate its strength. Table 6.14 lists AASHTO's

TABLE 6.13
Characteristics of ASTM Gradations Necessary for Evaluation with Filter Criteria

	Particle size (inch)		
ASTM Gradation	D_{15} (Size at which 15 Percent by Weight of the Material is Smaller)	D_{85} (Size at Which 85 Percent by Weight of the Material is Smaller)	D_{50} (Size at Which 50 Percent by Weight of the Material is Smaller)
No. 5	0.528 to 0.688	0.917 to 0.982	0.722 to 0.857
No. 57	0.388 to 0.450	0.812 to 0.929	0.437 to 0.679
No. 6	0.375 to 0.469	0.667 to 0.732	0.484 to 0.607
No. 67	0.208 to 0.328	0.625 to 0.812	0.354 to 0.562
No. 7	0.187 to 0.258	0.438 to 0.487	0.307 to 0.400
No. 78	0.125 to 0.241	0.425 to 0.487	0.281 to 0.400
No. 8	0.117 to 0.200	0.335 to 0.375	0.241 to 0.287
No. 89	0.059 to 0.156	0.312 to 0.362	0.168 to 0.268
No. 9	0.055 to 0.100	0.164 to 0.187	0.109 to 0.143
No. 10	< 0.0059 to 0.018	0.148 to 0.187	0.058 to 0.102

Calculated from data in ASTM D 448.

TABLE 6.14
Filter Criteria for Geotextiles

Soil to be Separated Out	Geotextile Opening Size for Which 95 Percent of the Holes are Smaller (O_{95})	Geotextile Apparent Opening Size (AOS)
Relatively coarse-grained soil ($\geq$50 percent passing No. 200 sieve)	< 0.59 mm	$\geq$ No. 30 sieve
Relatively fine-grained soils (>50 percent passing No. 200 sieve)	< 0.30 mm	$\geq$ No. 50 sieve

AASHTO (1991), cited in Rollings and Rollings (1999), p. 13.

TABLE 6.15
Holders of Registered Trademarks Mentioned in This Chapter

Registered Trademark	Holder	Headquarters Location
Arkalite	Arkalite	West Memphis, Arkansas
Buildex	Buildex	Ottawa, Kansas
Carolina Stalite	Carolina Stalite	Salisbury, North Carolina
Envirotac	Environmental Products and Applications	Palm Desert, California
Gravelite	Big River Industries	Alpharetta, Georgia
HydRocks	Garick	Cleveland, Ohio
Lehigh Cement	Lehigh Cement	Union Bridge, Maryland
Lelite	Lehigh Cement	Union Bridge, Maryland
Norlite	Norlite	Cohoes, New York
Northeast Solite	Northeast Solite	Saugerties, New York
PaveCryl	Rohm and Haas	Philadelphia, Pennsylvania
Penzsuppress	American Refining Group	Bradford, Pennsylvania
PermaTill	Carolina Stalite	Salisbury, North Carolina
PolyPavement	PolyPavement	Los Angeles, California
Profile	Profile Products	Buffalo Grove, Illinois
Road Oyl (Resin Pavement)	Soil Stabilization Products	Merced, California
Sierra Cascade	Sierra Cascade	Chemult, Oregon
Soil-Sement	Midwest Industrial Supply	Canton, Ohio
Solite	Solite	Richmond, Virginia
Stabilizer	Stabilizer Solutions	Phoenix, Arizona
Stalite	Carolina Stalite	Salisbury, North Carolina
T-NAPS	George L. Throop Co.	Pasadena, California
Tri-Texas Industries	Tri-Texas Industries	Dallas, Texas

filter criteria for separation geotextiles. In certain circumstances, a geotextile also provides some tensile strength that inhibits deformation in the pavement layer and helps distribute traffic load onto the underlying subgrade, as described in Chapter 3.

TRADEMARKS

Table 6.15 lists the holders of registered trademarks mentioned in this chapter.

ACKNOWLEDGMENTS

A 1996 fellowship from the Landscape Architecture Foundation supported travel to the mid-Atlantic area in 1996, which led to the recognition of unbound aggregate as a valid stand-alone porous paving material. During that trip Michael J. Lenny of Cherry Hill, New Jersey cordially permitted the author to visit his office during his absence.

The following persons generously provided information used in this chapter: Alphonse Anderson of Hanson Aggregates (Athens, Georgia); Scott Beveridge, University of Georgia Physical Plant; Steven Bopp of Austin Tao and Associates, St. Louis; Jim Dykes of Dykes Paving (www.dykespaving.com); Frank Elmer FAIA of Lincoln Street Studio (www.lincolnstreet.com); George Howard and Charles Machemiel of the Georgia Crushed Stone Association; Tom Masten, Solite (www.solitecorp.com); Byron McCulley of Amphion Environmental, Oakland, California; Chere Peterson of Petrus UTR (www.petrusutr.com); Charles Pryor of the National Stone, Sand and Gravel Association (www.nssga.org); Bob Randolph and Samuel Randolph of Soil Stabilization Products (www.sspco.com); Robyn Ritting of the Expanded Shale, Clay and Slate Institute (www.escsi.org); staff of APAC — Pan American Construction (Miami, Florida); Bruce Thompson, USDA-NRCS, Amherst, Massachusetts; Kim Thompson, Conrad Yelvington Distributors, Orlando, Florida; Peter Tourtellotte, Orchard Nursery (www.orchardnursery.com); Diep Tu of Florida Concrete and Products Association (www.fcpa.org); Malcolm Wells, architect, formerly of Cherry Hills, New Jersey; Scott Wilson of North-East Trees, Los Angeles, California (www.northeasttrees.org).

The following persons constructively critiqued early drafts of this chapter: Tom Masten of Solite (www.solitecorp.com); Charles Pryor of the National Stone, Sand and Gravel Association (www.nssga.org); and John Ries of the Expanded Shale, Clay and Slate Institute (www.escsi.org).

REFERENCES

American Association of State Highway and Transportation Officials (AASHTO) (1985). *Materials*, Washington, D.C.: American Association of State Highway and Transportation Officials.

American Association of State Highway and Transportation Officials (AASHTO) (1993). *AASHTO Guide for Design of Pavement Structures 1993*, Washington: American Association of State Highway and Transportation Officials.

Barksdale, Richard D. (1991). *The Aggregate Handbook*, Washington: National Stone Association.

Bohnhoff, William (1999). *Porous Paving Design Compared to Impervious Pavement Standards*, Aurora, Colorado: Invisible Structures (www.invisiblestructures.com).

Bolen, Wallace P. (2002). *Construction Sand and Gravel — 2002 Annual Review*, Mineral Industry Surveys, Reston: U.S. Geological Survey.

Consumer Products Safety Commission (no date). *Playground Surfacing Materials*, Document No. 1005, Washington: Consumer Products Safety Commission (downloadable at www.kidsource.com).

Eaton, Robert A., and Peter C. Marzbanian (1980). *Structural Evaluation of Porous Pavement Test Sections at Walden Pond State Reservation, Concord, Massachusetts*, Special

Report 80–39, Hanover, New Hampshire: U.S. Army Cold Regions Research and Engineering Laboratory.

Ferguson, Bruce K. (1992). Soils and Aggregates, in *Materials, Volume 4 of Handbook of Landscape Architectural Construction,* pp. 1–18, Scott S. Weinberg and Gregg A. Coyle, eds., Washington: Landscape Architecture Foundation.

Ferguson, Bruce K. (1994). *Stormwater Infiltration,* Boca Raton: Lewis Publishers.

Gourley, Elizabeth (2001). Paving Alternatives, *Landscape Architecture* 91, 24–26 and 99.

Gustaitis, Rasa (2001). Los Angeles River Revival, *California Coast & Ocean* 17, 2–14.

Hammatt, Heather (2002). Retreating through History, *Landscape Architecture* 92, 64–71 and 89–90

Huhta, Richard S. (1991). Introduction to the Aggregate Industry, in *The Aggregate Handbook*, Chapter 1, Richard D. Barksdale, Ed., Washington: National Stone Association.

Keating, Janis (2001). Porous Pavement, *Stormwater* 2, 30–37.

Keating, Janis (2003). Chemical Soil Stabilization, *Erosion Control* 10, 40–47.

Moulton, Lyle K. (1980). *Highway Subdrainage Design,* Report No. FHWA-TS-80-224, Washington: U.S. Federal Highway Administration.

Nichols, Frank P., Jr. (1991). Specifications, Standards, and Guidelines for Aggregate Base Course and Pavement Construction, in *The Aggregate Handbook*, Chapter 15, Richard D. Barksdale, Ed., Washington: National Stone Association.

Rollings, Marian P, and Raymond S. Rollings, Jr. (1996). *Geotechnical Materials in Construction,* New York: McGraw-Hill.

Rollings, Raymond S., and Marian P. Rollings. (1999). *SF-Rima, A Permeable Paving Stone System,* Mississauga, Ontario: SF Concrete Technology.

Shergold, F.A. (1953). The Percentage Voids in Compacted Gravel as a Measure of Its Angularity, *Magazine of Road Research* 5, 3–10.

Tepordei, Valentin (2002) *Crushed Stone — 2002 Annual Review,* Mineral Industry Surveys, Reston: U.S. Geological Survey.

Thompson, J. William (2001). Roots Over the River, *Landscape Architecture* 91, 62–69.

Thompson, Marshall R. (1991). Aggregate as a Structural Product, in *The Aggregate Handbook*, chapter 11, Richard D. Barksdale, Ed., Washington: National Stone Association.

Tian, Ping, Musharraf M. Zaman, and Joakim G. Laguros. (1998). Variation of Resilient Modulus of Aggregate Base and Its Influence on Pavement Performance, *Journal of Testing and Evaluation* 26, 329–335.

Uzan, Jacob. (1999). Granular Material Characterization for Mechanistic Pavement Design, *Journal of Transportation Engineering* 125, 108–113.

Wagner, David L. (1991). Decomposed Granite, *California Geology* 44, 243–249.

Webster, L. F. (1997). *The Wiley Dictionary of Civil Engineering and Construction,* New York: John Wiley and Sons.

Wells, Malcolm B. (1974). An Underground Office, *Progressive Architecture* 55, 112–113.

White, Thomas D., Aggregate as a Component of Portland Cement and Asphalt Concrete, in *The Aggregate Handbook*, Chapter 13, Richard D. Barksdale, Ed., Washington: National Stone Association.

7 Porous Turf

Ronald B. Sawhill

CONTENTS

Relationship between Turf and Traffic 243
Turfgrass Species and Varieties 245
    Warm-Season Grasses 246
        Bahia Grass 246
        Bermuda Grass 246
        Buffalo Grass 247
        Centipede Grass 247
        St. Augustine Grass 247
        Seashore Paspalum 247
        Zoysia Grasses 248
    Cool-Season Grasses 248
        Bentgrasses 249
        Bluegrasses 249
        Tall Fescue 249
        Perennial Ryegrass 250
Turf Construction 250
    The USGA Profile 250
    The California Profile 251
    Sand-Based Root Zone Mixtures 252
    Turfgrass Installation 254
    Base Course for Drainage, Stability, and Rooting 255
Turf Maintenance 257
    Irrigation 257
    Mowing 258
    Fertilization 259
    Aeration 259
    Topdressing 260
    Overseeding 260
    Insect Control 261
    Disease Control 261
    Weed Control 261
    Traffic Control 261
    Snow Removal 262
    Combined Effect of Maintenance Operations 262

Turf Infiltration Rate 262
    Infiltration Rate in Sand-Based Root Zones 263
    Infiltration Rate in Fine-Textured Root Zones 264
Unreinforced Turf for Pedestrian Traffic 264
    Missouri Botanical Garden 265
    Scott Amphitheater, Swarthmore College, Pennsylvania 265
    Amphitheaters in Other Climates 266
Unreinforced Turf for Vehicular Traffic 268
    Big Top Flea Market, Florida 268
    Mall of Georgia 269
    Muirfield Village, Ohio 271
Turf Reinforcement 272
    Overlaid Meshes 272
    Embedded Mats 274
    Integral Fibers 275
Reinforced Turf for Vehicular Traffic 277
    Lino Lakes State Bank, Minnesota 277
    Troy Burne Golf Club, Wisconsin 278
    Urban Ventures Soccer Field, Minnesota 278
Trademarks 280
Acknowledgments 280
References 281

Living turf is a pavement, or a pavement surrogate, in places where it is used for the purpose of bearing traffic. Turf can be selected, designed, and installed to meet limited traffic requirements in a stable and durable manner, while it simultaneously maintains the health and appearance of living grass. For this, the surface soil layer must be a viable rooting medium, and should be able to withstand a compactive traffic load without sacrificing the porosity and permeability that are essential to turf's continued health.

The subject of this chapter is uniform turf growing in a surface soil, with or without synthetic reinforcement in or on the soil medium. Turf is also a component in the composite surfaces of geocell, block, and grid pavements, which support traffic through their solid ribs while protecting turf in their joints or cells. Those composite surfaces are covered in later chapters. This chapter supports them by introducing the turf component.

Table 7.1 distinguishes alternative structural conditions in turf and composite surfaces. An unreinforced turf surface has the structural benefits only of a careful selection of grass, planting medium, and drainage; it can bear the least traffic. A reinforced turf surface bears traffic equally directly, but contains synthetic reinforcement that assists the turf in resisting wear and compaction, so it can bear somewhat heavier traffic. A protected turf surface is located slightly below the level of a structural grid, block, or geocell, so it does not directly bear the full traffic load; the composite surface can carry still heavier traffic. In all structural conditions, living grass roots add some tensile strength, resisting horizontal and vertical movement with flexible, resilient anchoring.

TABLE 7.1
Alternative Turf Structural Conditions

Condition	Turf Surface Position	Surface Components
Unreinforced turf	Directly contacted by traffic	Grass and soil only
Reinforced turf	Directly contacted by traffic	Grass, soil, and synthetic reinforcement
Protected turf	Below top of protective structure	Grass, soil, and protective structure such as geocell, open-celled grid, or open-jointed block

Based on distinctions made by Samuel Randolph, personal communications (2002 and 2004).

Turf's living dynamism distinguishes it from all other traffic-bearing surfaces. Turf absorbs carbon dioxide from the atmosphere and emits oxygen. It actively cools by transpiration; no other surfacing material can compete with it in counteracting the urban heat island (Asaeda and Ca, 2000). It has the appearance of "green space," making it look as if no pavement were present. It captures migrant organic and inorganic particles and builds them into the soil structure. It reduces glare by absorbing light, and noise by absorbing sound. Healthy turf is inherently porous and permeable, so it increases stormwater infiltration and reduces runoff. The turf ecosystem captures and treats infiltrating pollutants.

An inherent cost of all living turf is that it must be regularly maintained. Maintenance routines can include mowing, irrigation, fertilization, aeration, topdressing, overseeding, and managing of insects and diseases.

Traffic-bearing turf technology has grown greatly in recent decades, driven originally by developments in golf greens and sports fields. It is aimed at supporting healthy, uniform grass while resisting traffic damage. Applications to pedestrian and vehicular loads benefit from the demanding specifications previously established in golf and sports venues. Table 7.2 defines some terms with distinctive application to turf.

Those seeking basic background information in turf science and its contemporary applications are referred to manuals by Beard (2001), Puhalla et al. (2002), Turgeon (1999), and White (2001). Table 7.3 lists examples of sources of updated information on turfgrass selection, suppliers, installation, and maintenance.

RELATIONSHIP BETWEEN TURF AND TRAFFIC

Turf's vehicular applications have included seldom-used or overflow parking areas, emergency access lanes, golf-cart paths, helipads, and aircraft taxiways. For pedestrians, turf has been used in amphitheaters, public event venues, sports fields, golf courses, and garden pathways. A uniform turf surface can be smoother and more accessible than many composite surfaces involving grids or cells. When a turf surface is not in use for its primary traffic, it is available for multiple uses such as band practice, music concerts, and passive and active recreational activities.

Determining a site's prospective traffic level is essential to a decision to use turf surfacing, and to the selection of an appropriate variety of grass, soil profile, and reinforcing material.

Excessive traffic injures grass directly by tearing and crushing the foliage, and indirectly by compacting the root zone soil (Craul, 1992, pp. 219–220; Shearman,

TABLE 7.2
Some Terms with Distinctive Application to Turf

Term	Definition
Black layer	Layer of anaerobic organic matter just below a turf surface
Blade	Leaf of grass
Bunch grass	Grass that tends to grow in a clump without rapid vegetative spreading
Cool-season grass	Grass adapted to growth in cool weather, and subject to dormancy or decline in hot weather
Crown	Junction of stem and root in a grass plant
Cultivar	A plant variety developed by breeding or selection and having a designated name
Divot	Puncture in a turf surface and the severing of the turf from the underlying soil
Grass	Narrow-leafed herbaceous plants; or members of the Gramineae botanical family
Lawn	Ground covered with fine, closely mowed grass
Overseeding	Sowing of seed into existing turf
Plug	Small piece of turf or sod
Rhizome	Horizontal underground stem or branch that produces roots and shoots
Rootzone	Soil-like material for turfgrass rooting, or the layer in which it is placed
Shoot	Sending out of new growth
Sod	Turf, especially turf cut into pieces or sections for grassing a lawn
Spreading grass	Grass that grows horizontally by sending out rhizomes, stolons, or both
Sprig	Small division of grass used for propagation
Stolon (runner)	Horizontal above-ground stem or branch that produces roots and shoots
Thatch	A layer that accumulates on the soil surface, composed of grass roots and shoots intermingled with dead and decaying organic material
Topdressing	Application of granular material onto existing turf
Turf	Mowed vegetation and the stratum of intermingled soil, roots and shoots in which it is rooted
Turfgrass	A grass species or cultivar maintained as turf
Verdure	Green growing vegetation
Warm-season grass	Grass adapted to growth in warm weather, and subject to dormancy in cold weather
Wear	Injury directly effected upon grass by traffic, such as tearing, crushing and abrasion of blades, stems, crowns, or stolons

1999). The overall effects of both forms of injury are reduced turf growth, loss of density, and premature senescence.

Grass wear is the immediate result of excessive traffic: the abrasion, tearing, and stripping of foliar tissue. Crushed, torn tissue quickly desiccates and turns straw-brown; growth is inhibited. Tissue injuries invite insect and fungal attack; the thinning of ground cover invites weed invasion. A grass's resistance to wear injury seems to come as much from flexibility and resilience as from strength. Wear damage can be prevented and recovery effected by grass selection, correct maintenance for health and vigor, and traffic control. Carrow and Johnson (1989) found that golf carts caused less turf damage when turning at broad 22-foot radii than when making sharp reverse turns at the same speed. The results suggest that, in laying out parking areas, angle parking and aisle alignments that minimize turning could help minimize wear.

TABLE 7.3
Examples of Sources of Updated Information on Turfgrass Selection, Suppliers, Installation, and Maintenance

Name	Contact Information
American Lawns	www.american-lawns.com
Golf Course Superintendents Association	www.gcsaa.org
Grounds Maintenance Magazine	www.grounds-mag.com
Lawngrasses.com	www.lawngrasses.com
National Turfgrass Evaluation Program	www.ntep.org
Sports Turf Research Institute	www.stri.co.uk
Turfgrass Information Center	www.lib.msu.edu/tgif
Turfgrass Producers International	www.turfgrasssod.org
UC Guide to Healthy Lawns	www.ipm.ucdavis.edu
US Golf Association Green Section	www.usga.org/turf

Soil compaction is the longer-term or more chronic effect of excessive traffic. It reduces a soil's porosity, aeration, drainage, and root penetrability. Low oxygen and poor porosity and permeability inhibit root growth. With less root growth, foliar growth and quality decline. The above-ground visible signs of grass stress resulting from soil compaction can often be confused with those of other environmental stresses such as heat and drought. Compaction damage can be prevented and reversed by the construction of a compaction-resistant root zone, correct irrigation and aeration, and traffic control.

A grass's tolerance to traffic is a composite of its tolerance to wear and soil compaction. Spreading grasses send out rhizomes or stolons, which send new roots down while filling in bare spots in the turf carpet. The multiple root locations give some spreading grasses an advantage in recovering from traffic stress.

Where grass vegetation thins or completely fails, it exposes bare soil to surface crusting under further traffic and rainfall. Although a rainfall-crusted layer is less than one inch thick, its low permeability further deprives the underlying soil of water and air.

TURFGRASS SPECIES AND VARIETIES

A species is a botanically related unit of interbreeding. Varieties (cultivars) are produced by managed breeding and selection, and given designated names. The U.S. Department of Agriculture and related organizations have been assessing and breeding grasses for lawn use since the early twentieth century. There are over 10,000 grass species, about 50 of which are used in lawns; only a few important examples are discussed here.

Selection of a turfgrass must take into account microclimatic conditions such as temperature, moisture, and shade. Daily parking casts shade; different varieties of grass have different sensitivities to shade. Low areas and exposed hilltops may have different enough temperatures to influence turfgrass selection. Soil moisture can be influenced by topography, groundwater conditions, and the provision of soil drainage in turf installation.

WARM-SEASON GRASSES

Warm-season grasses are those that can thrive in warm weather, so they are used mostly in warm southern climates. In cold winter weather they go dormant, turning yellow-gray. They may also go dormant in dry summers if not irrigated. During dormancy they have little resilience to traffic stress. Winter-dormant warm-season grasses are often overseeded with cool-season, traffic-tolerant perennial ryegrass, to improve their seasonal appearance and traffic tolerance. Table 7.4 compares some warm-season grasses.

Bahia Grass

Bahia grass (*Paspalum notatum*) is native to South America. It spreads by rhizomes and grows well in sandy soils. Bahia grows rapidly, to the extent that it is considered an invasive plant in southern ecosystems. It thrives in abundant moisture and can stand brief periods of inundation, but is intolerant of cold. Frequent mowing stimulates dense cover. Three commercial varieties are Argentine, Pensacola, and Tifton 9. Most Argentine varieties have wide blades. Pensacola has narrower leaves; it is more tolerant of cold and drought. Tifton 9 has even finer blades.

Bermuda Grass

Bermuda grass (*Cynodon dactylon*) spreads quickly by both rhizomes and stolons. It can provide dense, high-quality cover, but under frequent traffic, requires large quantities of fertilizer to maintain its condition. The runners elongate rapidly, but require frequent mowing to form a thick continuous carpet. Bermuda grass survives a wide range of temperatures, soil types, and fertility, and is tolerant, to a degree, of salinity. It requires sun and thrives in hot weather as long as it has sufficient water; in drought it goes dormant.

When Bermuda is green, it is one of the most traffic-tolerant of turfgrasses. Among its many cultivars, some of the most wear-tolerant are Texturf 10, Texture 1F, NK 78098, Tiflawn, Ormond, and Tifway (Beard et al., 1979, pp. 24–26).

TABLE 7.4
Examples of Warm-Season Turfgrasses

Name	Form	Relative Traffic Tolerance when Green
Bahia grass	Spreading	Good
Bermuda grass	Spreading	Excellent
Buffalo grass	Spreading	Poor
Centipede	Spreading	Poor
St. Augustine	Spreading	Fair
Seashore paspalum	Spreading	Good
Zoysia grasses	Spreading	Fair: resistant to damage, but slow to recover

Traffic tolerances partly from Shearman (1999).

When Bermuda is dormant, however, it is very susceptible to traffic wear. Injured stolons are not replaced until growth resumes. Traffic is detrimental even during initial early-spring growth, because of the tenderness of the young shoots. Overseeding Bermuda grass during winter dormancy with cool-season ryegrass improves wear resistance and appearance. In the spring, Bermuda may be reseeded over the rye.

Buffalo Grass

Buffalo grass (*Buchloe dactyloides*), native to the western and southwestern U.S., spreads by stolons, forming a dense, deep sod. Like other warm-season grasses, it may benefit from winter overseeding to protect dormant stolons. Buffalo grass requires water during establishment, after which it is drought-resistant. It requires sun.

Buffalo grass seems to be intolerant of sandy soil, which would inhibit its use in sandy traffic-bearing root zone materials like those that will be discussed later in this chapter. If an adequate soil mixture can be found and traffic tolerance can be assessed, this water-conservative native will deserve increased use in traffic-bearing turf.

Centipede Grass

Centipede grass (*Eremochloa ophiuroides*) is native to southeastern Asia. In North America it is used in warm southern climates. Common centipede has long been commercially available. Other, newer varieties include Oaklawn, Tennessee, Hardy, and Tifblair. Centipede covers the ground with elongating stolons and wide blades, suppressing weeds and forming a thick, low carpet. Because of its low stature, it requires relatively little mowing.

Centipede is more shade-tolerant than Bermuda. Its traffic tolerance has not been rigorously studied; it is probably less tolerant than Bermuda and zoysia.

St. Augustine Grass

St. Augustine grass (*Stenotaphrum secundatum*) is native to America's Gulf Coast and the Caribbean. It forms a thick, dense carpet of stolons and wide leaf blades and is salt-tolerant, and moderately shade-tolerant. Its rapid growth suppresses pests and weeds; it is particularly weed-free in the shade where many weeds do not grow. St. Augustine is not as traffic-tolerant as Bermuda grass or zoysia, but it has been used successfully under infrequent traffic.

Seashore Paspalum

Seashore Paspalum (*Paspalum vaginatum*) is indigenous to the Gulf Coast, the Caribbean, and tropical coastal areas worldwide. Botanically, it is a cousin of Bahia grass. Unlike its cousin, it is supposed to be noninvasive.

Seashore Paspalum spreads by both rhizomes and stolons, forming a prostrate, dark green mat. It tends to go dormant in the fall later than other warm-season grasses, and to green out later in the spring. It thrives in moist soil but is also drought-tolerant where it has been properly irrigated to promote deep rooting. It is distinctively salt-tolerant: effluents and ocean water can be used to irrigate it, as long as salts are not allowed to accumulate.

Seashore Paspalum has numerous diverse naturally occurring varieties. Relatively wear-tolerant cultivars include 223, Boardwalk, Durban CC, EE1, Salam, SeaIsle 1, and SeaIsle 2000 (Trenholm et al., 2000). Research on Seashore Paspalum's traffic tolerance is ongoing.

Zoysia Grasses

Zoysia grasses were introduced into North America from southeastern Asia about the turn of the twentieth century. They spread slowly by both rhizomes and stolons, making dense turf. The blades of zoysia grasses tend to be narrow and stiff. They can be grown in many kinds of soils and are tolerant of salt, but intolerant of poor drainage. They are somewhat tolerant of shade and, being deep-rooted, respond slowly to drought and quickly to the return of adequate moisture. They are very resistant to wear damage, but recuperate slowly after injury happens. They can accumulate thatch quickly.

Korean velvet grass or Mascarene grass (*Zoysia tenuifolia*) is very fine-textured and very intolerant of cold.

Japanese (or Korean) lawn grass (*Zoysia japonica*) is a relatively coarse-textured Zoysia. It is more cold-hardy than other Zoysia species, so it is used a little farther north. Examples of varieties are Meyer, Belair, and El Toro.

COOL-SEASON GRASSES

Cool-season grasses thrive in cool weather and remain green through the winter, even under a blanket of snow. Hence they tend to be used in cool northern climates. They can take longer than warm-season grasses to develop roots fully and to stabilize sandy rootzones (three to five years, versus two to three years for warm-season grasses).

Some cool-season grasses can go dormant in summer drought; upon return of moisture they green up again. In warm summers they do not grow as densely as warm-season grasses, and may be susceptible to injury. They are commonly overseeded (with the same species) to increase density of cover. Reliable vigor and quality during hot, dry summers requires irrigation.

Table 7.5 compares some cool-season grasses.

TABLE 7.5
Examples of Cool-Season Turfgrasses

Name	Form	Relative Traffic Tolerance
Bentgrasses	Spreading	Good
Kentucky bluegrass	Spreading	Medium resistance, recovers quickly
Supina bluegrass	Spreading	Good; stolons aid rapid recovery
Tall fescue	Bunch	Excellent wear tolerance, but low compaction tolerance
Perennial rye grass	Bunch	Excellent

Traffic tolerances partly from Shearman (1999).

Bentgrasses

Bentgrasses are a group of species that include colonial bentgrass (*Agrostis tenuis*), creeping bentgrass (*Agrostis palustris*) and velvet bentgrass (*Agrostis canina*). Bentgrass spreads by stolons, forming a dense turf despite its characteristically thin blades. It is used throughout northern U.S., with greatest success in New England and the Pacific Northwest. With careful management, certain creeping bentgrass cultivars have also been used on golf greens in the South.

Bluegrasses

Bluegrasses are a group of spreading cool-season species.

Kentucky bluegrass (*Poa pratensis*) was imported from Europe during America's early colonial period. It is adapted to cool, humid areas of the U.S. It spreads by rhizomes. Kentucky bluegrass' many cultivars differ in their adaptation to shade, moisture, and other conditions. Some of the more traffic-tolerant are A-20, A20-6, A-20-6A, A-34, H-7, PSU-173, T-13, Sydsport, Trenton, Wabash, Cheri, and Glade (Minner et al., 1993). Some varieties are rather aggressive spreading grasses that can bear shade and wear. Some are tolerant of deicing salt. Cultivar selections for a specific project should be verified with local turf experts or agricultural extension services. It is common to plant a mixture of varieties together, so favorable varieties in the mixture can take advantage of site-specific conditions.

Supina bluegrass (*Poa supina*) was introduced relatively recently from Europe. It spreads rapidly by both rhizomes and stolons. It is tolerant of traffic, and recovers from wear quickly.

Tall Fescue

Tall fescue (*Festuca arundinacea*) is a coarse-textured bunch grass. It has been in the U.S. for nearly 200 years. It grows rapidly and requires frequent mowing in spring and fall. However, individual bunches tend not to persist for more than a few years; annual overseeding replaces declining bunches and maintains turf density. Tall fescue is considered an invasive plant in some native ecosystems. Except for mowing, it requires less maintenance than many other cool-season grasses and tolerates low fertility. It tolerates drought and can remain green into hot, dry weather longer than some other cool-season grasses.

This grass is adapted to a wide range of soil conditions. It thrives in moist (but well-drained) fine-textured soil; it is also drought-tolerant once established. On sandy soils, frequent, thorough irrigation is required to maintain density.

Tall fescue is shade-tolerant and in hot climates performs best when it receives afternoon shade. However, in heavy shade, conscientious maintenance is required to maintain density.

Tall fescue is very wear-tolerant, unlike most other cool-season grasses (Carrow, 1980; Shearman, 1999). Nevertheless, it is intolerant of soil compaction, and thins dramatically under prolonged traffic. Regular overseeding and soil aeration are used to maintain turf quality under traffic. Minner et al. (1993) gave high traffic-tolerance ratings to the varieties Rebel, Rebel II, Jaguar, Mesa, and Tribute.

Perennial Ryegrass

Perennial ryegrass (*Lolium perenne*) is native to Eurasia and has been carried around the globe. It is often used in mixtures that include Kentucky bluegrass or fescue. This grass is also used for overseeding winter-dormant warm-season turfs. It has hundreds of cultivars.

Perennial ryegrass has an excellent track record under traffic, having an excellent combination of wear and compaction stress tolerance (Shearman, 1999). Minner et al. (1993) rated the varieties Diplomat, SR4000, Gator, Ovation, and Prelude highly for traffic tolerance. Although perennial ryegrass is generally considered a bunch grass, it sometimes becomes stoloniferous, and this may enhance its traffic tolerance.

TURF CONSTRUCTION

The development of contemporary turf construction owes much to the golf industry. Golf courses are stringent testing grounds for traffic-bearing turf: many public golf greens bear 50,000 to 140,000 pedestrians per year, while maintenance vehicles track over them five to seven times per week (Davis et al., 1990, p. 1). Wear and compaction stresses are high.

The ideas developed for golf applications provide for a series of layers of materials which together provide drainage, aeration, rooting habitat, and resistance to the effects of compaction. They are clearly specified and well supported by research and experience. They make a foundation for understanding the functions of contemporary traffic-bearing turf construction and a checklist for evaluating any site-specific proposal.

Prior to 1940, many recreational turfs were constructed with native soil planting media. Loamy soils made fine turfgrass growing media; the moisture-holding capacity of fine-textured soils compensated for the lack of irrigation. Under compaction, however, fine-textured poorly drained root zones deteriorated rapidly.

After World War II, the use of recreational facilities increased, bringing compaction problems to the fine-textured, poorly drained, often wet and plastic, planting media. At the same time, irrigation became more available, making the turf's water-holding capacity less important. Since the early 1950s, the United States Golf Association (USGA) and other organizations have been supporting research into new types of traffic-bearing turf media.

THE USGA PROFILE

In 1960, USGA and researchers at the Texas A&M University first codified the use of a notably sandy root zone supported by a drainage layer. This construction profile has now been used on golf greens around the world. It is commonly called the USGA or Texas-USGA method; USGA updates its specifications from time to time (USGA, 2004). Figure 7.1 summarizes the construction profile.

The surface layer is the principal rooting layer for grass and the direct support for traffic. The layer is commonly called a root zone; in pavement parlance it could be called a surface course. It is distinctively sandy, in order to maintain porosity and permeability under traffic, and to remain well-drained and aerated even in wet weather. Sand's characteristically low water-holding capacity is commonly

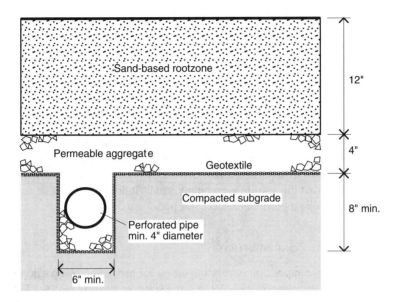

FIGURE 7.1 USGA turf construction profile (data from USGA, 2004).

compensated for by irrigation. Its low nutrient-holding capacity is compensated for by careful fertilization.

An underlying layer of permeable aggregate drains the root zone and conducts excess water to drainage pipes. Thorough drainage empties the root zone's pores for aeration, and prevents saturation that would weaken the root zone under load. The aggregate material is matched to a given project-specific root zone mixture so that the combination of materials meets filter criteria like those described in Chapter 6. The required aggregate may turn out to be a special specification, and not a standard gradation stockpiled by local aggregate producers. A geotextile fabric may be added as a separator between the drainage aggregate and the subgrade.

If the available root zone and aggregate materials do not meet filter criteria, an additional layer of coarse sand may be inserted between the drainage aggregate and the root zone to a depth of 2 to 4 inches. In conventional pavement parlance, this layer is called a "choke" layer or a "filter" layer; USGA merely calls it an "intermediate" layer. The material is selected to meet the filter criteria of both the root zone above and the drainage aggregate below.

The subgrade is compacted. Perforated drainage pipes are placed along the turf area's low edges, in any water-collecting depressions, and elsewhere throughout the turf area no more than 15 feet apart. They are installed in trenches cut into compacted subgrade deep enough to establish their drainage gradient.

THE CALIFORNIA PROFILE

Researchers at the University of California have been exploring sandy root zone media and turf construction profiles since the mid-1960s. Davis et al. (1990) and

Harivandi (1998) summarized the outcome. The approach is commonly called the "California" approach; it is applied in California and elsewhere as a relatively low-cost alternative to the USGA profile.

Figure 7.2 summarizes the construction profile. A sandy root zone rests directly on the subgrade. The sand is drained by entrenched pipes like those in the USGA profile. There is no intervening aggregate drainage layer; its omission reduces construction cost compared with the USGA profile. The profile's performance depends strongly on the rootzone layer and its composition, so the sand gradation must be selected carefully (Ferro and Otto, 2001). Uniformly coarse sands can result in a drought-like effect; very fine sands can hold too much water. To further reduce cost, the root zone has been made as shallow as 6 inches in some installations. Some practitioners speculate that this profile's simplicity makes it possible for small construction errors to have disproportionate effects on performance.

SAND-BASED ROOT ZONE MIXTURES

A correct root zone composition is very important for producing compaction resistance and thorough drainage while holding some water to support rooting. The vital component in contemporary root zones is open-graded (single-sized) sand. Open-graded sand has open voids for air and water circulation. Sand is nonplastic, so its voids tend to remain open under compaction, maintaining the root zone's infiltration, percolation, and aeration rates. The sand particles must be strong and durable; they must not decompose or be crushed under traffic.

Figure 7.3 shows the required gradations in the USGA and California root zone mixtures. The chart shows the range of quantity allowed from each size range; for example, USGA requires that 0 to 20 percent of the mixture be from the 0.15 to 0.25 mm range, and 60 to 100 percent from the 0.25 to 1.0 mm range; California requires that 82 to 100 percent be from the 0.1 to 1.0 mm range. Both mixtures emphasize

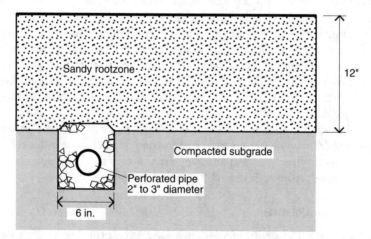

FIGURE 7.2 California turf construction profile (after Davis et al., 1990, Figure 2).

Porous Turf 253

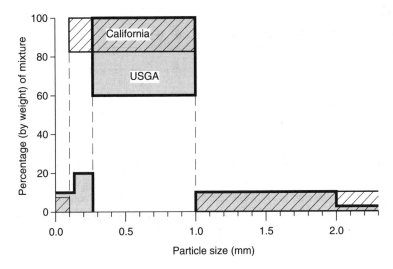

FIGURE 7.3 Allowable size ranges in the USGA and California root zone gradations (data from USGA, 2004, Table 3, and Davis et al., 1990, Table 2).

the 0.1 to 1.0 mm range, and allow very little material outside that range, so both of them can be considered open-graded (single-sized) gradations.

Material complying with the USGA specification is commonly referred to as "USGA sand." USGA allows the mixture to be made up by blending sand together with amendments such as peat, soil, compost, zeolite, or perlite from various sources, so long as the result has the specified gradation. A common mixture is made by blending sand, peat, and a topsoil of sand, loamy sand or sandy loam. Blending in limited amounts of fine-grained amendments can improve the mixture's water-holding capacity and nutrient exchange capacity. Any proposed mixture can be tested for compliance with USGA's specifications for physical characteristics and performance.

USGA (2004, Table 4) specifies root zone performance criteria which must be met simultaneously with the gradation criteria. The mixture must have a total porosity of at least 35 percent, and saturated hydraulic conductivity of at least 6 inches per hour. It is possible for a mixture that has an allowable gradation to have an unacceptably low infiltration rate, so it is essential for a prospective mixture to be tested. In effect, the gradation requirement is only a first screen for selecting mixtures for further laboratory testing; a selected mixture must comply with both the gradation and infiltration rate requirements. Table 7.6 lists tests standardized by ASTM (www.astm.org) for compliance with specifications such as those of USGA.

As an alternative, the California guideline emphasizes a root zone made of unamended sand, or "straight sand." In California, there are major deposits of aeolian sand readily available that produce the necessary properties. Adequate material is inexpensive and available from many producing companies, so California's root zone specification is aimed at finding the "right" sand, not at blending materials together. The idea is that carefully selected sand can have favorable water retention and infiltration rate, so amending it with organic matter or loam to improve these properties ought not to be

TABLE 7.6
Examples of ASTM Standards for Root Zone Testing

Number	Title
F 1632	Standard Test Method for Particle Size Analysis and Sand Shape Grading of Golf Course Putting Green and Sports Field Root zone Mixes
F 1647	Standard Test Methods for Organic Matter Content of Putting Green and Sports Turf Root Zone Mixes
F 1815	Standard Test Methods for Saturated Hydraulic Conductivity, Water Retention, Porosity, Particle Density, and Bulk Density of Putting Green and Sports Turf Root Zones

necessary. Amendments to improve ion exchange capacity are equally unnecessary, where low capacity is made up for with a balanced fertilization program. The California guideline does not specify a particular minimum infiltration rate, although it emphasizes the importance of an adequate infiltration rate (Davis et al., 1990, p. 4).

TURFGRASS INSTALLATION

There are three major methods for installation of grass on a root zone: laying sod, sprigging, and seeding. In all cases, a sample of the root zone material should be sent to an agricultural testing lab, and the soil treated with lime or fertilizer as recommended for turfgrass establishment.

Sod comes as a pregrown layer of grass, most typically with an adhering layer of soil up to about 1 inch thick. Sodding is the fastest method of establishing a nonerodible surface and a usable traffic-bearing turf. Establishment is a matter of extending the sod's roots into the root zone. Thin sod promotes particularly rapid rooting. When schedule permits, a specified cultivar can be grown at a sod farm to fit a project's requirements precisely. Sod can be grown with reinforcing materials embedded in its adhering soil. The attached soil must be at least as coarse-textured as the root zone upon which it will be placed; if the attached soil is too fine-textured, it may be inconsistent with the specification and purposes of the root zone. Although sod is more expensive than other planting methods, it is highly reliable in establishment, weed control, and subsequent turf quality. Turfgrass Producers International (1995) maintains standard specifications for sod preparation, fertilization, material, transplanting, installation, and maintenance.

An alternative is soil-free sod. It is soil-free as a result of washing, or of being grown in a soil-free medium (for example, the BAyr Root process, www.strathayr.com). Soil-free sod is light and thin; it can be installed in large units with few seams. Upon placement it tends to suffer relatively little root trauma, speeding its establishment in its new setting. It avoids the hazard of layering of sod soil upon a different underlying base soil.

Sprigging is used for spreading grasses where the cost of sod would be prohibitive, or where turf is being installed into the cells of open-celled grids. Establishment requires at least one full growing season following planting, during which the area should not be trafficked. If planting is done late in the season, the establishment period may need to extend into the next growing season. Maintenance during grow-in focuses on generating a complete cover; additional sprigs may be needed to fill in slow-growing patches.

Seeding is the least expensive and least reliable installation method. Seeds can be stored for much longer period of time than sprigs or sod. Seed sources should be chosen to avoid unsuitable levels of weed seed. Seeding is routine for fescue and buffalo grass. For other species, the number of varieties available in seed form is limited but growing. Seed can be placed by spreading, drilling, or hydroseeding. The major concern following spreading is to get the seed into proper contact with the soil, which can be achieved by rolling, raking, or pulling a steel mat.

Establishment commonly requires at least one full growing season. Seeding of cool-season grasses in the fall gives the turf a chance to establish its root system prior to experiencing heat stress. During germination and establishment traffic must be prohibited, and the surface must be protected from erosion by mulch or erosion-control blankets. Weed management is necessary at installation and throughout the establishment period. Reseeding may be necessary in thinly grassed patches, particularly for bunch grasses.

BASE COURSE FOR DRAINAGE, STABILITY, AND ROOTING

In pavement parlance, an aggregate layer like that specified by USGA is a base course. Like base courses elsewhere, a base course under a turf root zone can be designed to perform multiple vital functions.

In the USGA concept, the layer's vital function is root zone drainage. Drainage of excess water leaves the root zone's voids clear for aeration, and avoids the hazard of soil plasticity and softness. For effective drainage, the layer must be of permeable aggregate and at least several inches thick. It can be supplemented with water-storage chambers like those described in Chapter 4. Perforated pipe like that specified by USGA is necessary where the layer is not otherwise adequately drained.

The same layer can also be designed to enhance the profile's stability under traffic load. A base course's structural role is to distribute the load widely over the subgrade by increasing total profile thickness. Required thicknesses were discussed in Chapter 3. Some of the manufacturers of turf-reinforcing products (to be discussed later in this chapter) provide guidelines for layer thicknesses when using their products. For structural stability the layer must be made of durable, angular aggregate, and compacted in place.

The same layer can be designed simultaneously to enlarge the rooting volume and moisture reservoir, like the "structural soils" described in Chapter 5. Although structural soils' application to turf has not been researched as thoroughly as their application to trees, the same principles presumably apply. Structural soils are based on "skeletons" of durable single-size aggregate; the soil is blended into the aggregate's void space, where the aggregate skeleton protects it from compaction. The structural soils described in Chapter 5 for application to tree rooting had limited soil content so the soil would not over-fill the aggregate's void space, become compacted, and preclude aeration and drainage. However, for application to turf, the layer must support a sandy root zone, so the filling of voids to a greater extent may be justified to prevent migration of the root-zone material into the voids.

TABLE 7.7
Material Proportions in Turf Base-Course Structural Soil Suggested by Carolina Stalite

Item	Percent of Total Volume	Percent of Aggregate Volume
ESCS aggregate, 3/8 inch	75	100
Sandy loam, USGA specification	20	27
Compost	5	7

www.permatill.com.

Table 7.7 lists components for a turf base-course structural soil suggested by Carolina Stalite, one of the producers of ESCS aggregate discussed in Chapter 6. The producer listed material proportions as a percentage of the mixture's final volume; the table also shows those proportions as percent of the aggregate volume, because the soil materials are mixed into the aggregate's void space without an increase in total volume. The aggregate's suggested 3/8-inch gradation is open-graded; it has about 35 percent void space between particles plus internal void space in the particles. The soil materials add up to 34 percent of the aggregate volume, which comes close to filling the voids between particles. The aggregate producer believes that the mixture will still have 10 to 15 percent porosity, presumably much of which will be within aggregate particles. The mixture has been installed as a base course in emergency access lanes, under a one-inch layer of USGA root zone. If sod is placed over this type of base, the sod's attached soil should be sandy.

The ESCS-soil mixture is related conceptually to the ESCS mixtures for tree rooting mentioned in Chapter 5. The idea for both applications can be traced back to the work of James Patterson (1976; Patterson and Bates, 1994), who in 1971 rotary-tilled expanded-slate aggregate 8 inches deep into the turf surfaces of a playground and a softball outfield in Washington, DC. The sites' native silt loam and sandy loam soils became the soil components of the ESCS-soil mixtures. Although the aggregate amounted to only 20 to 33 percent of the soil volumes, after 22 years of pedestrian traffic the aggregate-soil mixtures had higher porosities and faster infiltration rates than untreated native soils, and were penetrated by more abundant grass roots. This result encouraged further experimentation, leading to the development of mixtures with higher aggregate contents, such as the one described in the preceding table and those described in Chapter 5.

Patterson's experiment mixed aggregate into a surface root zone to seek more stability, durability, and root viability. However, surface aggregate is susceptible to being kicked up by mowers or damaging mower blades and, in certain settings, may be hazardous to running or falling people. Nevertheless, to this day, the U.S. Federal Aviation Administration (1995, p. 130) provides for a surface of "aggregate-turf pavement" on landing areas used by light aircraft (up to 12,500 pounds). The mixture is intended to remain firm during wet weather (CBR of 20 or more) while viably supporting living grass roots. It is produced by blending open-graded aggregate into native soil. Unfortunately FAA does not specify the required proportions of aggregate and soil.

Under a structural-soil base course the subgrade, too, could be made available for additional rooting and infiltration by the scrupulous prevention of compaction during construction, and the omission of any geotextile at the bottom of the pavement structure.

TURF MAINTENANCE

Traffic-bearing turf requires regular maintenance as does any other lawn. The upkeep of a sandy, well-drained root zone and the counteraction of traffic stress require particularly conscientious maintenance. Potential management operations include irrigation, mowing, fertilization, aeration, topdressing, overseeding, disease control, insect control, weed control, and traffic control. Although these practices require ongoing expense, they are common and can be accomplished using available equipment and well-known techniques. Table 7.8 lists ASTM standard guides for turf maintenance.

Conscientious, informed turf maintenance is readily available in settings managed by parks departments and other organized landscape maintenance organizations. It is less reliable in settings maintained by street-sweeping crews.

Maintenance efforts should be most concentrated where traffic is greatest, and at times when the greatest traffic is expected.

To free turf surfaces for regular maintenance, they must be installed only in settings where they will be free of traffic on a predictable schedule. Examples are parking areas for stadiums, churches, movie theaters, and shopping centers. Turf can also be used in emergency access lanes, which are very seldom trafficked at all. In contrast, residential parking areas could have some traffic almost any hour of any day, so regular grass maintenance would be difficult, and turf surfacing should usually be avoided there.

IRRIGATION

A reliable water supply assures turf's greenness and uniformity. It compensates for the low water-holding capacity of sandy, well-drained root zones. It enhances traffic tolerance by assuring vigorous growth, thereby allowing the grass to recover from incipient wear (O'Malley, 2001).

Almost any turf requires irrigation during the establishment period: approximately the first two years after planting, before grass roots have grown deep enough to exploit the water-holding capacity of the entire rooting zone. During this period, temporary irrigation could be supplied when necessary by a truck or "hose-end" sprinkler system.

TABLE 7.8
Examples of ASTM Standards for Turf Maintenance

Number	Title
F 2060	Standard Guide for Maintaining Cool Season Turfgrasses on Athletic Fields
F 2269	Standard Guide for Maintaining Warm Season Turfgrasses on Athletic Fields

For long-term water supply, a permanent irrigation system distributes water through sprinkler heads or subsurface pipes. Permanent irrigation is likely to be required in warm climates with low or irregular rainfall in the growing season. It is less likely to be required in cool, reliably moist climates, especially where bases and subgrades are designed to provide moisture-holding capacity. Sandy root zones require more frequent watering cycles than fine-textured soils.

A permanent irrigation system can be significantly expensive to install and to operate (the cost was illustrated in a case study of the Mitchell Center Arena in Chapter 2). Whether to bear this cost is an important question and one that depends on the desired turf quality, and also whether turf will be selected at all in preference to nonliving surfacing alternatives.

Theoretically, a sandy root zone overlying a permeable aggregate layer, like that in the USGA profile, can hold suspended water in its small capillary pores, with excess water draining out only when the soil is saturated. Moisture stored in the root zone would support turf growth between irrigation cycles or rainfall events. Not all observations have confirmed however, that sandy root zones in fact suspend significant moisture at their interface with drainage layers (Prettyman and McCoy, 2003), so it is prudent to assume that a sandy root zone will require regular water replenishment.

Irrigation operations should be scheduled based on a turf's actual week-by-week water requirements. Too much water fills the root zone pores, preventing air circulation. Too little water can cause turf to wilt, making it susceptible to wear damage. Enough water should be applied at each operation to encourage deep rooting, reducing the significance of temporary surface drought. In dry climates, salts can accumulate in irrigated soils; periodic flushing can remove them.

Subsurface turf irrigation is a relatively new technology which may offer advantages compared with sprinkler irrigation. An example of a proprietary subsurface system is from Evaporative Control Systems (www.ecsgreen.com).

Mowing

All turfgrasses require routine mowing. A neat lawn may need to be mowed weekly in the growing season. A meadow-like setting may need to be mowed or harvested only two or three times per year. In the eighteenth century, lawns were kept short by scythes and grazing animals; mechanical mowing started in the early nineteenth century.

Lawn grasses can be mown beneficially because, unlike other plants, they grow from the base or crown, below mowing level. Each mowing removes part of the leaves, reducing the plants' ability to photosynthesize. The plants naturally compensate by producing additional leaves, shoots, stolons and rhizomes, thereby building a thicker, more continuous lawn surface.

High mowing leaves tall grass, which is relatively resistant to traffic stress (Youngner, 1962; Shearman, 1999). In tall grass verdure, rhizomes and stolons can develop fully. Thick verdure cushions the crowns, reduces wear injury, and to a degree insulates the soil from compaction. Frequent mowing encourages horizontal spreading. Mowing patterns should be altered from time to time in order to limit wear from repetitive wheel action.

FERTILIZATION

Turf nutrients need to be supplied, replenished, and maintained by correct fertilization. Balanced nutrition helps maintain turfgrass stress tolerance (Shearman, 1999). Sandy root zones have low nutrient (ion) exchange capacities, requiring particularly conscientious nutrient monitoring and maintenance.

Mayer (1998) recommended the following specific components of a nutrient management program on traffic-bearing sandy turf: (1) annual or semiannual agronomic soil testing, (2) fertilization schedule based on the test results, (3) fertilizer types and schedules that maintain fertility levels continuously rather than in surges, and (4) careful management of secondary and minor elements, because there is a very narrow range within which these elements are acceptable in sand profiles. A soil test is an extremely economical investment for planning correct fertilization. Experienced turf managers make a soil test at least once every other year, to plan fertilization in the coming year (O'Malley, 2001).

Nitrogen is the most abundant nutrient, and the most important for traffic tolerance. Insufficient nitrogen produces weak, sparse growth; excessive nitrogen produces tender, traffic-sensitive growth (Canaway, 1975). The ideal is a controlled nitrogen release that consistently provides neither more nor less than the turf actually needs.

Traffic tolerance is almost equally sensitive to potassium nutrition. Sand does not hold potassium well, so on sandy root zones potassium should be applied frequently in light doses or in controlled-release formulas.

Phosphorus is not as transient as other nutrients; it can be built up in the soil. Initial frequent applications can be performed to build up soil phosphorus levels.

AERATION

Root zone air circulation can decline with traffic compaction, thatch accumulation, blockage of pores by fine particles, and water-logging (Chong et al., 2001). Aeration is mechanical penetration of the thatch and soil to allow air circulation. The operation goes from 1 inch to 6 inches deep. It is a fundamental response to traffic compaction (Shearman, 1999). Well-aerated plants use water and fertilizer efficiently, grow vigorously, crowd out weeds, and outgrow disease.

Coring is probably the most effective type of operation. Coring removes narrow cores or plugs of sod, leaving holes that admit air into the thatch and soil and that can be filled with topdressing material. It breaks open slowly permeable thatch layers and readmits air to soil pores; it counteracts compaction, improves soil permeability, and hastens thatch decomposition. Subsequent mowing chops up the soil cores left lying on the surface.

Other types of aeration are slicing, spiking, and high-pressure injection.

Warm-season grasses should be aerated during the summer growing season; cool-season grasses should be aerated during their fall and early spring growing periods. Turf that bears significant traffic, or that becomes thatched, could be aerated twice per growing season, and more often where traffic is heavy.

Topdressing

Topdressing is a surface application of sand, soil or other granular substance. It encourages thatch decomposition, and hence aeration, by establishing a loose, aerated, soil-like environment where moisture is suspended and microbiota can decompose organic debris. It can simultaneously be used to smooth off slightly traffic-rutted areas. The topdressing material should be at least as coarse and open-graded as the root zone to which it is added, so that it does not retain surface water that would reduce aeration.

The material should be applied in small increments during the growing season, and not in thick layers. The amount of material applied at any one time should be less than 1/16 inch; after being settled by one irrigation operation, the turf should not appear to have been topdressed (Davis et al., 1990, p. 21).

Topdressing is often done following aeration to fill cores and slices with loose permeable material. It is also done in conjunction with overseeding to establish seed-to-soil contact. It can be added in further increments during growing periods.

Recent research has suggested the possibility of topdressing with crumb rubber: small rubber particles from shredded tires (Rogers et al., 1998). Crumb rubber is an alternative to angular sand grains, which have caused wear injury to near-surface portions of turf plants especially where traffic is intense. Round crumb-rubber particles cause much less tissue damage. High turf quality has come from the use of crumb-rubber particles between 0.05 and 2.0 mm in size, applied at 0.75 to 1.5 pounds per square foot. The presence of crumb rubber raises soil temperatures; this has improved turf growth in a cool northern climate, but it is not clear whether this would be a problem in warm southern climates. Crumb rubber will be described further in Chapter 13.

Overseeding

Overseeding protects turf from wear and assists its recovery from incipient wear by thickening the protective turf mat and supplying fresh growing grass that can bear further traffic. It is done as often as five times per year. It is preceded by a close mowing to obtain seed-to-soil contact, and often by aeration to roughen the soil surface and further increase seed-to-soil contact. It is followed by topdressing to obtain still more soil contact, and adequate irrigation to support seed germination.

In some cases, the overseeded grass matches the preexisting grass to thicken the lawn cover. For example, if cool-season Kentucky bluegrass appears stressed after a hot, dry summer, it can be overseeded with more bluegrass in the autumn to build new green cover for the winter and the coming year.

In other cases the overseeded grass is a different grass, to create a versatile mixture. For example, warm-season Bermuda grass is commonly overseeded in autumn with cool-season ryegrass to make the lawn green and vigorous during the winter. In the spring, the warm-season grass returns to growth and dominates the lawn.

Insect Control

Many insects live in turf all the time. Healthy turf tends to resist insect infestation and damage, but certain insect populations can become excessive given the right environmental conditions. They feed on leaves or roots, which produces brown spots on the turf, giving it an uneven appearance.

When any turf damage is evident, the cause must first be determined to distinguish insect damage from uneven irrigation or other environmental stresses, which can generate many of the same symptoms. A plug of turf containing soil, roots, crowns, and shoots can be taken from the damaged area, and examined to identify insect pests visually. Plugs can be shipped to agricultural labs for expert analysis. Examples of insects that can cause turf damage are grubs, cutworms, sod webworms, and chinch bugs.

When infestation occurs, competitive organisms such as ladybugs and ground beetles can be used for biological control. Alternatively, pesticides can be applied to suppress many types of insects.

Disease Control

Most turf diseases are fungal infections. Fungi access grass tissue through open wounds caused by mowing, traffic damage and insects. Hence disease occurrence can be limited by management of mowing frequency and blade sharpness, control of traffic, and monitoring of insects. Sprinkler irrigation should be scheduled near dawn, and not at night, so the turf surface dries rapidly without fostering fungal growth. Fungicides can be applied during turf establishment for disease prevention, and after establishment to treat specific types of infection.

Weed Control

Weeds compete with turfgrass plants for water and compromise a lawn's uniformity and neatness. Weeds can develop rapidly where the turfgrass cover has been thinned by traffic or poor maintenance. All aspects of conscientious turf management inhibit weed development by encouraging a thick cover of healthy, competitive grass. When weeds occur, they can be suppressed with herbicides. Some weeds can be controlled with organic herbicides, such as certain corn meal products, fatty acids, herbicidal soap, vinegar, and lemon juice.

Traffic Control

Traffic control can have an enormous effect on turf quality and health at essentially no cost (O'Malley, 2001). Traffic routes and use areas can be spread out or rotated among different portions of a turf area, reducing the traffic intensity or duration on each portion.

Planned traffic control can head off overuse before it begins, giving the turf preventive "resting" periods between uses. Golf-course superintendents frequently relocate cart-path access points to distribute traffic over time and allow turf recovery before traffic is returned to the same place. The turf parking area at Westfarms Mall (described in Chapter 8) uses movable ropes and posts to rotate access points and parking-stall use from day to day or week to week. In pedestrian areas, wide footpath corridor layouts invite pedestrians to spread out, dispersing traffic and reducing potential stress.

Alternatively, traffic control can divert traffic away from places that have begun to show wear, giving them "renovation" periods long enough for recovery and regrowth. To make a diversion without inconvenience, the turf area layout and its management program must allow the restriction or redirection of traffic when signs of stress appear.

SNOW REMOVAL

In cold climates, when snow has to be cleared from turf areas, it must be removed in ways that do not damage the turf surface. On small areas, snow blowers remove snow by lifting rather than pushing. On larger areas, snowplow blades can be equipped with skids or rollers that keep the blades one or two inches above the surface. Where deicing salt will be used, grass varieties should be selected for salt tolerance. When turf is winter-damaged, spring recovery involves the usual maintenance combination of aeration, overseeding, and topdressing.

COMBINED EFFECT OF MAINTENANCE OPERATIONS

Many types of maintenance operations work together to maintain turf aeration and moisture levels. Their interaction and the importance of carrying them out correctly are necessitated by "black layer," a problematic result of careless maintenance.

Black layer is a glue-like, foul-smelling layer that is formed just below the soil surface by the respiration of anaerobic bacteria. It is a visible symptom of prolonged hypoxia in sandy root zones. The waterlogged, impermeable layer aggravates the underlying soil's hypoxia; the sulfide respiratory products themselves may be chemically toxic. When a rooting zone becomes hypoxic, turfgrass roots cease functioning and die; the grass turns yellow, wilts, and thins.

The vicious cycle of black-layer hypoxia may begin with the accumulation of silt and clay particles from topdressing material that is too fine-grained, and of undecomposed thatch that blocks air circulation by suspending moisture in its small pores, all in combination with overirrigation and perhaps unbalanced fertilization. Water that evaporates from the surface does not infiltrate and reach the roots.

Black-layer hypoxia can be prevented or responded to with coring to penetrate the layer for air and water circulation, topdressing with correctly coarse-grained material to encourage thatch decomposition, and irrigating with no more water than actually needed.

TURF INFILTRATION RATE

Living grass actively maintains its root zone's infiltration rate and controls, to a degree, the infiltration rate's change over time (Craul, 1992, p. 148; Ferguson, 1994, pp. 187–191). The vegetative cover intercepts raindrops, protecting the soil surface from crusting and sealing. Networks of elongating and branching roots entwine soil particles, binding them together and connecting their pores in a three-dimensional

Porous Turf

matrix where water and air can move. Turf's infiltration rate, like that of nonliving surface types, can decline over time under certain conditions. However, unlike nonliving surface types, turf's infiltration rate can increase under certain other conditions.

INFILTRATION RATE IN SAND-BASED ROOT ZONES

USGA (2004, Table 4) specifies that a sand-based root zone has saturated hydraulic conductivity of at least 6 inches per hour. This criterion applies to the root zone mixture in the laboratory before it is placed, grassed, or compacted by traffic. It can also be seen as a minimum target for sand-based rooting zones at any time during a project's life. California (Davis et al., 1990, p. 5) suggests, but does not require, at least 20 inches per hour in the laboratory, and then expects a reduction to as low as 12 inches per hour upon placement in the field.

Figure 7.4 shows root zone infiltration rates measured in the laboratory by Prettyman and McCoy (2003) in alternative turf profiles. They measured two root zone mixtures complying with USGA criteria for gradation and other properties. The "high-permeability" mixture had sand and sphagnum in a 9:1 ratio by volume; the "low-permeability" mixture had sand and composted biosolids in a 2.7:1 ratio. The mixtures were installed in 12-inch-thick root zones in both USGA and California profiles. In the USGA profile, the root zone was drained by a continuous aggregate layer which fed excess water to perforated pipes. In the California profile, the root zone was placed on an impermeable base, so excess water had to travel laterally through the lower part of root zone layer to reach the drainage pipes. The root zones were lightly compacted and grassed with 15-month-old creeping bentgrass. Both mixtures had faster infiltration than the USGA and California mimina. The USGA profile gave higher infiltration rates than the California profile, because rapid root zone drainage by the USGA aggregate layer freed the root zone to receive further infiltrating water.

In the field, infiltration rates can be lower than those in the laboratory due to traffic compaction and unpredictable variations in construction and maintenance.

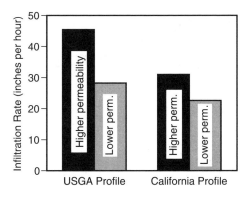

FIGURE 7.4 Surface infiltration rates measured in experimental turf profiles (data from Prettyman and McCoy, 2003).

Crawley and Zabcik (1985) measured infiltration rates on four golf greens on three different courses in California, and found infiltration rates of 9.5 to 25.2 inches per hour. The greens were two to seven years old; infiltration rate did not vary systematically with age. The root zones were sand amended with perlite. The greens' construction profiles were not reported.

INFILTRATION RATE IN FINE-TEXTURED ROOT ZONES

Fine-textured root zones, not complying with the USGA or California root zone specifications, produce lower infiltration rates.

Hamilton (1990, pp. 47–53) measured infiltration rates on loamy-soiled residential lawns in Pennsylvania. The lawns had mixtures of Kentucky bluegrass, perennial rye grass, and fescues. The lawn with the lowest infiltration rate (0.2 inches per hour) was the youngest; its installation two years before had involved soil disturbance and compaction. The lawn with the highest infiltration rate (3.9 inches per hour) was six years old, and had experienced no compaction during installation.

On similar residential lawns in the same locale, Partsch et al. (1993) observed the systematic changes over time shown in Figure 7.5. Compacted turf had low infiltration rates immediately after construction, but with time, as the grass roots and accumulating organic matter penetrated and aggregated the soil, the infiltration rate rose to a moderate plateau of about 2 inches per hour. In contrast, uncompacted turf (sod on uncompacted base soil) initially had high infiltration rates, but with time, as the artificially loose soil settled into place, its infiltration rate converged on the same moderate plateau as that of other mature lawns.

UNREINFORCED TURF FOR PEDESTRIAN TRAFFIC

The following case studies illustrate the application of unreinforced turf to the bearing of pedestrian traffic in a variety of settings and climates.

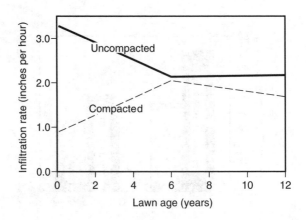

FIGURE 7.5 Lawn infiltration rates on loamy soils in Pennsylvania (data from Partsch et al., 1993, Table 3).

MISSOURI BOTANICAL GARDEN

Figure 7.6 shows turf in a rather common use: on a garden path at the Missouri Botanical Garden in Saint Louis. This path is representative of the garden's numerous secondary routes: greenswards bounded by the loosely sweeping curves of planting areas. The path is 4 feet wide at the narrowest place. On one side, brick edging defines a botanical display area; on the other side careful, frequent trimming of grass and adjacent ivy maintains the distinction between path and planting bed. This path carries about 100 pedestrians per day. The turf is thick and uniform under the garden's superb maintenance. The garden's more heavily trafficked routes are surfaced with stones or with fixed pavements.

SCOTT AMPHITHEATER, SWARTHMORE COLLEGE, PENNSYLVANIA

Figure 7.7 shows the turf floor of the Scott Amphitheater, on the campus of Swarthmore College in Pennsylvania. The turf supports pedestrian seating and walking during the college's orientation and commencement ceremonies, prayer and memorial services, and summer concerts. The amphitheater was constructed in 1942 according to a design by landscape architect Thomas W. Sears.

The arboretum space is unforgettable. It is entered from above via stone steps and paths that wind down a wooded, flowered slope. It is comfortably enclosed by low stone walls and by large rhododendrons and other woodland plants, the green of which is consonant with the amphitheater's grass. At the bottom of the space is a flat, grassy proscenium floor, wide enough to "seat" the amphitheater on the slope. The proscenium is backed and spatially further "stabilized" by a grass-floored stage

FIGURE 7.6 Turf walking area in the Missouri Botanical Garden, Saint Louis, Missouri.

FIGURE 7.7 The turf-floored Scott Amphitheater on the campus of Swarthmore College in Pennsylvania.

elevated by low stone walls. Sixteen informally located tulip poplars rise in colossal columns to frame the arboretum's high "ceiling."

Ten curving retaining walls descend the amphitheater's slope. They are made of native schist, split into slabs for laying. Although the tops of the walls are only 18 inches wide, they are in practice walkways; signs ask visitors to walk on the stone and not on the grass. Three stone steps are set into each 18-inch-high wall.

Each stone wall supports a level turf terrace. Each terrace is about $6^1/_2$ feet wide, which is wide enough for simultaneous seating and walking, and even for setting of portable chairs. The root zone is native fine-textured soil. Lines connecting steps across the terraces are implicitly walkways. Traffic along the lines has compacted the soil and worn the grass; after rain they are dark and muddy. Arguably these spots could be rehabilitated by replacing the native soil with a contemporary root zone of well-drained, uncompactible sand, and regrassing.

Amphitheaters in Other Climates

Amphitheaters in other parts of the country illustrate the beauty and comfort of turf in other climates. Figure 7.8 shows an amphitheater at the Heard Museum in Phoenix, Arizona. The amphitheater houses outdoor exhibits, events and performances. Its terraces are supported by concrete retaining walls, ascended by concrete steps. All the concrete is stained pink, in keeping with the desert setting. The deeply blue-green irrigated grass contrasts stunningly with the hot, hard planes of concrete, inviting arriving audiences to take comfort on its cool softness.

Porous Turf   267

FIGURE 7.8 Turf terraces at the Heard Museum amphitheater in Phoenix, Arizona.

FIGURE 7.9 Turf terraces at the amphitheater in the Village Homes residential community in Davis, California.

Figure 7.9 shows the turf amphitheater in the Village Homes residential community in Davis, California. The amphitheater is the site for the community's meetings and performances. The low concrete seating walls were economical to construct; their casually irregular lines are appropriate for the community's character. The grass terraces are maintained along with the other grass open spaces in the community's common areas.

UNREINFORCED TURF FOR VEHICULAR TRAFFIC

The following case studies illustrate the application of unreinforced turf to the bearing of vehicular traffic in a variety of settings and climates.

BIG TOP FLEA MARKET, FLORIDA

A parking lot at Big Top Flea Market near Tampa, Florida illustrates the use of turf in weekly scheduled parking. The market is in effect a special-purpose shopping center (www.bigtopfleamarket.com). It differs from other shopping centers in that it is open only two or three days a week, so the parking area bears no traffic four or five days each week. The 2500-stall parking lot surrounds a market building that holds more than 600 vendors. Figure 7.10 shows that the turf parking spaces are accessed by paved lanes. Parking is organized only by the orientations of access lanes and lines of bollards; individual spaces are not marked.

The soils are fine gray sands from the Millhopper and Tavares series, with substantial depth to bedrock and water table. This native substrate provides naturally durable porosity, drainage, and aeration for turf root zone and bearing of traffic.

FIGURE 7.10 Turf parking area at Big Top Flea Market, Hillsborough County, Florida.

The grass is bahia, which spreads to form a surface mat. Visits in 1999 and 2001 showed that, the turf was in a fair-to-good condition. In the most-used parking stalls, very few spots were rutted into depressions where water stands following storms. Outside those spots, there was almost no sign of compacted wheel treads or poor drainage, even in wet summer periods.

Between parking rows are simple, sturdy wooden and concrete bollards 6 inches × 6 inches in size, five to eight feet apart. Rows of bollards orient parking spaces on the unmarked grass; clusters of them protect trees and mark intersections. They conveniently support raised irrigation heads, while protecting the heads from traffic. Trimming grass around the very numerous bollards presumably warrants a sizeable amount of labor.

Numerous secondary parking lanes are surfaced with dense-graded, recycled-concrete aggregate laid on geotextile. The lanes are 16 feet wide and marked for one-way traffic. The aggregate appears compacted from traffic, but there is no sign of water erosion. Where traffic is lightest, the lanes are partly overgrown with spreading grass. A few central entry lanes are paved with dense asphalt.

The overall site slopes naturally at 1 or 2 percent. Within that context, the site was graded to give stalls and lanes a consistent topographic pattern. The grass stalls slope noticeably down from the bollard lines to the aggregate lanes. The lanes drain gently toward the site's stormwater management basins.

The parking lot's hydrology is akin to that of a sandy meadow. The soil ecosystem treats oils and other pollutants from the numerous vehicles. The only routinely concentrated runoff is from the market building's rooftop. The site's stormwater management basins are much smaller than they would have to be if the huge parking lot had been given an impervious surface.

This site's soil and land use make turf surfacing highly successful. The market's intermittently scheduled operation limits the surface's long-term traffic load, permits routine maintenance operations such as mowing and irrigation, and sets aside regular periods for turf recovery following wear. Continuous turf bays make huge rooting zones for trees. The cost of the turf is the long-term regime of mowing, irrigation, and other maintenance operations.

Mall of Georgia

Turf parking at the Mall of Georgia near Atlanta illustrates the use of turf in seasonal overflow parking. The mall is the biggest shopping center in southeastern U.S. (www.simon.com). Its peripherally located grass parking area is frequently full of cars during the few weeks of the Christmas shopping season, but lies empty the rest of the year. Figure 7.11 shows the area in the spring three years after installation.

The area can hold about 100 cars. Grass parking stalls are accessed by dense asphalt lanes. The lanes' edges are unreinforced, but are nevertheless in good condition. A raised concrete curb separates the area from adjacent grass landscape areas. Parking is organized only by the paved lanes, the peripheral curbs, and flowering trees between interior parking rows; there are no wheel stops or other markings to define individual stalls. Overflow runoff drains to an inlet at the area's low corner.

FIGURE 7.11 Turf overflow parking area at The Mall of Georgia in Buford, Georgia.

The grass is Bermuda, which spreads to form a mat. When the area is used for parking the grass is winter-dormant and seldom protected by frozen ground or snow cover. Nevertheless, after three years, the turf surface was generally continuous and healthy. The only visible wear and rutting are along the central entry lane, where numerous vehicles veer across small grass corners. The area is empty of cars during the entire growing season, when the turf has to be mowed, irrigated, and otherwise maintained, and it has time to recuperate from traffic stress. The turf bays give trees large, viable rooting zones.

The Bermuda sod is rooted in a one-inch-thick sand layer cut from the sod farm where it was grown. The sod layer rests on a thick base of fine sand. Whether a drainage layer or perforated pipes underlie the base is not known. The sandy layers are extremely important to this unreinforced turf's success, because the native subgrade is red clay. If grass were planted directly in the native soil, the clay would be moist and plastic during the winter parking season, and easily susceptible to rutting and compaction.

Pop-up sprinkler heads are located at the corners and edges of the turf bays. These locations are correct for efficiently uniform water distribution, but vehicles drive over the heads when veering at corners or entering parking stalls. At the most heavily trafficked corners, some heads have been knocked out of alignment and others' plastic edges have been chipped off. Heads located at the turf's periphery, where there is no traffic, have not been touched. In future turf installations, pop-up irrigation heads could be protected from traffic by concrete or plastic collars, or they could be located between bays where there is no traffic, or irrigation could be by subsurface tubing when that technology has been proven and is available.

Muirfield Village, Ohio

Parking fields in Muirfield Village, Dublin, Ohio illustrate the use of turf in intense but infrequent event parking. The fields hold most of the parking for a once-per-year summer professional golf tournament (www.thememorialtournament.com). Shuttle vans transport spectators from the fields to the golf venue. Figure 7.12 shows the fields in June, 2001, exactly one week after that year's tournament. Parking is on unreinforced turf; access lanes are paved with aggregate.

The fields are large enough to hold many thousands of cars. They were set aside in James Bassett and Associates' master plan for the Muirfield Village community and its golf course, and are accessed from Crossgate Drive. They are separated from the public street by grassy berms, and from surrounding homes by hedgerows of trees and shrubs. The alignment of lanes and parking bays is smooth and regular. The fields have no significant use other than for annual tournament parking, although they are physically eligible to be considered for other occasional uses.

The lanes are of dense-graded, angular, tightly packed limestone aggregate. They are about 12 feet wide, which is wide enough for one-way travel; presumably, parking attendants direct traffic during events. Their nearly impervious surfaces occupy about 25 percent of the fields' area. Their driving condition was good after several years of use.

The parking stalls are in coarse but uniform meadow grass planted directly in the native clay till soil. There is no reinforcement of the turf, and no permanent irrigation system. There are no wheel stops or stall markings; parking is organized only by the orientation and spacing of the lanes. The fields' fundamentally smooth surface and healthy turf suggest that the field is mowed several times per growing

FIGURE 7.12 Parking field at Muirfield Village in Dublin, Ohio.

season, and occasionally leveled and overseeded in the spring before the summer golf event.

A small number of special parking spaces are set aside for busses and large recreational vehicles. These spaces are made of aggregate, have electric hookups, and are numbered for reservations.

In the month preceding the 2001 tournament, there had been extraordinarily high rainfall, making the clay soil plastic and unusually vulnerable to traffic. After the tournament, some low spots in the parking bays were wet and rutted, and wheels had tracked a clay veneer onto much of the grass. Rutting could accumulate over time if it were not for conscientious maintenance. Outside the low areas, the turf was in good condition.

The potential for rutting in wet weather could be reduced by selective retrofitting in wet spots, amounting to perhaps 5 or 10 percent of the turf area. In these places, adding perforated pipes, fiber reinforcement, or a sandy root zone would improve soil drainage and support. In future installations, it will be possible to analyze a field site before installation, and to direct drainage and reinforcement to low, wet areas.

Muirfield's turf parking is economical of construction materials, and consistent with the low-impact stormwater objectives of the community's comprehensive plan. The fields' maintenance is an economical extension of the large, well-organized landscape management of the surrounding golf club and community.

TURF REINFORCEMENT

Numerous proprietary meshes, mats, and fibers are available for reinforcing turf root zones. Traffic continues to contact the turf directly, but the reinforced turf resists compaction and rutting better, thereby maintaining its porosity, permeability, aeration, drainage, and appearance. Manufacturers have invented a variety of reinforcement configurations; they are categorized here as overlaid meshes, embedded mats, and integral fibers. Some of the individual products have considerable technical credibility, being supported by published research and detailed technical guidelines.

OVERLAID MESHES

Table 7.9 lists examples of overlaid mesh products. They are placed at or near the root zone surface, giving strength and protection to near-surface roots and crowns. They help distribute traffic loads laterally, reducing the point load intensity, but they cannot completely compensate for inherently soft plastic soil. Some are intended for vehicular traffic; they provide tensile or beam-like strength to spread load across the turf surface. Others are only for pedestrian traffic; they may provide a relatively continuous mat-like surface to improve pedestrian accessibility or make cart travel smoother, while providing some lateral strength. Some specific products are described below.

GrassTrac is a flexible steel mesh placed on or near a turf surface to distribute vehicular traffic load. The mesh's wires are spaced approximately 3 inches × 4.5 inches. Its Belgian manufacturer had experience with this product in Europe before introducing it in North America. The mesh can be installed on a root zone before

Porous Turf

TABLE 7.9
Examples of Overlaid Mesh Products

Company	Product	Contact Information
Bekaert	GrassTrac	www.grasstrac.com
Mat Factory	Safety Deck	www.matfactoryinc.com
Netlon Turf Systems	TurfGuard	www.netlon.co.uk
Netlon Turf Systems	NetPave 25	www.netlon.co.uk
Tenax	LBO SAMP	www.tenax.net

seeding or sodding. Alternatively it can be installed over existing turf, as long as the grass is first cut as short as possible; after installation the grass grows up through the wire mesh. The wire covers only a small percentage of the turf's surface; the surface appearance and permeability are essentially those of the turf. The mesh can reputedly be used in place of a base course for small vehicles such as passenger cars; a base is required for heavier vehicles.

TurfGuard is a polyethylene mesh with small grid apertures and high tensile strength. It is laid on a root zone surface before grassing, or on top of preexisting turf after close mowing. The flexible mesh is rolled out over the surface and staked in place. A sandy topdressing is applied to cover the mesh ribs, and the area is overseeded. Grass plants grow through the mesh and intertwine with it. TurfGuard was used at a universally accessible archery range in Britain, where the surface reinforcement and closely spaced ribs made the turf effectively wheelchair-accessible, while protecting the turf crowns from traffic damage (Netlon Turf Systems, 2000).

LBO SAMP is a polypropylene grid with distinctive tensile strength. It is intended for pedestrian and lightweight vehicular traffic. On preexisting turf, the mesh is laid down at the beginning of the grass growing season following a close mowing. The area is then overseeded and topdressed. Grass grows through the mesh; grass roots intertwine with the mesh filaments. With complete grass cover, the mesh is nearly invisible, and the area has the appearance of a simple lawn.

Netpave 25 is in effect a thin version of the Netpave 50 geocell that will be described in Chapter 8. Unlike Netpave 50, it is intended to be installed over preexisting turf. It is made of recycled polyethylene; it comes in interlocking panels 20 inches square and one inch thick. Prior to placement, the turf should be cut as low as possible and any depressions filled with a sandy topdressing. The area can be used immediately after placement, but its condition improves as the turf integrates with the mesh. It has been used for overflow parking areas. Due to the size and depth of the openings, it should probably not be used in recreational areas or where narrow heels are likely to be worn. It can be used temporarily and then removed, for example in places, where the use of a turf area for infrequent events would not warrant a permanent installation. It could also prove useful for reinforcing patches of preexisting turf that have become muddy and rutted. At the Wellingborough Golf Club in the U.K., NetPave 25 was used to reinforce overflow parking without having to go through the permit process that would have been required for other types of pavement (Netlon Turf Systems, 2002).

EMBEDDED MATS

Table 7.10 lists examples of embedded mat products. These are three-dimensional wire matrices in the interstices of which sandy root zone soil is placed. Soil particles and growing roots interlock with the mats' filaments. The composite soil-root-mat structure combines the compressive strength of sand with the tensile strength of the roots and mat. After being compressed by a load, some types of mats return resiliently to their original shape and thickness, resisting rutting and compaction in wet conditions. They can be characterized using tests such as those listed in Table 7.11.

Enkamat is a springy mat about 3/4 inch thick, composed of long nylon filaments, bent and interwoven in a complex pattern and welded where they intersect. It contains 95 percent open space for soil and roots. The mat is placed on root zone material and filled with further material, and the surface is grassed. Growing roots entwine with the mat's filaments to make a composite structure that distributes weight and resists compaction and rutting. The turf can be grown into the reinforcing mat on a sod farm; the sod is then cut and placed with the reinforcing mat built in.

Tensar Mat is made of polyethylene filaments. It has a flat base layer and a cuspated upper surface that absorbs lateral movement. The mat is placed on root zone material, filled with further material, and grassed. Roots entwine with the mat's filaments, making a composite structure that resists compaction.

SportGrass has polypropylene fibers arranged vertically like grass blades and attached to a woven synthetic backing. It is installed on sandy root zone material; further material is added to fill the space between the standing fibers. Grass grows up alongside the projecting parts of the fibers, forming a reinforced canopy. Grass roots entwine with the fibers, and may be able to grow downward through the backing into the underlying root zone material (McNitt and Landschoot, 1998). SportGrass-reinforced sod can be grown at a sod farm, and lifted away for transplanting.

TABLE 7.10
Examples of Embedded Turf-Reinforcing Mat Products

Company	Product	Contact Information
Colbond Geosynthetics	Enkamat	www.colbond-usa.com
Hummer Turfgrass	Grasstiles	www.usaturf.com
Motz Group	TS-II	www.themotzgroup.com
SportGrass	SportGrass	www.sportgrass.com
Tensar	Tensar Mat	www.tensar.co.uk

TABLE 7.11
ASTM Standard Tests for Turf Reinforcement Mats

Number	Title
D 6454	Standard Test Method for Determining the Short-Term Compression Behavior of Turf Reinforcement Mats (TRMs)
D 6524	Standard Test Method for Measuring the Resiliency of Turf Reinforcement Mats (TRMs)

The TS-II mat has a backing of jute and plastic mesh, with attached polypropylene fibers. The mat is placed onto a root zone layer. Further root zone material is added to fill the space between the vertically oriented fibers, and grassed. Grass blades grow among and above the fibers, creating a living surface reinforced by the standing parts of the fibers. Roots intertwine with the lower parts of the fibers, which are fibrillated (slit) so the roots can entwine with them intimately. The combination of backing, fibers, and living grass makes a composite structure.

Grasstiles mats make portable reinforced-turf modules. Vertically oriented nylon fibers are attached to an 85 inch × 85 inch polyethylene drainage pad. A 2-inch layer of sandy planting medium fills the space between the fibers, and is grassed. The entire sod unit can be lifted out with the drainage pad as a pallet. Permanent installations are placed on sandy root zone material; examples are the playing fields at Hershey Park Stadium in Hershey, Pennsylvania, and the PSINet Stadium in Baltimore, Maryland. Temporary installations can be placed on any solid base with enough porosity for drainage. Portability enables replacement of turf sections as necessary.

INTEGRAL FIBERS

Table 7.12 lists examples of integral fiber products. They consist of fibers or small segments of fiber mesh that are tilled, injected, or sewn into the root zone in great numbers. They add tensile strength while stabilizing networks of open pathways for air and water circulation.

Advanced Turf is a fabric-like mesh segment. Each segment is about the size of a playing card. Large numbers of segments are mixed into sandy root zone material. The flexible, resilient mesh interlocks with sand particles and growing grass roots to make a springy structure that resists compaction and rutting, while maintaining porosity, infiltration, and aeration. The manufacturer provides guidelines, summarized in Table 7.13, for root zone and base thicknesses that can bear specific traffic loads. After the mixed, reinforced root zone is placed, the surface is seeded and top-dressed, or bare-root sod may be placed. Figure 7.13 shows infiltration rates maintained by Advanced Turf's fiber structure under light traffic for three years; the fibers maintained some infiltration rate even in plastic clay loam.

Reflex Mesh is a mesh segment with a bowed or pre-flexed form that rebounds when the soil is compressed. This mechanical action maintains pore spaces; sandy planting medium containing flexing mesh structures drains notably rapidly. It greatly increases the soil's load-bearing capability. The resilience of the mesh structures

TABLE 7.12
Examples of Turf-Reinforcing Integral Fibers

Company	Product	Contact Information
Desso DLW	GrassMaster	www.dessodlw.com
Netlon Turf Systems	Reflex Mesh	www.netlon.co.uk
Netlon Turf Systems	Advanced Turf	www.netlon.co.uk
Stabilizer	TurfGrids	www.stabilizersolutions.com

TABLE 7.13
Summary of Manufacturer's Guidelines for Turf Profiles with Root Zones Reinforced by Advanced Turf Product

Traffic	Rootzone Thickness	Base	Subgrade Treatment
Pedestrian and very light vehicular traffic	< 8 inches	None required	Cultivated for infiltration, then lightly recompacted
Heavier traffic	≥ 8 inches	Required; choke layer may be required to meet filter criteria	Compacted

Netlon (no date).

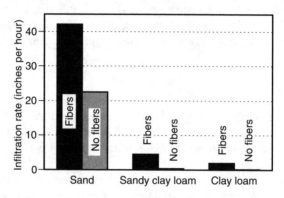

FIGURE 7.13 Infiltration rates in lightly trafficked turf with and without Netlon Advanced Turf fibers (average of four measurements per year over a three-year period, data from Beard and Sifers, 1993).

maintains a surface that is safe for athletic use. The manufacturer suggests specific base, soil, and drainage specifications for various levels of traffic loading.

GrassMaster uses long polypropylene fibers that are injected or sewn into the root zone with needles. Each fiber is injected vertically to form a "u" shape like an inverted staple. It is injected about 8 inches into the soil, leaving its two ends emerging about 3/4 inch above the soil surface. Growing roots intertwine with the fibers to make a composite soil structural matrix, while the emergent parts of the fibers protect the grass crowns from wear. The fibers can be injected into a sandy root zone prior to seeding, or into a completed sod.

Turfgrids are polypropylene strands 1/2 to 2 inches in length, manufactured in packed bunches. During incorporation into the soil each bunch opens and expands into a net up to 6 inches in length, interlocking with soil and growing roots. When used with a sandy planting medium, shear strength can increase 10 to 100 percent; bearing capacity (CBR) can increase 50 to 100 percent (Stabilizer, 1996). The fibers are spread on a layer of root zone material at a rate of 1 pound per 10 square feet and rotary-tilled into the top 4 inches. Alternatively, soil and fibers can be mixed before placement. The reinforced soil is compacted and grassed. To retrofit preexisting turf,

Porous Turf   277

a 5- to 6-inch layer of fiber-reinforced planting medium can be placed on top of the preexisting surface; this approach was used for a turf parking and event-staging area at the Louisville Waterfront Park in Louisville, Kentucky.

REINFORCED TURF FOR VEHICULAR TRAFFIC

A series of parking lots installed in the Minneapolis area by the Glenn Rehbein Company illustrate the successful use of reinforced turf for limited vehicular traffic. They also illustrate the successful use of traffic-bearing turf as a porous pavement in a notably cold climate.

All three sites have similar construction (Jay Hudson and William McCully, personal communications 2002). The surface layer is a 4- to 6-inch root zone of sand reinforced with Advanced Turf fibers and grassed with bare root sod. The root zone soil is 80 percent (by volume) washed sand and 20 percent sandy loam. The grass is a mixture of Kentucky bluegrass varieties.

LINO LAKES STATE BANK, MINNESOTA

Overflow parking at Lino Lakes State Bank on Otter Lake Road in Lino Lakes, Minnesota illustrates fiber-reinforced turf surfacing for an integrated combination of overflow parking, stormwater infiltration, and snow storage purposes. Figure 7.14 shows the area two years after installation. The grass cover is very uniform. It is big enough to hold 13 cars. Because it is located at the site's periphery, it is seldom used for parking, so it is usually free for its other purposes such as snow storage and

FIGURE 7.14 Mesh-reinforced turf parking at Lino Lakes State Bank in Lino Lakes, Minnesota, in November 2002.

stormwater control. It is separated from a dense asphalt turning lane by a concrete strip. The bank's other parking stalls are of dense asphalt.

The turf parking bay receives both direct rainfall and the runoff from two thirds of the bank's impervious roof and pavements. Perimeter concrete curbs collect drainage and concentrate it on the turf. The grass bay forms a shallow swale to hold infiltrating stormwater and meltwater. Presumably water stands here briefly during intense storms and snowmelt periods. Overflow runoff drains to an inlet at one end of the swale.

The grass bay is plowed for snow removal, like the rest of the parking lot. A flexible rubber snowplow blade is used to avoid rutting the turf, or a skid is attached to a metal blade to keep it one inch above the grass surface. The turf's fiber reinforcement helps resist plow rutting. A portion of the grass is used for snow storage; when the snow melts the water infiltrates into the turf.

The turf is a filter for the bank's runoff; suspended solids that come to rest in it are either built into the soil structure by the living grass, or are disposed with grass clippings after mowing. This experience indicates that an adequately constructed traffic-bearing turf surface can routinely receive and infiltrate substantial amounts of runoff from adjacent areas, unlike most other porous surfaces which must ordinarily drain away in all possible directions to avoid clogging.

TROY BURNE GOLF CLUB, WISCONSIN

Overflow parking at Troy Burne Golf Club in Hudson, Wisconsin illustrates the use of fiber-reinforced turf in highly maintained overflow parking. Figure 7.15 shows the parking area three years after installation. The area can hold approximately 90 grass parking stalls along dense asphalt lanes; the main parking area is entirely in dense asphalt. Both areas are located near the clubhouse building's main entrance.

The area's central turf bay forms a gentle swale, draining to the center where a surface inlet removes overflow runoff. The sandy root zone is well drained; the surface is dry soon after rain while puddles remain on the asphalt lanes. The turf's rooting zone is a 6-inch layer over granular subgrade soil. Pop-up irrigation heads are located near the center of the bay; they are well protected from traffic by their locations and their low placement amid dense turf.

The grass is *Poa supina*, mixed with Kentucky bluegrass. The grass is very neatly maintained, and notably smooth and uniform. The turf showed no traffic damage except at one small corner, where all traffic enters at high speed and cuts across the grass. The turf's excellent maintenance is an economical extension of the golf club's routine procedures.

The turf is unused for parking mid-winter, so snow accumulations are not removed. However, deicing salt is used on the asphalt lanes in late fall; briny runoff presumably makes its way into the grass. For fast recovery from winter conditions, the turf is sometimes overseeded in early spring with perennial ryegrass.

URBAN VENTURES SOCCER FIELD, MINNESOTA

Event parking at the Urban Ventures soccer field at Fifth Avenue and 29th Street in Minneapolis illustrates the visible effects of topdressing in a traffic-bearing turf.

FIGURE 7.15 Mesh-reinforced turf parking at Troy Burne Golf Club in Hudson, Wisconsin, in November 2002.

FIGURE 7.16 Mesh-reinforced turf parking at the Urban Ventures soccer field in Minneapolis, in November 2002.

TABLE 7.14
Holders of Registered Trademarks Mentioned in This Chapter

Registered Trademark	Holder	Headquarters Location
Advanced Turf	Netlon Turf Systems	Blackburn, United Kingdom
Bayr Root	StrathAyr	Seymour, Victoria, Australia
Enkamat	Colbond Geosynthetics	Enka, North Carolina
Evaporative Control Systems	Evaporative Control Systems	Reno, Nevada
GrassMaster	Desso DLW	Oss, The Netherlands
GrassTiles	Hummer	Lancaster, Pennsylvania
GrassTrac	Bekaert	Kortrijk, Belgium
LBO SAMP	Tenax	Wrexham, UK
Netpave 25	Netlon	Blackburn, United Kingdom
ReFlex Mesh	Netlon Advanced Turf Systems	Blackburn, United Kingdom
Safety Deck	Mat Factory	Mesa, California
SportGrass	SportGrass	McLean, Virginia
Tensar Mat	Tensar	Blackburn, UK
TS-II	Motz Group	Newtown, Ohio
TurfGrids	Stabilizer	Phoenix, Arizona
TurfGuard	Netlon Turf Systems	Blackburn, United Kingdom

Figure 7.16 shows the parking area four years after installation. It is small, and irregularly shaped to fit into a narrow urban space. This is the home soccer field for the Urban Stars Athletic Club; it is sometimes called the Honeywell Home Field. Youth activity here is supported by Urban Ventures and other nonprofit organizations.

The root zone is very sandy. The grass is healthy, spreading supina bluegrass. In the winter, the parking surface is neither plowed for snow removal nor sanded for traction. The surface is routinely aerated and topdressed. Aeration is by coring or piercing; it is only 1 inch deep, so it penetrates little more than the surface thatch or crust. It hurts the integral fibers little, so it does not alter the turf's performance.

The photo was taken shortly after an aeration treatment and an application of topdressing sand. The sand was clean and open-graded; the grains were loose and noncohesive. It made a thin, uniform layer barely touching the soil. Almost all of the grass blades remained standing through the sand layer. If the sand has any effect on driving, it is to assist traction and prevent slipping that would otherwise abrade the grass and soil. In the spring, more living blades can sprout through the sand, while the sand grains are built into the living soil structure.

TRADEMARKS

Table 7.14 lists the holders of registered trademarks mentioned in this chapter.

ACKNOWLEDGMENTS

The following persons generously provided information used in this chapter: James Bassett of James Bassett and Associates (www.bassettassociates.com); Don Bright and Lisa Frasier of Tensar (www.tensarcorp.com); Douglas Fender of Turfgrass Producers International (www.turfgrasssod.org); Charles Friedrich and Barry Gemberling of Carolina Stalite (www.permatill.com); Jay Hudson, Mike Kelly, and William McCully of Glenn Rehbein (www.rehbein.com); Samuel Randolph of Soil Stabilization Products Company (www.sspco.com); and William Siler, a former graduate assistant at the University of Georgia.

The following persons constructively critiqued early drafts of this chapter: William McCully of Glenn Rehbein (www.rehbein.com) and Samuel Randolph of Soil Stabilization Products Company (www.sspco.com).

REFERENCES

Alexander, Ron (2002). Compost and the Landscape Architect, *Landscape Architecture* 92, 42–46 and 98.

Asaeda, Takashi, and Vu Thanh Ca (2000). Characteristics of Permeable Pavement During Hot Summer Weather and Impact on the Thermal Environment, *Building and Environment* 35, 363–375.

Beard, James B. (2001). *Turf Management for Golf Courses*, 2nd ed., New York: John Wiley and Sons.

Beard, James B., and Samuel I. Sifers (1993). *Stabilization and Enhancement of Sand-Modified Root Zones for High Traffic Sport Turfs with Mesh Elements*, College Station, Texas: Texas Agricultural Experiment Station.

Beard, James B., S.M. Batten, and A. Almodares (1979). *An Assessment of Wear Tolerance Among Bermudagrass Cultivars for Recreational and Sports Turf Use*, PR-3836, College Station: Texas Agricultural Experiment Station.

Canaway, P.M. (1975). Turf Wear: A Literature Review, *The Journal of Sports Turf Research Institute* 51, 9–40.

Carrow, R.N. (1980). Influence of Soil Compaction on Three Turfgrass Species, *Agronomy Journal* 72, 1038–1042.

Carrow, R.N., and B.J. Johnson (1989). Turfgrass Wear as Affected by Golf Car Tire Design and Traffic Patterns, *Journal of the American Society of Horticultural Science* 114, 240–246.

Chong, She-Kong, Richard Boniak, Chang-Ho Ok, Sam Indorante, and F. Dan Dinelli (2001). How Do Soils Breathe? *Golf Course Magazine*, reprinted at www.grasspave.com.

Clancy, Tony (1996). Infiltration Measurements Surprise, *Turf Management* March–April, p. 37.

Crawley, Wayne, and David Zabcik (1985). Golf Green Construction Using Perlite as an Amendment, *Golf Course Management* July.

Craul, Phillip J. (1992). *Urban Soil in Landscape Design*, New York: John Wiley and Sons.

Davis, William B., Jack L. Paul, and Daniel Bowman (1990). *The Sand Putting Green: Construction and Management*, Publication No. 21448, Davis: University of California Division of Agriculture and Natural Resources.

deShazer, S.A., T.P. Riordan, F.P. Baxendale, and R.E. Gaussoin (2000). *Buffalograss: A Warm-Season Native Grass for Turf*, EC92-1245, Lincoln: Nebraska Cooperative Extension (www.ianr.unl.edu).

Ferguson, Bruce K. (1994). *Stormwater Infiltration*, Boca Raton: Lewis Publishers.

Ferro, Sam, and Duane Otto (2001). Drainage Problems? Sand-Based Root zones Help Golf Courses and Athletic Fields Find "Soil Solutions," *Grounds Maintenance* June.

Hamilton, George W. (1990). *Infiltration Rates on Experimental and Residential Lawns*, M.S. thesis, University Park: Pennsylvania State University.

Harivandi, M. Ali (1998). Golf Green Construction—A Review of the University of California Method, *California Turfgrass Culture* 48, 17–18.

Henderson, J.J., J.R. Crum, T.F. Wolff, and J.N. Rogers, III (2001). *Athletic Field Root Zone Mixes: What Is the Best Mix for Your Field?*, East Lansing: Michigan State University Department of Crop and Soil Sciences (http://www.lib.msu.edu).

Mayer, Eugene (1998). Compaction Resistance and Drainage: The Driving Force Behind Sand Based Root Zones, *Turf Magazine* June.

McNitt, A.S. and P.J. Landschoot (1998). *The Effects of Soil Inclusions on Soil Physical Properties and Athletic Field Playing Surface Quality*, Final Report, University Park: Pennsylvania State University Department of Agronomy.

McNitt, A.S., P.J. Landschoot, and A.R. Leonard (1999). Evaluation of the Agronomic and Sports Turf Quality of a Modular Turf System Installed over Black Top, in *1999 Annual Research Report*, University Park: Pennsylvania State University Center for Turfgrass Science (http://turf.cas.psu.edu).

Minner, D.D., J.H. Dunn, S.S. Bughara, and B.S. Fresenburg (1993). Traffic Tolerance Among Cultivars of Kentucky Bluegrass, Tall Fescue, and Perennial Ryegrass, in *International Turfgrass Society Research Journal 7*, R.N. Carrow, N.E. Christians, and R.C. Shearman, Ed., Overland Park, Kansas: Intertec Publishing.

Netlon Turf Systems (2000). *The Keppleway Holiday and Activity Centre, Lake District; Archery Range and Maze*, Case Study 5, www.netlon.co.uk.

Netlon Turf Systems (2002). *Wellingborough Golf Club, Northants; Overspill Parking*, Case Study 7, www.netlon.co.uk.

Netlon Turf System (no date). *NetPave 25*, www.netlon.co.uk.

O'Malley, Penelope (2001). Maintaining Turf on Golf Courses and Playing Fields: Management Is the Key to Combating High Traffic and Overuse, *Erosion Control* 8, 80–85.

Partsch, C.M., A.R. Jarrett, and T.L. Watschke (1993). Infiltration Characteristics of Residential Lawns, *Transactions of the American Society of Agricultural Engineers* 36, 1695–1701.

Patterson, James C. (1976). Soil Compaction and Its Effects upon Urban Vegetation, *Better Trees for Metropolitan Landscapes*, pp. 91–102, Frank S. Santamour, Henry D. Gerhold, and Silas Little, Ed., U.S. Forest Service General Technical Report NE-22, Upper Darby, Pennsylvania.

Patterson, James C., and Christine J. Bates (1994). Long Term, Light-Weight Aggregate Performance as Soil Amendments, *The Landscape Below Ground: Proceedings of an International Workshop on Tree Root Development in Urban Soils*, pp. 149–156, Gary W. Watson and Dan Neely, Ed., Savoy, Illinois: International Society of Arboriculture.

Prettyman, G.W., and E.L. McCoy (2003). Profile Layering, Root Zone Permeability, and Slope Affect on Soil Water Content during Putting Green Drainage, *Crop Science* 43, 985–994.

Puhalla, Jim, Jeff Krans, and Mike Goatley (2002). *Sports Fields: A Manual for Design, Construction and Maintenance*, New York: John Wiley and Sons.

Richards, Carolyn W. (1994). *Report on the Effects of "Turfgrids" on Soil Physical Properties of Sand Rootzones under Laboratory Conditions*, West Yorkshire, U.K.: Sports Turf Research Institute.

Rogers, John N. III, J. Timothy Vanini, and James R. Crum (1998). Simulated Traffic on Turfgrass Topdressed with Crumb Rubber, *Agronomy Journal* 90, March–April, 215–221.

Shearman, Robert C., 1999, Managing Intensively Used Turfs, *University of Nebraska Center for Grassland Studies Newsletter* 5, 2–5.

Shearman, Robert C., E.J. Kinbacher, and T.P. Riordan (1980). Turfgrass-Paver Complex for Intensely Trafficked Areas, *Agronomy Journal* 72, 372–374.
Stabilizer, Inc. (1996). *Turfgrids Fibers Benefits and Lab Performance Testing*, Phoenix: Stabilizer Inc.
Stier, J.C., J.N. Rogers, III, and J.C. Sorochan (1997). Development of Management Practices for Supina Bluegrass (*Poa supina* Schrad.), in *67th Annual Michigan Turfgrass Conference Proceedings*, Vol. 26.
Trenholm, L.E., R.R. Duncan, and R.N. Carrow (1999). Wear Tolerance, Shoot Performance, and Spectral Reflectance of Seashore Paspalum and Bermudagrass, *Crop Science* 39, 1147–1152.
Trenholm, L.E., R.N. Carrow, and R.R. Duncan (2000). Mechanisms of Wear Tolerance in Seashore Paspalum and Bermudagrass, *Crop Science* 40, 1350–1357.
Turfgrass Producers International (1995). *Guideline Specifications to Turfgrass Sodding*, Rolling Meadows, Illinois: Turfgrass Producers International (www.turfgrasssod.org).
Turgeon, A. J. (1999). *Turfgrass Management*, Englewood Cliffs, New Jersey: Prentice Hall.
U.S. Federal Aviation Administration (1995). *Airport Pavement Design and Evaluation*, Advisory Circular AC 150/5320-6D, Washington, DC: FAA (www.faa.gov/arp/150acs.htm).
U.S. Golf Association Green Section Staff (UGSA) (2004). *USGA Recommendations for a Method of Putting Green Construction*, Far Hills, New Jersey: U.S. Golf Association (www.usga.org).
U.S. Soil Conservation Service (1989). *Soil Survey of Hillsborough County, Florida*, Washington: U.S. Soil Conservation Service.
White, Charles B. (2001). *Turf Managers' Handbook for Golf Course Construction, Renovation and Grow-In*, Chelsea, Michigan: Sleeping Bear Press.
Youngner, Victor B. (1962). Wear Resistance of Cool Season Turfgrasses: Effects of Previous Mowing Practices, *Agronomy Journal* 54, 198–199.

8 Plastic Geocells

CONTENTS

Geocell Installation 286
Maintenance Cost 287
Geoblock 289
    Ritz-Carlton Huntington Hotel, Pasadena, California 290
    Public Health Laboratory, Anchorage, Alaska 291
    Remote Trails, Alaska and Wisconsin 291
Grasspave^2 295
    Orange Bowl, Miami, Florida 297
    Westfarms Mall, Farmington, Connecticut 298
    Trace Pointe, Clinton, Mississippi 301
    Blue Cross Blue Shield, Southfield, Michigan 302
Grassroad and Tufftrack 303
    Hyatt Regency Hotel, Long Beach, California 304
    California JPIA, La Palma, California 305
Grassy Pavers 306
    Medgar Evers Building, Jackson, Mississippi 307
    Residence, Santa Monica, California 308
Gravelpave^2 309
    Burger King Restaurant, Henderson, North Carolina 309
    Grand Canyon Trust, Flagstaff, Arizona 310
    Frostburg State University, Frostburg, Maryland 312
Netpave 314
    Nigel's Pitch, Middletown, Rhode Island 314
Other Models of Plastic Geocells 317
Flexible Plastic Webs 318
    Stone Mountain State Park, Georgia 318
Trademarks 319
Acknowledgments 319
References 321

The geocell family of pavement materials is defined by manufactured plastic lattices, the cells of which can be filled with aggregate or a planting medium for turf. The term "geocell" comes from Koerner (1998, p. 62; Robert Koerner, personal communication 2001); in his terms geocells are plastic mattresses or panels forming networks of box-like cells that are filled with earth material.

Geocells for paving seem to have originated in the U.S. about 1977, perhaps inspired by precedent in Australia. At that time a number of companies independently developed proprietary models.

The lattices come in panels a few feet square, or in rolls that can be spread out to cover large areas. In many models the lattice is about 1 or 2 inches thick; the lattice's ribs define cells a few inches wide. In some models the ribs are joined at the bottom by a perforated plastic sheet which bears on the underlying base course. Tabs or pins connect adjacent panels to each other.

The resulting pavement surface can be structurally flexible. A flexible lattice filled with aggregate or soil can absorb modest surface undulations without compromising appearance or function, so it is tolerant of uncompacted, swelling, and frost-susceptible subgrades. Plastic lattice materials tend not to absorb moisture or to be affected by chemicals or freeze-thaw cycles. The lattice spreads traffic load over its base course through the base sheet's continuity, the panel's rigidity, and the strength of connection of one panel to another.

Healthy grass that spreads over a geocell's ribs can form a splendidly uniform green surface. It can be as accessible for pedestrians as any other uniform grass surface. The open area at the bottom of a geocell panel is important for rooting of grass; openings in the walls between cells may assist spreading of certain grasses.

The models discussed in this chapter are 88 to 98 percent open at the surface, so the porosity, permeability, appearance and heat-island effect of a finished pavement are almost entirely dependent on the cells' fill and vegetation.

Many manufacturers supply guidelines for installation of their products. However, the geocell-producing industry does not have the benefit of an industrial association to set uniform standards of comparison or to educate potential users about appropriate applications. Manufacturers' reports of strength and other characteristics are too often based on tests that are inconsistent between one manufacturer and another, and between geocells and other types of paving material. In the absence of uniform measures of performance, potential users are left to rely on experience with specific models in specific types of settings. An impartial industrial association would give guidance to users and credibility to suppliers. The formation of an industrial association or ASTM committee to formulate uniform standards for plastic paving geocells is called for.

GEOCELL INSTALLATION

Because industry-wide construction guidelines have not been produced, typical components of construction must be inferred from general knowledge of pavement construction and the guidelines of individual manufacturers for their specific products.

Like any other pavement surfacing, geocells require a firm base course to bear traffic load. Over the base a setting bed of smaller aggregate may be required to make a uniform surface to receive the geocell units.

The light weight of plastic geocells makes them easy to transport, to put into position and, where necessary, to remove and replace for utility access or pavement repair. However, they may similarly be jarred easily by moving traffic, so some models, in some applications, need to be anchored to the base. Anchors can be metal or plastic spikes, pins, or rods. Placement of boulders, logs, or wheel stops over the geocell layer can similarly secure it. Edge restraint is not required for panels that are anchored or firmly interlocked panel to panel.

In installations with aggregate fill, the aggregate must be angular in order to resist displacement. It must be open-graded in order to be permeable; a typical size is ASTM No. 8 or No. 89. It is compacted into place with a vibrating plate or roller; additional aggregate is then swept in for leveling.

In installations with vegetation, a sandy planting medium resists compaction and maintains aeration and drainage better than does fine-textured soil or organic matter. The medium can be settled into the cells through vibrating or watering. Irrigation may be important at least in the first year or two because the small pockets of soil in the cells have little water-holding capacity, especially if the medium is sandy. Plastic geocell materials tend to be nonabsorbent; they neither "steal" moisture from the planting medium nor store it for later use. Traffic should be allowed on the surface only after the vegetation has established itself.

Sod can be installed by pressing it into empty cells. This type of installation is known to be feasible with thin-walled geocells; the thick ribs of other models may inhibit pressing in of the sod. Sod installed this way should be grown in noncohesive sandy soil, free of reinforcing netting materials, and cut at a thickness equal to the depth of the cell. It is theoretically possible to grow grass in a plastic geocell at a remote site before installation, and then when ready install the geocell with its contained grass as one would a layer of sod. Anchoring of geocells to inhibit movement may protect growing grass roots. In turn, deep roots may add tensile strength to a pavement structure. Following installation, mechanical aeration of turf must be avoided: probing or coring can damage plastic geocell material.

The base course can be made part of the rooting zone if the base course material (and setting bed, if any) is open-graded so that it is penetrable by roots, aerated, and well drained. The aggregate material can be made more viable by blending into it a limited volume of planting medium. It is very important to limit the quantity of medium in order to preserve void space and prevent compaction; the resulting base mixture should be like those for tree rooting described in Chapter 5. If the aggregate is limestone, the grass type must be tolerant of alkaline soil. Geotextiles must be omitted from the upper layers of the pavement so roots can grow unobstructed into the base course; where necessary an open-graded aggregate filter layer might be used in place of a geotextile.

MAINTENANCE COST

Table 8.1 illustrates the potential maintenance costs of geocell pavements. It lists estimated annual costs for a parking lot at Mitchell Center Arena in Mobile, Alabama, the layout of which was introduced in Chapter 2. The parking lot holds 450 cars in a 2.9-acre area. It combines surfaces made with the Grasspave^2 and Gravelpave^2 models of geocell, both of which will be described later in this chapter. The university's grounds department (Lindsey, 1999) estimated geocell maintenance

TABLE 8.1
Estimated 20-year Maintenance Costs for the Grasspave^2–Gravelpave^2 Parking Lot at Mitchell Center Arena, and of a Dense Asphalt Lot in the Same Location

Item	Maintenance Cost ($ per square yard per year as of 1998)
Grasspave^2-Gravelpave^2 Combination	
Mowing (0.052 labor + 0.019 mower)	0.070
Fertilizer & lime	0.007
Irrigation maintenance	0.002
Irrigation water	0.112
Painting & stripping	0.001
Box blading & leveling of Gravelpave^2	0.103
Total	0.296
Dense Asphalt	
Resurfacing (0.126) & sealing (0.088)	0.214
Painting & stripping	0.148
Pothole repair	0.131
Crack filling	0.044
Total	0.497

Source: Lindsey, 1999.

cost for a 20-year life span, and compared it with that of uniform dense asphalt (the installed size may have been slightly smaller than that assumed in the estimate, with greater proportion of Grasspave^2 in the area).

The estimator based the asphalt maintenance costs on the university's recorded costs for asphalt parking pavements in the previous five years. Forty-three percent of the asphalt's cost would be seal coating every five years.

In the geocell parking lot Grasspave^2 with Bermuda turf occupies the parking stalls. The grass is irrigated and seasonally overseeded with annual ryegrass. Because the university had no historical maintenance data for Grasspave^2 surfaces, the estimates in the table are based on the university's experience with maintenance of unreinforced turf. The biggest component of the estimated cost is irrigation water. The estimator assumed that no cost would be necessary for repair or replacement of the Grasspave^2 geocell during the 20-year estimate period (observations in 2002 suggest that this assumption may be correct).

Gravelpave^2 with aggregate fill occupies the turning lanes. Because the university had no historical maintenance data for Gravelpave^2 surfaces, the Gravelpave^2 cost in the table is hypothetical. It consists of an assumed 4 hours per month of box blading and leveling. The estimator assumed that no cost would be necessary for repair or replacement of the Gravelpave^2 geocell during the 20-year estimate period (observations in 2002 suggest that some such cost may become necessary in heavily trafficked portions of the turning lanes, as described in Chapter 2).

According to these estimates, the maintenance of the porous pavement would cost 40 percent less than that of dense asphalt. Table 8.2 differentiates the costs attributable to the Grasspave^2 and Gravelpave^2 areas.

TABLE 8.2
Costs Attributable to the Separate Types of Porous Pavement in the Mitchell Center Arena Parking Lot

	Grasspave²	Gravelpave²	Total
Area (acres)	1.3	1.6	2.9
Area (% of total)	45	55	100
Cost ($ per acre per year)	2080	910	1433
Cost ($ per year)	2700	1,455	4155
Cost (% of total)	65	35	100

Calculated from Lindsey (1999).

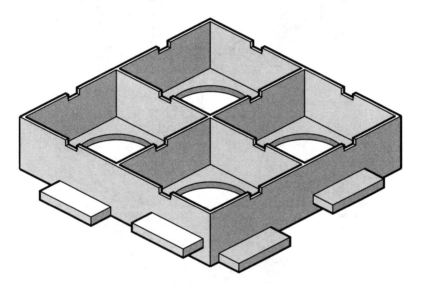

FIGURE 8.1 Portion of a Geoblock panel.

GEOBLOCK

The Geoblock model of geocell is manufactured by Presto Products (www.presto-geo.com). Geoblock is intended for use with either grass or aggregate. It has also been used without any fill at all, for wide-wheeled off-road vehicles to drive directly on the geocell ribs. The Presto company and some regional installers make useful technical literature available to designers.

Geoblock can be recognized in the field by the distinctively square arrangement of its ribs as shown in Figure 8.1. Each panel is 2 inches thick and 20 inches × 40 inches in area. Each open cell is 3 1/8 inches × 3 1/8 inches in size. The ribs are notably rigid and sturdy; they are 1/8 inch thick on the interior of the panel and 50 percent thicker on the periphery. Peripheral tabs hold adjacent panels 5/8 inch apart, making the joint in effect into another open cell. The unit is open 92 percent at the top and 60 percent at the bottom (measured from sample provided by manufacturer).

FIGURE 8.2 Geoblock emergency access lane at the Ritz-Carlton Huntington Hotel in Pasadena, California.

Geoblock is a heavy-duty model. Its rigidity and peripheral tabs cause weight to distribute across a panel and from panel to panel. Laying the panels with staggered joints integrates the geocell layer structurally. The interlock of adjacent panels at the peripheral tabs is flexible; where necessary screws can be driven through the tabs to prevent panels from separating.

Geoblock is made partly from recycled plastic. A large portion of Presto's business is the making of plastic bags; the company recycles scrap material from the manufacturing process into the geocell product. Geoblock is black in color.

RITZ-CARLTON HUNTINGTON HOTEL, PASADENA, CALIFORNIA

An emergency access lane at the Ritz-Carlton Huntington Hotel in Pasadena, California exemplifies the use of Geoblock in a common type of application (Erika Pryor, personal

Plastic Geocells

FIGURE 8.3 Geoblock emergency access lane at the Alaska Department of Health and Social Services Public Health Laboratory, Anchorage, in 2003.

communication 2001). Figure 8.2 shows what appears to be a wide strip of lawn at the base of the building. To make an access lane the turf is reinforced by a Geoblock layer resting on a firm aggregate base. The meticulously maintained turf makes a thick living surface which completely hides the geocell's black plastic ribs. The grassy surface is used for formal or semiformal events such as wedding parties and is visually prominent at all times beneath the hotel's rooms and terraces.

PUBLIC HEALTH LABORATORY, ANCHORAGE, ALASKA

A Geoblock emergency access lane at the Alaska Department of Health and Social Services Public Health Laboratory in Anchorage illustrates the use of geocells in a notably cold setting. Figure 8.3 shows the lane three years after installation. The grass is a mixture of (in order, beginning with the biggest component) red fescue, Bering hairgrass (*Deschampsia beringensis*), smooth brome, and five other local grasses (planting plan by Land Design North, courtesy of Tamas Deak, April 2003). In this cold setting both growth and weed invasion are slow; mowing is expected to be required very seldom. The naturalistically coarse-textured turf layer thoroughly hides the black ribs of the Geoblock reinforcement except for a few square inches adjacent to a concrete walk. Beneath the Geoblock is an aggregate base course.

REMOTE TRAILS, ALASKA AND WISCONSIN

Geoblock's great sturdiness has invited its use for stabilization of trails used by all-terrain vehicles (ATVs) and off-road bicycles. The trails illustrate the use of Geoblock in a distinctive type of construction and in challenging climatic environments.

An ATV trail on the wet, low-lying Palmer Hay Flats Game Refuge in Alaska inevitably crosses areas of soft soil and standing water (Hunt, 2002; Kevin Meyer, personal communications 2002 and 2003). In the late 1990s, ATVs on the unstabilized trail had eroded muddy ruts, then veered into new paths, trampling wide swaths of sedges and grasses. In Alaska, ATVs are a practical way to access remote country,

FIGURE 8.4 Geoblock ATV trail at Palmer Hay Flats State Game Refuge, Alaska.

but damage to vegetation and wildlife habitat from ATV traffic is a public concern (Komarnitsky, 2001; Zencey, 2001). In 2000, the Alaska Department of Fish and Game and the National Park Service's Rivers, Trails and Conservation Assistance program experimented with various stabilization materials, and then using the results stabilized the most sensitive 800 feet of the trail with Geoblock.

The Geoblock panels were laid directly on the fine-grained, vegetated soil without a base course. Adjacent panels' peripheral tabs were fastened together with screws and plastic cable ties to make a secure trail 6.4 feet wide. Where the trail goes through seasonally standing water, to prevent the panel from floating, the cells were filled with native gravel 3/4 inch to 2 inches in size. In drier areas there is no gravel fill because of the difficulty of transporting fill to the remote trail location; the ATVs ride directly on the geocell ribs.

Figure 8.4 shows the installation in March 2003, three years after installation. Indigenous grasses had regenerated under the geocell mat and grown up through the empty cells from below. Growing ice crystals have lifted soil into the cells; the soil partly fills the cells and supports dense grassy vegetation (dormant and low at the time of the photograph). The gravel-filled units support little vegetation. The Geoblock units were in excellent condition, aligned panel-to-panel with no crushing of the ribs by traffic. However, in warm weather, some of the panels have buckled upward because they expand with the heat and are locked rigidly together by the screws at their joints (Kevin Meyer, personal communication 2003).

The National Park Service has built similar Geoblock ATV trails on other lands in Alaska. Meyer (2002) and Allen et al. (2000) provided detailed guides to this type of construction. The geocells prevent soil erosion by the vehicles, distribute the traffic load over the soil, and pass naturally diffuse surface and near-surface flows of water (Meyer, 2002, pp. 25–29). The exposed plastic ribs are not considered slippery for ATVs even when wet (although they make a surface that is only fair for pedestrians and wildlife, and poor for horses).

Figure 8.5 shows that the Geoblock panels can be assembled into trails of various widths. In each arrangement staggered joints distribute the load from panel to panel while keeping the panels in alignment.

In the Park Service's practice, following leveling of the trail ground as necessary and elsewhere the shearing of tussocks of grass at the base, the panels are laid directly on vegetated soil (Meyer, 2002, pp. 25–29). Vegetation has regrown through 70 to 90 percent of the unfilled cells within a couple of years after installation. The regrowth stabilizes the installation, restores some of the site's biological productivity, and visually integrates the installation into the environment. On extremely wet and soft subgrade, geotextiles or mats are added to keep the plastic soil from squeezing into the geocell's unfilled cells.

Geoblock panels that are rigidly fastened together and exposed to the sun may expand in warm weather up to 1 percent by length of the panel (Meyer, 2002, p. 44; Kevin Meyer, personal communication 2003). To make "expansion joints," panels of Netpave (a flexible model of geocell described later in this chapter) replace panels of Geoblock at intervals of approximately 40 feet along the trail. Because Netpave does not have a bottom sheet that distributes loads over soft subgrade, it is placed on pockets of firm soil or over supplemental mats or geotextiles.

In the late 1990s, the U.S. Bureau of Land Management used Geoblock to confine ATVs to a single, well-defined and stable path on the Quartz Creek trail in Alaska's White Mountains National Recreation Area. The trail traverses a hillside of alpine tundra with shallow, organic, easily erodible soils. At the trailhead Geoblock panels were fastened together with screws into 6.4-foot-wide units to be carried to the site. The trail route was leveled with shovels and, where necessary, drained at the side with a small ditch. The Geoblock units were laid over sheets of plastic netting to help resist sinking into the mud. At intervals Netpave "expansion joints" were attached to the Geoblock panels by trimming projecting tabs off both units and driving in screws. The cells were filled with soil to anchor the panels. Project staff observed that the soil fill reduced potential chipping and breaking of the cell walls, blended the trail into the tundra environment, and reduced possible animal entrapment.

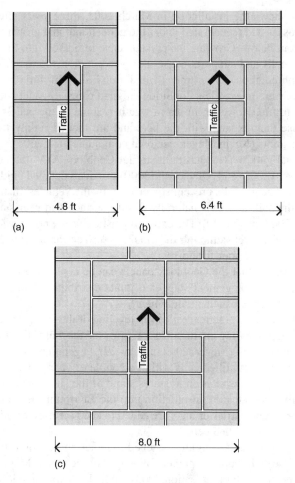

FIGURE 8.5 Arrangement of Geoblock panels to form trails of various widths (adapted from Meyer, 2002, p. 40).

In Wisconsin, Geoblock was used to stabilize trails designated for off-road recreational bikes in Kettle Moraine State Forest (*Erosion Control*, 1995; Bjorkman, 1996). Thirty thousand bikers per summer season had eroded the unstabilized trails, especially on steep slopes. Bikers swerved around the eroded spots, in effect widening the trails and inviting further erosion. In the mid-1990s the Wisconsin Department of Natural Resources installed Geoblock on steep (usually 25 percent and steeper) sections of trail 100 to 350 feet in length, amounting to 5 or 10 percent of the total trail length. The panels were placed on a 2-inch base of 1/4 inch aggregate and fastened together with screws. Traction was of greater concern here than in the Alaskan ATV trails because of the steep slopes and the different type of vehicle, so all cells were filled with soil. In shady areas and areas of intense traffic vegetation did not thrive in the cells; the soil wore down to the tops of the Geoblock ribs, which then resisted further erosion. Nevertheless, monitoring showed that on a section of

TABLE 8.3
Web Sites Where Information about Trail Stabilization with Geoblock and Other Materials May Be Updated

Subject	Web Address
Palmer Hay Flats project	www.prestogeo.com
White Mountains project	http://wwwndo.ak.blm.gov/WhiteMtns/html/projects/geogrid/default.html
Trail construction	National Off-Highway Vehicle Conservation Council, www.nohvcc.org

Geoblock-stabilized trail with 62 percent slope, after three months and five rainfall events, soil loss from the stabilized trail was less than 1 percent of that from an equally steep section of unstabilized trail.

Table 8.3 lists web sites where information about projects of this type may be updated.

GRASSPAVE^2

The Grasspave^2 model of geocell is manufactured by Invisible Structures Inc. (ISI, www.invisiblestructures.com), which developed the original version of the model in 1982. ISI intends this model for use with grass, in contrast to the company's Gravelpave^2 model which is meant for aggregate (Gravelpave^2 will be described later in this chapter). ISI is one of the largest manufacturers of geocells in North America; its web site makes product information and installation case studies abundantly available.

In ISI's geocell products the ribs are in the shape of 1-inch-high rings as shown in Figure 8.6. At the bottom the rings are bound onto a sheet of thin, flexible connecting ribs. The flexible geocell structure conforms smoothly to an undulating base surface. It can be cut with pruning shears. It comes in the form of rolls 3 feet wide; tabs at the edges anchor adjacent rolls to each other. Gravelpave^2 is open 98 percent at the top and 92 percent at the bottom sheet (measured from sample provided by manufacturer).

In the field, ISI's geocell products are easily identified by the distinctive ring pattern shown in Figure 8.7. Grasspave^2 can be distinguished from Gravelpave^2 by Grasspave^2's absence of an attached geotextile at the base, its distinctive dark color, and ring walls which are slightly thinner than Gravelpave^2's.

The circular configuration gives the thin ribs individual stability. However the thin, flexible bottom ribs neither stiffen the rings further, nor significantly distribute traffic load from the rings across an underlying base layer. Although the thinness and flexibility of the material make Grasspave^2 a relatively light-weight model, Grasspave^2 has been structurally successful in practice because the grass placed in its cells limits its applications to settings with infrequent traffic.

Because Grasspave^2's rings occupy so little of the surface area, the surface's hydrology is fundamentally determined by the porous material that fills the cells, and not by the impervious plastic. This was demonstrated in monitoring by Brattebo and Booth (2003) as described in Chapter 4. Two parking spaces were surfaced

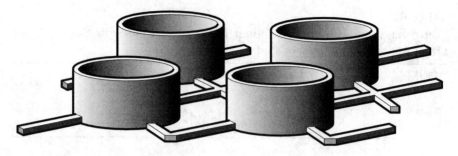

FIGURE 8.6 Portion of a Grasspave² panel.

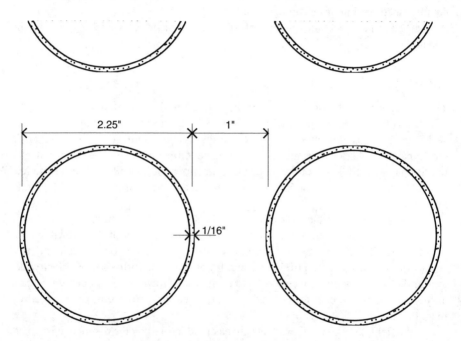

FIGURE 8.7 The Grasspave² model of geocell as seen from above (measured from sample provided by installer).

with Grasspave²; the cells were filled with sand, and grassed. Although the surface ended up with inconsistent grass cover, in the area's gentle rainfall it generated next to no runoff, unlike a parallel panel of dense asphalt where runoff closely mimicked the rate of rainfall. The Grasspave² panel generated a small amount of runoff during one unusually intense storm of 1.65 inches of rain in 14 hours, but the runoff coefficient was still only 0.02. Runoff occurred also during other, less intense, events when cars occupied the Grasspave² parking stalls, as water sheeted off the roofs and hoods of the cars and temporarily saturated parts of the Grasspave² surface.

ORANGE BOWL, MIAMI, FLORIDA

A parking lot at the Orange Bowl stadium in Miami, Florida exemplifies the use of Grasspave^2 for event parking (Thompson, 1996; White, 1996b). The stadium is owned by the City of Miami and used for major sports events. It is located in an old part of the city, in the midst of a dense residential neighborhood. In the mid-1990s the stadium needed more parking space and improvements in its drainage and external appearance. In political meetings neighborhood residents advocated an appearance of "open space," and not of a "pavement," while the limited capacity of the city's old storm sewers dictated that at least some water had to be disposed on-site. The combination of concerns prompted the use of a porous grassy surface.

In 1995, 1745 parking spaces were installed in Grasspave^2. Figure 8.8 is an interior view during the 2001 football season. Grasspave^2 parking bays occupy 65 percent of the 9-acre parking field. The 36-foot-wide bays are double-loaded and hold up to 20 parking spaces in each row. Driving lanes of dense (impervious) asphalt bear heavy moving traffic; short painted lines on the asphalt edges delineate individual Grasspave^2 parking spaces. There are no curbs or wheel stops anywhere in the parking lot; perimeter parking is confined by bollards. (Parking spaces for daily use by stadium employees are in separate parking lots made of dense asphalt.)

The Grasspave^2 geocells were placed on a 6-inch-thick base of compacted crushed limestone similar to customary road bases in the area (Enrique Nunez, personal communication July 2000). They were laid so the elevation at the top of the rings matched the surface elevations of the adjacent asphalt lanes.

The grass is Argentine bahia, a "tough," drought-tolerant grass that is commonly used in highway projects in the region. It was installed in the form of sand-grown sod, cut with soil 1 inch thick to match the depth of the Grasspave^2 cells. The sod sections were laid on top of the rings, then rolled with cylinders to embed them in the cells (Enrique Nunez, personal communication 2000; White, 1996b). The friability of the sand, the resilient roots of the grass, and the narrowness of Grasspave^2's ribs allowed the sod to be embedded without harm. Additional sand was spread on top to fill any remaining joints between sod sections. The elevation of the finished sod surface matched that of the ring structure.

Surface runoff from the asphalt driving lanes drains onto the grass parking bays, where it is treated and absorbed. Each bay forms a shallow swale with a drop inlet at one end for overflow runoff. Although the parking lot is only 4 to 5 feet above sea level and the seasonal high water table is within 12 inches of the surface, only the excess runoff generated during infrequent large storms reaches the drainage inlets and the city's storm sewers.

The parking lot's vehicular and pedestrian traffic is intense during events, many of which involve day-long tailgating. Between events are long periods with little or no traffic; only a limited number of events occur per year.

The same trained maintenance team that cares for the turf inside the stadium also maintains the parking lot (White, 1996b). The grass is not irrigated (Enrique Nunez, personal communication 2003). Six years after installation the grass was in uniformly excellent condition, with essentially no evidence of rutting five days after a major college football game. The grass cover completely hid the Grasspave^2 ribs. If

FIGURE 8.8 Grasspave^2 parking at the Orange Bowl, six years after installation and five days after heavy use during a college football game.

any sediment had been tracked onto the field by vehicles, the living grass had built it into the porous soil matrix. The grass was visually in better shape than the adjacent asphalt, which showed signs of patching and repainting.

If there were to be a contest for the best-looking parking lot in North America, of any surface material, the Orange Bowl's Grasspave^2 parking lot would be a candidate. In the long periods when the lot is vacant the appearance is that of a large, well-defined lawn. Neighborhood residents think of it as open space, not as a parking lot (Thompson, 1996). Its attractiveness is produced by a number of factors of design, maintenance, and use. The appropriately trafficked and splendidly maintained grass is green, uniform, and neat. The grass environment is sensibly cooler than "hard" pavements in other nearby parking lots. The gently curving asphalt lanes and grass panels define the field's internal order; their rhythmical repetition defines distance across the space. The perimeter of well-kept and color-coordinated bollards, fencing, and trees reinforces the definition of distance and the shape of the ground plane within which the lanes and panels lie like an orderly patterned carpet.

WESTFARMS MALL, FARMINGTON, CONNECTICUT

A parking lot at Westfarms Mall in Farmington, Connecticut illustrates Grasspave^2's use in seasonal overflow parking. Westfarms Mall is a large, busy shopping mall with well-maintained grounds. The Grasspave^2 overflow area was installed during enlargement of the mall in 1995; Figure 8.9 shows the area seven years after installation.

The use of porous grass-based pavement was prompted by a municipal requirement for stormwater control (Thompson, 1996; White, 1996a; Vince McDermott, personal communication 2002). The crowded shopping center site had no room for

FIGURE 8.9 Grasspave² overflow parking at Westfarms Mall in Farmington, Connecticut, in 2002.

enlarged detention reservoirs. Although Grasspave²'s cost was higher than that of dense asphalt, it was less than half the projected cost of a $1-million detention and infiltration structure that would have been required if an equivalent amount of dense pavement had been installed (*Environmental Building News,* 1994) and it protected the quality of groundwater by soil filtration (Shawn McVey, personal communication 2002). The local municipality also advocated leaving a portion of the site in "green space."

The 4.7-acre Grasspave² area is located in a corner of the shopping center site distant from building entrances. It holds 400 parking spaces for overflow use in the peak holiday shopping season (Kevin Keenan, personal communication 2002). The remainder of the shopping center's parking is on dense asphalt or in parking garages.

A 6-inch-wide concrete band separates the Grasspave^2 from adjacent asphalt paving; a raised curb separates it from unreinforced grass outside the parking area. In the grassy field, Grasspave^2 covers both the parking stalls and the aisles. Parking bays are indicated only by curbed planted "islands" at the ends of the bays and lines of trees between them; there are no wheel stops or continuous internal curbs.

The Grasspave^2 geocells were placed on an aggregate base course. The cells were filled with topsoil and hydroseeded (White, 1996a). However, two successive large thunderstorms washed away large areas of seed. With no time left for establishment of seeded grass before it would be used in the peak holiday shopping season, the third and final attempt at planting was with sod (Kevin Keenan, personal communication 2002). The sod is a blend of bluegrass varieties. It was rolled to press it onto the rings, but because the cells had earlier been filled with soil some of the turf remained above the rings. The soil in the sod is fine-textured; its infiltration rate is moderate.

During the one-month peak holiday shopping season the Grasspave^2 area is used mostly for employee and valet parking (Kevin Keenan and Vince McDermott, personal communications 2002; White, 1996a). Although the area is rarely completely filled there is traffic into it all day long. Seasonally employed staff mark the grass with athletic-field paint or place ropes between the trees to delineate parking bays. To reduce local wear they sometimes alternate the driveways into the lot from one day to another.

Burial of the Grasspave^2 rings beneath the sod surface has left tires in direct contact with unreinforced soil. Rutting has occurred at the main entry into the grass area where traffic is most concentrated; essentially no similar damage has occurred in the less-trafficked parts of the area. The plastic soil is especially vulnerable when it is wet and soft; less damage is done when it is firmly frozen (Kevin Keenan, Shawn McVey, personal communications 2002).

In years with early winter snow the area is snowplowed up to a few times per parking season. To prevent the plow from damaging the turf, maintenance staff attached a roller to the back of a plow blade which keeps the blade 1/2 inch to 1 inch above the surface (Kevin Keenan, personal communication 2002; White, 1996a).

During the growing season the field is empty of cars, and its appearance is that of a large, well-maintained lawn, where red maple trees and a few curbed "islands" give definition to the space. It presents a green foreground for the view of the mall from a nearby street. Turf maintenance is scheduled as a routine part of the mall's grounds care. The entire growing season is available for turf restoration and growth. Summer visits six and seven years after installation showed that thick continuous grass covered the geocell ribs everywhere except in the tire tracks at the heavily trafficked entrance.

The grass is irrigated. It is given more water than other nearby turf to compensate for the small water-holding capacity of the soil in the cells (Kevin Keenan, personal communication 2002). The irrigation water is harvested from a detention pond that had been installed during the shopping center's original development.

The trees rooted in the parking field were large and healthy seven years after planting. Presumably the permeable material all around them provides abundant rooting space. There is no perceptible evidence of heaving by roots or by frost; the flexible geocell structure would be tolerant of heaving, were it to happen.

Plastic Geocells

FIGURE 8.10 Drainage from the Westfarms Mall Grasspave² area several hours after the end of a substantial rain in May 2002.

Near the low edge of the area the soil is sometimes wet. Evidently, excess water drains through the base course to this low edge and then saturates the soil from below. As shown in Figure 8.10, after a substantial rain water overflows visibly from the grass onto adjacent dense asphalt, long after the impervious surfaces have dried off, visibly reflecting the contrast between rapid runoff from the dense pavement and detention in the porous pavement.

Trace Pointe, Clinton, Mississippi

A Grasspave² emergency access lane at Trace Pointe in Clinton, Mississippi exemplifies geocell placement on swelling soil. Trace Pointe is a large retirement facility that opened in 1999. Figure 8.11 shows the access lane at the rear of the building two years later. Under the Grasspave² surface is an aggregate base course on compacted subgrade.

Grass surfacing was selected for its appearance in the midst of the residents' walkways and sitting areas. The lane gets essentially no traffic except that of landscape maintenance vehicles.

The grass in the access lane is centipede sod. It is irrigated from a permanent sprinkler system. It was more or less well established after one year, with the black Grasspave² rings almost completely hidden.

The subgrade soil is silt loam in the Loring series, derived from the notoriously swelling Yazoo clay. Although there was no noticeable deformation of the surface two years after installation, if swelling were to happen it would cause a gentle undulation

FIGURE 8.11 Emergency access lane at Trace Pointe Retirement Community, Clinton, Mississippi.

in the Grasspave^2 which would not compromise the use or appearance of the flexible grassy surface.

BLUE CROSS BLUE SHIELD, SOUTHFIELD, MICHIGAN

A Grasspave^2 emergency access lane at Blue Cross Blue Shield of Michigan exemplifies the beauty that grass-based paving can achieve with careful detailing in the most common type of application. The setting is the grounds of a large office building in Southfield, Michigan, adjacent to a wooded stream corridor. Figure 8.12 shows the access lane winding prettily at the edge of the trees behind the building.

The lane is wide, well maintained, and lit by bollard lamps. Its character is more that of a safe, well-maintained footpath than of a utilitarian access lane. On one side is the wooded valley with the sound of a stream. On the other side the base of the building opens into a large terrace from which landscape edges arc artfully into the winding lane. The sequence from building, to terrace, to terrace-side landscape, to grass, to wooded valley is a gradient from man-made to natural. The grassy lane is a logical step in this sequence; its visibly evident maintenance makes the lane and even the edge of the woods part of the cared-for landscape (Nassauer, 1997).

At the time of a visit in April 2002 the lane's uniform grass hid the black Grasspave^2 rings. A few small spots were slightly worn where saturation at the end of the snowmelt season had softened the soil under the wheels of maintenance vehicles, and where snow plows pushing snow from an asphalt road onto the grass had scraped away a small area of turf and beaten up the plastic ribs.

Plastic Geocells 303

FIGURE 8.12 Grasspave[2] emergency access lane at Blue Cross Blue Shield of Michigan, Southfield, Michigan.

Curbs would be out of keeping with the nonstructural character of the lane. Instead the grassy path is lined with trees, plants, and bollard lamps that define the path's alignment while blending it into a composition with the terrace and the wood's edge. The bollards in particular are obviously man-made markers. They are spaced about 50 feet apart on each side of the lane and placed alternately from side to side, so they are only 25 feet apart on the two sides together. The topography of the lane further defines it: the slope is downward to the valley on one side, and to the terrace area or a swale on the other side. The plantings, bollards, and topography simultaneously define the lane and integrate it into the landscape sequence from building to valley.

GRASSROAD AND TUFFTRACK

The Grassroad geocell is manufactured by NDS (www.ndspro.com). The full name of the model is Grassroad Pavers[8]. In 1999, NDS acquired the previous manufacturing company, Bartron, which had been making Grassroad since 1984 (John Bosanek, personal communication January 2002).

Grassroad comes in panels 24 inches × 48 inches in area and 1 1/4 inches thick. Each panel has 240 hexagonal cells like those shown in Figure 8.13. In the field Grassroad can be distinguished from other hexagonal geocells by the indentations at the tops of the ribs. The ribs are joined at the bottom by a perforated sheet. Tabs lock adjacent panels together. Cleats projecting 5/8 inch from the bottom inhibit lateral

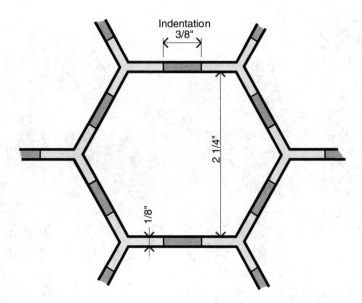

FIGURE 8.13 The Grassroad geocell as seen from above (measured from sample provided by manufacturer).

movement. The panels are made partly from recycled material. They can be cut with a saw. They are either green or black.

The indentations at the tops of the ribs might allow grass to spread from cell to cell protected from traffic compaction. Openings in and around the base sheet permit horizontal and vertical extension of grass roots. The panel area is open 94 percent at the top and 61 percent at the bottom sheet (measured from sample provided by manufacturer).

NDS also manufactures the similar Tufftrack model, having acquired the previous manufacturing company ADS (John Bosanek, personal communication January 2002). Tufftrack has hexagonal cells similar to Grassroad's but without indentations at the top of the ribs. Some users believe that the smooth upper surface improves pedestrian accessibility. Tufftrack is 1/4 inch thicker, has different panel-to-panel latching, is more open at the bottom, and has no bottom cleats. It comes in 24 inch x 24 inch panels, and only in black. Like Grassroad, it is made partly from recycled material.

HYATT REGENCY HOTEL, LONG BEACH, CALIFORNIA

An emergency access lane at the Hyatt Regency Hotel in Long Beach, California exemplifies Grassroad's use in a common geocell application. Figure 8.14 shows that the lane is set very visibly at the base of the tall hotel building and adjacent to a heavily used public park. In this carefully maintained landscape the turf is thick and uniform; the geocell ribs are completely hidden. The arrangement of trees and the topography indicate the alignment of the lane.

Plastic Geocells 305

FIGURE 8.14 Grassroad emergency access lane at the Hyatt Regency Hotel in Long Beach, California.

CALIFORNIA JPIA, LA PALMA, CALIFORNIA

Grassroad parking stalls at the California Joint Powers Authority (JPIA) in La Palma, California make a complex and neatly detailed landscape. JPIA (www.cjpia.org) occasionally has large training events, but the office is in a crowded urban setting with little space for parking, so this installation is for overflow parking and occasional turning of large vehicles. Figure 8.15 shows the installation in JPIA's dense but campus-like setting.

About one third of the campus' parking spaces are in Grassroad-reinforced grass. The permeable surfaces may enlarge the rooting zones of the campus' many trees, which thrive under the landscape's scrupulous maintenance. The Grassroad spaces are on the parking lot's perimeter. On the outside of the stalls is a raised concrete

FIGURE 8.15 Grassroad overflow parking spaces at California JPIA in La Palma, California.

curb. On the inside is a concrete strip which stabilizes the edge of an adjacent asphalt pavement. During peak-use periods cars double up for parking back-to-front on the grassy spaces and adjacent asphalt spaces. Wheel stops indicate individual parking stalls.

GRASSY PAVERS

The Grassy Pavers geocell is manufactured by RK Manufacturing (www.rkmfg.com). The model is often referred to by the company's initials, RKM.

Grassy Pavers come in panels 15 inches × 20 inches in area and 1 3/4 inches thick. The ribs make a hexagonal lattice similar to that of Grassroad, joined by a bottom sheet. In the field, Grassy Pavers can be distinguished by the crenellations on its ribs: as shown in Figure 8.16, at the top of each rib are nine low ridges, and at each rib intersection is a low round knob. Although each panel is rigid, tabs connect adjacent panels flexibly. Laying the panels with staggered joints integrates the geocell layer structurally. Perforations in and near the base permit horizontal and vertical extension of roots and percolation of water. Cleats projecting 3/4 inch from the bottom inhibit lateral shifting. The panels can be cut with a saw. They are colored dark green. The manufacturer makes colored plastic inserts for placing in the cells to delineate edges or parking stalls. The unit is open 94 percent at the surface and 42 percent at the bottom sheet (measured from sample provided by manufacturer).

Plastic Geocells

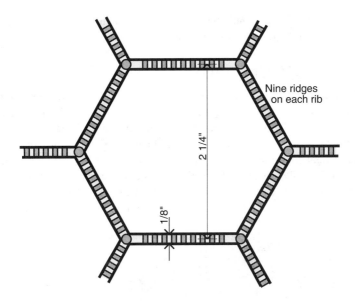

FIGURE 8.16 The Grassy Pavers ("RKM") model of geocell as seen from above.

RKM also manufactures a more flexible variety of Grassy Paver, called Equiterr. It is intended for use in horse paddocks and stalls.

Medgar Evers Building, Jackson, Mississippi

A loading area at the Medgar Evers Building in Jackson, Mississippi illustrates the use of Grassy Pavers with grass and the placement of porous pavement on swelling soil. Medgar Evers is the central post office facility for the city.

The Grassy Pavers cells are filled with fine-textured soil and planted with centipede grass. The soil was placed to the top of the plastic ribs, exposing it to traffic compaction without protecting from the ribs. Neither centipede nor fine-textured soil has a good reputation for tolerating traffic. In this casually installed driveway, even infrequent vehicular use has been sufficient to compact the soil and suppress the grass in the wheel tracks. In the summer growing season, the dark-green Grassy Paver ribs are exposed in the sparsely vegetated wheel tracks but are visually unobtrusive in the generally green grassy surface. During the dormant winter season, when centipede thins and turns yellow-gray, the plastic lattice may be visually more prominent.

The Grassy Pavers panels are in good condition even where the soil is compacted. The plastic ribs protrude above eroded soil without apparent damage.

Below the geocells is an aggregate base course on compacted subgrade. The subgrade soil is derived from the notoriously swelling Yazoo clay. There is no visible deformation of the pavement surface; evidently, the compaction of the subgrade and the weight and firmness of the base have been sufficient to inhibit soil swelling.

If swelling were to occur despite these features, the flexible geocell pavement could absorb surface undulation without damage.

Residence, Santa Monica, California

A small residential parking area in Santa Monica, California illustrates the use of Grassy Pavers with aggregate fill, and distinctive design details in a small urban space. The city installed the parking for a small multifamily residence as one of its experiments in urban rehabilitation and reduction of nonpoint pollution (Neal Shapiro, personal communication August 2001). Figure 8.17 shows the parking area soon after its installation in 2001, with construction signs still up and rye grass beginning to sprout. The residential building faces a confined alley space in a dense urban neighborhood. Grassy Pavers adjacent to the building reinforce two aggregate

FIGURE 8.17 Residential parking spaces of Grassy Paver with aggregate fill in Santa Monica, California, soon after construction.

parking spaces and small adjoining grass areas. There are no curbs. At the time of the visit to the installation in 2001 it had not been in operation long enough to show any signs of use; only its design features allowed speculative critique and prognosis.

The aggregate is angular, open-graded, 1/4 inch to 1/2 inch in size, and highly permeable. It is tan in color. The parking spaces are delineated by bright white aggregate held in rows of Grassy Paver cells. Over time, displacement of the aggregate might obscure the gravel parking stripes; surface drainage might bring clogging matter from a downspout on one side, and from the dense alley pavement on the other.

Small areas of soil and grass fill the cells in the installation's periphery. The turf's edges and corners are vulnerable to compaction by vehicles straying off the asphalt pavement. Reddish concrete paving blocks 12 inches x 12 inches in size rest in cut-away portions of the grass-filled geocells, presumably on the same base as the geocells. If the portions of the grass later become compacted by vehicles, those portions could be removed and replaced by similar blocks cut into the surface.

GRAVELPAVE^2

Gravelpave^2 is made by Invisible Structures, the same company that produces Grasspave^2 (www.invisiblestructures.com). The model was developed in 1993. It is intended to be filled with aggregate. It is made of the same type of recycled plastic as Grasspave^2 and formed into flexible rolls with the same distinctive ring structure. In the field Gravelpave^2 can be distinguished from Grasspave^2 by the attachment of a geotextile to the underside, and by slightly thicker ring walls (3/32 inch instead of 1/16 inch). The purpose of the geotextile is to prevent small aggregate particles in the cells from collapsing into the voids of potentially large open-graded aggregate in the base below. Gravelpave^2 comes in different colors to match the aggregate that will be placed in it. Figure 1.19 showed the laying out of Gravelpave^2 rolls on a base course of ASTM No. 57 aggregate blended with sand.

Gravelpave^2's aggregate fill has invited applications under traffic frequency much greater than that to which Grasspave^2's turf is limited, such as the main entry aisles of busy parking lots. However, Gravelpave^2's thin ribs have been structurally challenged by the concentrated traffic loads to which they have been subjected. Portland cement has been used to stabilize the aggregate fill in some settings (Chere Peterson, personal communication 2000). The cement is mixed into wet aggregate at 10 to 15 percent by volume, and the mixture is compacted into the cells. The surface may then take on the color of the cement.

BURGER KING RESTAURANT, HENDERSON, NORTH CAROLINA

A Burger King restaurant in Henderson, North Carolina exemplifies the use of geocells in a retail setting. The restaurant is located on U.S. Business Rt. 1 at Parham Road. Gravelpave^2 covers about half of the restaurant's parking stalls, to provide the restaurant's commercially required customer parking capacity while satisfying a city ordinance limiting impervious cover (Shannon Britt and Chris Spelic, personal communications 2000 and 2001). The driving lanes and other parking stalls are of dense asphalt. Figure 8.18 shows two of the restaurant's parking bays a couple of years after construction.

FIGURE 8.18 Gravelpave² parking at a Burger King restaurant in Henderson, North Carolina.

The aggregate is rounded river gravel, uniformly sized at about 3/8 inch. It is extremely permeable. It is tan and white in color, and held in tan-colored Gravelpave² rings.

Most of the Gravelpave² spaces are on the parking lot's perimeter. On the outside of the stalls is a raised concrete curb. On the inside the aggregate abuts the dense asphalt of the driving lanes. There is no marking of individual parking spaces. Vehicles park properly where the curbs are configured for 90° parking, but tend to park equally at 90° in bays shaped for diagonal parking.

In bays sized for only 5 to 10 cars, parking is limited to a dead-end configuration, the slow-moving cars displace little aggregate, and the geocell is in good condition. In contrast the small patch of unsightly displacement shown in Figure 8.19 developed in a long bay marked for boat and RV parking. Many large vehicles had parked lengthwise along the bay and then, upon restarting, drove along the bay as if it were a driving lane before turning out at high speed. The moving, turning traffic eroded the rounded gravel particles and beat down the plastic rings.

This experience shows that geocell-reinforced aggregate is suitable for dead-end parking stalls even in commercial settings. However, it also affirms the requirement for the aggregate to be angular in order to resist displacement. To further inhibit displacement, aggregate should be used only in parking bays where through traffic is prohibited by perimeter location, limited size, or wheel stops.

Grand Canyon Trust, Flagstaff, Arizona

A Gravelpave² parking lot at the Grand Canyon Trust in Flagstaff, Arizona exemplifies the use of geocells with aggregate in a cold, snowy setting (Keating, 2001). The

FIGURE 8.19 Displacement of rounded gravel by moving vehicles at the Henderson Burger King.

trust is a regional conservation group (www.grandcanyontrust.org); it shares a small office facility with a local office of The Nature Conservancy. The elevation of the site is 7000 feet. Although Flagstaff gets only 19 to 20 inches per year of precipitation (measured as liquid water), in the winter it occurs in the form of 80 to 90 inches of snow (Ruffner and Bair, 1977).

The trust installed its 6000 square foot parking lot in 1998. It is used daily by about a dozen automobiles and an occasional delivery truck. Gravelpave^2 covers both the parking stalls and a central lane. Because of the small size of the lot, traffic is limited even in the lane. The cells are filled with decomposed granite 3/16 inch to 1/4 inch in size. Concrete wheel stops define individual parking stalls. The trust's staff monitor their porous parking lot through their daily use of it and adjust detailed maintenance as needed (Rick Moore, personal communication 2003).

This was one of the first installations of Gravelpave^2, and during the first year some small corrections were made to the installation (Keating, 2001). Anchoring had been omitted. At the entrance concentrated traffic tended to bounce up the edge of the geocell. In the stalls, slush under parked cars could freeze, and when the cars moved again they could pull up some aggregate and lattice. In both places retrofit anchors (12-inch spikes) were placed and a small amount of aggregate was replaced.

In subsequent years a hand shovel, rake, and wheelbarrow have been used to move displaced aggregate back into position at entries where moving traffic concentrates (Keating, 2001; Rick Moore, personal communication 2003). The labor amounts to 4 to 6 man-hours per year. There are next to no weeds even in low-traffic corners, perhaps reflecting a scarcity of weeds in Flagstaff's dry, cold climate.

The plastic lattice was in good condition five years after installation (Rick Moore, personal communication 2003). The surface is free of dust. Rainwater visibly infiltrates; runoff is unknown.

In the first year snowplows caught on the lattice and lifted up small sections. Subsequently the plow operators were told to plow only if the snow got deeper than 4 inches and to keep the blade a couple of inches off this special surface (an alternative would be a roller attached to the blade, like that at the Grasspave^2 installation at Westfarms Mall described earlier in this chapter). Since then plowing has caused no abrasion (Rick Moore, personal communication 2003). Cars pack the remaining snow down into the aggregate surface. When the snow melts, the meltwater drains into the aggregate without standing puddles.

Walking on the aggregate is comfortable to the trust's staff, who tend to dress for outdoor conditions, but is bothersome to occasional visitors wearing high heels (Rick Moore, personal communication 2003). Aggregate particles are sometimes tracked into cars.

This experience indicates that aggregate reinforced with geocells is suitable for snowy climates where traffic is light and plowing for snow removal is controlled. It points out a special need for Gravelpave^2 anchoring in parking stalls in cold climates.

FROSTBURG STATE UNIVERSITY, FROSTBURG, MARYLAND

A parking lot on the campus of Frostburg State University in Frostburg, Maryland exemplifies the use of Gravelpave^2 in a larger parking lot with correspondingly greater traffic in the central driving lane. The lot is called the Sand Spring lot; it is located on University Drive, adjacent to Cambridge Residence Hall. It is 14,800 square feet in area; this is large enough to hold about 40 cars. It is used daily in season by dormitory residents, day students, and visitors to nearby sports fields. Figure 8.20 shows the parking lot about two years after installation.

The porous Gravelpave^2 surface was installed in about 1998 to protect an adjacent stream while stabilizing a formerly unpaved surface which had become muddy and rutted. The layout invites 90° parking on both sides of a central lane.

Plastic Geocells

Gravelpave[2] covers both the stalls and the lane. Vehicles park properly at 90° despite the absence of noticeable markings for individual stalls.

The surface fill is open-graded crushed limestone about 3/8 inch in size, colored purple and tan. Its permeability is very high. The aggregate-filled geocell lies on a base of crushed limestone 1 inch to 1 1/2 inches in size.

A visit in July 2000 showed that aggregate particles had been displaced in the heavily trafficked part of the central lane near the lot's entrance, leaving the aggregate surface as much as 1/4 inch lower than the top of the rings. In a few patches the tops of the exposed rings had deformed under the traffic or been torn away leaving deeper pits in the surface. Fine particles ground up and eroded from the pits covered nearby rings in shallow sheets, reducing the surface infiltration rate.

Two years later university staff had added a 1-inch layer of aggregate on top of the original surface. The new material was a mixture of 3/8 inch crushed stone and sand. However, the new aggregate had no geocell reinforcement and was subject to the same concentrated traffic. In the driving aisle pits continued to erode, alternating with "dunes" of displaced aggregate and wide sheets of crusted, slowly permeable fine material. One can infer that the central lane always needed something more stable than an aggregate type of surface, because the traffic is inherently too intense for it.

In the parking stalls there was very little gravel displacement from the beginning. The surface has been uniformly smooth and stable enough for easy walking, and highly permeable. The geocell ribs are almost unnoticeable, being uniformly filled and covered with gravel. After the addition of the new aggregate layer, gentle traffic

FIGURE 8.20 The Sand Spring parking lot on the campus of Frostburg State University, Frostburg, Maryland.

compaction only smoothed the surface. The sand in the new layer washed down into the older aggregate layer, leaving uniformly coarse particles at the surface.

The Frostburg parking lot remains porous and permeable. However, the experience shows that concentrated traffic in a busy parking aisle requires a cement binder or a structurally heavy-duty type of porous surfacing. In contrast, the light traffic of dead-end parking stalls gives great flexibility in choice of porous paving material.

Where a pitted geocell is repaired, the geocell material should be cut out of the broken area and replaced, perhaps with heavier-duty material, level with the original surface. A layer of aggregate on top of a geocell layer is as displacable as one without any geocell reinforcement at all. Repair can be economically limited to the torn areas; the addition of a uniform aggregate layer is mostly wasted.

NETPAVE

The Netpave model is distributed in North America by licensees of the British company Netlon (www.netlon.co.uk). It is manufactured in Germany by the Schwab company, where it is called Schwabengitter 2000 (www.horst-schwab.de; Ronnie Doctor, personal communication July 2002). Figure 8.21 shows a portion of the Netpave grid as seen from above. It can be recognized in the field by the distinctively curved ribs between rigidly square grids. All the ribs are notably thick and sturdy. Each panel is 20 inches × 20 inches in area. The Netpave 50 model is 2 inches thick. Netlon also makes Netpave 25, a thinner model intended for placement on top of existing grass.

The curved ribs make a Netpave panel flexible. It conforms to uneven ground and absorbs temperature changes without buckling. Adjacent panels connect with lugs and slots. The panels are manufactured from recycled plastic. They are black in color. They can be cut with a saw. The unit is 88 percent open at the top. White plastic cell inserts are available for marking parking stalls or lane edges.

Although the individual ribs widen slightly at the bottom, there is no bottom sheet to distribute traffic load over the layer on which the geocell sits, so it is vital to place Netpave on a firm base that can bear the load directly from the ribs (Meyer, 2002, pp. 44–45). Figure 8.22 shows the thickness of aggregate base course suggested by Netlon. In addition to the base thicknesses shown, for installations with aggregate fill Netlon suggests a 1-inch bedding layer of 2 to 6 mm (0.08 to 0.24 inch) aggregate, and the same aggregate in the cells (Netlon, no date). For installations with soil and grass, Netlon suggests a 2-inch compacted bedding layer of 70 percent sand and 30 percent other soil, and the same medium in the cells to within 1/4 or 1/2 inch of the surface before seeding.

Because Netpave is flexible and heavy-duty, it has been used as an "expansion joint" in Geoblock trails as described earlier in this chapter. It is not used for the entire trail, because its lack of a base sheet prevents it from spreading out load over soft, wet soils.

NIGEL'S PITCH, MIDDLETOWN, RHODE ISLAND

One of the very few Netpave installations in North America to date is a parking lot at a soccer field called Nigel's Pitch in Middletown, Rhode Island. The town

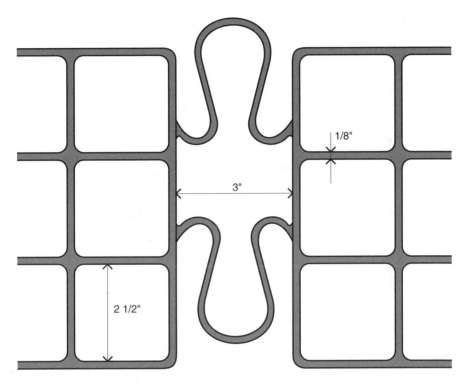

FIGURE 8.21 Portion of a Netpave 50 panel as seen from above.

constructed Nigel's Pitch on Wyatt Road in 2000 (Mike Embry, personal communication 2002). The purpose of porous pavement on the site is to protect the Maidford River watershed, which is a designated drinking-water protection area for the City of Newport. The installation illustrates the durability of Netpave under traffic. However, it was poorly installed, and exemplifies the consequences of an inadequate base course.

The layout invites parking on both sides of a central lane. Netpave 50 covers both the parking stalls and the central lane. Only a fence on one side and a hedgerow on the other indicate the parking arrangement; individual parking spaces are unmarked. In the summer the parking lot is heavily used during soccer events. There is no winter maintenance, whether plowing, sanding or salting.

The subgrade is fine-textured, plastic, slowly permeable till. Much of it is fill resulting from leveling of the soccer field. The fill in the geocells is the same native soil, planted with grass. This is an appropriate place for a grassy surface, because the seasonal use limits annual traffic load, the scheduled use allows maintenance of the grass, and the grassy surface suits the pastoral setting of farms, homes, and hedgerows.

However, two years after construction, traffic in the central lane had badly compacted the soil and suppressed the grass, and even pressed down the geocells to make an irregular surface. Crusts of compacted clay alternated with washes of

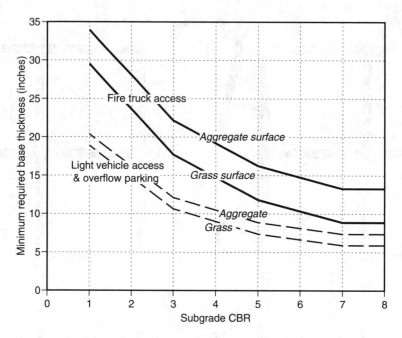

FIGURE 8.22 Suggested thickness of firm aggregate base under Netpave 50 (data from Netlon, no date).

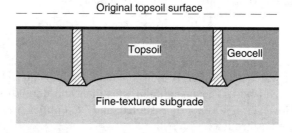

FIGURE 8.23 Inferred subsidence of geocell into soft subgrade at Nigel's Pitch, Middletown, Rhode Island.

coarser particles; infiltration rate was low. On the two sides grass remained in the parking stalls; infiltration rate was adequate.

The pavement had not been given an adequate base course (Arthur Erhardt, personal communication 2002). Figure 8.23 shows an interpretation of what had happened. Without a firm base course, traffic in the central lane pressed the Netpave ribs and fill against and into the underlying subgrade. The inadequacy of the base course caused the fine-textured topsoil in the cells to be compacted and subsequently crusted by rainfall when the grass was lost, making the installation a failure both structurally and hydrologically.

Plastic Geocells **317**

The installation did not do justice to the Netpave product. The geocell showed no sign of breakage or deformation even in the spots with the most concentrated traffic. The flexibility of the lattice had maintained the continuity of the surface layer even while the lattice was pressed into the underlying soil.

The product has recently been installed at Schmelz Countryside Volkswagen in Maplewood, Minnesota, and Spirit Hills Center in Lino Lakes, Minnesota (William McCully, personal communication 2003).

OTHER MODELS OF PLASTIC GEOCELLS

Other models of plastic geocells are intended for use in pavements and reputedly have been used with grass fill in residential driveways and commercial emergency access lanes. However, specific installations have not been located.

The NeoTerra Lawn Sav'r model of geocell is manufactured by Matéflex (www.mateflex.com). According to the manufacturer's project literature its ribs form interlocking cloverleaf shapes. The ribs are connected by a perforated bottom sheet. The cells are approximately 2 inches wide. Each panel is nominally 20 inches × 40 inches in area and 1.75 inches thick. It is green in color; white inserts are available for indicating the edges of lanes and parking stalls.

The PermaTurf model of geocell is manufactured by the PermaTurf company (www.permaturf.com). Its ribs form a hexagonal pattern as shown in Figure 8.24. The ribs are attached to a bottom sheet with a hexagonal perforation at each cell. Each panel is 1.5 inches thick and 13 inches × 13 inches in area; it is open 94 percent at the top and 33 percent at the bottom. It is green and made from recycled plastic. Adjacent panels interlock firmly but flexibly.

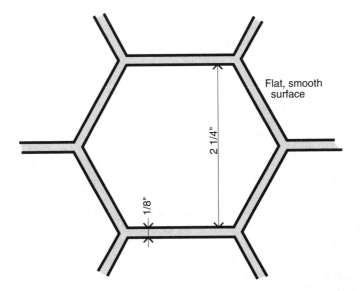

FIGURE 8.24 The PermaTurf geocell as seen from above (measured from sample provided by manufacturer).

TABLE 8.4
Examples of Flexible Plastic Webs for Cellular Stabilization of Aggregate

Product	Manufacturer	Contact Information
Envirogrid	GeoProducts	www.envirogrid.com
Geolok	Fluid Systems	www.fluidsystems.com
Geoweb	Presto Products	www.prestogeo.com
TerraCell	Webtec	www.webtecgeos.com

FLEXIBLE PLASTIC WEBS

Flexible plastic webs are available for strengthening aggregate layers in base courses and certain surface applications. The webs come in "mattresses" several inches in thickness. The mattresses are pulled to open the cells, and the cells are filled with soil or aggregate. Table 8.4 lists examples of this type of product. Essentially, all of them can be recognized in the field by the almost fabric-like flexibility of the cell walls. Not all of them are intended for use in surface courses.

The tensile strength of a plastic web inhibits aggregate from shifting laterally, so the aggregate distributes load and reduces the intensity of pressure on the subgrade. The addition of a web may reduce a pavement's cost by permitting reduced thickness of the aggregate base layer, or avoidance of removal and replacement of soft subgrade. The use of a web may be combined with the use of a geotextile at the bottom of the aggregate layer.

In Washington State, the U.S. Forest Service (Gusey, 1991, pp. 6, 19) tested Geoweb's stabilization of recreational trails used by hikers, motorcyclists, and equestrians. The stabilized areas were on soft soils and steep slopes. The cells were filled with native sandy soil from alongside the paths. Geoweb was considered easy to transport to the remote stabilization sites, but the installations did not blend into the woodland settings visually. In traction and durability they were rated "good."

STONE MOUNTAIN STATE PARK, GEORGIA

Figure 8.25 shows a Geoweb-stabilized road at Stone Mountain State Park, Georgia. Vehicles use the road to ascend the mountain for maintenance of tourist facilities and transmission equipment. Although the mountain has numerous hikers, the road's alignment hides it from pedestrians, and its purpose is utilitarian.

The road is made of crushed granite aggregate up to 1 1/2 inches in size. The material is open-graded with large pores and very high permeability. It is placed directly on the mountain's bare, steeply sloping granite surface without excavation for leveling. In some places the aggregate structure is held on the downhill side by a low retaining wall of cement-filled sandbags. Geoweb stabilizes the uppermost few inches of the aggregate material. Where the Geoweb is exposed at the surface, traffic has beaten the fabric down to the level of the aggregate. On parts of the driving surface cement was added to inhibit displacement by vehicles. Most of the mountain's surface runoff drains through the road's aggregate without concentration. Culverts carry

Plastic Geocells 319

FIGURE 8.25 Aggregate maintenance road stabilized with Geoweb at Stone Mountain State Park, Georgia.

concentrated water where it occurs. There is no sign of erosion or displacement of the aggregate anywhere on the road.

TRADEMARKS

Table 8.5 lists the holders of registered trademarks mentioned in this chapter.

ACKNOWLEDGMENTS

The following persons generously provided information used in this chapter: Gary Bach, Bart Bartman, Linda Lecker and Patricia Stelter of Presto Products

TABLE 8.5
Holders of Registered Trademarks Mentioned in This Chapter

Registered Trademark	Holder	Headquarters Location
Envirogrid	GeoProducts	Houston, Texas
Geoblock	Presto Products	Appleton, Wisconsin
Geolok	Fluid Systems	Belle Chase, Louisiana
Geoweb	Presto Products	Appleton, Wisconsin
Grasspave[2]	Invisible Structures	Golden, Colorado
Grassroad Pavers[8]	NDS	Lindsay, California
Grassy Pavers	RK Manufacturing	Jackson, Mississippi
Gravelpave[2]	Invisible Structures	Golden, Colorado
NeoTerra Lawn Sav'r	Matéflex	Utica, New York
Netpave	Netlon	Blackburn, UK
TerraCell	Webtec	Charlotte, North Carolina
Tufftrack	NDS	Lindsay, California

(www.prestogeo.com); Steven M. Bopp of Austin Tao & Associates (http://austin-tao.com/); John Bosanek of NDS (www.ndspro.com); Shannon Britt and Chris Spelic of Invisible Structures (www.invisiblestructures.com); Leta Bryan of Gary Graves Landscapes, Jackson, Mississippi; Cal Callahan of MCCA (Destin, Florida); Tamas Deak of Koonce Pfeffer Bettis (www.kpb-architects.com); Ronnie Doctor of Netlon (www.netlon.co.uk); Michael Embry of the town of Middletown, Rhode Island; Arthur Erhardt of GridTech (www.gridtech.com); Debbie Flanagan of The Nature Conservancy (www.tncarizona.org); Daryl Gusey of the U.S. Forest Service, Lakewood, Colorado; John Hechtel, Alaska Department of Fish and Game; Jay Hudson and William McCulley of Glenn Rehbein Company (www.rehbein.com); Tom Hunt of the University of Wisconsin at Platteville; Kevin Keenan of Westfarms Mall; Robert Koerner of Drexel University; Dan Larsen of the Brock White Company (www.brockwhite.com); Tim Lawrence of Ohio NEMO (htp://nemo.osu.edu); Lisa Lawliss of the Rhode Island Department of Environmental Management; Vince McDermott and Jim MacBroom of Milone and MacBroom, (www.miloneandmacbroom.com); Shawn McVey of the U.S. Natural Resources Conservation Service, Storrs, Connecticut; Kevin Meyer of the National Park Service Rivers & Trails Conservation Assistance Team, Anchorage, Alaska; Rick Moore and Steele Wotkyns of the Grand Canyon Trust (www.grandcanyontrust.org); Enrique Nunez of the City of Miami Planning Department; Chere Peterson of Petrus UTR (www.petrusutr.com); Christopher Ponzani and Brian Kralovic of Ponzani Landscaping (St. Clairsville, Ohio); Erika Pryor of Soil Stabilization Products (www.sspco.org); Samuel Randolph of Soil Stabilization Products (www.sspco.com); Christine Sassenberg, PermaTurf (www.permaturf.com); Brad Stringer of Cooke Douglass Farr Lemons, Jackson, Mississippi (www.cdfl.com); Neal Shapiro of the City of Santa Monica Environmental Programs Division; Devi Sharp of Wrangell-St. Elias National Park, Alaska; Lance Uhde of Lance Construction (Athens, Georgia).

The following persons constructively critiqued early drafts of this chapter: Bill Bohnhoff of Invisible Structures (www.invisiblestructures.com); Chere Peterson of

Petrus UTR (www.petrusutr.com); David Nichols of the University of Georgia School of Environmental Design (www.sed.uga.edu); and Samuel Randolph of Soil Stabilization Products (www.sspco.com).

REFERENCES

Allen, J.L, K.E. Shea, W.M. Loya, and P.J. Happe (2000). *All-Terrain Vehicle (ATV) Trail Mitigation Study: Comparison of Natural and Geosynthetic Materials for Surface Hardening*, Research Report No. 00-1, Copper Center, Alaska: U.S. National Park Service, Wrangell–St. Elias National Park and Preserve.

Bjorkman, Alan W. (1996). *Off-Road Bicycle and Hiking Trail User Interactions: A Report to the Wisconsin Natural Resources Board*, Madison, Wisconsin: Wisconsin Department of Natural Resources, Bureau of Research.

Brattebo, Benjamin O., and Derek B. Booth (2003). Long-term Stormwater Quantity and Quality Performance of Permeable Pavement Systems, *Water Research* 37, 4369–4376.

Environmental Building News (1994). Paving with Grass, in *Environmental Building News* 3, downloadable at www.buildinggreen.com.

Erosion Control (1995). Erosion Controllers Hit the Trail, *Erosion Control* January/February, p. 15.

Gusey, Daryl L. (1991). *Trail Hardening Test*, Project No. ORV-86-3P, U.S. Forest Service, Wenatchee National Forest.

Hunt, Tom (2002). Hard Trails in Alaska: Responding to ATV Impacts on the Last Frontier, *Land and Water* 46, 31–34.

Keating, Janis (2001). Porous Pavement, *Stormwater* 2, 30–37.

Koerner, Robert M. (1998). *Designing with Geosynthetics*, 4th ed., Upper Saddle River, New Jersey: Prentice Hall.

Komarnitsky, S.J. (2001). Wear Repair, *Anchorage Daily News* July 17, p. B1.

Lindsey, Andy (1999). *Comparisons of Maintenance*, Mobile: University of South Alabama Grounds Department; available at www.invisiblestructures.com.

Logsdon, Alva D. (2001). Window is Inspiration for Award Winning Campus Plaza, *Landscape Architect and Specifier News* 17, 18–22.

Meyer, Kevin G. (2002). *Managing Degraded Off-Highway Vehicle Trails in Wet, Unstable, and Sensitive Environments*, 2E22A68-NPS OHV Management, Missoula, Montana: U.S. Forest Service Technology and Development Program.

Nassauer, Joan Iverson, 1997, Cultural Sustainability: Aligning Aesthetics and Ecology, *Placing Nature: Culture and Landscape Ecology*, pp. 66–83, Joan Iverson Nassauer, Ed., Washington: Island Press.

Netlon (no date). *Netpave 50 Grass Paving System, Design and Installation Guidance for Gravel Surfaces and Grassed Surfaces*, Blackburn, U.K.: Netlon Turf Systems (www.netlon.co.uk).

Nichols, David (1995). Comparing Grass Pavers, *Landscape Architecture* 85, 26–27.

Nonpoint Education for Municipal Officials (no date). *West Farms Mall*, University of Connecticut Nonpoint Education for Municipal Officials (http://nemo.uconn.edu).

Presto Products Company (1998). *Geoblock System General Design Package*, Appleton, Wisconsin: Presto Products Company.

Rabb, William (1999). Paved with Good Intentions, *Mobile Register* March 7.

Ruffner, James A., and Frank E. Bair (1977). *The Weather Almanac*, 2nd ed., Detroit: Gale Research.

Sipes, James L, and John Mack Roberts (1994). Grass Paving Systems, *Landscape Architecture* 84, 31–33.
Thompson, J. William (1996). Let That Soak In: Landscape Architects Are Finding Ways of Putting Storm Water to Good Use in "Green" Parking Lots, *Landscape Architecture* 86, 60–67.
Warren, John (1997). Friends University Uses Porous Pavement System for Renovation of Davis Hall, *Midwest Contractor* May 12.
White, Patrick (1996a). A Whole Lot of Turf: Permeable Paving Permits Mall Expansion in Connecticut, *Turf Magazine* February 1996, reprinted at www.invisiblestructures.com.
White, Patrick (1996b). Parking on the Grass: Miami's Orange Bowl Gets a Turf Parking Lot, *Turf Magazine* October 1996, reprinted at www.invisiblestructures.com.
Zencey, Matt (2001). Managing ATVs: High-Tech Trail Mats May Help Reduce Conflicts, *Anchorage Daily News* July 30, p. B4.

9 Open-Jointed Paving Blocks

CONTENTS

The Concrete Paving Block Industry 324
Configuration and Use of Block Surfaces 326
    Blocks for Open-Jointed Surfaces 326
    Relief and Use of Open-Jointed Surfaces 329
Construction of Open-Jointed Block Pavements 329
    Base and Bedding Materials 330
    Edge Restraint 331
Joint-Fill Materials and Their Implications 332
    Infiltration Rate 332
    Contrast in Potential Aggregate Gradations 332
    Aggregate Clogging and Restoration 334
    Vegetated Joints 336
Factors in Stability and Infiltration 337
    Factors in Block Layer's Contribution to Stability 337
    The Interaction between Block-Layer Stability and Infiltration 338
    Total Pavement Thickness 339
Drainstone Concrete Block 339
    English Park, Atlanta, Georgia 340
Ecoloc Concrete Block 342
    Annsville Creek Paddlesports Center, Cortland, New York 345
Eco-Logic Concrete Block 347
    Pier A Park, Hoboken, New Jersey 348
Eco-Stone Concrete Block 349
    Mickel Field, Wilton Manors, Florida 350
    Giovanni Drive, Waterford, Connecticut 353
    Castaic Lake Water Conservatory, Santa Clarita, California 355
    Harbourfront Fire Station, Toronto, Ontario 356
SF-Rima Concrete Block 358
    Alden Lane Nursery, Livermore, California 360
    Forest Hill Apartments, Wilmington, North Carolina 362

Paving Blocks of Other Materials 363
    Other Models of Concrete Block 363
    Manufactured Concrete Slabs 363
    Concrete Rubble Slabs 364
    Blocks of Porous Material 366
    Wooden Blocks 366
    Metal Plates 367
    Natural Stone 368
    Bricks 370
Open Patterns with Rectangular Paving Blocks 372
Trademarks 373
Acknowledgments 373
References 376

Open-jointed paving blocks are solid units shaped or placed to leave open spaces in the joints between adjacent units. The blocks are manufactured from concrete or clay, or cut from natural stone or wood. The joints are filled with porous aggregate or vegetated soil, or left vacant. Together, the blocks and joints form a composite surface that can be porous and permeable while bearing traffic loads through the structure.

A related kind of pavement material is the open-celled paving grid, which will be described in the following chapter. Grids differ from blocks in that grids have open cells cast into each paving unit; the porous portion of a pavement surface comprises the cells within the units, as well as any open joints between units.

Properly constructed block pavements are capable of bearing heavy traffic loads and can be remarkably durable in a wide variety of conditions. Many block products are manufactured in controlled factory conditions to meet industry-standardized requirements. Different available colors and shapes give versatility in appearance and freedom in aesthetic design; a pavement's appearance can be customized for a specific site and given a high level of detail (Smith, 2004). If soils swell or subside, the blocks can be lifted off, the base releveled, and the same blocks replaced. Access to subsurface utilities is unobstructed and is fast, quiet, and simple; waste is small. In many projects the blocks' potentially long service life offsets their high initial cost by lowering life-cycle cost.

Table 9.1 lists some terms that have particular application in open-jointed block construction. Paving blocks are sometimes referred to as "stones," but in this book that term is used to designate only natural mineral material.

THE CONCRETE PAVING BLOCK INDUSTRY

Blocks have been used to make pavements for many centuries. Originally, almost all the blocks were of stone and brick. After World War II, European cities rebuilt themselves while replicating the prewar character of their brick streets; for this purpose European companies developed the first concrete paving blocks (Borgwardt, 1998, pp. 19–21; Fischmann, 1999; Rollings and Rollings, 1992a, pp. 8–9; Shackel, 1990, pp. 1–12; Smith, 1999). The uniformity, durability, and versatile shapes of the concrete units got them quickly adopted. Simultaneously, in North America, most paving

TABLE 9.1
Some Terms with Particular Application in Open-Jointed Block Construction

Term	Definition
Bed (bedding, setting bed, laying bed, bedding layer)	Layer of relatively small aggregate on which block units are placed
Block	Solid piece of material used as a construction unit
Brick	Block made from fired clay
Chamfer	Beveled edge at the intersection of two faces
Flag (flagstone)	Thin flat piece of stone
Joint	Space or interface between adjacent blocks
Open-jointed	Having joints wide enough to produce or permit production of significant porosity and permeability
Paver	Any block, brick, grid, panel, tile, or lattice used for paving; or narrowly, a paving block of relatively compact dimensions, unlike a slab
Pavior	Craftsman who lays pavers
Precast	Cast, as concrete, before placement in final position
Segmental pavement	Pavement assembled from modular blocks or grids
Slab	Piece that is relatively broad in area
Stone	Natural mineral matter

focused on other materials such as asphalt and concrete due to those materials' comparatively low initial costs and America's great emphasis on automobile-oriented development where the only requirement was a smooth riding surface. But in the early 1960s German and Dutch companies developed efficiently mechanized concrete block manufacturing equipment; the technology was soon transferred to Britain, Japan, New Zealand, and South Africa, and in the 1970s to North America, for use in multifunctional urban renewal projects. At the same time, German companies developed open-jointed blocks. At least some of them were modifications of preexisting models of tight-jointed interlocking blocks; and because they were based on already-proven block shapes and installation techniques it was easy for them to catch on in Europe (Shackel, 1997).

The industry has been well supported by research and technical guidelines, examples of which are cited in the references at the end of this chapter. The research by Brian Shackel and his colleagues at Australia's University of New South Wales has been enormous. In North America, Rollings and Rollings researched block structures first in the 1980s for the Army Corps of Engineers, and then for manufacturers' associations. Theses done at the University of Guelph under the direction of William James (published in the proceedings of Computational Hydraulics International, www.chi.on.ca) have produced valuable new quantitative knowledge. An excellent contemporary summary manual is that by Smith (2001).

Today, open-jointed models are a burgeoning part of North America's concrete paving block industry. The industry is characterized by technical soundness and innovation, with new models becoming available for refined fulfillment of multiple pavement purposes, and a growing knowledge of design, installation and performance. The international licensing of German models and the activities of the Interlocking Concrete Pavement Institute (www.icpi.org) overcome distance and

TABLE 9.2
Associations and Sources of Information in the North American Concrete Paving Block Industry

Name	Contact Information
ASTM International	www.astm.org
Interlocking Concrete Pavement Institute (ICPI)	www.icpi.org
National Concrete Masonry Association (NCMA)	www.ncma.org

language to make continuing international developments available. Designers have the option of using stock models supplied by manufacturers, or of devising new models with custom shapes, colors and imprints (Campbell, 1993). The associations and sources of information listed in Table 9.2 advocate uniform standards and make available a wealth of design and construction guidelines and contractor certification.

ASTM C 936, *Standard Specification for Solid Concrete Interlocking Paving Units*, specifies for concrete blocks the constituents, strength, absorption of water, resistance to damage from freezing and thawing, and resistance to abrasion. Units conforming to the standard are very strong and durable; in effect they are made from a "no-slump" concrete mix, compacted into shape under pressure and vibration to give them an average compressive strength of at least 7200 psi. Models of blocks conforming to ASTM C 936 are little damaged by frost, and can withstand the scraping of snow plowing.

Trucks can economically transport concrete blocks up to 200 or 300 miles from manufacturing plants to construction sites, so the plants tend to be geographically distributed to supply regional markets. Even block models that are nationally and internationally licensed tend to be produced by regionally distributed licensees. Table 9.3 lists some examples of regional manufacturers. Additional manufacturers can be identified from member lists of the national industrial associations.

In choosing a concrete block model for a specific project, a first step may be to identify manufacturers within 200 to 300 miles of the project site and the models they have available. Alternatively, if a specific model is desired, regional sources for it might be identified from a list of licensed manufacturers on the licensing agency's web site.

CONFIGURATION AND USE OF BLOCK SURFACES

Block pavements are sometimes referred to as "segmental" pavements to distinguish them from poured panels of asphalt and concrete. They are assembled from modular units; their surfaces are mosaics with systematic pattern and relief. The various shapes given to different block models are attempts to balance in different ways the competing demands of porous area, solid traffic-bearing area, and stable interlock.

BLOCKS FOR OPEN-JOINTED SURFACES

Figure 9.1 delimits the range of block shapes that are discussed in this chapter under the heading "open-jointed blocks." As defined in this book, open-jointed blocks are

TABLE 9.3
Examples of Manufacturers or Suppliers of One or More Models of Open-Jointed Concrete Paving Blocks

Supplier	Contact Information
Abbotsford Concrete Products	www.pavingstones.com
Acker Stone	www.ackerstone.com
Anchor Concrete Products	www.anchorconcreteproducts.com
Angelus Block	www.angelusblock.com
Balcon	phone 410-721-1900
Basalite	www.basalitepavers.com
Belgard	www.belgard.biz
Borgert Products	www.borgertproducts.com
Capitol Ornamental Concrete	www.capitolconcrete.com
Castle Rock Pavers	phone 888-347-7873
Grinnell Pavers	www.grinnellpavers.com
Hessit Works	phone 812-829-6246
Ideal Concrete Block	www.idealconcreteblock.com
Interlock Paving Systems	www.interlockonline.com
Kirchner Block	www.kirchnerblock.com
Mutual Materials	www.mutualmaterials.com
Navastone	www.navastone.com
Nicolock	www.nicolock.com
Paveloc	www.paveloc.com
Paver Systems	www.paversystems.com
Pavestone	www.pavestone.com
Tremron	www.tremron.com
Unilock	www.unilock.com
Westcon	www.westconpavers.com
Willamette Graystone	www.willamettegraystone.com

those with indented edges, integral spacers, or laying patterns that produce open areas between adjacent blocks, while other parts of the blocks are in contact with each other for structural interlock. "Completely tight-jointed" blocks are those with little or no open space between blocks. "Widely spaced" blocks are those placed with large open areas between them, such that adjacent blocks have no significant structural contact.

In the figure, open-jointed "blocks with indentations" have narrow joints on a substantial part of the perimeter, while indentations create open areas at intervals. The narrow portions of the joints can produce significant block-to-block interlock for stability; some models have complex shapes that nest the blocks tightly in the paving pattern. At the same time the indentations' drainage openings can occupy a significant portion of the pavement surface.

Open-jointed "blocks with spacers" have projecting spacers that maintain uniformly wide joints. The wide joints commonly produce total open area comparable to that produced by blocks with indentations. Because the joints are uniformly wide, they invite joint fill to be of large, open-graded aggregate for rapid infiltration.

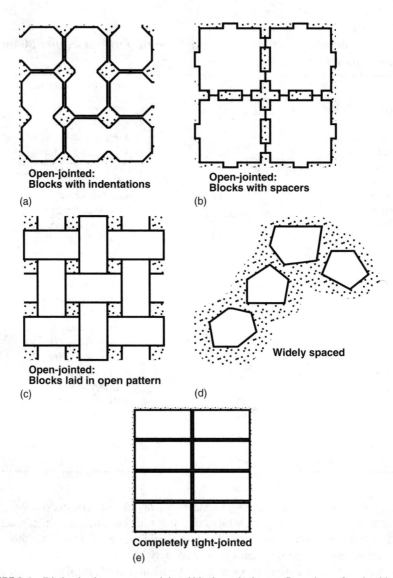

FIGURE 9.1 Distinction between open-jointed blocks and other configurations of paving blocks.

For significant porosity and permeability, joints or indentations wide enough to be filled with highly permeable open-graded aggregate are essential. In this book only those openings 1/4 inch and wider are considered to add significantly to the potential porosity and permeability of a pavement.

The consideration of opening width eliminates from use as porous pavements the many available block models with integral spacers 1/16 inch to 1/8 inch in size on all sides. In this book those models are considered tight-jointed as are those with no spacers at all. The purpose of such small spacers is to tighten the structural interlock between blocks by admitting dense-graded sand into the joint. Interlock and stability in such pavements can be very sound. However, infiltration through joints is very slow because of the narrowness of the channels between the sand grains and the adjacent block faces. Hade (1987) confirmed that blocks with exclusively narrow, sand-filled joints have high runoff coefficients, except under rain of the least possible intensity during which water may have been absorbed as much by the concrete block material as by pores in the sand joints.

RELIEF AND USE OF OPEN-JOINTED SURFACES

Open-jointed pavement surfaces have relief. To inhibit chipping, almost all block models have chamfers around the upper edges which produce relief of about 1/4 inch in the pavement surface. In the joints, aggregate fill is commonly recessed below the chamfer by up to an additional 1/4 inch. Thus, in many installations, the total surface relief is between 1/4 inch and 1/2 inch. This raises questions about, and opportunities for, applications of open-jointed paving blocks.

For pedestrian accessibility, the question of whether a given type of surface produces a required degree of accessibility is, with the knowledge we possess today, often a judgmental one which must be decided by the individual designer and owner. The *ADA Standards for Accessible Design* (U.S. Department of Justice, 1994, pp. 22–23) set an early judgmental guideline:

> "Changes in level up to 1/4 inch may be vertical and without edge treatment. Changes in level between 1/4 inch and 1/2 inch shall be beveled with a slope no greater than 1:2. Changes in level greater than 1/2 inch shall be accomplished by means of a ramp...."

According to that judgement, surfaces of chamfered blocks with slightly recessed aggregate may be accessible. Such published standards may be revised in the future. It has been speculated that block models with large flat surfaces (perhaps 8 inches × 8 inches and larger) present platforms for secure stepping, no matter how much relief there is in the joints. Narrow joint width (for example, 1/4 inch) and minimal chamfer may be another approach to ensure pedestrian accessibility. Ongoing research is defining quantitatively the terms in which accessibility can be evaluated, particularly for persons with disabilities (Cooper et al., 2002).

The movement of automobile wheels across a block surface produces a modest vibration which can be sensed by drivers and is audible to pedestrians. This is desirable where the vibration is used to calm traffic, and the sound to warn pedestrians of oncoming vehicles. In a different type of setting, the noise could be objectionable, for example, on busy city streets surrounded by residences. In those instances it is possible to reduce the noise by selecting blocks with narrow joints and little or no chamfer at the upper edges, and placing their long dimensions diagonal or parallel to the direction of traffic (Shackel and Pearson, 1997).

CONSTRUCTION OF OPEN-JOINTED BLOCK PAVEMENTS

Figure 9.2 shows typical components of an open-jointed block installation. Smith (2001) presented guidelines for this type of construction; his guidelines are aimed specifically at open-jointed blocks, so they should be used in place of any guidelines

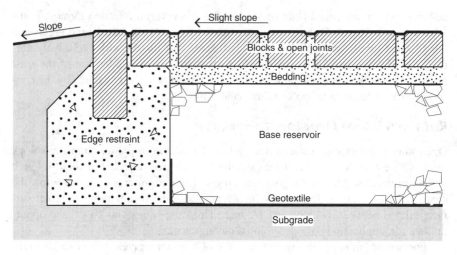

FIGURE 9.2 Typical components of open-jointed block pavement construction (adapted from Borgwardt 1992; Borgwardt 1997, Figure 2; and Smith 2001, pp. 8–9).

aimed at tight-jointed construction (examples of the many available guidelines for tight-jointed construction are those of Interlocking Concrete Paving Institute (ICPI) 1995 and National Concrete Masonry Association (NCMA) 1995).

BASE AND BEDDING MATERIALS

An aggregate base course or base reservoir spreads traffic load on the subgrade. No. 57 is an open-graded gradation commonly used for this purpose. The base aggregate is compacted as it is placed in lifts. A geotextile may be placed under the base following criteria such as those described in Chapter 3. Some conservative installers place geotextile under every block pavement, without regard to soil type.

A layer of relatively small open-graded aggregate is laid over the base as level bedding for the blocks. The bedding material must be angular and durable. Smith (2001, p. 28) suggested No. 8 gradation, which meets filter criteria over No. 57 and therefore does not require geotextile separation. The "bedding layer is a construction necessity but a structural weakness" (Rollings and Rollings, 1992a, pp. 39–40). It allows blocks to be placed and leveled, but under traffic it may be susceptible to compaction and rutting. Susceptibility to rutting increases with increasing thickness of the bedding layer, so the effective thickness is commonly limited to 1 inch; at this thickness rutting is not noticeable. On an open-graded base, ordinarily about 1 inch of the bedding material is compacted into the base's large voids, and then another inch is spread out, without compaction, as the bedding layer. Upon placement and vibration of the blocks, the bedding material works into the joints from below.

In relatively small projects the blocks are placed on the bedding by hand. On large projects specialized machines economically speed installation by picking up and placing many blocks at once. Joint-fill material is spread into the joints, and the block surface is vibrated to settle the blocks, bedding, and joint fill into a level, firm

Open-Jointed Paving Blocks

position. Thoroughly compacting (vibrating) the blocks into the bedding layer limits future rutting under traffic. A slight slope on the block surface discharges overflow runoff and surface debris to the pavement's edge; limiting the slope to no more than a few percent inhibits washing of aggregate out of the joints and promotes surface infiltration of water.

EDGE RESTRAINT

The provision of some form of edge restraint is imperative. Without firm edging, horizontal shifting of blocks and migration of bedding material from beneath them produces spreading joints, block rotation, and surface settlement (Burak, 2002a). Strong edge restraint is most particularly needed on the low sides of significant slopes and where traffic comes near or over the edge.

The form of restraint is a site-specific question almost independent of the construction of the rest of the pavement. The example of edge restraint in the previous figure was a block unit laid on end in concrete to form a flush curb. Figure 9.3 shows a raised curb set in concrete such as might be used for vehicle control between a street and a walkway (a raised curb does not block overflow drainage if it is located on the upslope side of a pavement). The subgrade or base material under a curb's concrete footing should be compacted, even if the subgrade under the adjacent base reservoir is not compacted.

The edge restraint shown in Figure 9.4 is a manufactured strip of metal or plastic held in place by a stake or rod. Examples of strips manufactured for this purpose were listed in Chapter 3. Edging strips must be placed on compacted subgrade or base material, and not on the bedding (Burak, 2002a). A specific edging strip must be selected for the traffic load and type of block where it will be used.

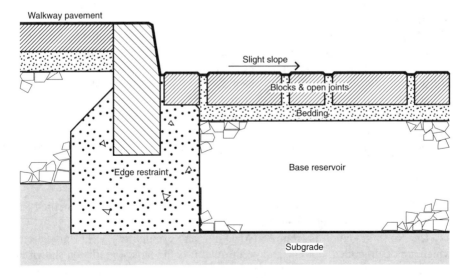

FIGURE 9.3 Example of block construction with raised curb (adapted from Borgwardt, 1992).

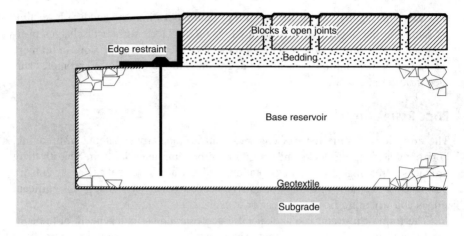

FIGURE 9.4 Example of block construction with edging strip of metal or plastic.

JOINT-FILL MATERIALS AND THEIR IMPLICATIONS

Whereas most block models are manufactured to fixed standards everywhere they are licensed, the fill for the joints must be specified and approved for each separate project. It is commonly the same material as that in the bedding. Proper joint fill need not be expensive; it must only be specified decisively, and contractors must be prevented from putting deficient material in its place.

INFILTRATION RATE

Figure 9.5 illustrates the importance of open-graded joint fill for pavement infiltration rate. It shows rates measured by Borgwardt (1997) in pavement surfaces containing different gradations of material. The designations "0/2 mm," "1/3 mm," and "2/5 mm" indicate ranges of particle size in millimeters. The 2/5 mm aggregate was the largest and most open-graded of the aggregates sampled, so it had the largest open pores and produced the highest permeability. (ASTM No. 8 aggregate, which is discussed at several points in this chapter, is even larger and more open, and can presumably produce even higher infiltration rates.) Smaller fill material produced lower permeabilities. Dense-graded topsoil material was used in joints planted with vegetation; its very fine particles greatly limited the infiltration rate; the growing vegetation presumably maintained as much permeability as the soil was capable of.

CONTRAST IN POTENTIAL AGGREGATE GRADATIONS

Figure 9.6 contrasts some of the gradations of aggregate that could be specified for joint fill in block pavement using North American standards.

The No. 8 gradation is from ASTM D 448 and exemplifies ASTM gradations which are open-graded, porous, and permeable. It is the largest gradation shown in the figure. Its large voids admit water rapidly and are relatively unsusceptible to clogging, so it is suitable for use in pavements for which high infiltration rate is a

Open-Jointed Paving Blocks 333

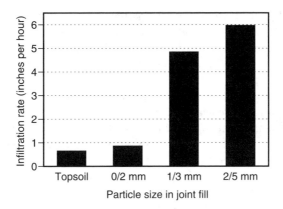

FIGURE 9.5 Average infiltration rates of block pavements with open joints occupying about 15 percent of the surface area, a few years after installation (after Borgwardt, 1997).

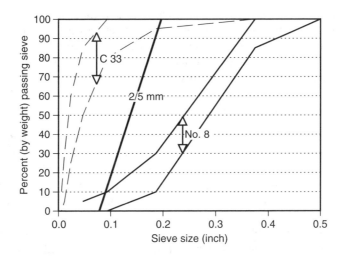

FIGURE 9.6 Selected contrasting aggregate gradations (data partly from ASTM C 33 and D 448).

priority. The particles are large enough not to collapse into the voids of a No. 57 base course. It is the gradation suggested by Smith (2001) for open-jointed pavements. When used with blocks with spacers that make uniformly wide joints, No. 8 aggregate comfortably fits into the joints and gives decisively high and durable infiltration rates. However, if No. 8 is used with blocks that have some narrow joints, it is unlikely to completely fill the narrow joints, and this might be a structural disadvantage under substantial traffic.

The 2/5 mm gradation has been used commonly in Germany for porous pavements. Although this gradation's particles are smaller than those of No. 8, its steep curve indicates a narrow range of particle sizes and freedom from fine particles. For blocks having combinations of indentations and narrow joints, this gradation's relatively small particles can fit into the narrow joints to assist structural interlock. Its voids are open enough to produce adequate infiltration rate, but they are small enough to trap debris and invite rooting of moss and other vegetation, and possibly result in a decline in its infiltration rate over time.

The C 33 gradation is defined in ASTM C 33, *Standard Specification for Concrete Aggregates*; it is commonly referred to as "concrete sand." It is very familiar in American block pavements; many guidelines for tight-jointed block pavement construction recommend this gradation as joint fill. However, it is a dense-graded material; its curve rises steeply at the left side of the chart, indicating a high content of fine particles. Its permeability is inevitably very low; it must never be used in pavements that are intended to be porous and permeable. Other fine-graded sands such as "mason sand" and "mortar sand" must similarly not be used for joint fill in open-jointed blocks, no matter how accustomed local contractors are to using them for tight-jointed pavements. A case study of the Castaic Lake Water Conservatory presented later in this chapter illustrates the consequences of using C 33 concrete sand in an open-jointed block pavement.

Joint-fill aggregate must be durable, angular, and, for rapid infiltration, open-graded. During construction, if the specified aggregate particles seem too large to penetrate and fill the joints, some contractors brush sand over the surface to fill the joints. They must be explicitly instructed not to do so, if the porosity and permeability of the pavement are to be preserved.

AGGREGATE CLOGGING AND RESTORATION

Table 9.4 and data presented in Chapter 4 show that the infiltration rates of many installations with 2/5 mm (0.08 to 0.2 inch) joint fill have declined over time. This may result from some combination of settling of the aggregate in place and the accumulation of exogenous particles in the pores. Theoretically Smith's recommended gradation of No. 8 (0.1 to 0.4 inch) aggregate should produce infiltration rates higher and more durable than those listed in the table.

James and Gerrits (2003) found that infiltration rates in an eight-year-old block pavement could be significantly increased by removing about 1/2 inch of fill from the joints. It is possible to remove the clogged or crusted surface material with a vacuum sweeper and replace it with new, clean, open-graded material. If a sweeper does

TABLE 9.4
Measured Infiltration Rates in Open-Jointed Block Surfaces with Joints Occupying about 15 Percent of the Surface Area, Filled with 2/5 mm Aggregate

Surface Character	Infiltration Rate (inch/h)	Reference
Newly installed	8.5–9.23	Borgwardt, 1997; Shackel and Pearson, 1997
2–5 years after construction; joint fill contained some clogging sediment	2.8–5.7	Borgwardt, 1994

not adequately remove the fill, then lifting up an area of blocks and resetting with new joint fill is possible.

James and Gerrits' tests were conducted in a parking lot made of Eco-Stone, an open-jointed block described later in this chapter. The joint fill was "clear washed stone gravel", but it may have been much smaller than No. 8. The pavement was located on the campus of the University of Guelph in Ontario (its construction was discussed in Chapter 3). Like other campus pavements, it had been routinely plowed for snow, sanded for winter traction, and swept with rotating brushes (William James, personal communication 2002). Experimental Eco-Stone panels occupied several parking stalls and half of a driving lane. The parking stalls were used almost daily by automobiles. The driving lane carried approximately 60 vehicles per day, and for one year had borne heavy vehicular traffic during the construction of a nearby building. The parking lot's gentle slope directed surface drainage from the parking stalls to a drop inlet in the center of the driving lane. The researchers measured the infiltration rates in different areas of the pavement surface, and then measured them again after incremental depths of aggregate fill were removed from the joints.

Figure 9.7 contrasts the results from three types of location in the parking lot. The parking stalls, with the lightest traffic and at the top of the drainage, had maintained the highest infiltration rate over time, and required little excavation to achieve still higher infiltration rates. The driving lane, with the heaviest traffic and at the bottom of the drainage where fine particles accumulated, had lost much of its infiltration rate, and was so deeply clogged that its infiltration rate benefited little from excavation. The margin between the lane and the stalls, with intermediate traffic and in the middle of the drainage sequence, had lost much of its infiltration rate, but had dramatic improvement from excavation.

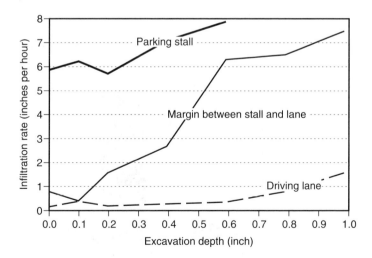

FIGURE 9.7 Recovery of infiltration rate with removal of joint-fill aggregate from 8-year-old pavement at the University of Guelph (after James and Gerrits, 2003, Figure 22.3; infiltration rates higher than 8 inches per hour were not measured).

The researchers attributed the clogging, especially in the busy driving lane, to deposition of fine material from vehicles, winter sand, and the general environment; in-place generation of fine particles by vehicular grinding; and in-place settlement and compaction. It is also likely that runoff had washed fine particles contained in the original aggregate material and in winter sand from upper areas in the parking stalls to lower areas in the driving lane.

These results show that portions of a pavement with intermediate levels of traffic, drainage, and clogging can be most effectively rehabilitated. Areas high in the drainage or with little traffic already have high infiltration rates and require little rehabilitation. Areas low in the drainage or with more traffic may be too deeply clogged to benefit much from excavation; low, heavily trafficked areas may benefit most from regular cleaning to prevent deep and irreparable clogging.

VEGETATED JOINTS

For rooting vegetation in the joints, the joint-fill planting medium should be as sandy and open-graded as the vegetation can tolerate. As described in Chapter 7, sandy root zone material has durable drainage and aeration, and is resistant to compaction and crusting. In the bedding and base course, a limited quantity of planting medium could be mixed into open-graded aggregate to deepen the rooting zone. In Germany crushed-lava aggregate has been used to supply very high porosity for plant roots (Soenke Borgwardt, personal communication 2003); the ESCS aggregate described in Chapters 6 and 7 could presumably supply similarly high porosity. Vegetation type must be suited to the traffic level and to the blocks' potential heat-island temperature regime.

Figure 9.5 showed that topsoil such as that sometimes used to support vegetation has a rather low infiltration rate. However, the same vegetation can delay surface runoff in time. Borgwardt (1997) found that standing vegetation forms a small dam around each block, retaining water on the block surface up to about 3/4 inch deep. Figure 9.8 shows that, among the pavements in Borgwardt's sample, runoff from a pavement with vegetated joints began almost 30 minutes after the start of rainfall.

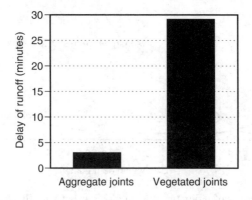

FIGURE 9.8 Average delay of runoff after start of rainfall from open-jointed block pavements (after Borgwardt, 1997).

Shearman et al. (1980) researched grass performance in open-celled grids. Their results are probably applicable also to open-jointed blocks. The study is described in Chapter 10.

FACTORS IN STABILITY AND INFILTRATION

The blocks and joint-fill aggregate make up a composite layer which determines the pavement's surface infiltration rate, and contributes, together with the base course, to pavement stability.

Factors in Block Layer's Contribution to Stability

A block pavement with aggregate fill and bedding is a structurally flexible pavement. The block layer's contribution to pavement stability comes from the interlock between blocks. Under traffic, adjacent blocks interlock when they attempt to rotate in place and wedge against each other and any aggregate particles jammed between them. Interlock resists a pavement's vertical deformation ("rutting") by sharing traffic load among adjacent blocks and spreading the load out on the underlying base. It resists horizontal "creep" under severe horizontal loads such as vehicular braking and turning (Rollings and Rollings, 1992a, pp. 33–36). The degree of interlock is determined by the blocks' interlocking shapes, the tightness of the aggregate joint fill, and the thickness throughout which adjacent blocks are in contact (Shackel and Lim, 2003).

One factor contributing to interlock is block shape. Blocks with complex shapes and narrow joints (for example, joints no wider than 1/8 in.) on many sides wedge themselves together and interlock in multiple directions to a greater degree than simple rectangular blocks (Rollings and Rollings, 1992a, pp. 33–36; Shackel and Lim, 2003). The simplicity of rectangular blocks can be compensated for to a degree by a herringbone laying pattern, which encourages more complex interlock than running bond (Rollings and Rollings, 1992a, p. 5). However on the whole, with the knowledge we have at the present time, block shape contributes to load distribution and rutting resistance only in a relative, "qualitative sense, and does not support a specific quantitative reduction in required pavement thickness" (Rollings and Rollings, 1992a, p. 36).

A second factor contributing to interlock is block dimensional proportions. ASTM C 936 limits the size of a block's exposed face to 100 square inches and the length to no more than 4 times the thickness. The purpose is to avoid broad, thin units that could crack easily under asymmetrical load or produce little interlock along their sides in proportion to the leverage of traffic load imposed from above. In practice units outside these dimensional limits are available; Burak (2003) suggested that they be used only for pedestrian traffic, not vehicular.

A third factor contributing to interlock is block thickness. Great thickness gives blocks leverage as they attempt to rotate under traffic load and to transfer compressive load laterally from one block to another (Rollings and Rollings, 1992a, pp. 20–22, 36–37). For many block models, block thickness can be specified at the time of ordering. Block thicknesses are specified as commonly in millimeters or

TABLE 9.5
Equivalent Block Thicknesses

Millimeters	Centimeters	Inches	Inches
60	6	2.36	2 3/8
80	8	3.15	3 1/8
100	10	3.94	3 15/16
120	12	4.72	4 3/4

centimeters as in inches; Table 9.5 lists conversions of some common thickness measurements. Load-spreading ability of the block layer and the pavement's resistance to rutting increases notably as block thickness increases from 6 to 8 cm; further thickening produces relatively little further improvement. Consequently, 6 cm (2 3/8 inch) blocks tend to be used for pedestrian traffic and 8 cm (3 1/8 inch) blocks for moderate vehicular traffic. Blocks 10 to 12 cm thick tend to be called upon only for extraordinary loads or as an additional safety factor.

THE INTERACTION BETWEEN BLOCK-LAYER STABILITY AND INFILTRATION

Depending on site-specific traffic loads and hydrologic objectives, some pavements might be designed with stability as a priority, and others primarily for infiltration. With our current knowledge, it may be difficult to maximize both stability and infiltration rate in the same block layer; choices of block model and joint-fill aggregate are sometimes made to maximize one at the expense of the other.

Blocks with narrow joints, or with combinations of narrow joints and wide indentations, are capable of producing great stability, where the narrow joints are filled by aggregate small enough to penetrate the joints (Shackel, 1997; Shackel and Lim, 2003). Across narrow joints, small aggregate particles contribute to interlock by transmitting compressive rotational force from block to block. However, small particles may inhibit infiltration. Where high stability is a priority, these blocks require a careful selection of small but open-graded aggregate that fits into the narrow joints but does overly sacrifice infiltration rate. An alternative approach for pavements where stability is less critical is illustrated later in this chapter in a case study of Mickel Field; the case study seems to demonstrate that, on an adequately stable base course, it is possible to use a relatively large, open-graded, highly permeable aggregate which simply leaves the narrow joints largely vacant even though it produces less than the maximum possible interlock.

Blocks with uniformly wide joints are capable of producing at least some interlock through direct contact of blocks with thick spacers. These blocks are also capable of producing very rapid infiltration, where the joints are filled with large, open-graded, highly permeable aggregate. Where high, durable infiltration rate is a priority over stability, one can select strictly open-graded aggregate and forego the maximum stability that would be produced by packing narrow joints with small particles (Borgwardt, 1994; Shackel, 1997, Figures 9 and 10). Less-than-maximum stability in the block layer can be compensated for by a sturdy base course and total pavement thickness.

TOTAL PAVEMENT THICKNESS

A block surface with its bedding layer can be conservatively (safely) equated structurally to an equivalent thickness of asphaltic concrete, aggregate base course, or a combination of both (Rollings and Rollings, 1992a, pp. 28–29, 33–36). This equivalency is consistent with approaches to flexible pavement design that do not distinguish between various flexible materials, but instead treat the entire pavement thickness as a single elastic unit. Equivalent thickness is a convenient method of adapting well-known flexible (asphalt) pavement design methods to include block pavements.

All more or less interlocking block-and-bedding layers distribute load to this degree. Complex block shapes, narrow joints, and dense joint fill probably contribute more to pavement interlock and stability than do other block-and-bedding features. Wide joints, large open-graded aggregate, and vacant joints may contribute less. However, the quantitative degrees to which layers with those features replace greater or lesser thicknesses of other flexible materials is as yet unknown. Until they are known "the most conservative approach is to simply let each inch of block and [bedding layer] be equal to 1 inch of required flexible pavement thickness" (Rollings and Rollings, 1992a, p. 77).

Accordingly, total flexible-pavement thicknesses such as those cited in Chapters 3 and 12 can also be used for block pavements. With the knowledge we possess currently, this conclusion applies to all block pavements, whether the surface layer is constructed for maximum infiltration or for maximum stability. Block shape contributes to load distribution and rutting resistance only in a relative, "qualitative sense, and does not support a specific quantitative reduction in required pavement thickness" (Rollings and Rollings, 1992a, p. 36).

Further quantitative research is called for on the effects on required pavement thickness of uniformly wide joints filled with open-graded aggregate, and of narrow joints left vacant of fill. Vacant joints and large open-graded aggregate are attractive for their very high infiltration rates. Their structural limits must be positively known so they can be designed confidently together with their base courses to bear given traffic loads.

Software is available to help determine minimum required pavement thickness. Uni-Group USA (www.uni-groupusa.org) distributes design software developed by Brian Shackel called *Lockpave Pro: Structural Design of Interlocking Concrete Block Pavements*. It determines base thickness appropriate for a given traffic level, subgrade frost susceptibility, and subgrade CBR. It can derive the base thickness required to carry traffic using any aggregate material, including open-graded aggregate, provided the designer enters appropriate design information such as the material's modulus (an index of response to external stress). It has preset settings for block models produced by Uni-Group members, and can deal with any other type of block provided the designer enters the appropriate data (Brian Shackel, personal communication 2004).

DRAINSTONE CONCRETE BLOCK

The Drainstone model of concrete block was developed in the 1990s by Bosse Concrete Products, a manufacturer of blocks in the Belgard line (www.belgardhardscapes.com; Dolph Bosse and Alan McLean, personal communications

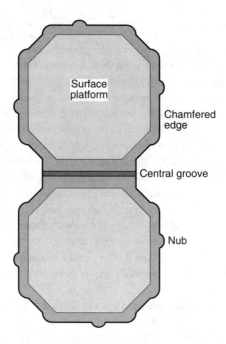

FIGURE 9.9 The Drainstone block as seen from above.

2001 and 2003). The block is in the shape of two attached octagons as shown in Figure 9.9. Indentations in the block's sides and corners leave open areas in the joints between adjacent units. A chamfer 1/4 inch wide around all the sides defines octagonal surface platforms.

In the field Drainstone can be distinguished from other octagonal models by the central groove between the two surface platforms and the symmetrical octagonal shape of the openings between blocks. The central groove and the octagonal openings give the overall paved surface an appearance of uniformly symmetrical octagonal units.

The block fits within a total area of 4 inches × 8 inches. The rectangular module allows the block to be laid in patterns such as those as shown in Figure 9.10, and in patterns together with the many other 4 inch × 8 inch models of paving blocks available in North America.

Table 9.6 lists the open areas in a Drainstone surface. The 1 1/2 inch × 1 1/2 inch indentations between adjacent units are sufficiently large to accept open-graded aggregate such as No. 8, but not large enough to allow the tires of vehicles to enter the opening and compact or erode the aggregate. On the other faces of the block, staggered nubs make joints 1/16 inch wide. The total area of open joints is 15 percent of a pavement's surface.

ENGLISH PARK, ATLANTA, GEORGIA

A parking lot at English Park in Atlanta illustrates careful detailing of the Drainstone laying pattern, and the use of a porous pavement's base as a stormwater detention

Open-Jointed Paving Blocks

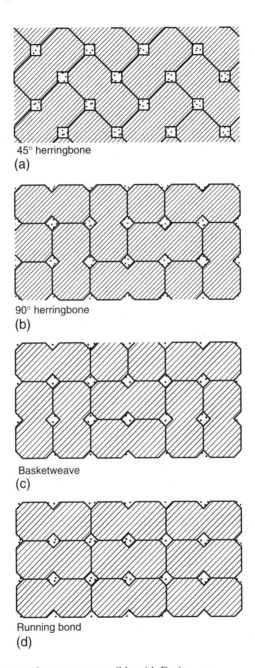

45° herringbone
(a)

90° herringbone
(b)

Basketweave
(c)

Running bond
(d)

FIGURE 9.10 Some paving patterns possible with Drainstone.

reservoir. The park is a City of Atlanta neighborhood park; the Drainstone parking lot is accessed from Bolton Road.

The lot has eight stalls and a turning lane. Drainstone covers both the stalls and the lane. The perimeter is restrained by concrete curbs, flush with the pavement

TABLE 9.6
Open Areas in a Drainstone Pavement Surface

Item	Size	Area
Large indentations	1 1/2 inch × 1 1/2 inch	14% of pavement area
Narrow joints	1/16 inch wide	1% of pavement area
Total open area		15% of pavement area

Measured from sample provided by manufacturer.

surface except where raised curbs separate vehicles from a pedestrian sidewalk. The Drainstone joints are filled with aggregate equivalent in size to the German 2/5 mm specification; its appearance is that of small but open-graded aggregate. A few months after installation the aggregate surface was about 1/2 inch below the surface of the blocks.

The details of construction are simple but admirably integrated. One row of blocks (a "soldier course") parallels the curving perimeter curb, visually emphasizing the edge. A running bond fills the remainder of the pavement surface. The parking stalls are aligned in exact coordination with the paving pattern as shown in Figure 9.11.

Figure 9.12 shows the parking lot's construction. The void space in the aggregate base reservoir is large enough to store the water from a 100-year storm. An impervious membrane at the bottom of the structure makes the base course into a detention reservoir without percolation into the subgrade (it is arguable that the membrane could have been omitted to allow some stormwater infiltration). Visits to the installation after large storms have confirmed the reservoir's detention effect, as drainage out of the pavement's discharge pipe continues at a low rate long after the rain has stopped (William Brigham, personal communication 1998). The cost per square yard of paving was greater with these open-jointed blocks than it would have been with dense pavement, but no park land had to be claimed for an off-pavement stormwater detention basin.

ECOLOC CONCRETE BLOCK

The Ecoloc model of concrete block is manufactured under license from the von Langsdorff group (www.langsdorff.com), which holds an applicable U.S. patent (Barth et al., 1994). The von Langsdorff group originated in Germany; in North America it is associated with the Uni-Group USA manufacturers' association. Regional distributors of Ecoloc and other Uni models can be identified on the Uni-Group web site (www.uni-groupusa.org).

Figure 9.13 shows that the block has a complex shape, which combines multidimensional structural interlock of adjacent blocks with open-joint indentations. It is organized into three square surface pads, each effectively 4 inches × 4 inches in area. Around the pads is a chamfered rim that holds various nubs and indentations. In a completed pavement the square pads appear to be arranged in a complex herringbone pattern.

Table 9.7 lists the open areas in an Ecoloc pavement surface. The total open area is 12 percent.

Open-Jointed Paving Blocks 343

FIGURE 9.11 Painted dividing lines aligned with Drainstone paving blocks at English Park, Atlanta, Georgia.

TABLE 9.7
Open Areas in an Ecoloc Pavement Surface

Item	Size	Area
Large indentations	1 1/4 inch x 1 1/4 inch	10% of pavement area
Narrow joints	1/8 inch wide	2% of pavement area
Total open area		12% of pavement area

Measured from sample provided by manufacturer.

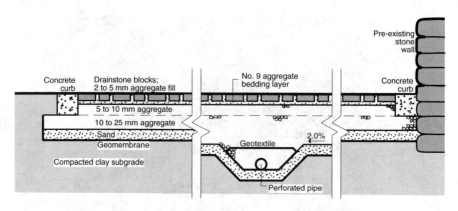

FIGURE 9.12 Construction of the Drainstone parking lot at English Park (after project drawing by Atlanta Parks Department, courtesy of William Brigham).

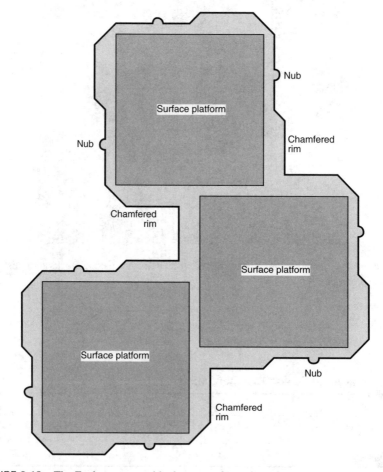

FIGURE 9.13 The Ecoloc concrete block as seen from above.

For block models like Ecoloc with combinations of open and narrow joints, some practitioners have specified combinations of large and small aggregate material, in the hope that the small particles would pack the narrow joints for stability while the large particles would occupy the large indentations for porosity and permeability. Shackel (1997) found that such blending of particle sizes does little good for either stability or infiltration rate. Figure 9.14 shows the results of his infiltration test. Relatively open-graded 0.4 inch aggregate produced rapid infiltration, although it produced only limited structural interlock because its large particles left the narrow joints unpacked. Small-particled sand made structural interlock firm by packing the narrow joints, but it produced slow infiltration because the small pores between sand particles were the only infiltration paths through the pavement surface. A blend of the two particle sizes had no faster infiltration than that of the sand alone, although it had larger average particle size, because the small sand particles clogged the voids between the large particles. The blend produced structural interlock as limited as that of the large aggregate alone, because the large particles prevented sand from completely packing the joints. This result confirms that to produce significantly high infiltration rates, there is no alternative to using open-graded (single-size) aggregate material, as large as possible, foregoing the very high strength that would come from the packing of narrow joints with fine particles.

Ecoloc's indentations are large enough to accept No. 8 open-graded aggregate, but No. 8's large particles would leave the narrow joints mostly empty. To fill both the indentations and the narrow joints, a smaller open-graded aggregate such as the German 2/5 mm gradation is required.

ANNSVILLE CREEK PADDLESPORTS CENTER, CORTLAND, NEW YORK

An Ecoloc parking lot at Annsville Creek Paddlesports Center in Cortland, New York exemplifies use of porous pavement in brownfield reclamation and reuse

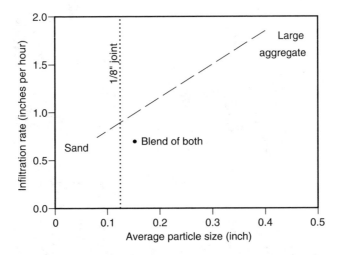

FIGURE 9.14 Infiltration rates in an Ecoloc pavement surface with joints filled with aggregate of different gradations (data from Shackel, 1997).

(Allen 2002). The center was installed on the Hudson River shore in late 2001 and early 2002 as part of Hudson Highlands State Park. It is accessed from Rt. 202. Figure 9.15 shows the completed parking lot in early 2002, as construction of the surrounding walkways and earthen berms continued.

The site was formerly a maintenance yard of the state transportation department. The subgrade is artificial fill. Before conversion of the site for park use, to improve the quality of the site's drainage into the river, petroleum-based pollutants were removed from the soil, and all dense pavement was removed. In the site's new use as a paddlesports center, the direct flow of pavement runoff into the river was prohibited. The new pavement is smaller than the old DOT pavement, and porous. The Ecoloc model was considered a heavy-duty model suited to the center's expected traffic.

Ecoloc covers parking stalls and turning lanes alike. The perimeter is stabilized by raised concrete curbs. Many perimeter blocks are cut to fit the smoothly curving

FIGURE 9.15 Ecoloc parking lot at Annsville Creek Paddlesports Center, Cortland, New York.

Open-Jointed Paving Blocks

edge. The bedding and joint fill are angular open-graded aggregate equivalent to the German 2 to 5 mm size. As of early 2002, the aggregate was about 1/4 inch below the block surface. The pavement's structural condition was excellent despite the recent passage of heavy construction equipment; no displacement of blocks was evident. The complex texture of the block surface created a light rumbling vibration when vehicles drove slowly over the surface.

The infiltration rate was high except at a few spots on the perimeter where excavation for walkways had produced sediment that clogged the pores. The pavement's low corners are on the inland side of the park; at these points curb cuts release potential overflow runoff into highway swales for further treatment and infiltration before the runoff reaches the river.

ECO-LOGIC CONCRETE BLOCK

The Capitol Ornamental Concrete Specialties Company (www.capitolconcrete.com) developed the Eco-Logic concrete block in the late 1990s in response to specifications by Arnold Wilday (a joint venture of landscape architects Henry Arnold and Cassandra Wilday) for Pier A Park in Hoboken, New Jersey (Michael Grossman, personal communication 2001). Having developed the manufacturing capability for this block and understanding its environmental advantages, Capitol gave the model its name and continues to make it available.

The model is shaped in rectangular blocks resembling bricks in 4 inch × 4 inch and 4 inch × 8 inch sizes. From the combination of sizes and two or more colors a designer can compose a huge variety of paving patterns and effects. Figure 9.16 shows the model.

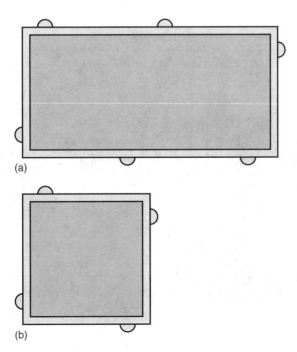

FIGURE 9.16 The two shapes of the Eco-Logic concrete block, as seen from above.

Spacers create joints 1/4 inch wide on all four sides. This is sufficiently wide to receive No. 10 aggregate fill. It is arguably the narrowest width that joints could have and still be significantly porous and permeable when filled with any aggregate. Table 9.8 lists the open areas produced by each of the two block sizes. A pavement combining the two sizes would have an intermediate porous area.

Pier A Park, Hoboken, New Jersey

The purpose of the porous Eco-Logic pavement at Pier A Park was the provision of viable rooting space for trees in the midst of considerable pedestrian traffic, as described in Chapters 2 and 5. Many trees are planted in the block-paved areas; many others are planted in areas surfaced with turf or aggregate.

Figure 9.17 shows that Eco-Logic forms 18-foot wide walkways along the park's waterfront perimeter. The walkways are punctuated by paved plazas and numerous benches looking out to the Hudson River and Manhattan. Where the pavement adjoins lawns, the edge is held by a metal strip. Blocks of various colors and sizes are laid in meticulous patterns to form carpet-like panels aligned with water edges,

TABLE 9.8
Areas of Open Joints in Eco-Logic Concrete Block Pavement Surfaces

Block Size	Joint Width	Joint Area
4 inch × 4 inch	1/4 inch	12.5% of pavement area
4 inch × 8 inch	1/4 inch	9% of pavement area

FIGURE 9.17 Eco-Logic walkways at Pier A Park in Hoboken, New Jersey.

planting beds, benches, and intersecting walkways. The pavement bears the load of very heavy pedestrian traffic and occasional park maintenance vehicles.

The base material is ESCS aggregate of ASTM C 330 1/2-inch gradation, into which planting medium is blended to make it an "air-entrained soil." The blocks are laid directly on the base course, without a separate setting bed (specification by Arnold Wilday, provided by Capitol Ornamental Concrete Specialties).

The joint-fill aggregate is angular crushed traprock. Table 9.9 lists the gradation; it is generally consistent with ASTM's No. 10 gradation. One year after installation the aggregate was 1/4 inch below the block surface. In the warm season, moss grows in the joints where tree shade prevents the surface from drying out completely. Although the aggregate is small, it is open-graded, with small but open pores. Infiltration through the aggregate joints is rapid. The porous, permeable surface provides air and water to tree roots growing in the air-entrained soil below.

ECO-STONE CONCRETE BLOCK

Eco-Stone is the most widely used of the several open-jointed concrete models developed by the von Langsdorff organization (www.langsdorff.com). For Eco-Stone the group holds an applicable patent (Barth and von Langsdorff, 1989). The von Langsdorff group licenses its models to North American manufacturers through Uni-Group USA. Regional distributors of Uni models can be identified on the Uni-Group web site (www.uni-groupusa.org).

Uni has admirably supported research into the performance and design requirements of Eco-Stone and other Uni models, and made the results available to design professionals. Examples are the research publications of James and his colleagues cited in the references at the end of this chapter, technical design guidance by Rollings and Rollings (1993) and Cao et al. (1998), and software for hydrologic and structural design. Uni has published an easy-to-use summary of Eco-Stone research and experiences (Uni-Group USA, 2002b).

Eco-Stone's shape is a predecessor of Drainstone's, consisting of two attached octagons that form rectangular openings in the indentations between adjacent units. Figure 9.18 shows the shape. Eco-Stone can be distinguished in the field by the nubs on only one of the two attached octagons, and the rectangular, not square or octagonal, joint opening. The surface fits within a 4 1/2 inch × 9 inch rectangle. Spacers create joints 1/8 inch wide on all the sides where blocks abut.

TABLE 9.9
Gradation of Joint-Fill Aggregate in Eco-Logic Blocks Specified at Pier A Park

Sieve Size	Percent (By Weight) Passing
No. 4	85–100
No. 40	25–50
No. 100	10–30

Henry Arnold, personal communication (2001); project specifications by Arnold Wilday provided by Capitol Ornamental Concrete Specialties.

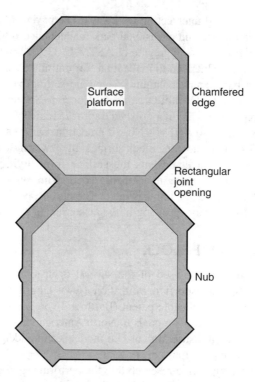

FIGURE 9.18 The Eco-Stone block as seen from above.

TABLE 9.10
Open Areas in an Eco-Stone Pavement Surface

Item	Size	Area
Rectangular indentations	7/8 inch × 2 inch	10% of pavement area
Narrow joints	1/8 inch	2% of pavement area
Total open area		12% of pavement area

Measured from sample provided by manufacturer.

Table 9.10 lists the open areas in an Eco-Stone pavement surface. The indentations in the joints are large enough to receive open-graded aggregate such as No. 8. The total open area is 12 percent of the pavement surface.

Figure 9.19. shows the placing of Eco-Stone in a parking bay. The blocks are cut to meet a previously installed concrete curb and placed on a bed of No. 89 crushed granitic rock mixed with sand.

MICKEL FIELD, WILTON MANORS, FLORIDA

An Eco-Stone parking lot at Mickel Field in Wilton Manors, Florida illustrates the successful use of open-jointed block for rapid surface drainage, and affirms the role of strictly open-graded joint-fill aggregate in achieving that end. Figure 9.20 shows the parking lot in 2001, six years after installation.

Mickel Field is a sports complex maintained by the city of Wilton Manors. It is located in the midst of a residential neighborhood at Northwest 7th Avenue and Northwest 27th Street. The parking lot holds about 80 cars.

The use of porous pavement was motivated by the need for surface drainage (Donna DeNinno, personal communication 2001). There is no local storm sewer system, the topography is very level, and there is little outfall for surface drainage amid the surrounding streets and homes. For years, after every storm, water had stood on a dense asphalt pavement in the park's parking lot. Although open-jointed block was an expensive kind of pavement, porous pavement to infiltrate stormwater into the site's sandy soil or discharge it laterally through the bedding and joints was

FIGURE 9.19 Eco-Stone being placed in a parking bay.

FIGURE 9.20 Eco-Stone parking lot at Mickel Field, Wilton Manors, Florida.

a less expensive option than reconstruction of the city's storm sewer system. In 1995, the city replaced the old asphalt pavement with 0.7 acre of Eco-Stone.

The blocks are installed on a 1-inch-thick bed of sand over a base of crushed limestone meeting conventional Florida DOT specifications. They were placed rapidly using mechanized equipment. The edging is concrete grout troweled into place after laying of the blocks, so no cutting of edge blocks was required. There are no raised curbs, and almost no drainage onto the surface from off-pavement. The bulk of the area is in light gray blocks; darker charcoal-gray blocks indicate lines between parking stalls. Wheel stops of recycled plastic are anchored by metal bars. Six years after installation the blocks showed no displacement horizontally or vertically. The joint-fill aggregate was 1/4 inch below the block surface.

The joint fill is angular limestone, single-sized, approximately 3/8 inch in size with only a small amount of finer particles. It is a truly open-graded, porous, permeable aggregate. It comfortably fills Eco-Stone's joint indentations. However, the narrow joints are almost empty of fill, as the aggregate particles are too large to fit into them. Six years after installation, the pavement's infiltration rate was extremely high; the rapidity with which water disappeared from the surface was reminiscent more of unbound aggregate surfaces than of other open-jointed block installations.

Park employees had not seen aggregate particles moving around on the pavement surface, and had not replenished the fill in any way since construction. The only maintenance requirement perceived by park personnel was the occasional suppression of weeds in the joints, for which they applied herbicide where needed from time to time in the course of their normal park maintenance rounds. Weeds sprouted most commonly at a place where runoff from an adjacent street washes in seeds and organic matter.

The highly permeable pavement has eliminated the standing-water problem that formerly existed in the park. The installation has been completely successful in that regard, "no matter how hard it rains" according to park personnel. Park personnel acknowledge that the initial expense was high, but the success at surface drainage and the expected long lifetime of the installation make it a "good investment." After the city succeeded with this initial paving experience, it installed a similar pavement surface in the nearby Highlands Park in about 2000 (Donna DeNinno, personal communication 2001).

GIOVANNI DRIVE, WATERFORD, CONNECTICUT

Giovanni Drive, a residential street in Waterford, Connecticut, was paved in Eco-Stone as an experiment in protecting the water quality of Long Island Sound (Phillips et al., 2002, Uni-Group USA, 2002a). Runoff has been monitored to compare with that from two other nearby suburban watersheds in the EPA-sponsored "Jordan Cove" research project. Table 9.11 lists web sites where information on the project may be updated as it continues.

Eco-Stone covers the last few hundred feet of the Giovanni Drive cul de sac, occupying 15,000 square feet and giving access to 12 homes. The installation exemplifies both the hydrologic success of open-graded joint-fill aggregate, and the necessity to protect the pavement from on-draining sediment to preserve its infiltration rate. Figure 9.21 shows the street in 2002, one year after installation.

The street's construction is carefully detailed. A 6-inch-wide concrete band stabilizes the perimeter. A parallel "soldier course" of gray rectangular blocks visually re-emphasizes the edge. A panel of Eco-Stone occupies the remainder of the street

TABLE 9.11
Web Sites Giving Information on the Jordan Cove Research Project

Agency	Address
Nonpoint Education for Municipal Officials University of Connecticut	http://nemo.uconn.edu www.canr.uconn.edu/nrme/jordancove

FIGURE 9.21 Eco-Stone pavement on Giovanni Drive, Waterford, Connecticut.

surface. The Eco-Stone blocks are in subtle variations of red and gray, randomly placed in the panel. Many blocks are cut to fit the curving street edge. A vegetated swale receives overflow runoff and protects the pavement from on-drainage of sediment, except at the uppermost point of the street where the swale is absent. The surface has been snowplowed without difficulty; the snow is pushed freely over the flush curb into the broad vegetated swale. One year after installation the structural condition was excellent: no displacement of blocks was evident despite the recent load of construction traffic. The joint fill is single-sized (about 1/4 inch) angular aggregate, essentially even with the block surface; the bedding is the same material. The base is 12 inches of 1 1/2-inch gravel with low fines content. The subgrade soils are well drained.

Eco-Stone also covers the residential driveway aprons, from the street's concrete band to the beginning of each private lot. A double course of gray blocks (presumably laid on concrete) outlines each apron, distinguishing it from the concrete-edged street. A linear drain across the driveway diverts runoff from the private lot into the roadside swale, preventing the lot's runoff from entering the street pavement and permitting separate monitoring. Beyond the diverting drain each home has a different pavement type: Eco-Stone, unbound aggregate, or dense asphalt.

At the time of a visit in May 2002, it was seen that the Eco-Stone infiltration rate was high in the driveways, where the joint-fill aggregate had visibly open voids. However, the infiltration rate was low in the street, where the aggregate surface had a solid, dense appearance, and beneath the surface the aggregate's voids were substantially filled with fine-graded sand.

The sediment clogging the street surface originated on one or two residential lots at the uppermost point of the street, where contractors had used an unpaved, unvegetated area for storage (Donna DiNinno and Jack Clausen, personal communications 2002 and 2004). Eroded sediment had visibly drained onto the pavement without a swale for diversion, construction trucks visibly tracked mud onto the pavement, and additional vehicles tracked the sediment further throughout the pavement. In contrast, the downward slope of the driveway aprons toward the street had protected them from the spreading sediment. The street's infiltration rate could be restored by removing the top inch or so of aggregate from the joints and replacing it with clean open-graded material.

This experience confirms that high infiltration rate is possible in open-jointed block surfaces. It also indicates that a finished street serving various buildings or lots must be protected from tracking and washing of sediment during the subsequent construction of individual properties. The properties should be separated from the street by a continuous swale. Construction storage should be confined to clean pads of open-graded aggregate, perhaps in the locations of proposed driveways. Another approach would be to leave the street only in subbase aggregate during construction and to complete the pavement surface only after all the properties are completed and stabilized.

CASTAIC LAKE WATER CONSERVATORY, SANTA CLARITA, CALIFORNIA

An Eco-Stone parking lot at the Castaic Lake Water Conservatory in Santa Clarita, California exemplifies both an attractive parking lot layout, and the surface crusting and slow infiltration that can occur where a block surface's joint-fill aggregate contains fine material. The conservatory is a public-education facility of the Castaic Lake Water Agency (www.clwa.org). It is used by educational groups and for public meetings and formal events. The purpose of porous pavement in the parking lot is to demonstrate materials that could infiltrate water for use by plants in the nearby watershed, where the average annual rainfall is only 13 inches. Figure 9.22 shows the installation in 2001, six years after construction. Projecting planted islands divide the approximately 30 parking spaces into groups of four or five.

On the upslope side the edge is stabilized by raised concrete curbs; individual parking spaces are not explicitly marked. The downslope edge is stabilized by a flush

FIGURE 9.22 Eco-Stone parking lot at the Castaic Lake Water Conservatory in Santa Clarita, California.

concrete band 24 inches wide. Wheel stops anchored in the band designate individual parking stalls. Overflow runoff drains across the band into a planted median.

Painted lines designate three spaces for the handicapped in the Eco-Stone surface. However, formal events have raised an unforeseen accessibility concern for open-jointed blocks: an employee complains of difficulty with high heels; they get "stuck in the cracks."

Six years after construction the blocks were in excellent horizontal and vertical alignment. They are placed on a setting bed of ASTM C 33 "concrete sand" (Uni-Group USA, 1998); as described earlier in this chapter, C 33 concrete sand is a notably dense-graded aggregate. The base is a layer of crushed recycled concrete 7 inches thick. The subgrade is compacted silty and clayey sand.

To make joint-fill aggregate the concrete sand was blended with 2 to 5 mm angular aggregate (Uni-Group USA, 1998), making a mixture similar to that studied by Shackel in Ecoloc pavements. Six years after construction the infiltration rate of the parking lot as a whole was moderate. At the upslope edge the aggregate had a more or less open appearance between the approximately 1/8-inch particles, and the infiltration rate was adequate. Near the low edge, closer to the building, the aggregate had a substantial amount of fines, even dust-size particles; washes of fine particles formed a surface crust; the infiltration rate was low. Evidently rain-splash had washed fines from upper areas into lower areas, where traffic near the building had further ground and dispersed the fine particles. Employees had seen "mud" washing out of joints in wet weather. The cleanliness and infiltration rate of the lower areas could be improved by removing the top inch or so of aggregate and replacing it with clean single-sized aggregate.

HARBOURFRONT FIRE STATION, TORONTO, ONTARIO

An Eco-Stone drive at the Harbourfront Fire Station in Toronto, Ontario exemplifies the ability of permeable open-jointed block pavements to bear heavy traffic loads,

Open-Jointed Paving Blocks 357

and to do so in a climate with deep winter frost. The station is located on a reclaimed industrial site on Queens Quay West, adjacent to a bay of Lake Ontario. The purpose of porous pavement is to protect the bay's water quality. Figure 9.23 shows the drive in 2002, two years after construction. The pavement is an entry drive for fire trucks and a parking area for employees.

Five Eco-Stone panels, each 20 feet × 20 feet in area, form the entry drive. The panels are defined by concrete strips 20 inches wide between panels and 36 inches wide at the periphery. The geometry of bands and panels extends at the side of the drive to define incidental planting and parking areas. In front of the fire station is a circular Eco-Stone turning area 60 feet in diameter, edged by a concrete band and a parallel course of Eco-Stone blocks.

FIGURE 9.23 Eco-Stone drive at Harbourfront Fire Station, Toronto.

In winter the pavement is plowed and salted when necessary; it is not sanded. Because there are no raised curbs, plows push snow freely off the edges of the pavement.

The pavement's structural condition was excellent after two winters, with no visible displacement. The exception was a small area where several blocks had subsided by a fraction of an inch relative to an adjacent concrete strip; subsidence could be attributed to incomplete compaction of the base course adjacent to the strip.

The joint fill is crushed limestone aggregate 3/8 inch in size, with a small amount of finer particles. The aggregate was level with the block surface over most of the area. The infiltration rate was high except at one edge where mud had washed in from an adjacent unpaved drive.

Dense-graded, slowly permeable aggregate that would pack the joints has evidently not been necessary to enable the pavement to bear the load of fire trucks. The block has withstood scraping by snowplows. Well-made open-jointed concrete block on a thick, sound base course seems to be a favorable porous paving material for heavy traffic in cold northern climates.

SF-RIMA CONCRETE BLOCK

The Rima model was developed by the SF Cooperative of Germany. In North America Rima is licensed to manufacturers through SF Concrete Technology (www.sfconcrete.com). Figure 9.24 shows the Rima block's square shape.

The block's surface is 7 3/4 inches × 7 3/4 inches in area. The relatively large level platform, like a "stepping stone," may make the block surface relatively accessible for pedestrians with high heels. Thick nubs project 1/2 inch from all four sides. The blocks can be rotated to make joints either 1/2 inch or 1 inch wide, as shown in Figure 9.25. Spaces on all sides give Rima less interlock than complex shapes such as Ecoloc.

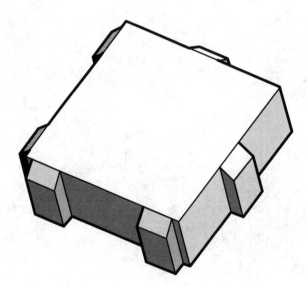

FIGURE 9.24 The SF-Rima block.

Open-Jointed Paving Blocks

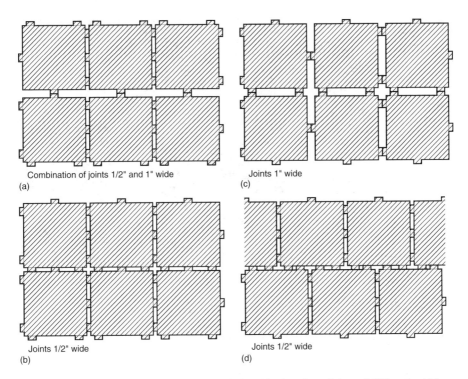

FIGURE 9.25 Alternative arrangements of Rima block to form joints of different widths.

TABLE 9.12
Open Areas in SF-Rima Pavement Surfaces

Type of Opening	Area
1 inch joints	21.5% of pavement area
1/2 inch joints	12% of pavement area

Measured from sample provided by manufacturer.

Both joint widths are sufficiently wide to receive open-graded aggregate such as No. 8. The 1-inch joints are wide enough for rooting of vegetation. Table 9.12 lists the open areas in Rima surfaces using each joint width. A pavement with a combination of joint widths would have an intermediate open area.

Figure 9.26 shows Rima blocks being laid on a bedding layer of No. 89 aggregate. The bedding layer rests on a base reservoir of No. 57 aggregate. The blocks are laid in an office building's entry lane in a pattern with 1/2 inch joints. The joints are filled with further No. 89 aggregate.

Table 9.13 lists infiltration rates measured in Rima pavement surfaces. Drewes' measurements were conducted in a laboratory with clean aggregate materials. Borgwardt's were carried out in the field and included some areas with weeds growing in the joints. These results confirm that the infiltration rate of an open-jointed block surface can be high as long as the aggregate material is decisively open-graded.

FIGURE 9.26 SF-Rima being laid on a bedding layer of No. 8 aggregate.

TABLE 9.13
Infiltration Rates Measured on SF-Rima Pavement Surfaces

Joint Width	Joint-fill Aggregate	Infiltration	Conditions	Reference
1 inch	2/5 mm, clean	No runoff at rainfall intensity up to 21.3 in/hr	Freshly built, in laboratory	Drewes, 1989
1 inch	2/5 mm, with clogging sediment	No runoff at rainfall intensity up to 21.3 in/hr	Freshly built, in laboratory	Drewes, 1989
1/2 inch	2/5 mm	2.0 in/hr and higher	Parking lots and storage areas up to 4 years old	Borgwardt
1/2 inch	0/2 mm	0.3 in/hr	Parking lot 5 years old	Borgwardt

"Borgwardt" is Borgwardt 1995, cited in Rollings and Rollings 1999, p. 6; and Soenke Borgwardt, personal communication 2003.

ALDEN LANE NURSERY, LIVERMORE, CALIFORNIA

Rima pavements at the Alden Lane Nursery in Livermore, California illustrate the use of porous pavement in tree preservation. In 2001, the nursery constructed a new retail building and main entry. The layout of the new construction was organized around two very large landmark oak trees. A Rima panel overlies the rooting zone of each tree. One panel is the new entry driveway; the other forms a pedestrian courtyard at the new building.

Figure 9.27 shows the pedestrian courtyard one year after construction. The perimeter is stabilized by a flush concrete strip which is compatible with pedestrian accessibility and free surface drainage. Planters and furniture emphasize the edge between the courtyard and an adjacent asphalt drive. In the center is a large unpaved area around the tree base where nursery staff place product displays. The Rima surface is several inches above the tree-base soil, which implies that the pavement structure rests on the preexisting root zone without excavation. The roots of the huge old tree surely extend through and beyond the courtyard area; the porous Rima panel is the surface of the tree's preexisting root zone.

The installation is very level and uniform. The blocks are placed with 1/2 inch joints and cut to fit the curving perimeter strip tightly. The joints are filled with small but open-graded aggregate, which was about 1/2 inch below the top of the blocks one year after construction. The infiltration rate was very high. Despite the great width and depth of the joints, the blocks give the impression of an easily accessible surface due to the great size of the Rima platforms and the very level alignment of the blocks.

At the nursery's entry driveway, a Rima pavement wraps around the other oak's generously wide tree-base area. The pavement surface drains downward to a city street on one side and to a drainage inlet on the other. A raised concrete curb confines vehicles to the Rima driveway pavement. The same curb forms a retaining wall, raising the Rima surface about 12 inches above the soil near the tree-base area. This implies that the pavement structure is placed on the surface of the preexisting root zone without excavation. The large root zone is unexcavated and perhaps uncompacted to support the tree's vitality despite bearing commercial traffic.

FIGURE 9.27 SF-Rima courtyard at Alden Lane Nursery in Livermore, California in October, 2002.

FOREST HILL APARTMENTS, WILMINGTON, NORTH CAROLINA

A Rima parking lot at Forest Hill Apartments in Wilmington, North Carolina exemplifies the attractiveness that can come from simple and consistent details of layout. The apartment complex is located on Darlington Avenue. Rima was installed in a long peripheral parking bay to meet municipal stormwater requirements during an expansion of the complex in 2001. The remainder of the parking lot is of dense asphalt. Figure 9.28 shows the installation one year after construction.

The total coverage of Rima is 14,000 square feet. Projecting planted islands divide the long line of parking stalls into seven bays of 10 to 15 spaces each. The rhythmical repetition of similar bays organizes the space and defines distance.

FIGURE 9.28 Rima parking stalls at Forest Hill Apartments in Wilmington, North Carolina.

Open-Jointed Paving Blocks 363

A raised concrete curb stabilizes the perimeter of the Rima panels; blue lines painted on the curb delineate individual parking spaces. A-6-inch wide concrete band separates the blocks from an asphalt turning lane. Surface drainage is from the Rima panels onto the asphalt lane.

The blocks are laid with 1/2-inch joints. Compared with 1-inch joints seen elsewhere, the relatively narrow joints give the pavement an appearance of relative solidity and consistency.

The joint fill is sand similar to nearby native soil. One year after installation the infiltration rate was moderate. The sand was 1/4 inch below the block surface and partly occupied by moss. Fine sandy sediment had eroded from an adjacent slope, clogging portions of the joints near the periphery. There had been some effort to stabilize the slope with mulch; the further natural development of vegetation will stabilize the slope more permanently.

PAVING BLOCKS OF OTHER MATERIALS

Numerous other kinds of materials are available for use as blocks in porous pavement surfaces.

OTHER MODELS OF CONCRETE BLOCK

Models of open-jointed concrete block other than those described earlier in this chapter exist, but they are relatively new and no specific pavement installations were identified or surveyed in time for this publication.

ECO I Paver is one such example. ECO I is available from E.P. Henry (www.ephenry.com; Jon Bowman, personal communication 2004). The block occupies a rectangular area of 5 1/2 inches × 8 1/4 inches, most of which is a central rectangular pad. On the sides are sturdy spacers reminiscent of those on SF-Rima, which make substantial joints between adjacent blocks. It has reportedly been installed at the Newark Conservancy and other sites in New Jersey and Pennsylvania. A second (ECO II) model was being considered as of early 2004.

Egra Stone was developed by Paveloc Industries (www.paveloc.com). Each block has a square surface approximately 9 inches × 9 inches in size. Two spacers on each side make wide joints between adjacent units. The units can be laid in either stack bond or running bond.

The InfiltraStone model was developed by Pavestone (www.pavestone.com) in 2002 for use in an "ecologically friendly" parking lot at the Living Desert Zoo and Gardens in Palm Desert, California (www.livingdesert.org; Jason Forster, personal communication 2003). The block is similar in dimensions to Drainstone and Eco-Stone. Indentations in its sides and corners make circular openings between adjacent units, which can be filled with porous aggregate.

MANUFACTURED CONCRETE SLABS

Simple concrete paving slabs are available for pedestrian settings. Numerous regional manufacturers make slabs of various sizes and thicknesses; there are no known standard models. Where the slabs are widely spaced, their open joints are potentially porous and permeable.

Figure 9.29 shows an example at a residential terrace in Ohio. The concrete slabs are 36 inches × 36 inches in area. Their surfaces are broom-finished for secure footing. The joints between them are 1 1/2 inches wide, so the slabs have no structural interlock. The joints are filled with topsoil supporting well-kept bluegrass. The grass forms soft mounds slightly higher than the concrete and spreads horizontally an inch or more onto each slab surface. The mounds form dikes that pond up a bit of rain water on each slab, slowing the runoff and enhancing potential infiltration through the topsoil joints.

Concrete Rubble Slabs

The demolition of concrete pavements and buildings produces irregular slabs of rubble that can be directly recycled as widely spaced paving units. Suppliers of recovered concrete slabs can be identified at Recycler's World (www.recycle.net).

Figure 9.30 shows slab-like fragments of recycled concrete in a pedestrian plaza at the Lyle Center for Regenerative Studies at California State Polytechnic University in Pomona (www.csupomona.edu/~crs/). Open-graded aggregate supports pedestrians who step off the slabs. The aggregate area is edged with concrete bands. This plaza was designed under the direction of the center's founder, John Lyle, around the opening of the center in 1994. The center is intended to demonstrate the integration of the needs of the community with the restoration of natural systems.

FIGURE 9.29 Widely spaced concrete slabs with turf joints at a residence in St. Clairsville, Ohio.

Open-Jointed Paving Blocks

FIGURE 9.30 Plaza of recycled concrete rubble at the Lyle Center for Regenerative Studies in Pomona, California.

This simple paving installation exemplifies the ingenious discovery of artful and useful possibilities in discarded materials that was part of Lyle's vision (Lyle, 1994).

The easiest concrete rubble to work with is that from concrete buildings. The walls and ceilings of buildings are formed with uniform thickness; upon recycling for paving the flat fragments are conveniently placed on a level aggregate setting bed. Almost equally uniform and easy to use are concrete paving slabs that were placed as overlays on preexisting level pavement surfaces.

Rubble from a concrete slab placed directly on subgrade or on an aggregate base course may be more difficult to use. The under-surface of a concrete slab takes on the irregular shape of the material on which it was placed. When it is later recycled as slab fragments, tedious adjustment of each piece on a setting bed may be required to make a uniform, level, stable pavement surface.

Blocks of Porous Material

It is possible to make paving blocks of porous material. This type of block need not have open joints to be permeable; networks of pores in the block itself are the conduits for fluids into the subsurface. Although no specific porous models are known to be manufactured in North America at this time, experience overseas and experimentation at home leave open the possibility of this kind of development in the future. Research on this development is called for.

Manufactured blocks of porous concrete have been tried in Germany and Japan. They are made with open-graded aggregate like that in the cast-in-place porous concrete which will be described in Chapter 11. Cast-in-place pavements of porous concrete have been successful in every regard when properly constructed. Porous blocks made of the same material ought to be capable of being equally successful.

Borgwardt (1997) found that the infiltration rates of pavements made of tight-jointed porous concrete blocks in Germany were greatly reduced by the use of fine-graded aggregate to fill the joints. The sand had filled the pores in the blocks as readily as the joints between them, clogging the entire surface. In Japan, porous concrete blocks are reputed to be easier to restore than other kinds of porous pavement, because the clogged units can be removed and replaced (Smith, 2000). Borgwardt recommended manufacturing blocks with joints sufficiently wide so that the joint-fill aggregate could be large enough not to enter and clog the blocks' pores. Another possibility is to shape the units with relatively narrow joints and omit joint-fill aggregate altogether, if the blocks can withstand direct contact and remain in alignment under a given traffic load.

Bosse Concrete Products, a manufacturer of blocks in the Belgard line (www.belgard.com) has considered making blocks of porous concrete for use in North America (Dolph Bosse and Jeff Speck, personal communication 2003). Various block shapes, joint widths, concrete mixtures, and joint-fill materials have been considered to address issues of interlock and permeability. It is acknowledged that blocks made of porous concrete should probably not be expected to meet the standards of strength and absorption specified in ASTM C 936.

It is theoretically possible to cut porous paving blocks from naturally porous rocks such as pumice and tuff. However, there is no known commercial source of such material in North America, nor has there been any documented experience of this kind in other parts of the world.

Wooden Blocks

Wooden blocks and segments of wooden logs are used as stepping pads in garden-like pedestrian settings to which their rugged natural appearance is suited. Some are of naturally durable wood species such as cypress, redwood, or western cedar; others are chemically preserved.

Figure 9.31 shows the typical construction of a pedestrian pavement with sections of logs as practiced in Germany. The large spaces between the circular sections can be filled with porous aggregate or soil.

Rectangular wooden blocks had a temporary place in the history of American street pavements (McShane, 1979). In the mid-nineteenth century wooden blocks

Open-Jointed Paving Blocks

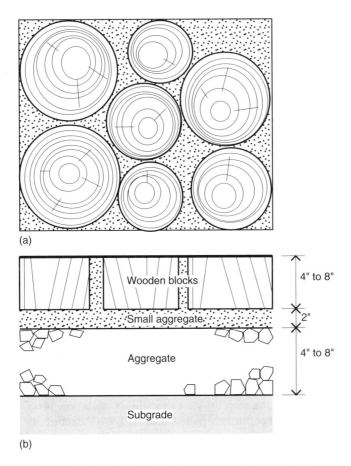

FIGURE 9.31 Plan and section of a pedestrian pavement made from log segments (after Baden-Württemberg, 1992).

preserved with creosote were relatively common on city streets because horses' shoes could find purchase in the cracks and joints while wagons rolled easily on the relatively uniform surface. The hooves and wheels were not excessively noisy on the wooden material; the block surfaces drained well. However, wooden pavements contributed to the spread of city fires, and were ultimately abandoned in favor of stone surfaces.

METAL PLATES

Heavy metal plates left behind in old industrial settings can be directly recycled as slab-like paving units. In Duisburg, Germany, the central platform of "Piazza Metallica" is surfaced with 49 cast steel plates, each 6 inches thick and 7 feet × 7 feet in area, set in small aggregate (Brown, 2001; www.latzundpartner.de).

The piazza is a feature in the Landschaftspark Duisburg-Nord, which is part of the regional Emscher Landscape Park. Duisburg-Nord is set in a reclaimed steel foundry; its theme is the direct and active reuse of the site's industrial relics. The piazza is a courtyard in the reclaimed foundry building, now used as a theater for outdoor movies and performances. The huge plates bear swathes of red-brown iron rust, visibly expressing both the powerful industrial forces that formed them and a concept of nature that includes spontaneous weathering and change.

The platform is a square with seven of the enormous slabs on each side. Adjacent plates are 9 inches apart. Each level plate forms a smooth floor panel. The surfaces of the plates are several inches above the aggregate bed, so one must step or jump carefully from plate to plate. Although the piazza's surface of industrial-strength paving blocks is irregular, it is massively firm.

Suppliers of recovered metal plates can be identified at Recycler's World (www.recycle.net).

Natural Stone

Many kinds of natural stone are quarried for use as paving blocks. There is no known national association concerned specifically with applications of stone in paving. However general information about stone materials can be found at web sites such as www.selectstone.com and www.plmarble.com. Some stone suppliers provide technical guidelines for installations of their products. Bartels (1992) illustrated variations in stone's design applications.

Cobblestone — rounded alluvial stone — was one of the first forms of hardened pavement in early America. Coastal cities took the material from natural alluvial deposits or the ballast of merchant ships. In the mid- to late-nineteenth century cobblestone set on a bed of sand was still the most common form of pavement on city streets where horses pulled heavy loads (McShane, 1979). Horses' hooves obtained traction from the irregular pavement surface, but iron horseshoes and iron-rimmed wagon wheels were notoriously noisy on the hard, irregular surfaces (Tarr, 1971).

Late in the nineteenth century, some of the cobblestone surfaces were replaced with rectangular brick or rectangular stone cut from quarries. Street cleaning on the flat units was easier than on the rounded cobbles (McShane, 1979).

Today, due to transportation costs, some kinds of stone are only regionally available. Many other kinds are marketed nationally and internationally for their special aesthetic or symbolic associations. Available types of stone differ in every characteristic. Some stone materials are stronger than concrete; others are much weaker. "Field stone" is used in the shape it was found in the ground, with its naturally irregular shapes and surfaces. Some other stones are polished for smoothness; still others are given systematic texture for traction. Suppliers of recycled stone can be identified at Recycler's World (www.recycle.net).

No known supplier makes available a stone block shaped to combine stone-to-stone structural interlock with open joints. Instead, stone paving units tend to come in simple blocks or slabs. These units can be placed with wide joints filled with porous soil or aggregate. Figure 9.32 shows a type of installation with vegetated joints practiced in Germany. Widely spaced units have no structural interlock, so they are ordinarily suited only for light pedestrian traffic.

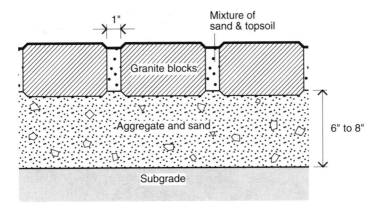

FIGURE 9.32 Installation of stone block with vegetated joints in Germany (after Baden-Württemberg, 1992).

Figure 9.33 shows an example with large stones in the Japanese Garden portion of the Missouri Botanical Garden in St. Louis. The native limestone rocks were selected for their flat shapes and their individual places in the composition. They are up to 3 feet × 5 feet in area. Such large stones require machinery for placement; they presumably rest on an aggregate base. The surrounding open-graded gravel infiltrates both direct rainfall and runoff from the rocks. Each rock bears one or two steps from a walking person. Because the rocks are laid with their length perpendicular to the direction of travel, they make a comfortably wide walkway. Nevertheless, they call for careful stepping. This path leads to the garden's tea house. In a traditional Japanese garden movement along a path toward a tea house is a preliminary stage of meditation (Holborn, 1978, pp. 76–78). The rough shapes of stones placed at irregular intervals draw one's attention as one walks; people places their feet with care. One is drawn away from the outer world, and moves toward the seclusion and calm of the tea house.

Sloan (2001) reported the use of even larger stones at the office campus of Sprint in Overland Park, Kansas. The campus' "Relaxation Courtyard" required a garden-like character for pedestrians, but also had to support heavy vehicles for emergency access. During mass grading of the site, large quantities of native limestone were excavated and stockpiled; much of it was in the form of natural slabs 18 inches thick. In the courtyard the slabs were used as surfacing for an emergency access lane. Heavy equipment placed the slabs on an aggregate base, over a compacted subgrade. The joints between the boulder-like rocks were filled with topsoil and sodded. To answer the local fire department's questions, before full installation, a full-size mockup was constructed and tested for stability with the department's equipment; the result showed full support. The great weight and breadth of the boulders and the firmness of their foundation gave the pavement stability under heavy load despite the lack of interlock between widely spaced units.

The Building Stone Institute provides considerable technical information on the types and uses of stone at its web site (www.buildingstone.org).

FIGURE 9.33 Limestone slabs at the Missouri Botanical Garden in St. Louis.

Table 9.14 lists examples of suppliers of paving stone. Some supply large quantities of bulk materials; others support custom craftwork. Other local suppliers can be found in all regions of North America. Searches for suppliers can be facilitated by web sites such as www.stoneworld.com, www.findstone.com, and www.traditional-building.com. *Building Stone* magazine describes stone products, applications, and quarrying operations, and displays advertisements from many suppliers.

BRICKS

Bricks are made by firing clay or shale in a kiln. Under intense heat the mineralogical structure of the material changes, and it becomes solid and firm. General introductions to bricks and guidelines for their paving applications are available in the literature of the Brick Industry Association (www.bia.org or www.brickinfo.org) and the book by

TABLE 9.14
Examples of Suppliers of Natural Paving Stone

Supplier	Contact Information
Archean Granite	www.archeangroup.com
B & H Stone	http://bhstonesupply.com
Bedrock Stone & Supply	www.bedrockstone.net
Buechel Stone	www.buechelstone.com
Camara Slate	http://camaraslate.com
Canadian Quarry Direct	www.canadianquarrydirect.com
Cantera Especial	www.cantera-especial.com
Carderock Stone	www.carderock.com
Champlain Stone	www.champlainstone.com
Cold Spring Granite	www.coldspringgranite.com
Columbia River Rock	www.columbiariverrock.com
Connecticut Stone	www.connecticutstone.com
Earthworks	www.earthworks1.com
Eden Stone Company	www.edenstone.net
Fletcher Granite Company	www.fletchergranite.com
Geo. Schofield	www.geoschofield.com
Granicor	www.granicor.com
Greenstone Slate	www.greenstoneslate.com
Halquist Stone	www.halquiststone.com
Herb Kilmer & Sons Flagstone	www.herbkilmerflagstone.com
Krukowski Stone	www.krukowskistone.com
Lompoc Stone	www.lompocstone.com
London Universal	www.londonuniversal.com
Luck Stone	www.luckstone.com
Milestone	www.milestonenm.com
North Carolina Granite	www.ncgranite.com
Pennsylvania Flagstone	www.pennsylvaniaflagstone.com
Polycor	www.polycor.com
Porphyry USA	www.porphyryusa.com
Santa Fe Stone	www.santafestone.com
Scotia Slate	www.scotiaslate.com
Siloam Stone	www.siloamstone.com
Stone Trade	www.stonetrade.com
Texastone Quarries	www.texastone.com

TABLE 9.15
Examples of Suppliers of Recycled Bricks

Supplier	Contact Information
Gavin Historical Bricks	www.historicalbricks.com
Schloss Paving	www.schlosspaving.com

Sovinski (2000). Brick suppliers can be found in most locales. Table 9.15 lists examples of suppliers of bricks recycled from old streets and buildings; further suppliers can be identified at Recycler's World (www.recycle.net).

Brick pavements on sand beds have been common in Holland and other European countries since the nineteenth century, and are still used there on streets with moderate traffic loads (Croney and Croney, 1998, p. 233). In Europe the art of laying bricks by hand is common enough that experienced teams of paviors routinely set and reset pavements that are, to American eyes, sensibly and interestingly patterned.

In North America, brick was first used for paving in the 1830s. Brick pavements were common for a time in the late nineteenth and early twentieth centuries (Croney and Croney, 1998, pp. 8–9). Manufactured bricks had the advantage of uniform size compared with the split stones that the bricks often replaced. By 1914, more than 6 million square yards of brick paving had been laid. Today, paving bricks are still used in portions of some cities, where they are valued for their historical associations and sense of craftsmanship.

For today's applications ASTM C 902 and C 1272 classify paving brick intended for different severities of exposure and traffic load. For each category they specify standards of freeze-thaw resistance, compressive strength, absorption of water, and abrasion resistance. Many paving bricks are 4 inches × 8 inches in area. They can be extremely strong and durable. The conventional thickness of paving brick for vehicular traffic is 2 5/8 inches; bricks for pedestrian traffic may be thinner. Although most bricks are reddish, the color can vary with source material and firing procedure and with the addition of coloring agents such as metal oxides.

Numerous brick models are available with spacers for narrow (1/16 inch to 1/8 inch-wide) joints, or no spacers at all. These joints are too narrow to make the pavements significantly porous and permeable. There are no known models of paving brick that are shaped for open permeable joints while interlocking for structural stability. Development of such models would make a competitive addition to the available options for porous pavements. The existing narrow-jointed models could form porous pavements where they are laid in open-jointed patterns as described below.

The attractiveness of bricks has led some cities to restore brick streets that had been installed in the early twentieth century, and later covered with asphalt (Schwartz, 2003). The restored streets are believed to revive the appearance and rejuvenate the economy of neglected downtown areas. Although the installation cost of new or restored brick pavements is substantial compared with that of dense asphalt, the material is believed to have low long-term maintenance costs.

OPEN PATTERNS WITH RECTANGULAR PAVING BLOCKS

Manufactured plastic spacers are available to widen the joints between blocks of common shapes, while maintaining some structural interlock between them. The one line of spacers known to be available in North America is Enviro-Pave, which originated in Germany and is distributed in North America by Pave Tech (www.pavetech.com). The line includes several different configurations, for creating joints from 3/8 inch to over 2 inches wide between blocks of rectangular and other shapes, in various laying patterns. Among available units that could be combined with these spacers to make open-jointed pavements are the many available models of otherwise tight-jointed concrete blocks, and simple rectangular paving bricks. Presumably, hand labor is necessary to place every block in proper arrangement with other blocks and the interconnecting spacers.

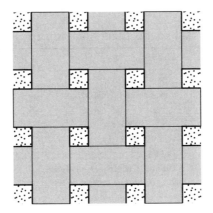

FIGURE 9.34 Open paving pattern with rectangular blocks.

Almost any rectangular blocks can be laid in an open pattern. Figure 9.34 shows a basketweave-like pattern with aggregate in the open joints. In this pattern any interlock between blocks comes from contact at their narrow ends. Because the interlock is limited, this type of pattern should be used only on a firm base course and in settings with light traffic load. If the length of the blocks is twice the width, then the open joint areas occupy 17 percent of the surface area. Blocks with shorter lengths relative to their width produce more continuous interlock and leave a smaller area in open joints. The Navastone company (www.navastone.com) has suggested the use of open patterns with its rectangular blocks in appropriate settings, and suggested a variety of complex patterns using blocks of different sizes and shapes.

TRADEMARKS

Table 9.16 lists the holders of registered trademarks mentioned in this chapter.

ACKNOWLEDGMENTS

Soenke Borgwardt of Borgwardt Wissenschaftliche Beratung, Norderstedt, Germany, and David R. Smith of the Interlocking Concrete Pavement Institute (www.icpi.org) have been extraordinarily generous in the provision of information and answers to specific inquiries. Marshalls Mono Ltd. of Halifax, England graciously permitted Borgwardt to send a copy of a sponsored report. Monica Ebenhoeh of the City of Vienna, Austria, translated literature from German into English while a graduate student in landscape architecture at the University of Georgia.

The following persons generously provided additional information used in this chapter: Dexter Adams and his staff in the University of Georgia's Physical Plant Department; Henry Arnold, landscape architect of Princeton, New Jersey (www.arnoldassociates.com); Nancy Bakker of the Cement and Concrete Association of New Zealand; Dolph Bosse of Bosse Concrete Products (Morrow, Georgia); Cedar Wells Bouta of the City of Olympia, Washington; Joe Bowen and

TABLE 9.16
Holders of Registered Trademarks Mentioned in This Chapter

Registered Trademark	Holder	Headquarters Location
Abbotsford Concrete Products	Abbotsford Concrete Products	Abbotsford, British Columbia
Acker Stone	Acker Stone Industries	Corona, California
Angelus Block	Angelus Block	Sun Valley, California
Archean Granite	Archean Granite	Edison, New Jersey
B & H Stone	B & H Stone	Gunnison, Utah
Balcon	Balcon	Crofton, Maryland
Basalite	Basalite Concrete Products	Tracy, California
Bedrock Stone & Supply	Bedrock Stone & Supply	Lakewood, New Jersey
Belgard	Oldcastle	Atlanta, Georgia
Big River Industries	Big River Industries	Alpharetta, Georgia
Borgert Products	Borgert Products	St. Joseph, Minnesota
Buechel Stone	Buechel Stone	Chilton, Wisconsin
Camara Slate	Camara Slate	Fair Haven, Vermont
Canadian Quarry Direct	Canadian Quarry Direct	Burnaby, British Columbia
Cantera Especial	Cantera Especial	Sherman Oaks, California
Capitol Ornamental Concrete	Capitol Ornamental Concrete	South Amboy, New Jersey
Carderock Stone	Carderock Stone	Bethesda, Maryland
Castle Rock Pavers	phone 888-347-7873	New Orleans, Louisiana
Champlain Stone	Champlain Stone	Warrensburg, New York
Cold Spring Granite	Cold Spring Granite	Cold Spring, Minnesota
Columbia River Rock	Columbia River Rock	Kettle Falls, Washington
Connecticut Stone	Connecticut Stone	Milford, Connecticut
Drainstone	Oldcastle	Atlanta, Georgia
Earthworks	Earthworks	Perryville, Missouri
ECO I Paver	E. P. Henry	Woodbury, New Jersey
Ecoloc (Uni Ecoloc)	F. von Langsdorff Licensing Ltd.	Inglewood, Ontario
Eco-Logic	Capitol Ornamental Concrete Specialties	South Amboy, New Jersey
Eco-Stone (Uni Eco-Stone)	F. von Langsdorff Licensing Ltd.	Inglewood, Ontario
Eden Stone Company	Eden Stone Company	Eden, Wisconsin
Egra Stone	Paveloc Industries	Marengo, Illinois
F. von Langsdorff Licensing Ltd.	F. von Langsdorff Licensing Ltd.	Inglewood, Ontario
Fletcher Granite	Fletcher Granite Company	North Chelmsford, Massachusetts
Granicor	Granicor	St. Augustin, Quebec
Greenstone Slate	Greenstone Slate	Poultney, Vermont
Grinnell Pavers	Grinnell Pavers	Sparta, New Jersey
Halquist Stone	Halquist Stone	Sussex, Wisconsin
Herb Kilmer & Sons Flagstone	Herb Kilmer & Sons Flagstone	Kingsley, Pennsylvania
Hessit Works	Hessit Works	Freedom, Indiana
Ideal Concrete Block	Ideal Concrete Block	Westford, Massachusetts
InfiltraStone	Pavestone	Dallas, Texas
Interlock Paving Systems	Interlock Paving Systems	Hampton, Virginia
Kirchner Block	Kirchner Block	Bridgeton, Missouri
Krukowski Stone	Krukowski Stone	Mosinee, Wisconsin

TABLE 9.16 Continued

Registered Trademark	Holder	Headquarters Location
Lompoc Stone	Lompoc Quarries	Laguna Niguel, California
London Universal	London Universal	Monterey, California
Luck Stone	Luck Stone	Manakin, Virginia
Marshalls Mono Ltd.	Marshalls Mono Ltd.	Halifax, England
Milestone	Milestone	Santa Fe, New Mexico
Mutual Materials	Mutual Materials	Bellevue, Washington
Navastone	Navastone	Cambridge, Ontario
Nicolock	Nicolia Industries	Lindenhurst, New York
North Carolina Granite	North Carolina Granite	Mount Airy, North Carolina
Oberfield's	Oberfield's	Delaware, Ohio
Pave Tech	Pave Tech	Prior Lake, Minnesota
Paveloc	Paveloc Industries	Marengo, Illinois
Paver Systems	Paver Systems	Sarasota, Florida
Pavestone	Pavestone	Dallas, Texas
Pennsylvania Flagstone	Pennsylvania Flagstone	Roulette, Pennsylvania
Pine Hall Brick	Pine Hall Brick	Winston-Salem, North Carolina
Polycor	Polycor	Quebec City, Quebec
Porphyry USA	Porphyry USA	Cabin John, Maryland
Santa Fe Stone	Santa Fe Stone	Santa Fe, New Mexico
Schloss Paving	Schloss Paving	Cleveland, Ohio
Scotia Slate	Scotia Slate	Rawdon, Nova Scotia
SF Concrete Technology	SF-Kooperation GmBH	Mississauga, Ontario
SF Rima	SF-Kooperation GmBH	Mississauga, Ontario
Siloam Stone	Siloam Stone	Canon City, Colorado
Stone Trade	Stone Trade	East Greenwich, Rhode Island
Texastone Quarries	Texastone Quarries	Garden City, Texas
Tremron	Tremron	Miami, Florida
Unicon Paving Stone	Unicon Concrete	Durham, North Carolina
Uni-Group USA	Uni-Group USA	Palm Beach Gardens, Florida
Unilock	Unilock	Georgetown, Ontario
Westcon	Westcon	Olympia, Washington
Willamette Graystone	Willamette Graystone	Eugene, Oregon

others of Mutual Materials (www.mutualmaterials.com); Jon Bowman of E. P. Henry (www.ephenry.com); William Brigham, landscape architect with the City of Atlanta Public Works Department; Ian Chin of Wiss, Janney, Elstner and Associates in Northbrook, Illinois; Jack Clausen, University of Connecticut; Rick Cummings and Tim Holder of Unicon Paving Stone (www.uniconconcrete.com); Kwesi DeGraft-Hanson of the University of Georgia School of Environmental Design; Donna DeNinno of UNI-Group U.S.A. (www.uni-groupusa.com); Deborah Edsall of Edsall and Associates, Columbus, Ohio; Jason Forster and Robert Rayburn of Pavestone (www.pavestone.com); Suzanna Gordon of Basalite Concrete Products (www.basalitepavers.com); Michael Grossman of Capitol Ornamental Concrete

Specialties (www.capitolconcrete.com); James Hade of the Maryland State Highway Administration, Office of Environmental Design; William James of the University of Guelph and Computational Hydraulics International (www.chi.on.ca); Steve Jones of Pave Tech (www.pavetech.com); David Kimbrough, Castaic Lake Water Agency (www.clwa.org); Carol Krawcyck of the University of Delaware Department of Plant and Soil Sciences; R. Lawson of Pine Hall Brick (www.pinehallbrick.com); Natalie Capone Maresca of Porphyry USA (www.porphyryusa.com); Allan McLean, formerly of Belgard; Larry Nicolai of Ideal Concrete Block (www.idealconcreteblock.com); Jack O'Donnell of Oberfield's (www.oberfields.com); Helga U. Piro of SF Concrete Technology (www.sfconcrete.com); David Quinn of Angelus Block (www.angelusblock.com); Christopher Pratt and his staff at Coventry University (U.K.); Steve Samaha of the National Concrete Masonry Association (www.ncma.org); Jon Savelle, Seattle Daily Journal of Commerce (www.djc.com); Brian Shackel of the University of New South Wales (Australia); Jeff Speck of Big River Industries (www.bigriverind.com); Brian Trimble of the Brick Institute of America (www.bia.org); and Harald von Langsdorff of F. von Langsdorff Licensing Limited (www.langsdorff.com).

The following persons constructively critiqued early drafts of this chapter: Soenke Borgwardt of Borgwardt Wissenschaftliche Beratung, Norderstedt, Germany; Donna DeNinno of UNI-Group U.S.A. (www.uni-groupusa.com); David B. Nichols of the University of Georgia; Helga U. Piro of SF Concrete Technology (www.sfconcrete.com); David R. Smith of the Interlocking Concrete Pavement Institute (www.icpi.org); and Harald von Langsdorff of F. von Langsdorff Licensing Limited (www.langsdorff.com).

REFERENCES

Addis, R.R., R.G. Robinson, and A.R. Halliday, 1989, *The Load-Spreading Properties of Clay Brick Pavements*, Transport and Road Research Technical Report 234, Transport and Road Research, Crowthorne, U.K.

Allen, Kenneth J. (2002) Improving Public Access to New York State's Hudson River, *Landscape Architect and Specifier News* 18, 20–24.

Baden-Württemberg Umweltministerium (Baden-Württemberg Ministry of Environment) (1992). *Lebendige Wege*, Besser Leben Mit der Natur Folge 2 (*Living Paths*, Better Living with Nature Part 2), Stuttgart, Germany: Baden-Württemberg Umweltministerium.

Bartels, Elizabeth 1992. Cobbles and Other Stones, *Landscape Architecture* 82, March,. 84–87.

Barth, Gunther, and Fritz von Langsdorff (1989). *Paving Stone*, patent No. 4,834,575, Washington: U.S. Patent and Trademark Office (www.uspto.gov).

Barth, Gunther, Fritz von Langsdorff, and Harald von Langsdorff (1994) *Angular Paving Stone for Paving Areas*, patent No. 5,342,142, Washington: U.S. Patent and Trademark Office (www.uspto.gov).

Borgwardt, Soenke (1992). Alternative Methoden der Regenwasserentsorgung, Ein Ueberblick aus der Sicht der Freiraumplanung (Alternative Methods of Stromwater Treatment, A survey from the point of View of Landscape Planning), *Das Gartenamt*, part 1: 4, 251–258; part 2: 5, 336–344.

Borgwardt, Soenke (1994). *Expert Opinion [of Eco-Stone Permeability]*, Hannover, Germany: Institute for Planning Green Spaces and for Landscape Architecture.
Borgwardt, Soenke (1995). *Expert Opinion on In-Situ Permeability Test of Concrete Paving Stones Commissioned by SF-Kooperation GmbH*, Hannover, Germany: Institute for Planning Green Spaces and for Landscape Architecture.
Borgwardt, Soenke (1997). *Versickerungsfahige Pflastersysteme aus Beton: Voraussetzungen, Anforderungen, Einsatz (Infiltratable Concrete Block Pavement Systems: Prerequisites, Requirements, Applications)*, Bonn: Bundesverband Deutsche Beton- und Fertigteilindustrie.
Borgwardt, Soenke (1998). *Survey and Expert Opinion on the Distribution, Performance and Possible Application of Porous and Permeable Paving Systems*, Halifax, England: Marshalls Mono Ltd.
Borgwardt, Soenke, Alexander Gerlach and Martin Koehler (1994). *Permeable Pavements: Planning Options for Water-Permeable Paving Systems*, Bremen, Germany: SF-Kooperation.
Brown, Brenda J. (2001). Reconstructing the Ruhrgebeit, *Landscape Architecture* 91, 66–95.
Burak, Rob (2002a). Edge Restraints for Interlocking Concrete Pavements, *Interlocking Concrete Pavement Magazine* May, 16–21.
Burak, Rob (2002b). Joints and Joint Sand for Interlocking Concrete Pavements, *Interlocking Concrete Pavement Magazine* November, 14–27.
Burak, Rob (2003). Pavers versus Slabs — Making the Right Choice, *Interlocking Concrete Pavement Magazine* 10, 8–10.
Burton, Brian (2000). Mechanized Installation of Interlocking Concrete Pavers, *Landscape Architect and Specifier News* 16, 20–22.
Campbell, Craig (1993). Custom Precast Paving, *Landscape Architecture* 83, 84–86.
Cao, Su Ling, Daryl Poduska, and Dan G. Zollinger (1998). *Drainage Design and Performance Guidelines for UNI Eco-Stone Permeable Pavement*, Palm Beach Gardens, Florida: UNI-Group U.S.A.
Clark, A. J. (1981). *Further Investigations into the Load-Spreading of Concrete Block Paving*, Cement and Concrete Association Technical Report 545, C&CA, Wexham Springs, Slough, Bucks.
Cooper, Rory A., et al. (2002). *Evaluation of Selected Sidewalk Pavement Surfaces*, Pittsburgh: Veterans Administration Rehabilitation Research and Development Center, Human Engineering Research Laboratories.
Croney, David, and Paul Croney (1998). *Design and Performance of Road Pavements*, 3rd ed., New York: McGraw-Hill.
Drewes, U. (1989). *Pilot Experiment to Determine the Infiltration of Rain Amounts of ESKOO-Concrete Paver System Rima*, Braunschweig, Germany: Technical University Braunschweig, Leichtweiss Institute for Hydraulic Engineering.
Fischmann, Ron (1999). Unilock Successfully Introduced Paving Stones to North America, *American Builder* 4.
Hade, James D. (1987). Runoff Coefficients for Compressed Concrete Block Pavements, *Landscape Architecture* 77, 102–103.
Holborn, Mark (1978). *The Ocean in the Sand, Japan: From Landscape to Garden*, Boulder: Shambhala Publications.
Interlocking Concrete Pavement Institute, (ICPI) (1995). *Construction of Interlocking Concrete Pavements*, Tech Spec No. 2, Washington: Interlocking Concrete Pavement Institute.
Interpave (2003). *Permeable Pavements: Guide to the Construction and Maintenance of Concrete Block Permeable Pavements*, Leicester, Britain: Interpave (www.paving.org.uk).
James, William (2002). Green Roads: Research into Permeable Pavers, *Stormwater* 3, 48–50.

James, William, and Christopher Gerrits (2003). Maintenance of Infiltration in Modular Interlocking Concrete Pavers with External Drainage Cells, in *Practical Modeling of Urban Water Systems*, Monograph 11, pp. 417–435, William James, Ed., Guelph, Ontario: Computational Hydraulics International.

Kresin, Christopher, William James and David Elrick (1997). Observations of Infiltration through Clogged Porous Concrete Block Pavers, in *Advances in Modeling the Management of Stormwater Impacts*, Vol. 5, pp 191–205, William James, Ed., Guelph, Ontario: Computational Hydraulics International.

Lilley, A. (1991) *A Handbook of Segmental Paving*, London: Chapman & Hall.

Lyle, John Tillman (1994) *Regenerative Design for Sustainable Development*, New York: Wiley.

McShane, Clay (1979). Transforming the Use of Urban Space: A look at the Revolution in Street Pavements, 1880–1924, *Journal of Urban History* Vol. 5 (May 1979), 279–307.

Mearing, Marc, Glen Fuller, and Sam Harb (2000). Paving the Way for Success: Paving the Olympic Precinct, *Concrete: Journal of the Cement and Concrete Association of New Zealand*, 44, 21–25.

National Concrete Masonry Association, (NCMA) (1995). *Pavement Construction with Interlocking Concrete Pavers*, TEK 11-2, Herndon, Virginia: National Concrete Masonry Association.

Phalen, T. (1992). *Development of Design Criteria for Flood Control and Ground Water Recharge Utilizing UNI Eco-Stone and Ecoloc Paving Units*, Boston: Northeastern University.

Phillips, Robert A., John A. Clausen, John Alexopoulos, Bruce L. Morton, Stan Zaremba, and Mel Cote (2002). The Jordan Cove Project, *Stormwater* 4, 32–38.

Pillsbury, James (2003). Green Spotlight on Pervious Paving, *The Cornerstone* (Green Building Alliance, Pittsburgh) Summer, 2.

Rients, Jay (2004). Permeable Interlocking Paving Stone Pavements, *Elevation* (Illinois Chapter, American Society of Landscape Architects) 2, 4–6.

Rollings, Raymond S., and Marian P. Rollings (1992a). *Applications for Concrete Paving Block in the United States Market*, Palm Beach Gardens, Florida: Uni-Group USA.

Rollings, Raymond S., and Marian P. Rollings (1992b). *Review Comments on "Development of Design Criteria for Flood Control and Ground Water Recharge Utilizing UNI Eco-Stone and Ecoloc Paving Units" and "Development of Filter Protection for the UNI Eco-Stone and Ecoloc Paving Systems for Mitigation of Downstream Flooding and Ground Water Recharge"*, Professor Thomas E. Phalen, Vicksburg, Mississippi: Rollings Engineering; on file at Palm Beach Gardens, Florida: UNI-Group U.S.A.

Rollings, Raymond S., and Marian P. Rollings (1993). *Design Considerations for the Uni Eco-Stone Concrete Paver*, Palm Beach Gardens, Florida: Uni-Group USA.

Rollings, Raymond S., and Marian P. Rollings (1999). *SF-Rima, A Permeable Paving Stone System*, Mississauga, Ontario: SF Concrete Technology.

Santa Barbara Bicycle Coalition (2002). Castillo Patched, *Quick Release* January (www.sbbike.org).

Schwartz, Emma (2003). Bricks Come Back to City Streets, *USA Today* August 31 (www.usatoday.com).

Seddon, P.A. (1980). The Behaviour of Interlocking Concrete Block Paving at the Canterbury Test Track, *Tenth Australian Road Research Board Conference, August 1980*, Vermont South, Victoria: Australian Road Research Board.

Shackel, Brian (1990). *Design and Construction of Interlocking Concrete Block Pavements*, London and New York: Elsevier Science Publishers.

Shackel, Brian (1997). Water Penetration and Structural Evaluations of Permeable Eco-Paving, *Concrete Precasting Plant and Technology* 3, 110–118.

Shackel, Brian, and Alan R. Pearson (1997). Concrete Segmental Pavements, *Constructional Review* February, 42–47.

Shackel, Brian, and Daniel Oon Ooi Lim (2003). Mechanisms of Paver Interlock, presented at *Pave Africa 2003, Seventh International Conference on Concrete Block Paving*, October 12–15, 2003, Sun City, South Africa, on file at Concrete Masonry Association of Australia (www.cmaa.com.au).

Shearman, RC., E.J. Kinbacher, and T.P. Riordan (1980). Turfgrass-Paver Complex for Intensely Trafficked Areas, *Agronomy Journal* 72, 372–374.

Sloan, Kevin W. (2001). Modern Solutions via Ancient Principles: Roman Construction Solves Fire Truck Access, *Landscape Architecture* 91, 28.

Smith, David R. (1999). A Road to the Future: Tracing the History of Concrete Pavers, *Landscape Design* September, S19–S21.

Smith, David R. (2000). Straight A's, *Interlocking Concrete Pavement Magazine* November, 24.

Smith, David R. (2001). *Permeable Interlocking Concrete Pavements: Selection, Design, Construction, Maintenance*, 2nd ed., Washington: Interlocking Concrete Pavement Institute.

Smith, David R. (2004). *Backyards and Boulevards: A Portfolio of Concrete Paver Projects*, Atglen, Pennsylvania: Schiffer Publishing.

Sovinski, Rob W. (2000). *Brick in the Landscape, A Practical Guide to Specification and Design*, New York: John Wiley and Sons.

Tarr, Joel A. (1971). Urban Pollution — Many Long Years Ago, *American Heritage* 22, 65–69 and 106.

U.S. Department of Justice (1994). *ADA Standards for Accessible Design*, Excerpt from 28 CFR Part 36, dowloadable from www.usdoj.gov.crt/ada.

Uni-Group USA (1998). *Rio Vista Water Treatment Plant*, Palm Beach Gardens, Florida: Uni-Group USA.

Uni-Group USA (2002a). *Jordan Cove Urban Watershed Project*, Palm Beach Gardens, Florida: Uni-Group USA (www.uni-groupusa.org).

Uni-Group USA (2002b). Uni Eco-Stone: Guide and Research Summary, Palm Beach Gardens, Florida: Uni-Group USA (www.uni-groupusa.org/EcoResearch.pdf).

10 Open-Celled Paving Grids

CONTENTS

The Concrete Grid Manufacturing Industry	382
Construction of Open-Celled Grid Pavements	384
Grass Performance in Grid Cells	386
Ambergrid Concrete Grid	388
Villa Duchesne School, Saint Louis, Missouri	388
Checkerblock Concrete Grid	389
Sands Point Museum, Sands Point, New York	391
Grasscrete Concrete Grid Slab	393
Anderson School of Management, Los Angeles, California	394
Furman Office, Charlotte, North Carolina	395
Missouri Botanical Garden, Saint Louis, Missouri	397
Monoslab Concrete Grid	398
Christian Meeting Room, Gahanna, Ohio	399
Baltimore Zoo, Baltimore, Maryland	401
Turfstone Concrete Grid	402
San Luis Bay Inn, Avila Beach, California	404
Tumwater Historical Park, Olympia, Washington	405
Municipal Parking Lot, Dayton, Ohio	407
Blue Heron, Annapolis, Maryland	408
Other Models of Concrete Grids	410
Open-Celled Paving Bricks	410
Trademarks	411
Acknowledgments	413
References	414

Open-celled paving grids are blocks or slabs containing openings that penetrate their entire thickness. They differ from open-jointed blocks in that the porous area is in the cells cast within each unit in addition to any open space in the joints between units. The grids are made of concrete or brick. Most come as premanufactured units; one model is formed by casting concrete in place. Their applications have previously been reviewed by Nichols (1995), Sipes and Roberts (1994), and Southerland (1984).

The open cells are filled with porous aggregate or vegetated soil. The grids' solid ribs or pedestals transmit traffic loads and protect, to some degree, the fill in the cells from compaction and erosion. Together, the solid and open portions of the pavement form a composite surface that can potentially be porous and permeable, while bearing considerable traffic loads through the structure.

Open-celled grids are sometimes referred to as "grid pavers" or "turf pavers," or by generic use of one of the model names. Table 10.1 lists some terms that have particular application in open-celled grid pavements.

THE CONCRETE GRID MANUFACTURING INDUSTRY

Open-celled paving grids were first developed in Germany in 1961 (Southerland, 1984). The cells between the concrete ribs were filled with soil and grass to reduce the heat-island effect of large parking fields. Similar grids were soon used for a variety of environmental purposes in many parts of Europe. They were introduced to the U.S. in the late 1960s and early 1970s. Germany remains a center of creative development for concrete paving materials and the technology to manufacture and install them (Smith, 2001).

The North American concrete grid industry is supported by the industrial associations listed in Table 10.2, which advocate uniform standards and make available design and construction guidelines, contractor certification, and lists of regional manufacturers of grid products.

TABLE 10.1
Some Terms with Particular Application in Open-Celled Grid Pavements

Term	Definition
Bed (bedding, setting bed, laying bed)	Layer of relatively small aggregate on which grid units are placed
Block	Solid piece of material used as a construction unit
Cast in place	Cast, as concrete, in permanent position
Castellation	Form with alternate solid parts and open spaces
Cell	Open space or compartment between ribs or pedestals of a structure
Grid	Framework of ribs or pedestals in a rectangular pattern
Joint	Space or interface between adjacent grids
Lattice	Framework of crossed strips or ribs
Paver	Any block, brick, grid, panel, tile, or lattice used for paving
Pedestal	Column with a level top
Precast	Cast, as concrete, before placing

TABLE 10.2
Associations in the Concrete Paving Grid Manufacturing Industry

Name	Contact Information
ASTM International	www.astm.org
Interlocking Concrete Pavement Institute (ICPI)	www.icpi.org
National Concrete Masonry Association (NCMA)	www.ncma.org

Open-Celled Paving Grids

The industry has developed different models of grids to address the different combinations of structural, environmental, and cost objectives. Their configurations have implications for appearance, pedestrian accessibility, vegetation maintenance, the proportion of the surface in open cells, and the protection of cell fill from compaction and erosion.

ASTM C 1319, *Standard Specification for Concrete Grid Paving Units*, specifies concrete grid units' thickness, material, strength, and absorption. It specifies that a model's durability must be demonstrated by proven field performance for three years

TABLE 10.3
Examples of Manufacturers or Suppliers of Open-Celled Concrete Grids

Supplier	Contact Information
Abbotsford Concrete Products	www.pavingstones.com
Acker Stone	www.ackerstone.com
Anchor Concrete Products	www.anchorconcreteproducts.com
Angelus Block	www.angelusblock.com
Basalite	www.basalitepavers.com
Belgard	www.belgard.com
Bend Industries	www.bendindustries.com
Bomanite	www.bomanite.com
Cambridge Pavers	www.cambridgepavers.com
D'Hanis Brick & Tile	www.dhanisbricktile.com
Dura Art Stone	www.duraartstone.com
E. P. Henry	www.ephenry.com
Grinnell	www.grinnellpavers.com
Hanover Architectural Products	www.hanoverpavers.com
Hastings Pavement Co.	www.hastingspavers.com
Hessit Works	phone 812-829-6246
Ideal Concrete Block	www.idealconcreteblock.com
Interlock Paving Systems	www.interlockonline.com
Kirchner Block	www.kirchnerblock.com
Metromont Materials	www.metromontmaterials.com
Mutual Materials	www.mutualmaterials.com
Navastone	www.navastone.com
Nicolock	www.nicolock.com
Nitterhouse Masonry Products	www.nitterhouse.com
Oldcastle Architectural	www.oldcastle.com
Paveloc	www.paveloc.com
Paver Systems	www.paversystems.com
Pavestone	www.pavestone.com
Quick Crete Products	www.quickcrete.com
Stone Age Pavers	wwws.stonagepavers.com
Tremron	www.tremron.com
Unicon	www.uniconconcrete.com
Unilock	www.unilock.com
Westcon	www.westcon.com
Willamette Graystone	www.willamettegraystone.com

in the same type of environment and under the same general traffic volume as that expected for the model's intended use. Where a special feature such as surface texture or color is desired, a designer can specify it separately. Many concrete grid units are heavy, solid, durable paving materials. For those models with high initial cost, the cost may be compensated by long lifetime and correspondingly low total life cycle cost.

In choosing a grid model for a specific project, a first step may be to identify manufacturers in the region of the project site and the models they have available. Alternatively, if a specific model is desired, regional sources for it can be identified from the list of suppliers on a licensing agency's web site. Table 10.3 lists some examples of regional manufacturers or suppliers of concrete grids. Additional manufacturers can be identified from member lists of the national industrial associations.

CONSTRUCTION OF OPEN-CELLED GRID PAVEMENTS

Guidelines for pavement construction with premanufactured grid units are available from the two industry associations; an example of a useful publication is ICPI (1997) Tech Spec No. 8. Some manufacturers suggest additional guidelines for specification and installation of their products.

Figure 10.1 shows typical components. An aggregate base course or base reservoir spreads traffic load on the subgrade. No. 57 is an open-graded gradation commonly used for this purpose. The base aggregate is compacted as it is placed in lifts. A firm base course is essential to prevent rocking, cracking or settlement of grid units under traffic. A geotextile may be placed under the base following criteria such as those described in Chapter 3. Some conservative installers place geotextile under every grid pavement, without regard to soil type.

A layer of relatively small open-graded aggregate is laid over the base as level bedding for the grid units. The bedding material must be angular and durable. No. 8 aggregate is commonly used for this purpose; it meets filter criteria over No. 57 and therefore does not require geotextile separation. The effective thickness is commonly limited to 1 inch to inhibit rutting. On an open-graded base, ordinarily about 1 inch of the bedding material is compacted into the base's large voids, and then another inch is spread out, without compaction, as the bedding layer. The grid units are vibrated into place in the bedding material.

An edge restraint is mandatory where vehicles may go near or over the pavement edge, and on the downhill edge where the pavement is laid on a slope. In other types of settings the support from the base course and the great breadth and weight of the grid units may be sufficient to keep the units in place.

Figure 10.2 shows Turfstone grids being placed in this type of construction (Turfstone is a model of concrete grid that will be described later in this chapter). The grids rest on a setting bed of small aggregate. Beneath the setting bed is a base course of larger aggregate. The grid units are cut where necessary to meet a fixed curb. Machine assistance is commonly necessary to lift and place the heavy concrete units; examples of special machines for this purpose can be seen at Pave Tech (www.pavetech.com) and Eurotekno (www.optimasamerica.com).

The joints between adjacent grid units make an overall pavement surface structurally flexible although the individual units are rigid. Interlock between adjacent

Open-Celled Paving Grids

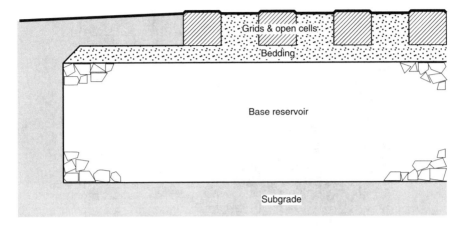

FIGURE 10.1 Typical components in installation of premanufactured grid units (adapted from ICPI 1997, Figure 19).

units is not counted upon for stability and there is no mention of interlock in the technical literature. In fact, NCMA (1996) recommended placing the units with a minimum joint spacing of 1/16 inch on the grounds that if the grids touch during repetitive traffic loading they may chip and spall. Instead, the great breadth and weight of the grid units give a layer of units its stability. Most pavements of this type are suited only for light or infrequent traffic such as that in emergency access lanes or lightly used parking stalls (David Smith, personal communication 2003).

For aggregate fill in the cells, ICPI (1997) suggested ASTM No. 8 gradation. The cells in most grid models are several inches wide, so they easily accept open-graded aggregate of a size sufficient to have large open pores and a high, reliable infiltration rate. In low-traffic areas mossy or weedy vegetation may germinate in the aggregate, perhaps requiring suppression or removal.

For installations with grass, a planting medium relatively resistant to compaction is helpful; under traffic a sandy soil tends to maintain greater pore size and rooting viability than does a clay soil. A common procedure is to place the medium into the cells within about 1/2 inch of the grid surface, dampen to settle, then add additional soil to bring it to the desired level. Leaving the soil about 1/4 inch below the surface may help avoid compaction, but may compromise pedestrian accessibility. Blending a limited amount of sand or other planting medium into the base course and bedding aggregate could form a "structural soil" that enhances the pavement's rooting volume.

Grass type must be suited to the traffic level and to the grids' potential "heat-island" temperature regime. Grass can be planted in the form of plugs. If the surface is seeded, the application rate for broadcasting seeds should be reduced to take into account the presence of the concrete ribs.

Grassed grids require the same maintenance as any other lawns: mowing, fertilizing, removal of thatch, topdressing of soil, replacement of dead grass, and applications of fertilizer (Sipes and Roberts, 1994). Aeration should be attempted only in

FIGURE 10.2 Placement of Turfstone grid units on an aggregate setting bed.

ways that avoid conflict with the concrete ribs or pedestals. Ribs or pedestals projecting above the soil surface might present rough travel for wheeled mowing machines; in such installations mowing by whip machines may be preferred. Irrigation may be critical in the first year or two to overcome the small natural moisture reservoir in the grid cells (Sipes and Roberts, 1994; Southerland, 1984).

GRASS PERFORMANCE IN GRID CELLS

A study by Shearman et al. (1980) in Nebraska illustrated how turf reacts to the environment in grid cells and to traffic wear in that environment. The grids were Monoslab, a concrete model described later in this chapter. Monoslab has concrete

Open-Celled Paving Grids

pedestals which leave 78 percent of the surface in open cells for vegetation. The grids were placed on a 4-inch sand setting bed directly over a silty clay loam subgrade soil. The same type of soil was used as planting medium in the open cells. The pedestals protruded 1/4 inch above the soil, but the large openings between pedestals presumably permitted some penetration by flexible tires. For comparison, turf without grids was planted nearby in similar soil. The researchers simulated heavy event-type traffic by driving a small truck 600 times over every sample, one day in the middle of each of two summers. Trained agronomists visually evaluated the turf's quality, injury, and recuperation.

In Figure 10.3 the "Turf quality" chart compares the average quality of grasses planted in the grids and those without grids throughout the two-year study period both before and after bearing traffic. Most of the grass varieties are marked near the diagonal line, indicating that their quality was little affected by the grids. Only bluegrass is notably below and to the right, indicating that its general quality was lower in the grids than in the unreinforced soil.

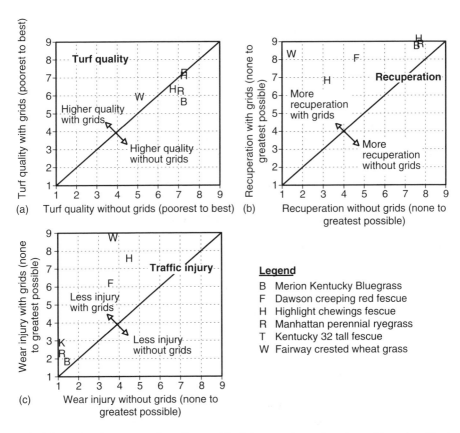

FIGURE 10.3 Average turf quality, traffic injury, and recuperation with and without Monoslab grids (average of two years' data from Shearman et al., 1980: turf quality from Table 3, traffic injury and recuperation from Table 5).

The "Traffic injury" chart compares the injury caused to grasses planted in the grids and those without grids three days after each traffic event. All grass varieties are marked to the upper left of the diagonal line, indicating that the grids protected all grasses to some degree. Wheatgrass and the two fescues suffered the most significant degrees of damage from the traffic events, but also benefited most from the protection of the grids.

The "Recuperation" chart shows the turf's recuperation in the two to three months following the traffic events. Taking into account all grass types, the recuperation was better for the grasses in grids than for those without grids, presumably because of the relatively low soil compaction in the grids following equal traffic application. The three grass types that had been most injured by the traffic and most beneficially protected by the grids were also the ones most beneficially assisted in their recuperation.

The results show that the grid environment did not greatly inhibit general grass quality, but that it protected grass from traffic injury and assisted its recuperation afterwards. Different types of grass benefited to different degrees from the grids' protection of grass and soil.

The soil was silty clay loam; it was therefore plastic and susceptible to compaction. Perhaps grass performance could have been still better had the grass been further protected from traffic by using a sandier planting medium less susceptible to compaction.

AMBERGRID CONCRETE GRID

The Ambergrid model of concrete grid was developed by Kirchner Block and Brick (www.kirchnerblock.com; Mark Wilhelms, personal communication 2001). Figure 10.4 shows the distinctively decorative shape of the precast unit as seen from above. The unit occupies a rectangular area nominally 12 inches by 18 inches in size. The concrete ribs enclose open cells both within and between the units.

Villa Duchesne School, Saint Louis, Missouri

An Ambergrid parking area at Villa Duchesne School in Saint Louis, Missouri illustrates the use of Ambergrid with grass. The area is accessed from the school's main entry drive and located within view of the school's main door, where it might be suitable for day-student parking. Figure 10.5 shows the parking area set among venerable old trees, lawns, and iron lamps. Two aisles of dense concrete give access to Ambergrid parking stalls on both sides of the aisles.

The impervious lanes drain onto the grassy Ambergrid stalls. Surface runoff on the stalls is confined by earthen berms. Water was standing at low corners in the stalls following significant rain in April, 2003.

The concrete units are in good condition except where they come in contact at narrow projecting points, some of which are chipped. Some blocks have settled, especially at the edges of the pavement, and this is attributable to local softness in the base.

The planting medium in the Ambergrid cells is fine-textured soil. It is within 1/4 inch of the surface over most of the area. The area is mowed like a lawn but supports vegetation which is only sparse and weedy. The Ambergrid cells are narrow enough

Open-Celled Paving Grids

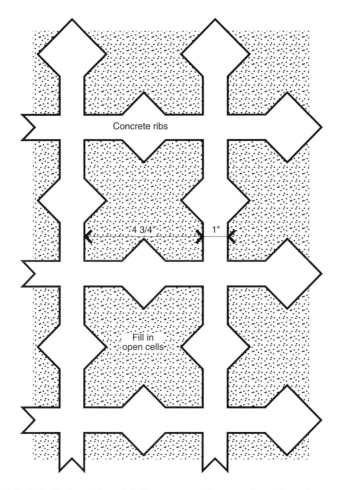

FIGURE 10.4 The Ambergrid model of concrete grid as seen from above (measured at Villa Duchesne School, Saint Louis, Missouri).

to protect the soil from intense compaction by car wheels. A sandier planting medium would have further helped vegetation by resisting compaction and maintaining drainage and aeration. The area could be rehabilitated by removing the uppermost soil from the cells, replacing it with a sandier planting medium, and reseeding grass.

CHECKERBLOCK CONCRETE GRID

Checkerblock is a proprietary model of the Hastings Pavement Company (www.hastingspavers.com). Hastings manufactures the grids or licenses them for manufacture at several locations in North America.

Checkerblock has a pedestaled or "castellated" shape consisting of concrete pillars held apart by low ribs, as shown in Figure 10.6. Around and between the ribs

FIGURE 10.5 Ambergrid parking stalls at the Villa Duchesne School, Saint Louis, Missouri.

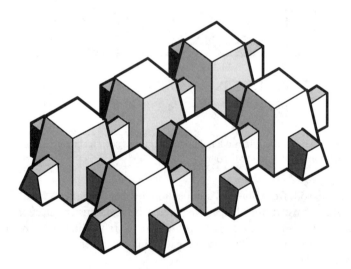

FIGURE 10.6 Portion of a Checkerblock grid unit.

and pedestals is a complex network of open cells. Each precast grid unit is 4 inches thick and 24 inches × 24 inches in area, and incorporates 16 pedestals. Checkerblock's slender ribs are strengthened with metal reinforcing.

After the cells are filled with soil or aggregate, all that is visible at the surface is the isolated square tops of the pedestals. Figure 10.7 shows the resulting pattern. The

Open-Celled Paving Grids

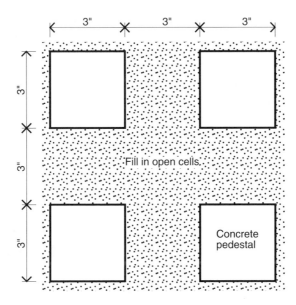

FIGURE 10.7 Surface pattern created by Checkerblock's concrete pedestals.

open cells occupy 75 percent of the pavement's exposed surface area (measured from sample provided by the manufacturer).

A spreading type of grass in the large, connected open area can produce the appearance of nearly continuous vegetative cover. In a comparison of several types of concrete grids and plastic geocells with Bermuda grass in a "low" maintenance regime, Nichols (1993) found that Checkerblock, the only pedestaled model in the comparison, had the best appearance according to ratings by trained landscape architects. The grass had spread easily around Checkerblock's pillars, producing a relatively uniform turf surface. Perhaps if maintenance had been at a higher level, the lattice-like grids and geocells would have had similarly continuous cover.

However, Checkerblock's wide, continuous cells admit the flexible tires of automobiles relatively easily, making the soil or aggregate in the cells relatively vulnerable to compaction, grinding or displacement. It is possible for the isolated concrete pedestals also to make pedestrian access difficult. Where the soil or aggregate in the cells is below the tops of the pedestals, it is hard for a pedestrian to find a place to step in the irregular composite surface; the tops of the pedestals are too small to be "stepping stones." Theoretically, this could be overcome by filling with a sandy, compaction-resistant planting medium to the tops of the pedestals and planting with a spreading type of grass.

SANDS POINT MUSEUM, SANDS POINT, NEW YORK

A Checkerblock parking lot at the Sands Point Museum on Long Island, New York illustrates both the durability of the Checkerblock grid and the necessity to lay out grassy pavements selectively and to maintain them adequately. The museum is part of the Sands Point Preserve, a historic estate in the town of Sands Point now maintained by the Nassau County Department of Recreation and Parks. If the parking lot was constructed during conversion of the site to public use in the mid-1970s, then it was approximately 25 years old at the time of a visit in 2002, making it is one of the oldest known extant porous pavements in North America. Figure 10.8 shows the installation in 2002.

The area of Checkerblock with grass can hold about 100 cars; overflow parking goes to peripheral areas of unreinforced grass. Traffic is seasonally heavy, including both autos and school busses. Checkerblock covers the parking stalls and lanes in a uniform field. The only indications of parking organization are rows of skinny, widely spaced bollards between parking bays and the parallel orientation of the Checkerblock grid. There are no curbs, trees, or wheel stops to reinforce the organization. Although

FIGURE 10.8 Checkerblock parking field at the Sands Point Museum on Long Island, New York.

cars park in alignment with the bays despite the sparseness of indicators, when they move many "cut" across the bays.

The Checkerblock grids are in good condition despite their age. The top surface of the pedestals is polished by traffic, exposing the concrete's aggregate, but the pedestal structures are solid and square. There is no visible displacement of the Checkerblock layer except immediately adjacent to an access road, where subsidence of some units is attributable to local softness in the base.

However, the soil and grass in the cells show the effects of long, heavy traffic and cursory maintenance. The soil is native sandy loam. The grass is not irrigated, which is a disadvantage for grass health on the rapidly draining soil. At the time of a visit in spring of 2002 most of the soil surface was bare of vegetation, particularly in the driving lanes.

At the center of the cells the soil is about 1/2 inch below the pedestal surface, forming channels of bare, compacted soil with washes of loose sand. Adjacent to each pedestal, where the soil is more protected from flexibly penetrating wheels, it was only lightly compacted and held scattered vegetation. The infiltration rate is surprisingly low considering the soil's sandy texture, even in the spots with some vegetation. The relief in the pavement surface makes a rumbling ride for vehicles and a walking surface that is less than universally accessible. The only solid patches of grass are in the portions of parking bays where only the front wheels of parked cars go and in peripheral parking bays distant from the museum entrance.

In the lanes the high level of traffic suppressed the grass, compacted the soil, and reduced the infiltration rate. Cutting of traffic across bays further injured the grass and soil in the parking stalls. Everywhere, the lack of irrigation accelerated the decline of the grass.

This experience confirms the necessity to select grassy pavements only for specific settings with notably infrequent traffic and as a corollary, to lay out grassy pavements to control traffic levels to fit the tolerance of grass. Sands Point's early porous parking lot did not have the benefit of prior experience with grassy pavements to guide distinctions between lanes and stalls, or between lightly and heavily used portions of a parking lot. In future installations of this type, for the turning lanes, some form of porous pavement that can tolerate the traffic should be selected. In the parking bays, cross-cutting traffic can be averted by wheel stops in the parking stalls, or by much more numerous and prominent bollards perhaps visually reinforced by rows of trees and curbed planting islands. In all grass areas, irrigation can encourage grass health.

GRASSCRETE CONCRETE GRID SLAB

Grasscrete is a concrete slab cast in place with special formwork to make an open-celled grid. It is a proprietary model of the Bomanite company (www.bomanite.com), which makes templates for patterned concrete, licenses their use, and certifies the installers. Figure 10.9 shows the distinctive surface pattern of rounded concrete pedestals that Bomanite's specialized formwork produces. The open cells occupy 53 percent of the pavement's exposed surface area (reported by Bomanite, www.bomanite.com).

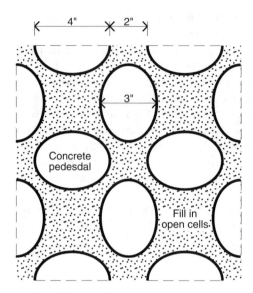

FIGURE 10.9 Surface pattern produced by Grasscrete's concrete pedestals (measured at the Anderson School of Management, University of California, Los Angeles).

The Grasscrete slab is ordinarily 5 1/2 inches thick. It is continuously reinforced with welded wire fabric or metal reinforcing bars. Because Grasscrete is a continuously reinforced monolithic slab, it has both compressive strength and flexural strength. It can be enormously strong. In a properly constructed installation, differential settlement should be unknown. Bomanite suggests guideline specifications to obtain this type of product. Where underground utilities exist, Grasscrete's monolithic structure could inhibit utility access.

ANDERSON SCHOOL OF MANAGEMENT, LOS ANGELES, CALIFORNIA

A Grasscrete emergency access lane at the Anderson School of Management on the UCLA campus in Los Angeles, California illustrates Grasscrete's use in a common type of application (Dallman and Piechota, 1999). Figure 10.10 shows the fire lane paralleling the south side of the school building.

In the center of the lane is a 22-foot-wide walkway of dense concrete. On each side is a 6-foot-wide strip of Grasscrete, making the lane 34 feet wide overall. The Grasscrete strips widen the firm area for emergency vehicles; because they are level and clear of obstructions they also perceptually widen the walkway. The cells are filled with soil and a bunch grass (perhaps fescue). Concrete bands 8 inches wide separate the Grasscrete from surrounding lawns and plantings. Access ports for buried utilities penetrate the Grasscrete at several places.

The walkway is heavily used by pedestrians and bicycles. Near walkway intersections the soil is compacted and bare where traffic cuts across the corners. Elsewhere conscientious maintenance keeps the grass green and dense, hiding the concrete pads and maintaining the soil's permeability.

Open-Celled Paving Grids 395

FIGURE 10.10 Grasscrete emergency access lane at the Anderson School of Management, University of California, Los Angeles.

FURMAN OFFICE, CHARLOTTE, NORTH CAROLINA

A Grasscrete parking lot at the office of David Furman Architecture in Charlotte, North Carolina illustrates Grasscrete's use with aggregate (David Furman and Will Siler, personal communications 2000 and 2003). The architect built the lot at 500 East Boulevard in 1989 to supply parking for his office while meeting a local restriction on impervious cover. Figure 10.11 shows the installation in 2003, 14 years after construction. Two Grasscrete parking bays hold 20 parking spaces. They are accessed by lanes of dense asphalt. The overall surface slopes at approximately 2 percent.

A Grasscrete panel occupies the central 6 feet × 14 feet of each stall. The panels are divided from each other and from the surrounding asphalt by concrete bands 24 to 36 inches wide. The bands implicitly define parking stalls 9 feet × 17 feet in

FIGURE 10.11 Parking spaces of Grasscrete with crushed-brick aggregate at the office of David Furman Architecture in Charlotte, North Carolina.

area and make very accessible walkways between panels. There are no wheel stops or internal curbs. Curbed planting islands define the ends of the bays.

The solid concrete bands are made with joints at all their intersections. The edges adjoining the Grasscrete panels are in a wavy pattern mirroring the outlines of the nearby pedestals.

Some of the Grasscrete pedestals have irregular edges and surfaces which could be attributed to hurried placement. They are in good condition except at the edges of some of the panels, where a few are broken or missing.

The aggregate in the cells is angular crushed brick. Its red color contrasts neatly with the Grasscrete's white concrete and matches the brick materials of nearby historic buildings. This recycled material is economically available in the area. The largest particles are 1/2 inch in size; the content of smaller particles is significant. The aggregate is level with the pedestals except at the edges of some of the panels, where some of the aggregate is missing and the aggregate surface is up to $1^1/_2$ inch below the concrete pedestals.

The missing pedestals and aggregate could be due to stress from vehicles' wheels near the edges of the Grasscrete panels. The same lateral force could both displace aggregate and break pedestals. Perhaps the hurried placement of the concrete left the pedestals more susceptible to breakage than those in other Grasscrete installations.

A visit during wet weather in May 2003, showed that water stood in the panels' low corners. Evidently, the voids in the surface course were full of water, and the base course is slowly permeable. This installation illustrates that Grasscrete can be placed

Open-Celled Paving Grids

selectively in complex patterns with solid concrete to make a useful combined composition. Although this installation was substantially more expensive than dense asphalt would have been, the architect-owner appreciates its attractive appearance.

MISSOURI BOTANICAL GARDEN, SAINT LOUIS, MISSOURI

Grasscrete parking bays at the Missouri Botanical Garden in Saint Louis illustrate a previous configuration of Grasscrete that is still extant on some sites, and the longevity of which well-made concrete grids are capable. The bays are located in portions of the garden's main parking lot, over the rooting zones of preexisting trees (the remainder of the parking lot is in dense asphalt). They were installed in 1982 (John Broderson, personal communication 2002). Figure 10.12 shows some of the slabs in 2003, 21 years after construction. The traffic on different parts of the installation varies; some portions are used almost daily by numerous garden visitors. Wheel stops indicate individual parking spaces.

Each Grasscrete panel is 24 feet × 24 feet in area, with a central open area at the tree base 8 feet × 8 feet in size. After two decades, the panels were in structurally excellent condition, in uniform alignment with no deformation or displacement. Cracks were rare and did not affect performance in any way. However, two decades of traffic had worn the cement matrix away from the concrete surface, leaving the aggregate particles standing up slightly above the rest of the surface.

The panels have cloverleaf-shaped perforations that distinguish this configuration of Grasscrete. Each perforation is 6 inches wide; adjacent perforations are

FIGURE 10.12 Grasscrete (old configuration) parking bays over the rooting zones of persimmon trees at the Missouri Botanical Garden, Saint Louis.

spaced 8 inches center-to-center. The fill in the open cells is angular aggregate 3/8 inch in size; it is open-graded with a low content of fine particles. The voids between aggregate particles are filled with soft, light, friable, porous organic matter which has dropped from the overhanging trees over the years and decomposed in place. The infiltration rate into the cells' mixture of aggregate and organic matter is very high.

The fill in the cells is up to 1/2 inch below the concrete surface. Its deep recession and the great width of the cells produce a noticeable rumbling vibration for cars driving slowly across the surface. Walking across the panels between perforations requires some care. Topdressing with additional 3/8 inch open-graded angular aggregate would bring the fill up to the concrete surface, reducing the vibration for cars and improving accessibility for pedestrians.

The trees that these slabs were built to protect have survived and thrived; they are notably large and healthy. They include bald cypress, swamp white oak, pin oak, and persimmon. They are visible above-ground evidence of the slabs' protection of large rooting zones with durable permeability to air and water despite the substantial and prolonged load of traffic.

MONOSLAB CONCRETE GRID

The Monoslab model of concrete grid is manufactured by Kirchner Block and Brick (www.kirchnerblock.com), which also licenses the model to other manufacturers. Some technical information about the model is available at www.ephenry.com.

Figure 10.13 shows Monoslab's shape. It has rows of concrete pedestals which are connected below the surface by intersecting ribs. A complex network of open cells wraps around the pedestals and over the ribs. The unit is 4.5 inches thick and 16 inches × 24 inches in area; it weighs more than 80 pounds.

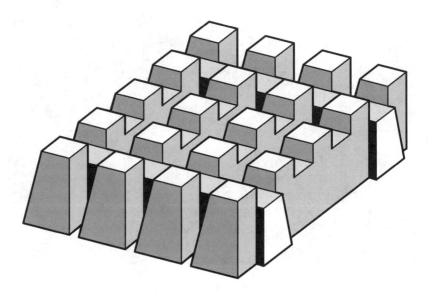

FIGURE 10.13 The Monoslab grid unit.

Open-Celled Paving Grids

In the field Monoslab can be distinguished from other pedestaled grids by the rectangular shape and spacing of the pedestals, shown in Figure 10.14. The open cells occupy 78 percent of the pavement's exposed surface area (calculated from data in the figure). Monoslab's large open areas, like those of Checkerblock, can affect accessibility, infiltration, and grass uniformity, in ways that depend on the character of the cell fill and site-specific traffic and maintenance.

CHRISTIAN MEETING ROOM, GAHANNA, OHIO

A Monoslab parking lot at the Christian Meeting Room in Gahanna, Ohio illustrates the ability of open-celled grids with vegetation to preserve grassy "green space." Although the church's postal address is on Sandburr Drive, that street is in a residential community, so the church has connected its access drive to the busier Wendler Boulevard. A row of trees screens the church building from the residences, leaving only a lawn fronting on to Sandburr. During weekly Sabbath meetings church parking overflows onto the exposed lawn. This area is reinforced with Monoslab grids. Figure 10.15 shows the parking area blending into the residential setting with trees and grass.

The Monoslab cells are filled with native clay soil and grassed. Only the tops of the concrete pedestals are visible amid the grass. Due to the slope of the ground the pedestals are entirely invisible from nearby residences. Parking is organized only by the orientation of a peripheral row of trees and the parallel rows of concrete pedestals on the ground surface; there are no wheel stops, lane markers, or perimeter edging.

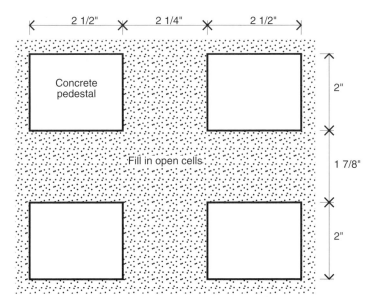

FIGURE 10.14 Surface pattern made by the tops of Monoslab's concrete pedestals (measured at Christian Meeting Room, Gahanna, Ohio).

FIGURE 10.15 Monoslab area for weekly parking at the Christian Meeting Room, Gahanna, Ohio.

At the time of a visit to the parking area in 2001, the concrete grids were in excellent condition, with no visible displacement or cracking.

The grass is thick and healthy over most of the area, especially in the areas implicitly used as parking stalls. The low frequency of traffic has allowed the grass cover to thrive, maintaining the soil's infiltration rate and blending the area into the residential setting.

The grass is thin and the soil compacted only where traffic concentrates at the vehicular entry point. Where numerous cars have followed the Monoslab cells' orientation, flexible tires have pressed onto the soil despite a degree of protection from the concrete pedestals. In this area sandy soil in the cells would be less susceptible to compaction.

Baltimore Zoo, Baltimore, Maryland

A Monoslab parking lot at the Baltimore Zoo illustrates both the durability of Monoslab grids under a substantial traffic load and the susceptibility of vegetated fill soil to compaction and erosion under the same traffic. The parking lot is large enough to hold several hundred cars. It receives correspondingly heavy daily use from the zoo's numerous visitors, especially in the summer growing season and in the portions of the parking lot closest to the zoo's entrance. Overflow parking goes to peripheral areas of unreinforced grass.

In the Monoslab-reinforced area the grids cover both the parking stalls and the driving lanes. Rows of timbers pinned to concrete bands define the bays; individual parking stalls are unmarked. The soil in the cells is fine-textured and loamy. At the time of visits between 2000 and 2003, roughly 15 years after the field was installed (Mark Reinberger, personal communication, January 2001), the concrete grids were in good condition; they were sound and smoothly aligned with each other horizontally and vertically, although in the most heavily trafficked areas the traffic had rounded off the edges of the concrete pedestals.

Figure 10.16 shows a portion of the parking lot that is used relatively infrequently because it is distant from the zoo's entrance. The soil in the cells is almost level with the concrete pedestals and the spreading grass is nearly continuous over soil and concrete alike. Pedestrian accessibility on the nearly uniform surface is good.

With increasing proximity to the entrance, the frequency of traffic increases and condition of the soil surface deteriorates. In the middle portion of the parking field the grass cover is good in the parking stalls, and especially in the parts of the parking stalls where only the front pair of wheels go. However, in the turning lanes the soil is compacted and the grass is thin; the grass–soil surface is 1/2 inch below the concrete pedestals.

Figure 10.17 shows an area close to the zoo's main entrance, where traffic is most heavy and frequent. Grass covers no more than 50 percent of the soil surface in the parking stalls and 20 percent in the turning lanes. The bare soil has been compacted and crusted by traffic and rainfall, and eroded by runoff; its surface is 1/2 inch or more below the concrete pedestals. The eroded soil has accumulated into a low bank at the downslope edge of each parking bay. The very uneven pavement surface causes car tires to make a rumbling noise and jostles the visitors' numerous baby carriages.

The failed parts of the surface could be restored by removing ungrassed soil from the cells and replacing it with a less compactible, less erodible fill such as angular open-graded aggregate. The Monoslab grids and their base course could be left in place, as they comprise a substantial portion of the pavement's cost and are in fundamentally good condition. (Other, firmly bound types of pavement materials could be selected for the most highly trafficked areas such as parking lot entrances.) In areas with only partially failed grass, perhaps the turf could be restored by loosening the soil surface, adding a sandy noncompactible planting medium to bring the soil surface up to level, and replanting. To minimize the cost of restoration, all treatments should be applied only in the portions of the parking lot that need them; the successfully grassed portions of the parking field should be left in their stable, healthily vegetated form.

FIGURE 10.16 Monoslab grids at the Baltimore Zoo in Baltimore, Maryland, in a portion of the parking lot distant from the zoo's main entrance.

Although this experience confirms that strong concrete grids on a firm base have good stability even under substantial traffic loads, it also affirms that turf fill should be used only in seldom-trafficked settings. In this heavily used parking lot, the grid pedestals have not prevented the heaviest traffic from killing the grass and compacting the soil. In future projects, analysis of site plans can distinguish settings that are suited for different types of pavement treatments, some grassy and others not.

TURFSTONE CONCRETE GRID

Turfstone is an unpatented and nonproprietary model of concrete grid (Smith and Hughes, 1981). Some manufacturers use other names for their versions of the model,

Open-Celled Paving Grids

FIGURE 10.17 A portion of the Baltimore Zoo parking lot near the zoo's main entrance.

such as Turfgrid, Turfblock, Turf-Slab, and Grass Grid. Numerous regional manufacturers make it probably the most widely manufactured grid model in North America. Some versions are manufactured to ASTM C 1319 standards; others may not be.

Figure 10.18 shows a typical Turfstone unit. It consists of a diagonal grid of solid ribs and square open cells. The generic Turfstone model can be identified in the field by the continuous lattice surface of the concrete ribs and their penetration by square cells. In one typical unit, the ribs are 2 1/8 inches wide. Its 3 1/2 inch $\times$ 3 1/2 inch openings occupy 34 percent of the pavement surface. The unit is 3 1/4 inches thick, occupies a 16 inch $\times$ 24 inch rectangular area, and is notably heavy.

Turfstone's rib lattice isolates the cells from each other. Pockets of soil in the cells have small water-holding capacity, making grass planted in the cells relatively susceptible to drought stress in the first year or so, before deep roots are established.

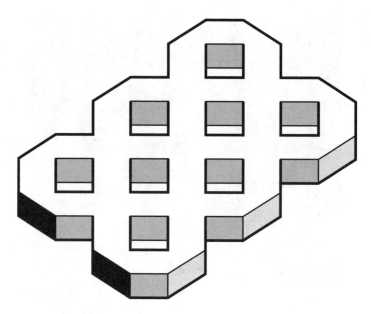

FIGURE 10.18 A typical Turfstone unit.

Some practitioners speculate that Turfstone's large concrete mass "steals" water from rooting soil in dry weather, and can have a local heat-island effect in which high afternoon temperature increases water stress. On the other hand, in an experimental installation in Renton, Washington, grass establishment was "easier and faster" in Turfstone than in a nearby plastic geocell; water that fell on Turfstone's ribs tended to concentrate in the cells, in effect giving the cells twice the natural rainfall; and the sturdy concrete ribs provided greater protection for the grass from the daily parking of vehicles than did the thin ribs of the geocell (Booth and Leavitt, 1999; Booth et al., 1997).

In well maintained and lightly trafficked settings where spreading grass covers the concrete ribs, the turf surface is as accessible for walking on as any other uniform lawn. In settings with more frequent traffic or less conscientious maintenance, the wide ribs are likely to be visible between grass-filled cells. In a low-maintenance regime, Nichols' (1993) comparisons found Turfstone to be relatively unsightly compared with pedestaled grids, because the surface was punctuated by unvegetated cells showing bare soil surrounded by white concrete ribs.

SAN LUIS BAY INN, AVILA BEACH, CALIFORNIA

A Turfstone emergency access lane at the San Luis Bay Inn in Avila Beach, California illustrates the use of Turfstone in a common type of application. The inn is a resort hotel overlooking the Pacific shore. The access lane is located at the foot of the building, forming the foreground in the ocean view. Figure 10.19 shows the grassy surface blending the lane visually into the hotel's landscaped setting.

Open-Celled Paving Grids

FIGURE 10.19 Turfstone emergency access lane at the San Luis Bay Inn in Avila Beach, California.

The grass is irrigated. Although it is a bunch grass (perhaps fescue) the turf cover is thick and uniform. The only area where the ribs are substantially exposed and visible is at the edge of the pavement, where grass spreads onto the rib surface from only one side. There is no visible edge restraint, even on the downhill edge of the surprisingly steep slope. Nevertheless, the surface is in excellent structural condition. It has probably never borne traffic except that of landscape maintenance vehicles and occasional pedestrians.

TUMWATER HISTORICAL PARK, OLYMPIA, WASHINGTON

A Turfstone parking bay at Tumwater Historical Park in Olympia, Washington illustrates the structural importance of a pavement's base course. The bay consists of a long line of parking stalls accessed by an asphalt access drive. Figure 10.20 shows the bay in 1999, five years or more after construction. Wheel stops indicate individual parking stalls; there are no other perimeter restraints. The park is heavily used for recreation, and the Turfstone bay is used for parking on many days of the year. Nevertheless, the predictable visitation schedule permits scheduled maintenance in the park's routine landscape maintenance program.

The cells are filled with fine-textured soil level with the grid surface. The grass cover is in good condition overall and easy to walk on. Near the perimeter of the parking spaces, where only the front wheels of cars pass, the grass is thick and covers the concrete ribs almost continuously. It is less continuous near the access lane, where all four wheels pass over, indicating that this is about as much traffic as the grass could

FIGURE 10.20 Turfstone grids with grass at Tumwater Historical Park, Olympia, Washington.

Open-Celled Paving Grids 407

stand without failing. The few spots with bare soil could be easily rehabilitated by replacing the planting medium with sandy, noncompactible soil, and replanting.

Near the asphalt drive a few Turfstone units have broken under the lateral pressure of numerous turning vehicles, but they remain aligned vertically. Some other units, located away from the concentrated traffic and surrounded by unbroken units, have broken and visibly settled relative to other nearby units. Such vertical displacement is attributable only to softness in the pavement's base or setting bed. Grid units that have settled or broken for any reason can be easily lifted out; the base could then be repacked and the units reset or replaced. A thick compacted base of strong material is necessary in all installations.

MUNICIPAL PARKING LOT, DAYTON, OHIO

A Turfstone municipal parking lot in downtown Dayton, Ohio illustrates the multiple environmental effects of Turfstone with grass. Although the parking lot no longer exists, its performance was quantitatively monitored and documented (four reports by Smith and others, 1981, and summary by Smith, 1984) and the results are still relevant today.

The parking lot was built in 1980 adjacent to municipal buildings at Third and Perry Streets. It held parking for 80 cars on 30,000 square feet of Turfstone with grass. It was used for municipal vehicles which were typically off the lot during the day and parked on it at night. Turfstone covered both the turning lanes and the parking stalls. Rows of solid blocks delineated the parking spaces, with panels of Turfstone between the lines of blocks. The Turfstone units were laid on an aggregate base and sand setting bed at a slope of 4 percent. The cells were filled with silty sand and seeded with a mixture of grass types selected for their prior success on lawns in the area or for their tolerance of deicing salt.

The grass was unmaintained except for mowing. In the first summer after installation, dry, dead and missing grass in certain parking stalls was attributed to summer drought and lack of irrigation during the establishment period, shade of parked cars, and oil dripped from vehicles. Cracked or misaligned grid units were noticed in parts of the turning lanes where numerous vehicles turned. Some units rattled loosely when ridden over by a vehicle; these were attributed to uneven settling of the uncompacted sand setting bed; the researchers concluded that in future installations the units should be vibrated into place on the bed. Mud and standing water occurred in the most heavily traveled parts of the turning lanes and at low spots in the surface during very wet spring weather. No difficulties were encountered in snow removal; this was attributed to the use of a plow blade with a flexible rubber edge. The salt-tolerant grass showed no ill effects from winter deicing salt.

Municipal employees who parked on the lot considered the surface fine for driving vehicles but difficult for walking; some reported heels stuck in the soft grass cells, and one reported twisting an ankle on the concrete ribs. The researchers suggested, providing walkways of smooth blocks along pedestrian paths for future Turfstone installations.

Figure 10.21 compares simultaneous temperatures at the Turfstone lot with those at a nearby parking lot of dense asphalt. Each point in the figure represents the

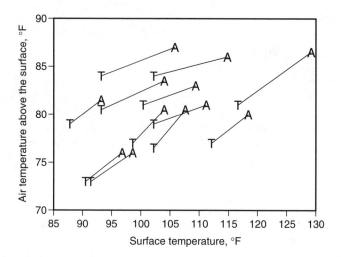

FIGURE 10.21 Selected pairs of simultaneous temperature observations on two parking lot pavements in Dayton, Ohio: T, Turfstone with grass; A, dense asphalt (data from Smith and Sholtis, 1981, p. 3).

average of readings at four places on each parking lot, in various settings with and without shade. The figure shows those 20 percent of the observations on bright or hazy days between 10:30 a.m. and 5:00 p.m. when the difference in air temperature between the two parking lots was the greatest. In these conditions Turfstone with grass was several degrees cooler than the asphalt, both at the pavement surface and in the air above. On cool, cloudy days the difference in temperature was not as great; on 2 percent of the occasions the Turfstone lot had higher temperatures than the asphalt lot.

Figure 10.22 shows the Turfstone's measured runoff coefficient during 11 natural rain events of varying intensity and duration in the late summer and early fall (Smith and Sholtis, 1981, pp. 6–9; two of the events are identical, so the chart seems to show only 10 points). During very small events the Turfstone generated no runoff. The highest runoff coefficient of 0.35 occurred during a storm that had been preceded by two days of wet weather. Evidently, the soil in the Turfstone cells stored moisture and produced relatively high runoff at a time of high antecedent moisture.

BLUE HERON, ANNAPOLIS, MARYLAND

A Turfstone drive and parking area at the Blue Heron residential community in Annapolis, Maryland illustrates the potential displacement of rounded alluvial gravel from the cells of a sloping grid pavement. Blue Heron was built on Boucher Avenue in 1991 or 1992. It serves five or six homes which generate only light traffic. The homes are connected by a Turfstone lane that expands into parking stalls at each home. The lane and stalls form a single flowing surface of Turfstone grids, edged by a raised concrete curb. Figure 10.23 shows a group of parking stalls about 9 years after construction.

Open-Celled Paving Grids   409

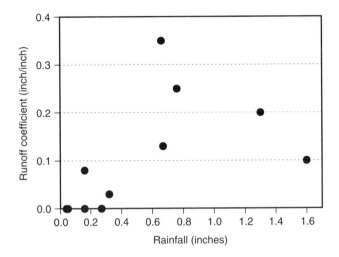

FIGURE 10.22 Runoff coefficients during natural rain events at the Dayton Turfstone parking lot (data from Smith and Sholtis 1981, p. 8).

FIGURE 10.23 Turfstone with gravel fill at the Blue Heron residential community in Annapolis, Maryland.

The grid units are structurally sound, showing little or no cracking or displacement. On the surface of the ribs traffic has worn away the concrete matrix, leaving a rugged residual surface of embedded aggregate; the selective erosion may be attributable to weak cement paste in the concrete material.

The fill in the cells is rounded river gravel. The maximum particle size is about 3/8 inch; the content of sand and smaller particles is substantial. The infiltration rate is moderately high. In the portions of the pavement that are level, the aggregate is within 1/8 inch of the grid surface. In a portion of the pavement that is steeply sloping and that receives occasional runoff channeled against the curb, the aggregate has washed out of the top 1/2 inch of the cells. The displaced particles are visible on the downslope where runoff and traffic have dispersed them on level parts of the pavement and in the public street. The total loss of particles over the years has been small and replacement has not been felt necessary (James Urban, personal communication January 2001).

This experience confirms that in order to resist displacement, unbound aggregate in pavement surfaces, even aggregate confined in the cells of paving grids, should consist of angular particles placed on a level or nearly level surface and free from concentrated flows of surface runoff. The washing out of particles in Blue Heron's steep channeled area could be inhibited by replacing the lost gravel with open-graded angular aggregate; such material would produce particle-to-particle interlock and absorb some of the runoff.

OTHER MODELS OF CONCRETE GRIDS

Table 10.4 lists additional models of concrete open-celled grids that are manufactured in North America. Although these units could be used for paving, no significant paving installations have been found that use them. Some of them have previously been applied in channel erosion control; others are very new models awaiting their first paving applications. All of them are available for potential paving applications in the future.

OPEN-CELLED PAVING BRICKS

Unlike the concrete grids and slabs described above, bricks are made from fired clay. The nature of clay paving bricks and their supply were introduced in the previous chapter.

Figure 10.24 shows that cored brick has been used as a type of open-celled paving grid. The open cells in cored brick can occupy up to 40 percent of the surface

TABLE 10.4
Other Models of Open-Celled Grids That May Be Used for Paving in North America

Model	Company	Contact Information
EcoGrid	Hanover	www.hanoverpavers.com
Grasstone	Pavestone	www.pavestone.com
Turf Pavers	D'Hanis Brick & Tile	www.dhanisbricktile.com
Uni-Green	F. von Langsdorff Licensing	www.langsdorff.com

Open-Celled Paving Grids

FIGURE 10.24 Cored brick used as an open-celled paving grid at the Southface Institute, Atlanta, Georgia (www.southface.org).

area. However, cored brick tends to be manufactured for use in building walls, not for the moisture and traffic conditions of pavements, so as a paving material it is appropriate only in undemanding situations like the storage-shed entrance shown in the figure.

The Pine Hall Brick company (www.pinehallbrick.com) has considered developing an open-celled model of brick specifically for use in pavements. This would be the first known model of clay brick intended to make pavements that are in any way porous and permeable. Development of this new configuration would provide a competitive alternative to concrete grid units and would allow the brick industry to participate directly in the production of porous pavement surfaces for the first time.

Figure 10.25 shows the unit's shape as it was being considered in late 2001. The unit is 4 inches × 8 inches in area and $2^{3}/_{4}$ inches thick, like many other models of paving brick. Nubs are placed asymmetrically to make joints 1/16 inch wide on all four sides. Three open cells provide porosity and permeability when filled with appropriate aggregate or soil. The cells occupy 19 percent of the surface area; they are large enough to receive No. 8 aggregate. However, aggregate small enough to fill the narrow joints would have a disappointingly low infiltration rate; perhaps those joints could be left vacant of fill.

TRADEMARKS

Table 10.5 lists the holders of registered trademarks mentioned in this chapter.

TABLE 10.5
Holders of Registered Trademarks Mentioned in This Chapter

Registered Trademark	Holder	Headquarters Location
Abbotsford Concrete Products	Abbotsford Concrete Products	Abbotsford, British Columbia
Acker Stone	Acker Stone Industries	Corona, California
Ambergrid	Midwest Products Group	Jefferson City, Missouri
Anchor Concrete Products	Anchor Concrete Products	Manasquan, New Jersey
Angelus Block	Angelus Block	Sun Valley, California
Basalite	Basalite Concrete Products	Tracy, California
Belgard	Oldcastle	Atlanta, Georgia
Bend Industries	Bend Industries	Appleton, Wisconsin
Bomanite	Bomanite Corporation	Madera, California
Cambridge Pavers	Cambridge Pavers	Lyndhurst, New Jersey
Checkerblock	Nicolia Industries	Lindenhurst, New York
D'Hanis Brick & Tile	D'Hanis Brick & Tile	San Antonio, Texas
Dura Art Stone	Dura Art Stone	Fontana, California
E. P. Henry	E. P. Henry	Woodbury, New Jersey
EcoGrid	Hanover Architectural Products	Hanover, Pennsylvania
Eurotekno	Eurotekno	Quackenbrück, Germany
Grasscrete	Bomanite Corporation	Madera, California
Grasstone	Pavestone	Dallas, Texas
Grinnell Pavers	Grinnell Pavers	Sparta, New Jersey
Hanover Architectural Products	Hanover Architectural Products	Hanover, Pennsylvania
Hastings Pavers	Nicolia Industries	Lindenhurst, New York
Hessit Works	Hessit Works	Freedom, Indiana
Ideal Concrete Block	Ideal Concrete Block	Westford, Massachusetts
Interlock Paving Systems	Interlock Paving Systems	Hampton, Virginia
Kirchner Block	Kirchner Block	Bridgeton, Missouri
Metromont Materials	Metromont Materials	Spartanburg, South Carolina
Monoslab	E. P. Henry	Woodbury, New Jersey
Mutual Materials	Mutual Materials	Bellevue, Washington
Navastone	Navastone	Cambridge, Ontario
Nicolock	Nicolia Industries	Lindenhurst, New York
Nitterhouse Masonry Products	Nitterhouse	Chambersburg, Pennsylvania
Oldcastle Architectural	Oldcastle	Dublin, Ireland
Pave Tech	Pave Tech	Prior Lake, Minnesota
Paveloc	Paveloc Industries	Marengo, Illinois
Paver Systems	Paver Systems	Sarasota, Florida
Pavestone	Pavestone	Dallas, Texas
Pine Hall Brick	Pine Hall Brick	Winston-Salem, North Carolina
Quick Crete Products	Quick Crete Products	Norco, California
Stone Age Pavers	Stone Age Pavers	Pompano Beach, Florida
Tremron	Tremron	Miami, Florida
Turf Pavers	D'Hanis Brick & Tile	San Antonio, Texas
Unicon Paving Stone	Unicon Concrete	Durham, North Carolina
Uni-Green	F. von Langsdorff Licensing	Inglewood, Ontario
Unilock	Unilock	Georgetown, Ontario
Westcon	Westcon	Olympia, Washington
Willamette Graystone	Willamette Graystone	Eugene, Oregon

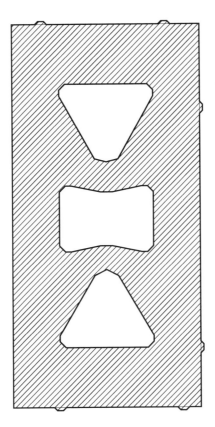

FIGURE 10.25 Surface view of open-celled paving brick being considered by Pine Hall Brick as of late 2001.

ACKNOWLEDGMENTS

The following persons generously provided information used in this chapter: Dexter Adams and his staff in the University of Georgia's Physical Plant Department; Steve Arnett of Air Vol Block (www.airvolblock.com); Steven M. Bopp of Austin Tao & Associates (http://austintao.com/); Soenke Borgwardt of Borgwardt Wissenschaftliche Beratung, Norderstedt, Germany; John Broderson of Ahal Contracting (www.ahal.com); Michael Difiore and Jim Panagos of Hastings Pavement Company (www.hastingspavers.com); Deborah Edsall of Edsall and Associates, Columbus, Ohio; David Furman of David Furman Architecture in Charlotte, North Carolina; Daryl Gusey, U.S. Forest Service, Lakewood, Colorado; Bill Hunt and John Stone of North Carolina State University (www.eos.ncsu.edu); R. Lawson of Pine Hall Brick (www.pinehallbrick.com); Jack O'Donnell of Oberfield's (www.oberfields.com); Jim Pillsbury of the Westmoreland Conservation District, Greensburg, Pennsylvania; Mark Reinberger and Scott Weinberg of the University of Georgia School of Environmental Design; Steve Riggs of Progressive Concrete (www.progressiveconcrete.com); Kellie Romero of Bomanite (www.bomanite.com); Will Siler, a former student at the University of Georgia, now in landscape architectural practice in North Carolina; David R. Smith of the Interlocking Concrete Pavement Institute (www.icpi.org); James Urban, landscape architect of Annapolis, Maryland.

Michael Lenny permitted access to his office in Cherry Hill, New Jersey. An AILA-Yamagami-Hope Fellowship from the Landscape Architecture Foundation supported research travel to porous pavement sites in 1996.

The following persons constructively critiqued early drafts of this chapter: David Nichols of the University of Georgia School of Environmental Design; Chris J. Pratt of Coventry University (UK); and David R. Smith of the Interlocking Concrete Pavement Institute (www.icpi.org).

REFERENCES

Booth, Derek B., and Jennifer Leavit (1999). Field Evaluation of Permeable Pavement Systems for Improved Stormwater Management, *Journal of the American Planning Association* Summer 1999, 314–325.

Booth, Derek B., Jennifer Leavitt, and Kim Peterson (1997). *The University of Washington Permeable Pavement Demonstration Project: Background and First-Year Field Results,* Seattle: University of Washington Center for Urban Water Resources Management.

Borgwardt, Soenke (1997a). *Leistungsfahigkeit und Einsatzmoglichkeit Versickerungsfahiger Pflastersysteme* (*Performance and Fields of Application for Permeable Paving Systems*), *Betonwerk + Fertigteil-Technik*, No. 2.

Borgwardt, Soenke (1997b). *Versickerungsfahige Pflastersysteme aus Beton: Voraussetzungen, Anforderungen, Einsatz* (*Infiltratable Concrete Block Pavement Systems: Prerequisites, Requirements, Applications*), Bonn: Bundesverband Deutsche Beton- und Fertigteilindustrie.

Dallman, Suzanne, and Thomas Piechota (1999). *Storm Water, Asset or Liability*, Los Angeles: Los Angeles and San Gabriel Rivers Watershed Council (www.lasgriverswatershed.org).

Day, Gary E. (1978). *Investigation of Concrete Grid Pavements*, Blacksburg: Virginia Polytechnic Institute and State University.

Day, Gary E. (1980). Investigation of Concrete Grid Pavements, in *Stormwater Management Alternatives*, pp. 45–63, J. Toby Tourbier and Richard Westmacott, eds., Newark: University of Delaware Water Resources Center.

Day, Gary E., D.R. Smith, and J. Bowers (1981). *Runoff and Pollution Abatement Characteristics of Concrete Grid Pavements*, Bulletin 135, Blacksburg: Virginia Polytechnic Institute and State University, Water Resources Center.

Interlocking Concrete Pavement Institute (ICPI) (1997). *Concrete Grid Pavements*, Tech Spec No. 8, Washington: Interlocking Concrete Pavement Institute.

National Concrete Masonry Association (1995). *Concrete Grid Pavers*, Spec Data No. 2, Herndon, Virginia: National Concrete Masonry Association, Concrete Paver Institute.

National Concrete Masonry Association (NCMA) (1996). *Concrete Grid Pavements*, TEK 11-3, Herndon, Virginia: National Concrete Masonry Association.

Nichols, David B. (1993). Open Celled Pavers: An Environmental Alternative to Traditional Paving Materials, in *Proceedings of the 1993 Georgia Water Resources Conference*, pp. 251–253, Kathryn J. Hatcher, Ed., Athens: University of Georgia Institute of Natural Resources.

Nichols, David B. (1995). Comparing Grass Pavers, *Landscape Architecture* 85, 26–27.
Pratt, C.J., J.D.G. Mantle, and P.A. Schofield (1989). Urban Stormwater Reduction and Quality Improvement through the Use of Permeable Pavements, *Water Science and Technology* 21, 769–778.
Pratt, C. J., J.D.G. Mantle, and P.A. Schofield (1995). UK Research into the Performance of Permeable Pavement, Reservoir Structures in Controlling Stormwater Discharge Quantity and Quality, *Water Science and Technology* 32, 63–69.
Shearman, RC., E.J. Kinbacher, and T.P. Riordan (1980). Turfgrass-Paver Complex for Intensely Trafficked Areas, *Agronomy Journal* 72, 372–374.
Sipes, James L, and John Mack Roberts (1994). Grass Paving Systems, *Landscape Architecture* 84, 31–33.
Smith, David R., and Mollie K. Hughes (1981). *Project Description, Part I of Green Parking Lot, Dayton, Ohio, An Experimental Installation of Grass Pavement*, Heritage Conservation and Recreation Service and on file at Interlocking Concrete Pavement Institute, Washington.
Smith, David R., and David A. Sholtis (1981). *Performance Evaluation, Part II of Green Parking Lot, Dayton, Ohio, An Experimental Installation of Grass Pavement*, Heritage Conservation and Recreation Service; and on file at Interlocking Concrete Pavement Institute, Washington.
Smith, David R. (1981). *Maintenance Evaluation, Part III of Green Parking Lot, Dayton, Ohio, An Experimental Installation of Grass Pavement*, Heritage Conservation and Recreation Service; and on file at Interlocking Concrete Pavement Institute, Washington.
Smith, David R. (1981). *Life Cycle Cost and Energy Comparison of Grass Pavement and Asphalt, Part IV of Green Parking Lot, Dayton, Ohio, An Experimental Installation of Grass Pavement*, Heritage Conservation and Recreation Service and on file at Interlocking Concrete Pavement Institute, Washington.
Smith, David R. (1984). Evaluations of Concrete Grid Pavements in the United States, in *Second International Conference on Concrete Block Paving*, Delft, April 10-12, 1984, pp. 330–336; and on file and Interlocking Concrete Pavement Institute, Washington.
Smith, David R. (2001). Another Attendance Record for the Bauma, *Interlocking Concrete Pavement Magazine* 8, 8–10.
Southerland, Robert J. (1984). Concrete Grid Pavers, *Landscape Architecture* 74, 97–99.
Wells, Malcolm (1981). *Gentle Architecture*, New York: McGraw-Hill.

11 Porous Concrete

CONTENTS

The Porous Concrete Paving Industry 418
Advantages and Disadvantages of Porous Concrete 421
    Cost of Porous Concrete 421
Hydrology of Porous Concrete 422
    Runoff 424
    Water Quality 425
    Surface Clogging and Rehabilitation 426
Constituents of Porous Concrete 426
    Aggregate 427
    Portland Cement and Its Supplements 428
    Reinforcement 429
    Admixtures 429
Construction of Porous Concrete Pavements 430
    Slab and Subbase 430
    Subgrade Compaction and Its Alternatives 431
Joints in Porous Concrete 432
    Control Joints 432
    Expansion Joints 434
Care Required in Porous Concrete Installation 435
    Installer's Qualifications 435
    Proportioning of Mixture 436
    Placement and Curing 437
    Testing for Quality Control 438
Porous Concrete in Successful and Unsuccessful Installations 439
    834 East Ocean Boulevard, Stuart, Florida 439
    827 S.E. Fifth Street, Stuart, Florida 440
Further Installations on Sandy Subgrades 441
    Hobe Sound Sprinklers, Hobe Sound, Florida 441
    Council on Aging of Martin County, Stuart, Florida 442
    Jones Ecological Research Center, Newton, Georgia 443
Porous Concrete on Fine-Textured Subgrades 444
    Southface Institute, Atlanta, Georgia 444
    Webb Bridge Park, Alpharetta, Georgia 447

Porous Concrete on the West Coast 448
Toward Porous Concrete in Cold Climates 450
Trademarks 452
Acknowledgments 452
References 453

Porous concrete is made by binding open-graded aggregate with Portland cement. Most of the volume of concrete is aggregate; the cement binds the aggregate particles together. Porous concrete is chemically identical to dense concrete; the defining difference is that porous concrete is made with open-graded aggregate, which creates the voids in the concrete structure.

Technically, the term "concrete" refers generically to any aggregate bound by a paste. Informally, the term is used to refer specifically to Portland cement concrete. This chapter follows that informal convention by using the terms "concrete" and "Portland cement concrete" interchangeably.

When Portland cement is mixed with water, the cement and water combine chemically (hydrate) to form a paste of calcium silicate hydrate. The concrete mixture is placed into formwork while in a fluid state. Upon completion of the chemical reaction ("curing"), the cement forms a hard shell around and between the aggregate particles. Over a period of days or weeks it produces a crystalline binder that turns the material into a rigid slab.

Previous introductions to porous concrete were those of the Florida Concrete and Products Association (no date) and the Portland Cement Pervious Association (2003). Other basic references such as the following supply further details on concrete and its pavement applications, as long as it is understood that they assume only a dense, nonporous concrete product: American Concrete Institute (1984), American Concrete Pavement Association (1996), Landscape Architecture Foundation (Hanson and Simeoni, 1992), Portland Cement Association (Kosmatka and Paranese, 1988), and *Time-Saver Standards* (Harris and Brown, 1998). Table 11.1 lists some terms with distinctive application in the construction of concrete, and of porous concrete in particular.

THE POROUS CONCRETE PAVING INDUSTRY

Concrete has been used in pavement surfaces since 1865, when dense concrete street pavements were first experimentally installed in Scotland (Croney and Croney, 1998, pp. 11–12). Initially, cities were hesitant to use concrete where it would inhibit access to underground utilities. In the early twentieth century, the material became more common where interest in concrete's notable durability exceeded concern about underground access. Today, concrete is second only to asphalt as a common paving material. Concrete suppliers and installers exist in most regions of North America.

Porous concrete was first used in pavements following World War II (Maynard, 1970). California used it to make drainage layers under dense concrete highway surfaces. Switzerland and other jurisdictions in Europe reportedly used it in surface courses. Subsequently it was used for overlays on dense concrete roadways to improve drainage and traction, and for well-drained tennis courts and highway shoulders (Monahan, 1981, pp. 9, 10, 20–21).

TABLE 11.1
Some Terms with Distinctive Application in Concrete Construction

Term	Definition
Admixture	An added ingredient in a concrete mixture
Air entrainment	Formation of microscopic air bubbles in a concrete mixture
Concrete	Any aggregate bound by a paste; used informally (in this chapter and elsewhere) to refer to Portland cement concrete
Consistency	Fluidity of a concrete mixture
Construction joint	The intersection between panels of concrete placed at two different times
Control joint (contraction joint)	Zone of weakness in a concrete slab formed to control the location of cracking
Curing	Hydration of Portland cement
Expansion joint (isolation joint)	Flexible gap between panels of concrete or between a concrete slab and a fixed structure
Fly ash	Fine airborne ash captured from the smokestacks of furnaces and incinerators during pollution control
Hydration	Chemical reaction combining Portland cement and water to form a rigid crystalline material
Isolation joint	An expansion joint between a concrete slab and a fixed structure
Joint	Part or space between two panels of concrete
Panel	Uniform area of concrete between joints
Portland cement	Hydraulic cement produced from limestone
Portland cement concrete	Concrete with Portland cement as the binder
Pozzolan	Any fine granular siliceous material which reacts chemically with Portland cement and water to supplement or enhance the cementitious binder
Proportioning	Establishment of desired ratios of aggregate, cement, water, and other constituents in a specific concrete mixture
Screeding	Striking or scraping of the surface of wet concrete with a board or blade for leveling, or the side forms that serve as a guide
Set time (setting time)	The time during which mixed concrete remains fluid and workable
Slump	Test of the consistency or workability of wet concrete based on the settlement of a mass of material after removing a cone-shaped form, or the result of this test
Subbase (base)	Any flexible pavement layer under a concrete slab
Workability	The ease with which a wet concrete mixture flows or can be pushed into forms

The use of porous concrete pavement specifically for its environmental effects started in the 1970s in Florida, where stringent requirements for stormwater retention had been established. A Florida engineer named John (Jack) Paine obtained jurisdictional "credit" for the material's stormwater retention capacity. The hydraulic storage capacity and Florida's characteristically sandy, permeable soils made porous concrete economically attractive where it reduced or eliminated the need for off-pavement stormwater management reservoirs. Table 11.2 lists some of the early Florida installations.

TABLE 11.2
Some Early Porous Concrete Parking Lot Installations in Florida

Name	Installation Date	Location	Surface Infiltration Rate In 1988
Royal Building	1976	Cape Coral	Very low (surface sealed during construction with cement paste)
1492 Colonial	1978	Fort Myers	0.96 inch/h (surface sealed during construction with cement paste)
Palm Frond	1980	North Fort Myers	745 inch/h
Witch's Brew Restaurant	1982	Naples	700 inch/h
Hampton Inn	1984	North Fort Myers	900 inch/h

Wingerter and Paine, 1989.

TABLE 11.3
Examples of National Associations in the Concrete Industry

Association	Contact Information
American Concrete Institute (ACI)	www.aci-int.net
American Concrete Pavement Association (ACPA)	www.pavement.com
American Society of Testing and Materials (ASTM)	www.astm.org
National Ready Mixed Concrete Association (NRMCA)	www.nrmca.org
Portland Cement Association (PCA)	www.cement.org
Portland Cement Pervious Association (PCPA)	temporary contact: www.petrusutr.com

With experience, Florida installers and engineers developed construction conventions suited to Florida's sandy soils, frost-free weather, and heavy rainfall. The Florida Concrete and Products Association assisted the technology's implementation with excellent guideline publications (Florida Concrete and Products Association, no date). The success of Florida's experience is reflected in the high infiltration rates shown in the preceding table for installations after 1980. Where failures occurred, they were associated with unqualified contractors who installed bad material. By 1990, more than 20 acres of the pavement had been laid in Florida, mostly in parking lots and lightly traveled drives.

Beginning in the mid-1990s, explorations were made into the use of porous concrete in other southern states and on the West Coast. At the same time national associations became interested in establishing industrial standards and advocating consistent product quality.

Table 11.3 lists some of the industry associations that today develop standards or information on concrete pavement in general or porous concrete in particular. In 2001 the American Concrete Institute formed a committee on Pervious Concrete (Committee No. 522) to compile technical knowledge about porous concrete on a national basis. In 2002, the Portland Cement Pervious Association (PCPA) was incorporated to develop standards for design and installation and to certify qualified porous concrete installers. Until PCPA develops its own web site it can be contacted through the Petrus UTR address listed in the table.

ADVANTAGES AND DISADVANTAGES OF POROUS CONCRETE

Porous concrete is one of the firmer and stronger materials covered in this book. The compressive strength of properly made porous concrete containing No. 89 aggregate is commonly 2000 pounds per square inch (psi) or more. In comparison, the compressive strength of dense concrete is commonly 3500 psi or more. Porous concrete's strength and durability have been adequate for the moderate traffic loads of numerous Florida parking lots where the material has been in place for more than 15 years. Porous concrete has also withstood greater traffic loads where it is installed in sufficient thickness.

The surface of porous concrete made with No. 89 aggregate is accessible by almost any measure. It is firm, regular, and well drained. It is adequately smooth for the movement of large-wheeled vehicles such as automobiles, bicycles, wheelchairs, and perhaps shopping carts, but probably not for the small wheels of roller blades. Traction is very high. It has been speculated (not observed in practice) that the surface's open pores could present a difficulty to walkers wearing very narrow high heels.

Concrete's color is light compared with that of asphalt. Light color makes the surface visible on dark nights. It improves driving safety. It permits a lighting reduction of 30 percent or more and corresponding cost reduction for lighting equipment and energy (Stark, 1986). For mitigating the effects of the urban heat island, porous concrete's light-colored surface has the theoretical advantage of low absorption of solar heat. However, porous concrete has been found as hot as dark asphalt on sunny summer days due to the low rate of radiant cooling (Asaeda and Ca, 2000).

Porous concrete's color can be controlled with dyes mixed into the material during construction, or stains applied later. It can be made either a brighter white or other desired colors or shades.

Porous concrete reduces street noise as does porous asphalt (*World Highways*, 1999).

COST OF POROUS CONCRETE

The cost of porous concrete today is commonly higher that that of dense concrete, because it is a special mixture with special installation procedures requiring specially experienced personnel. The difference in total cost depends on the relative costs of material and labor. The material from the supplier costs more because its cement content is high, and cement is the expensive component in concrete by weight. It takes longer to place a truckload of material, and some suppliers charge for the extra time. It might cost more or less in on-site labor. During placement labor is intensive and slightly slow because the wet material does not flow freely and has to be pulled and raked into place. But after placement there is no unproductive waiting for initial set and then surface finishing; labor is complete almost as soon as the last material comes out the chute.

Porous concrete, like other porous paving materials, is both a pavement and a component of site drainage. By reducing the need for off-pavement stormwater management, it can reduce total site development cost whether or not the cost of the pavement is greater than that of competing materials.

In 1987, porous concrete for a veterinary clinic's 15,000 square foot parking lot in Sanford, Florida cost $5000 more than dense concrete would have, but saved $25,000 on additional land purchase for a retention pond to meet local stormwater control requirements. The net saving was $20,000 (Bruce Glasby, personal communication 2000).

In Alpharetta, Georgia, the porous concrete roads at Webb Bridge Park, which will be described later in this chapter, cost 10 percent more than dense concrete would have (John Gnoffo, personal communication 1997). However, the porous material's runoff coefficient was low; it was allowed to be set equal to that of turf on the same soil type. Consequently, the park's stormwater conveyances and detention basins were correspondingly small. The net result of the porous material was a cost saving, because the porous road was properly counted as both a pavement structure and a part of the drainage system.

Properly constructed concrete pavement tends to have low long-term (life cycle) cost, because it has a long functional life and requires little routine maintenance. A survey of porous concrete installations in Florida (Wingerter and Paine, 1989) found that installations more than ten years old were structuraly in good condition and that subgrade density and permeability had not changed significantly since installation.

Concrete pavement has a long-term disadvantage in places where access to underground utilities will be required because concrete slab must be broken or cut away and afterwards patched with new material.

HYDROLOGY OF POROUS CONCRETE

The entire surface of porous concrete is porous and permeable, unlike composite surfaces such as those of open-jointed blocks and concrete grids. The void space in properly constructed porous concrete containing No. 89 aggregate is commonly between 11 and 21 percent; the exact percentage depends on such factors as the type of aggregate in the mixture (PCPA 2003, p. 5).

The measured surface infiltration rate of properly constructed material is from 55 inches per hour to many hundreds of inches per hour (Florida Concrete and Products Association, no date; Korhonen and Bayer, 1989, p. 16; Wingerter and Paine, 1989). Any decline of infiltration rate with time has not been measured but is believed to be slight as long as there is no significant exogenous source of clogging matter; concrete is a stable material. The production of a slab with high long-term infiltration rate is focused on the few hours when the wet concrete is being placed: proportioning and placement produce the full potential infiltration rate; improper mixture or excessive finishing can reduce surface infiltration rate to less than 15 inches per hour (Wingerter and Paine, 1989, p. 5).

The Southwest Florida Management District monitored runoff quantity and quality in areas containing some porous concrete at the Florida Aquarium in Tampa (Rushton and Hastings, 2001). The aquarium's parking lot was built in 1997. For comparative monitoring, different areas 0.23 to 0.26 acre in size were paved in different combinations of materials, and the runoff from each area drained to a separate outlet. Although the monitoring system did not quantitatively isolate porous

Porous Concrete 423

FIGURE 11.1 Porous concrete parking stalls and vegetated swale at the Florida Aquarium in Tampa, Florida.

concrete's effects from those of swales and dense concrete, it made porous concrete's types of effects visible.

Figure 11.1 shows the aquarium's partly porous concrete area in 1999. Monitoring equipment is visible at the end of the swale in the background. The porous concrete is in the parking stalls; the turning lane is of dense concrete. A gentle slope directs surface drainage from the center of the impervious lane across the porous stalls and into the swale. The surface infiltration rate of the porous concrete is high. The runoff monitored at the end of the swale is a composite of that from the dense concrete, the porous concrete, and the swale. The reference area for comparison of runoff had both turning lane and parking stalls in dense concrete, and a vegetated swale. The subgrade was sandy and had been modified over time by cycles of urban construction.

Runoff

Figure 11.2 shows the aquarium's runoff during an intense storm. It is evident that the porous concrete parking stalls reduced runoff. However, the effect is partly hidden by the runoff from the dense concrete turning lane, and the presence of swales in both drainage areas. Presumably if monitoring had focused exclusively on contrasting porous and dense pavement areas, the reduction of runoff by the porous material would be more visible.

Figure 11.3 shows the aquarium's runoff during a storm of low intensity. Note that the vertical scale is much lower than that in the previous figure. Runoff from the partly porous area was less than half that from the area containing only dense concrete, although again the effect of porous concrete is partly masked.

Table 11.4 lists the runoff coefficients of the aquarium's drainage areas over two years combining 66 storms of different intensities. The most typical runoff coefficients for the areas containing some porous concrete were about half those of the areas containing only dense concrete. Porous concrete was most effective at reducing runoff during small rain events (Rushton and Hastings, 2001, p. 20). If the runoff from the dense concrete is subtracted from the monitoring areas, the remaining median runoff coefficient from the porous concrete by itself was approximately zero. During the most intense events the coefficients of the two types of areas were almost equal.

Research to determine porous concrete's runoff coefficient directly is called for. Until that research is done, the runoff coefficient could logically be assumed similar to that of fresh porous asphalt, which is described in Chapter 4.

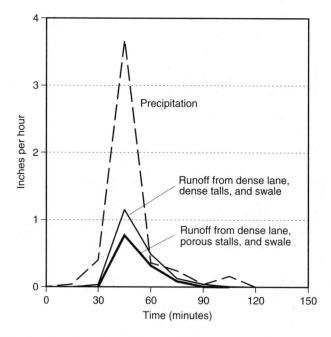

FIGURE 11.2 Precipitation and runoff at the Florida Aquarium during the short, intense rain of January 3, 1999 (data from Rushton and Hastings, 2001, Appendix E).

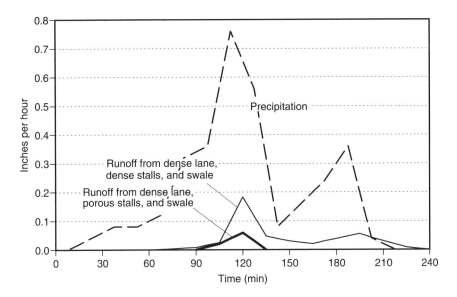

FIGURE 11.3 Precipitation and runoff at the Florida Aquarium during the long, soft rain of March 14, 1999 (data from Rushton and Hastings, 2001, Appendix E).

TABLE 11.4
Runoff Coefficients of Contrasting Drainage Areas at the Florida Aquarium.

	Dense Lane, Dense Stalls, and Swale	Dense Lane, Porous Stalls, and Swale
Average	0.23	0.13
Median	0.17	0.08
Maximum	0.75	0.72

Rushton and Hastings, 2001, Appendix C, data averaged over two years from two basins of each type.

WATER QUALITY

The large surface area in the pores of porous concrete provides the setting for treatment of water by attached microbial growth. Porous concrete has been used as a substrate for treatment of municipal secondary sewage, resulting in satisfactory reductions in organic carbon (Furukawa and Fujita, 1993).

Table 11.5 lists porous concrete's reduction in runoff pollutants observed at the Florida Aquarium. A constituent's total load is a product of both its concentration in runoff, and the total amount of runoff. For example, for total phosphorus, the average concentration from the partly porous concrete area was only 74 percent of that in the runoff from the area containing only dense concrete. The porous concrete area's combination of less runoff and lower concentration meant that the total phosphorus production was only 42 percent of that from the reference area. Phosphorus is a nutrient that can originate from automotive oils and is often carried on the surface of small solid particles. For total copper, the average concentration was 90 percent of that in the runoff from the dense concrete area, giving total production of only 47 percent of

TABLE 11.5
Runoff Quality in the Florida Aquarium's Partly Porous Concrete Area, as a Percentage of That in the Area Containing Only Dense Concrete

Constituent	Total Load
Total phosphorus	42 %
Total copper	47 %
Nitrogen	52 %
Iron	68 %

Calculated from data in Rushton and Hastings, 2001, Appendices F and J.

that from the reference area. Copper is representative of the metals produced from automotive components. The results for nitrogen and iron were similarly favorable (calculated from data in Rushton and Hastings, 2001, Appendices F and J).

The Florida Aquarium's results showed conclusively that the presence of porous concrete reduced the loads of all observed pollutants. However, the presence in both the porous concrete area and the reference area of vegetated swales and some dense concrete masked porous concrete's direct quantitative effect. Presumably, if observations were made in an area containing porous concrete alone, the measured reduction would be greater. Research to make such a direct comparison is called for.

Surface Clogging and Rehabilitation

An experiment on porous concrete in Florida showed that, if clogging with native Florida sand were to occur (simulated in the experiment by pressure-washing sand into the pores), rapid infiltration could be immediately restored with pressure washing with clean water and immediate brooming (Florida Concrete and Products Association, no date). Vacuuming without pressure washing could also be effective for sand.

In the Pacific Northwest, organic debris is a more likely clogging material. The cool, moist climate promotes net accumulation of organic matter, where there is a source of organic debris in overhanging vegetation (Brady, 1974, p. 156). In this type of environment it might be prudent to commit to a maintenance program before deciding to build a porous concrete pavement. On a porous concrete sidewalk in Olympia, Washington, after four years without maintenance, the surface pores were visibly clogged with debris from tree leaves and needles; moss was growing in the most abundantly clogged areas (Tomoseen, 2003). The city successfully used pressure washing to clean out the debris. The washing used a handheld, gas-powered pressure washer with potable water from fire hydrants. The washer had a "power head cone nozzle" to concentrate the water in a narrow rotating cone (other types of nozzles did not work as well). The sidewalk was 5.5 feet wide and 1,500 feet long, and took 41 man-hours to clean. Following washing, the surface pores were visibly clean and open.

CONSTITUENTS OF POROUS CONCRETE

Concrete's basic constituents are well known. Table 11.6 lists applicable ASTM standards.

TABLE 11.6
ASTM Specifications of Basic Constituents in Concrete

Number	Title
C 33	Standard Specification for Concrete Aggregates
C 150	Standard Specification for Portland Cement
C 260	Standard Specification for Air-Entraining Admixtures
C 494	Standard Specification for Chemical Admixtures for Concrete
C 618	Standard Specification for Coal Fly Ash or Calcined Natural Pozzolan for Use in Concrete
C 1116	Standard Specification for Fiber-Reinforced Concrete and Shotcrete

AGGREGATE

Aggregate occupies most of porous concrete's volume and is the principal load-bearing component. ASTM C 33 specifies limits of allowable abrasion, degradation in the magnesium sulfate test (a surrogate for freeze-thaw testing), and deleterious substances. The strength of porous concrete benefits from the interlock of angular particles such as those of crushed stone. The aggregate must be open-graded in order to produce voids.

Commonly used gradations are ASTM No. 8 and No. 89. No. 89 is probably the most commonly used because it is widely available from local suppliers. Its maximum particle size is about 3/8 inch, which is large enough to produce open pores and rapid infiltration. Figure 11.4 shows that No. 8 is more strictly single-sized, so where it is available it could contribute still more to a porous, permeable structure.

Larger No. 67 aggregate produces the hydrologic advantage of great pore size and permeability. Concrete made with No. 67 has less initial shrinkage and cracking that that with No. 89, but lower compressive strength because it has less internal surface area for aggregate-cement contact (Yang and Jiang, 2003). It is less "workable" for contractors during placement. It produces a coarse surface texture which is undesirable where accessibility and ease of walking are concerns.

Whether the aggregate in porous concrete must resist absorption of water is a controversial question. Some practitioners see absorption as a precursor to frost damage. However, the same micropores that absorb water also relieve pressure from expanding ice crystals, and many aggregates that have high absorption also have high durability (Rollings and Rollings, 1996, pp. 166–167).

In large parts of Florida the only economically available local aggregate is rather soft limestone or "limerock" which absorbs significant moisture. Florida contractors have used limerock in porous concrete parking lots with lasting success. Evidently, as long as absorptive limerock is selected to meet other requirements such as gradation and durability it is sufficient for the traffic loads of parking lots in Florida's frost-free climate (Diep Tu, personal communication 2001). In this environment, limerock's absorption of water may in fact be an advantage because the rock absorbs moist cement paste, contributing to a strong bond between cement and aggregate (Bruce Glasby, personal communication 2000).

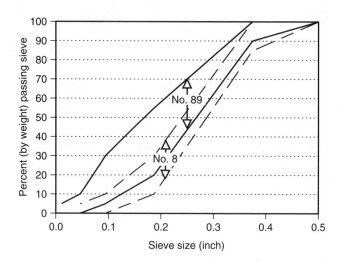

FIGURE 11.4 Gradations of ASTM No. 8 and No. 89 aggregate (data from ASTM D 448).

PORTLAND CEMENT AND ITS SUPPLEMENTS

ASTM C 150 specifies Portland cement's chemical composition, strength, and set time, and designates types with special characteristics. Portland cement is the most costly constituent of concrete, both financially and environmentally. To make it, limestone is fired at high temperature, with the consequent expense of energy and release of greenhouse gases. Porous concrete contains more cement than does dense concrete.

A "pozzolan" is any siliceous substance that reacts chemically with excess lime and water to supplement or enhance Portland cement's cementitious binder (King, 2000). By furthering the formation of complete calcium silicate hydrate crystals, pozzolans can enhance a concrete's strength and resistance to freeze-thaw conditions, deicing chemicals, high-sulfate soils, and salt water.

The most widely and abundantly available pozzolan in North America is fly ash. Fly ash has no direct energy cost in its production and would have to be landfilled if not used constructively (Shell, 2001). It is much less expensive than Portland cement. It has been used in small amounts as an admixture, and in large amounts to replace more than one quarter of the Portland cement in some concretes. Fly ash improves wet concrete's workability because its microscopic particles are naturally spherical and roll easily. It is a natural set retarder. ASTM C 618 defines types of fly ash and other pozzolans and specifies their reactivity with Portland cement.

The extent to which fly ash should be used in porous concrete is controversial. Its prolonging of the set time allows water to evaporate from the uncured material through the voids, so the material might cure inadequately and develop weaknesses (Bruce Glasby, personal communication February 2000).

Another pozzolan with significant occurrence in North America is blast-furnace slag, an industrial waste product for which addition to concrete is a constructive use.

Reinforcement

Reinforcement with metal or fiber increases concrete's strength, especially its tensile strength. It reduces a slab's deformation under stress, so it inhibits cracking caused by initial shrinkage or large temperature changes (ACI, 1984, p. 10). It holds incipient cracks tightly closed to keep segments of slab in alignment and transfer load between them.

Metal bars and meshes should seldom be used in porous concrete although they are commonly used to reinforce dense concrete. They do not bind well to the porous structure and are subject to corrosion where exposed in open voids. Where metal reinforcing must be used, coating it with cement paste before placement might improve bonding and durability.

An alternative reinforcement for porous concrete is polymer fiber. ASTM C 1116 defines types of fiber reinforcing and specifies their effects on strength and other characteristics of the concrete to which they are added. Fibers are easy to install because they are merely mixed into the fluid concrete mixture along with other components. However, their structural effect depends on the thoroughness of their mixing in the material and their bond with the concrete matrix; poorly mixed fibers make dense lumps in the mixture. Their total volume seldom exceeds 0.1 percent of the concrete material. The individual fibers are about 1 or 2 inches long, closely spaced, and randomly oriented within the concrete mixture (Portland Cement Association, 1991, p. 1). They provide tensile strength in all directions. They are relatively unsusceptible to corrosion or natural breakdown. The technology of polymer fiber reinforcing is relatively new and much remains to be learned about it. Some installers speculate that porous concrete with properly mixed fiber reinforcing is as strong as dense concrete without reinforcing.

Admixtures

An admixture is any material other than water, aggregate, cement, or reinforcement that is mixed into a wet concrete mixture to modify its properties. Chemical admixtures commonly come in liquid form. They are mixed into the wet concrete mixture in small quantities. ASTM C 494 defines admixtures with specific purposes such as water-reducing and retarding and specifies their effects on concrete's set time, strength, and shrinkage.

Water-reducing admixtures either increase the slump (fluidity) of concrete with a given water content, or maintain a given slump with a reduced amount of water. They can assist porous concrete installations during placement, because wet porous concrete is by nature stiff and zero-slump; it does not flow freely and must be raked into the forms. An admixture that increases slump without causing the paste to flow from the aggregate makes the material more workable.

Retarding admixtures delay setting of concrete. This keeps the concrete workable during placement. Many retarders also function as water-reducers (PCA, no date).

Air-entraining admixtures stabilize the millions of microscopic air bubbles that form in cement paste during mixing, making them a permanent part of the finished structure (ACI 1984, pp. 3 and 21–22; Kosmatka and Paranese, 1988, pp. 47–62).

Although the minute voids amount to only a few percent of the concrete's volume, individual voids are microscopically close to each other. Air entrainment helps concrete resist the effects of freezing and thawing, sulfate soil, sea water, and deicing chemicals. When water enters the micropores of cement and freezes, or chemical crystals start to grow, the nearby air-entrained pockets relieve the pressure in the material. Air entrainment tends to improve workability (fluidity) of wet concrete (ACI 1984, p. 3). ASTM C 260 specifies the effects of air-entraining agents on the strength, shrinkage, and other characteristics of the concrete to which they are added.

EcoCreto (www.ecocreto.com), Perco-Bond, Perco-Crete (www.percocrete.com), and StoneyCrete (www.stoneycreekmaterials.com) exemplify porous concrete products or mixtures containing proprietary admixtures. Research is called for to confirm the strength, infiltration rate, workability and other characteristics that these products may have. Perco-Crete (porous concrete created with the Perco-Bond admixture) seems to be able to make a porous and somewhat permeable mixture with distinctively fine aggregate. Perco-Crete has a distinctively smooth surface. Its numerous small pores hold water in suspension for a long time, making it available for evaporation that notably cools the pavement surface.

CONSTRUCTION OF POROUS CONCRETE PAVEMENTS

Figure 11.5 shows typical components of a porous concrete pavement. A porous concrete slab, like one of dense concrete, is structurally rigid; it distributes traffic loads over large subgrade areas. Its strength is principally in compression, not tension.

SLAB AND SUBBASE

Six inches is probably the minimum thickness for any porous concrete slab bearing light vehicular traffic, according to experience. Six inches of porous concrete can have the same strength as 1.5 inches of asphalt concrete with 5 inches of aggregate base (PCPA, 2003, pp. 3–4).

In Florida, porous concrete slabs for parking lots are built typically 6 inches thick over a 4-inch layer of open-graded aggregate (Paine, 1990; unpublished information from Florida Concrete and Products Association, Orlando, 1983). Heavier traffic loads require thicker slabs.

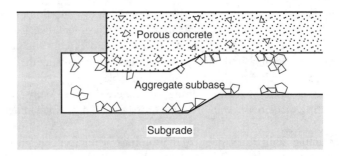

FIGURE 11.5 Typical components of porous concrete pavement construction.

Thicker slabs and subbases are being explored for heavier traffic loads. To carry numerous garbage trucks, a porous concrete slab 8 inches thick over a 6- inch aggregate base and compacted sandy subgrade was installed in 2000 at the Effingham County Dry Waste Collection Site in Guyton, Georgia (Chere Peterson, personal communication September 2003). To carry loaded concrete trucks, a thick porous concrete slab has been proposed for the entry drive of Quality Concrete in Salem, Oregon, in conjunction with the academic testing of material strength (Wilton Roberts, personal communication October 2003).

An aggregate layer under a rigid slab is referred to as a "subbase," perhaps because it does not interact with the slab to form a single structure, the way an aggregate base does under a flexible surface. On soft or scantily compacted subgrade a subbase makes a working platform for construction equipment placing the slab (AASHTO 1993, p. I-15). During curing it equalizes moisture conditions under the slab and insulates the slab from soil suction that would draw water out. In operation a subbase distributes load on the subgrade with relatively little expense, holds a slab in alignment across incipient cracks, and protects subgrade from frost penetration. An Army-Air Force manual (1992, p. 10-1) suggests an aggregate subbase under rigid slabs where the subgrade's Unified classification is OH, CH, CL, MH, ML, or OL, and on SM and SC soils where the water table is high or where soil drainage will be slow.

The thickness of subbase can be increased to compensate for great traffic load or soft subgrade, or to provide stormwater storage. In Florida parking lots the thickness of subbase has typically been determined more by hydraulic storage than by structural concerns as a result of Florida's combination of intense rainfall, stable soils, and shallow frost (Florida Concrete and Products Association, no date; Sorvig, 1993).

Extending the subbase beyond the edge of a slab supports the edge uniformly with the rest of the slab. Increasing the thickness of the slab further combats edge cracking where traffic may pass over or near the edge.

Grinding a hardened concrete surface to make it smooth could make it easier for walking in high heels and more resistant to damage from snowplows. In Florida, porous concrete swimming pool surrounds have reputedly been ground to make smooth, relatively skid-resistant, well-drained surfaces reminiscent of terrazzo. In Boone, North Carolina, grinding was reputedly done at a small basketball court where the rough-textured porous concrete surface had been wearing down shoes and balls. The surface was ground smooth for $0.25 per square foot and then vacuumed to remove dust from the pores. The smooth new surface has stopped wearing out shoes and balls. However, its skid resistance and noise absorption are lower and its reflective glare is higher.

Subgrade Compaction and Its Alternatives

In Florida the subgrade is commonly compacted to at least 90 or 95 percent Proctor density (unpublished information from Florida Concrete and Products Association, Orlando, 1983). Florida's sandy soils tend to retain significant infiltration even after compaction. If the subgrade's quantitative infiltration rate is consequential in the project's stormwater management, then, following compaction, the subgrade's infiltration rate is tested to confirm hydrologic calculations and the intended thickness of subbase or slab is adjusted to supply the necessary hydraulic storage.

In Savannah, Georgia, a testing lab recommended preserving some permeability in the sandy subgrade by limiting compaction to 95 percent standard Proctor and not to the 98 percent or more which is customary for dense pavements in the area (Joseph Whitaker, personal communication 2000). Where the subgrade is compacted less than 95 percent, greater thickness of subbase or slab could compensate structurally for the soft subgrade.

JOINTS IN POROUS CONCRETE

Concrete's rigid structural character makes it subject to cracking. Numerous tight cracks are benign; segments of slab can stay in alignment and transfer load across them. But if an isolated crack widens, one large segment of slab can become displaced relative to another and structural deterioration can be set in motion. Cracks in porous concrete develop in fundamentally the same ways as those in dense concrete and require essentially the same types of control.

In the first few weeks after a slab is placed numerous "initial shrinkage" cracks can result from Portland cement's shrinkage during hydration. Shrinkage of only a few percent is sufficient to crack most slabs. Shrinkage is a phenomenon of the paste; the aggregate has a restraining influence. Porous concrete shrinks only about half as much as dense concrete because its uniformly sized aggregate particles interlock more or less stone-to-stone with little cement between them. Consequently it develops correspondingly fewer and smaller initial cracks (Malhotra, 1976; Monahan, 1981, p. 5).

Other cracks can come from stress under traffic load and thermal deformation. Porous concrete's thermal expansion is about the same as that of dense concrete. With changing temperature a slab deforms with differential expansion or contraction at different depths in the slab.

A crack in porous concrete does not lead to saturation of an exclusive part of the subgrade as it does in dense concrete because the entire slab is already permeable; subgrade support is uniform in both wet and dry conditions. Porous concrete's uniformly sized aggregate particles interlock effectively, transferring load across incipient cracks and keeping segments of a cracked slab in alignment.

Control Joints

Control joints are deliberate zones of weakness in a slab, encouraging the formation of numerous small, evenly distributed, benign cracks in those locations and preventing the formation of a few large, randomly located, disruptive ones. They are sometimes called "contraction joints" because they control the numerous initial shrinkage cracks as well as any later warping cracks. Figure 11.6 shows examples of control joints. They can be added to a porous slab by forming with a roller before the concrete hardens. Alternatively, they can be cut into the surface with a saw after curing has set in, but cutting may loosen and detach adjacent aggregate particles. The joints typically penetrate one fourth to one third of the slab's thickness. The remaining

Porous Concrete

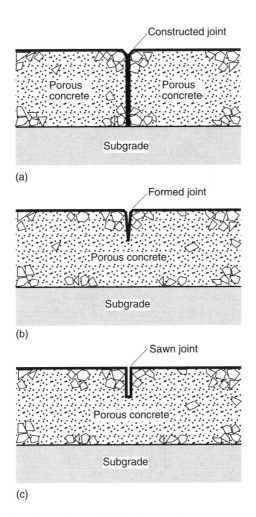

FIGURE 11.6 Section views of control joints in concrete.

three fourths of the slab maintains aggregate interlock across the crack, keeping the segments of slab in smooth alignment.

If work is stopped at a vertical face and later continued from the same point, the junction is a "construction joint." A construction joint may act as a control joint. Porous concrete needs no keying to keep slabs in alignment across a construction joint as does dense concrete; instead, the aggregate particles in the adjoining slabs interlock.

In porous concrete control joints should be spaced no farther apart than 20 feet (PCPA, 2003, p. 3). In comparison, control joints in dense concrete should be spaced no farther than 15 feet apart (ACI, 1984, p. 12). Porous concrete's control joints can be slightly farther apart because of porous concrete's relatively small initial shrinkage.

The arrangement of joints should form approximately square or circular panels. If a panel's ratio of length to width exceeds 1.5:1 cracking could occur (ACI, 1984, p. 15; PCPA, 2003, p. 3). This means that in one-lane driveways transverse joints end up about 10 to 12 feet apart and in many sidewalks only 5 feet apart. Longitudinal joints should be added to drives or roads that have two or more lanes. The pattern of joints can be planned to conform to the dimensions of a site and emphasize its geometry. Minor adjustments in the pattern can be made to meet incidental drainage inlets and manholes.

EXPANSION JOINTS

Expansion joints prevent cracking that would result from pressure during expansion of a slab or an adjoining structure in warm weather. Expansion joints are essentially flexible gaps in a large pavement, or between a pavement and another structure. Figure 11.7 shows typical examples. The gap is typically 1/2 inch to 3/4 inch wide and filled with flexible material to prevent clogging with sand or debris. Panels of concrete between expansion joints are of limited length and therefore of limited stress from temperate changes and warping (Croney and Croney, 1998, p. 43). The edge of each segment can be thickened along the joint as it can along a slab's outer edge.

Where an expansion joint is installed between a concrete slab and a fixed structure it may be called an "isolation" joint. It prevents buildup of pressure against the fixed structure. Additional isolation joints can be placed around slab inserts such as pipes, drains, hydrants, lamp posts, and column footings (ACI, 1984, p. 14).

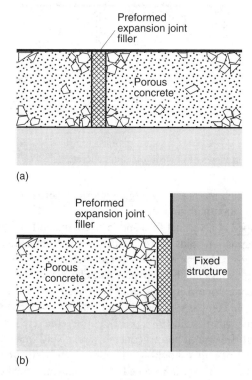

FIGURE 11.7 Section views of expansion joints in concrete.

Joint fillers should be preformed materials that can maintain their shapes during placement of the concrete material, but can be compressed under expansive pressure and then return to their original shapes (ACI, 1984, p. 15). They should be durable enough to resist long-term deterioration from moisture and other service conditions. ASTM D 1751, D 994, and D 1752 define preformed expansion joint fillers of materials such as cork, fiber, plastic, and sponge rubber and specify their absorption of water, the stress required to compress them, and the recovery of their shape when the stress is released.

Where expansion joints happen to be spaced no more than 20 feet apart they relieve the necessity for control joints. However, they should be used only where in fact necessary, because they can be difficult to maintain and if they become inflexible they can contribute to pavement failures (U. S. Departments of the Army and Air Force, 1992, pp. 15–16).

CARE REQUIRED IN POROUS CONCRETE INSTALLATION

Concrete is created by placing and curing on-site. It is not manufactured in standard models under uniform factory conditions. Porous concrete can be mixed by the same local concrete suppliers and delivered in the same concrete trucks as dense concrete. However, it is a special specification and requires an installer with special experience.

Traver et al. (2004) described the experience of placing the first installation in a locale where the contractors had no previous experience. The installers' early failures were caused by "not understanding the impact the porous concrete material properties had on construction practice." Many of the practices that were then developed by trial and error affirmed the special porous-concrete installation practices previously developed in Florida.

Although the proper placement of porous concrete is not difficult, it is different from that of dense concrete, and the difference is crucial to every aspect of the finished structure's success. The following are some of the special points that must be followed in mixing, placing, and curing porous concrete. PCPA's specifications stipulate in detail porous concrete's constituents, contractor qualifications, the submittal of test panels, proportions, subgrade preparation, formwork, mixing, placement, curing, jointing, and testing (PCPA 2003, pp. 7–14).

INSTALLER'S QUALIFICATIONS

The quality and performance of a porous concrete installation depend completely on the capability of the contractor who installs it. The only adequately knowledgeable installers are those with hands-on experience specifically with porous concrete, with at least a successful trial installation. No amount of experience placing dense concrete alone makes an installer qualified to place porous concrete. Contractors without experience and without the benefit of an experienced consultant should place several test panels to gain working knowledge and prove competence before completing the main portion of an installation.

The Portland Cement Pervious Association certifies installers on the basis of proven knowledge and experience. The certification process involves passing a proficiency exam and the completion of projects verified to be PCPA-compliant in terms of void content and structural condition (PCPA, 2003, p. 6). A given percentage of the work crew is required to hold Flatwork Technician or Finisher certification from the American Concrete Institute. If the number of PCPA-certified contractors grows, certification will become an important means for designers and owners to control quality in concrete installations.

Proportioning of Mixture

The required proportions of water, cement, aggregate, and admixtures in a concrete mixture are project-specific. Although experience shows that the proportions end up within a certain range, the correct quantitative proportions for a specific project depend on the character of the locally available aggregate and other variable conditions. Proportioning of constituents balances competing requirements such as strength, porosity, and workability (ACI, 1984, p. 5). The idea is to determine a practical and economical combination that produces a concrete with the required characteristics.

In a specific project, proportioning may be determined by the design specifier based on knowledge of local materials, by a testing lab analyzing several alternative combinations on behalf of the specifier or the installer, or by the installer at the project site based on experience with local materials and porous concrete mixtures.

PCPA's specifications require that the cement content in concrete containing No. 89 aggregate be not less than 600 pounds per cubic yard of material (PCPA, 2003, p. 10). With No. 67 aggregate, the cement content could be lower (unpublished information from Florida Concrete and Products Association, 1983; Joseph Whitaker, personal communication 2000).

The water content must be sufficient to allow the cement paste to coat the aggregate particles, but it is strictly limited by the danger of the overly wet paste draining through the material's pores. Drainage would leave aggregate particles in the upper part of the slab without adequate binding, and an accumulation of excess paste would clog the voids at the bottom. Years after installation, finished slabs placed with wrong water content or cured without protection from evaporative water loss can be recognized by ruggedly eroded patches where surface particles have come loose. Limited water content produces concrete of high strength with a given amount of cement (PCA, no date).

An indirect way of specifying water content is with the "slump test," which is described in ASTM C 143. The test indicates consistency by the settlement of a mound of wet mixture. For dense concrete, the expected slump is often about 5 inches (ACI 1984, p. 2). Most porous concrete is literally zero-slump. The low water content makes porous concrete "stiff" during placement; contractors say it has low "workability." This encourages unqualified installers to add more water, but the temptation must be resisted. On the other hand, with such low water content, there is the danger of not having enough water to hydrate the cement.

A qualified contractor checks the water content by careful visual inspection. Insufficient water produces a dull-appearing paste; too much water causes the paste to flow visibly from the aggregate. "Consistency of the product is controlled by

Porous Concrete

visual inspection;...the quantity of cement paste is considered sufficient when it coats the coarse aggregate with a shining film giving a metallic gleam" (unpublished information from Florida Concrete and Products Association, 1983).

Research is called for to develop a scientific, quantitative way to control water content in the small range between the minimum amount necessary to hydrate the cement, and an excessive amount that makes paste drain off the particles.

PLACEMENT AND CURING

The Florida Concrete and Products Association's installation guidelines are still very valuable in introducing porous concrete's placement and curing procedures (Florida Concrete and Products Association, no date).

Figure 11.8 shows the placement of porous concrete in progress. Because the mixture is stiff it is encouraged down the chute with shovels. It is raked into position 1 inch higher than the side forms.

After placement the concrete is rolled to compact it to its finished elevation. The roller shown in Figure 11.9 is a heavy steel pipe spanning between the side forms. Rolling consolidates and levels the material without smearing. It increases contact and bonding between surface aggregate particles. Compaction must be limited to preserve void space. A vibratory screed like that used for dense concrete has often been attempted but it visibly smears cement across the pores, sealing the surface.

Additional finishing operations such as floating and brooming, which are common on dense concrete, should not be done on porous concrete. They would seal the surface the same way screeding can. Table 11.2 showed that unnecessary finishing

FIGURE 11.8 Placing wet porous concrete.

FIGURE 11.9 Rolling porous concrete to level.

drastically reduced the surface infiltration rates of early installations; just below the surface the unsmeared material was open and permeable.

Because porous concrete has low moisture content it sets quickly. At the same time the small moisture content is vulnerable to rapid evaporation through the material's open pores. Consequently, it is vital to work quickly and to conserve the moisture in the wet mixture so it can complete the hydrating reaction in the cement paste (unpublished information from Florida Concrete and Products Association, 1983; Joseph Whitaker, personal communication 2000).

Immediately after rolling the material is covered with plastic sheeting to prevent drying. The sheeting remains for a number of days, held in place at the sides by boards or other clean weights.

Most "curing compounds" have no place in porous concrete. For dense concrete they supplement or replace covering with plastic sheets. They are sprayed or brushed onto the finished surface to form an impervious membrane that conserves moisture (ACI 1984, p. 26). They tend to be thick and sticky, so in porous concrete they could clog the surface while failing to coat the interiors of deep pores.

TESTING FOR QUALITY CONTROL

Testing laboratories can help assure proper installation (PCPA, 2003, p. 13; Joseph Whitaker, personal communication 2000). A laboratory can adjust a mixture's proportioning to accommodate the available aggregate. Prior to placement the subgrade can be evaluated for compaction or infiltration rate. After hardening, cores can be taken and tested for thickness, void content, and compressive strength.

Porous Concrete

At each stage the laboratory personnel must be aware of the special expectations for porous material and not get them confused with those for dense concrete to which they are accustomed. It may be desirable not to compact the subgrade fully; density and strength of porous concrete are not to be maximized as they are in dense concrete.

Placing and approving a test panel before proceeding with a full project assures that the material, the contractor, and the placement procedure are all working toward a successful project. After a test panel has thoroughly cured, it can be evaluated for infiltration rate, raveling, and general condition. Pressure-washing a test panel could quickly test whether any sections are weakly cemented or slowly infiltrating (Craig Tomoseen, personal communication 2003).

POROUS CONCRETE IN SUCCESSFUL AND UNSUCCESSFUL INSTALLATIONS

Two porous concrete parking lots only a block away from each other in Stuart, Florida exemplify contrasting successful and problematic installations in small, focused examples.

834 EAST OCEAN BOULEVARD, STUART, FLORIDA

The successful concrete parking for is that of a medical office at 834 East Ocean Boulevard. Figure 11.10 shows the terrace-like parking lot several years after installation. It forms the forecourt of the small office building. The parking stalls, the turning lane, and a handicap stall are all in porous concrete panels 12 feet wide by

FIGURE 11.10 Porous concrete parking lot at a medical office in Stuart, Florida.

15 to 18 feet long. White painted lines and concrete wheel stops define individual parking stalls.

The building and its parking lot are set 2 feet higher than the adjacent street. A small concrete retaining wall takes up the difference in elevation and makes a low parapet around most of the parking lot. A sloping driveway of porous concrete connects the parking lot terrace to the street. The entire parking lot is essentially level and accessible with respect to the building.

The parapet's color is consistent with the building's roof and painted details, unifying the terrace's perimeter. Because the forecourt terrace is clearly defined horizontally and vertically, the street seems not to "intrude" perceptually into it to the degree that the large quantity of traffic would otherwise make possible. The horizontal scale of the parking lot is in proportion with the vertical scale of the building. Planted islands projecting from the building penetrate into the parking surface.

One can tell at first glance that the lot's concrete surface is in good condition. The entire surface has uniformly open pores. There is no visible cracking, particle separation, or smeared-over areas. There is no significant source of sediment. Any surface overflow drains to the driveway. Entering cars drop sand from their wheels onto the driveway, whence it washes back to the roadside. This layout demonstrates the paradox that providing for surface runoff is one of the conditions to assuring high surface infiltration rate. It allows debris to escape the pavement's surface without being washed and ground into the pores.

The surface's infiltration rate is as excellent as one would expect from its open visual character. Water from a bottle disappears as fast as it is poured onto the surface. The exception is in the single stall marked for handicapped accessibility. This whole stall was painted blue, and the paint partly or wholly clogs many pores. In future installations the infiltration rates of accessible stalls could be better preserved by limiting paint to stripes or outlines and a central symbol.

827 S.E. Fifth Street, Stuart, Florida

A nearby but much less successful installation is a small parking lot at 827 S.E. Fifth Street, located behind 816 East Ocean Boulevard. This is another small medical facility. Figure 11.11 shows the main porous concrete parking lot at the side of the building (another, smaller area of porous concrete at the office door is reserved for handicapped access).

The concrete surface is visibly in poor condition. In many large patches particles have separated from the surface due to watery paste; in others the pores are visibly clogged solid by excess paste or smearing of the surface. This example affirms that, although cement paste binds porous concrete together, if it is in excess or watery it is simultaneously the enemy of the pavement's permeability, durability, and appearance. This concern about watery paste is relevant during the few hours when the material is being placed; if the problem is prevented when the installer places the concrete, then it will not exist for the life of the structure.

The parking lot is at the low point in the site's surface drainage. It is enclosed by raised curbs with no provision for overflow drainage. Some of the surface pores are visibly full of trapped sand and organic debris in addition to cement paste. The

Porous Concrete

FIGURE 11.11 Porous concrete parking lot at another medical office in Stuart, Florida.

infiltration rate is low in areas clogged by paste alone, and very low in areas clogged by both paste and debris. This experience affirms the general principle that porous pavements should not be located at the low point in a site's surface drainage.

FURTHER INSTALLATIONS ON SANDY SUBGRADES

The longest and most extensive experience with porous concrete pavement is on the sandy soils of Florida and adjacent parts of the southeastern Coastal Plain. Throughout the region the terrain is generally level; soils are mostly sandy; rainfall is plentiful; deep frost is rare. The following case studies from the region illustrate porous concrete's use in a variety of land use and traffic settings.

HOBE SOUND SPRINKLERS, HOBE SOUND, FLORIDA

A porous concrete parking lot at Hobe Sound Sprinklers in Hobe Sound, Florida bears the moderate traffic load of commercial trucks. Company trucks and other vehicles operate daily from the lot. Figure 11.12 shows the site. Although this installation is challenged by its traffic load it is successful structurally and hydrologically because it was well laid out and well constructed.

The concrete was placed in panels 12 feet × 20 feet in size. Roof downspouts discharge onto the pavement. The surface slopes gently from the building to the perimeter. There are no perimeter curbs; any overflow runoff drains freely off the edge of the pavement.

FIGURE 11.12 Porous concrete parking lot at Hobe Sound Sprinklers in Hobe Sound, Florida.

At the entry a short ramp of porous concrete slopes up from the street to the parking lot, buffering the parking pavement from any source of trackable sediment. There is some cracking of the concrete where fast-moving trucks exert great force as they "bounce" up the slight slope of the ramp.

The surface infiltration rate is extremely high, almost reminiscent of that of open-graded unbound aggregate. The surface appears uniformly open-pored, with no patches of paste-clogged pores or separating particles.

COUNCIL ON AGING OF MARTIN COUNTY, STUART, FLORIDA

A porous concrete parking lot at the Council on Aging of Martin County in Stuart, Florida, supports a viable rooting zone for trees. Figure 11.13 shows the parking lot and its trees about five years after installation. Parking stalls are defined by painted lines and concrete wheel stops. The concrete is placed in panels 12 feet × 30 feet in size. The traffic is moderate; it includes that of employees, meals-on-wheels volunteers, delivery for adult day care, and group parking for events. There is visible cracking on one concentrated traffic lane.

The surface pores appear generally open. The exception is at the edges of some panels where the particles are packed tightly together as a result of local over-compaction. There is no evidence of particle separation as would happen with watery paste. The pavement's surface infiltration rate is good overall.

A long entry drive of dense asphalt buffers the porous concrete from any tracking of sediment by vehicles. Surface drainage is away from the center of the lot to the periphery, so any sediment washes off. The organic matter dropping from the

FIGURE 11.13 Porous concrete surrounding palm trees at the Council on Aging of Martin County in Stuart, Florida.

internal trees is small in relation to the parking lot's area, and decomposes rapidly in the warm climate.

Palm trees are planted in small (3 feet × 3 feet) openings in the slab. The soil in the openings is sandy, as is all the subgrade soil in the vicinity. The trees were large and healthy five years after installation. Because this parking lot was well laid out and constructed, its performance is a success in every regard: permeability, durability, accessibility, and viability of tree rooting.

JONES ECOLOGICAL RESEARCH CENTER, NEWTON, GEORGIA

Drives and parking lots at the Jones Ecological Research Center near Newton, Georgia, illustrate the use of porous concrete in a nature preserve. The center (www.jonesctr.org) was constructed in the early 1990s; it was probably the first porous concrete pavement outside Florida. It is in a portion of the Coastal Plain where sandy soils overlie highly fractured sandy limestone. The construction was specified by designers from Atlanta who were resolving some of the contradictory advice about details such as control joints available from Florida, so it was a precursor to the installations outside the Coastal Plain which soon followed.

The research center used porous pavement because its environmental effects fit the center's ecological mission. The entry drive's curving alignment among the site's native longleaf pines was illustrated in Chapter 1. The entry drive, visitors' parking area, internal loop road, and walkways totaled 2.3 acres of porous concrete pavement. The slabs are 5 inches thick throughout the complex.

The construction cost was high compared with what dense concrete would have cost (Lee Tribble, personal communication 1998). In this rural location only one concrete supplier of limited means was available. To create the porous mixture the mixing plant had to be shut down and restarted with the special mixture.

The high cost of the paving material is counterbalanced by the low cost of drainage and water quality control. The pavement installation includes no curbs, no inlets, no drainage pipes, no swales, and no off-pavement stormwater reservoirs. Rainwater infiltrates through the pavement and into the soil as it does in the native forest floor throughout the site. Pavement overflow drainage (if any) disperses immediately into the site's sandy soil.

The pavement's infiltration rate is high despite the presence, in some spots, of a substantial amount of sand in the pores. Vehicles track sand from intersecting unpaved sand roads and from road verges, particularly at intersections and sharp curves. (Future tracking from road verges could be inhibited by installing bollards.)

Eight years after construction the pavement was structurally in excellent condition except for certain cracks. The internal loop road developed some longitudinal cracks soon after installation, in places where the pavement width exceeded 18 feet. Figure 11.14 shows one of them. The irregular cracks are unsightly and a slight nuisance to drivers. It is believed that the cracking was related to the absence of longitudinal joints in the wide slab.

POROUS CONCRETE ON FINE-TEXTURED SUBGRADES

Since 1996 porous concrete pavements have been installed on fine-textured subgrades outside the southeastern Coastal Plain. Although these pavements cannot percolate large quantities of rainwater into the soil during intense storms, they demonstrate that porous concrete pavements are structurally possible on fine-textured soil and that durable benefits of detention, water-quality amelioration, tree preservation, and gradual long-term infiltration are achievable.

SOUTHFACE INSTITUTE, ATLANTA, GEORGIA

A small driveway at the Southface Institute in Atlanta, Georgia, constructed in 1996, was the first installation of porous concrete on fine-textured subgrade. It exemplifies the physical success of porous pavements on clay subgrade and demonstrates a construction sequence that may be convenient in future installations elsewhere.

FIGURE 11.14 Crack in a two-lane porous concrete road at the Jones Ecological Research Center.

The Southface Institute (www.southface.org) is a nonprofit organization promoting resource conservation and sustainable development. The porous driveway shown in Figure 11.15 is one of many indoor and outdoor demonstration features at the institute's office. It bears light traffic of a few autos per day and an occasional truck. Some runoff discharges onto the pavement from roof downspouts. Overhanging hardwood trees supply plenty of organic debris. The slab's surface slopes gently downward and away from the building; any overflow runoff washes freely across pavement edges. The subgrade is red clay typical of the Piedmont region, modified over time by cycles of urban construction.

For several months before the placement of the concrete, the driveway site had been covered with a layer of large (3-inch) crushed-stone aggregate to provide a clean platform for construction equipment while the building was being erected.

FIGURE 11.15 Porous concrete driveway at the Southface Institute in Atlanta, Georgia.

Construction traffic pressed the aggregate into the clay soil, producing a very stable compacted stone-soil layer. Aggregate pads of this type are used commonly for temporary construction staging. When the building was complete the stone pad was left in place as a subbase for the driveway. A geotextile filter fabric was placed over the pad (it was arguably unnecessary over such a stable soil-aggregate mixture). The subbase was not counted in structural or hydraulic capacity; leaving it in place merely eliminated the cost of removal and produced the free benefit of a very stable and somewhat permeable subbase for the slab.

The concrete mixture followed the guidelines of the Florida Concrete and Products Association, with proportioning of constituents by a testing lab. The aggregate in the mixture is No. 89 crushed granite.

Five years after construction the concrete was in excellent overall condition. There is no deformation or general cracking.

At one edge two small (less than 1 square foot.) chips have broken away from the rest of the driveway under some great point load that exceeded the strength of the concrete edge, and of the subbase or subgrade into which the chips subsided. The subsidence of the chips suggests that the subbase did not extend to this edge of the driveway; the chips subsided directly into subgrade soil. Additional structural protection could have been provided, before placement, by examining the subbase and extending some sort of stable subbase to and beyond all planned slab edges.

Most of the surface has visibly open pores, but portions are smoothed over with smeared paste. The clogged-looking portions occur in parallel stripes. They were created during construction by screeding. Better construction would have used a roller rather than a board to level the surface. Presumably, under the screeded surface the interior of the slab is still open and porous, so although the partial smoothing reduces the surface's average infiltration rate, it does not reduce the slab's total pore capacity or contact of infiltrated water with the subgrade.

The pavement's surface infiltration rate is on the whole good despite the screeded portions. The screeded areas have low infiltration rates. Excess runoff from those spots flows a few inches onto areas with open pores, where it infiltrates. Surface drainage keeps the slab largely free of organic debris. Southface office workers say that the surface of the driveway remains conveniently dry even when rain is falling hard and water is flowing in the nearby street gutter so copiously that one is reluctant to try to cross the street.

Wooden strips 1 inch wide divide the concrete into five irregularly shaped panels from 10 to 35 feet wide. The wood was in good condition five years after construction. However, even pressure-treated wood can gradually disintegrate, so at some time it may need to be dug out and replaced with some other flexible material.

WEBB BRIDGE PARK, ALPHARETTA, GEORGIA

An access road of porous concrete at Webb Bridge Park in Alpharetta, Georgia, illustrates several points necessary for the construction of porous concrete on fine-textured soil, and for porous pavements of all kinds anywhere. The subgrade is deep red clay of the Madison series. Although the road was resurfaced with dense asphalt in 2003, the experience of the original porous concrete surface pointed out valuable lessons.

The park supports field sports (www.aspharetta.ga.us). It has considerable automobile traffic on seasonal weekends and evenings. The porous concrete entry drive and loop access road, totaling several acres of porous concrete, were installed in late 1997 and 1998. Earthwork was necessary to shape the loop road on the sloping ridgetop site. Figure 11.16 shows the resulting configuration. The road is marked for one-way traffic. The porous concrete mixture followed Florida guidelines. It was installed by a local contractor for $3.10 per square foot (Roy Keck, personal communication 1999).

A subbase of crusher-run aggregate was installed under most of the slab, at the contractor's discretion, to make a working platform on the irregular, fine-textured, and sometimes wet, subgrade. Where it was installed it helped prevent cracking by supporting the edges, equalizing moisture conditions during curing, and generally providing a uniform foundation for the slab. Where it was not installed the edges of

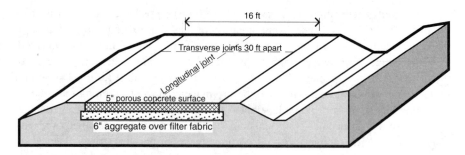

FIGURE 11.16 Configuration of the loop road at Webb Bridge Park in Alpharetta, Georgia (vertically exaggerated).

the slab have "rocked" and required patching with further porous concrete (Roy Keck, personal communication 1999). Although the dense-graded aggregate used in this project had low permeability, this experience confirmed the desirability of some type of aggregate subbase under porous concrete on fine-textured soil, and of edge support equal to that under the rest of the slab.

At one spot a large open longitudinal crack developed near the road's centerline; the adjacent concrete segments were visibly displaced against each other. The broken area was soon replaced with dense concrete. This crack could have been produced by the settlement of subgrade fill which was particularly deep at this spot on one side of the road. Settlement of fill results from incomplete compaction, and would displace any pavement placed at this location. Fill has to be compacted under porous concrete as it does under almost any other pavement. Alternatively, the crack could have been caused by excessively thick, impervious subbase material which washed out laterally during intense rainstorms and eliminated support for the concrete slab.

Apart from these details the concrete was in good condition several years after construction. Gentle undulations in the concrete surface were probably produced by irregular formwork during construction. The concrete's infiltration rate was moderate to high throughout the installation.

The loop road was properly configured to protect the pavement from on-drainage of sediment. A photograph in Chapter 2 showed the road soon after installation, with nearby earth actively eroding as grading of the park's sports fields continued. The road verges drain away from the pavement in every direction, including on the "uphill" side into a swale. Sediment drains naturally down eroding slopes without entering the pavement. Curbs are omitted so overflow drainage, when it happens, discharges freely into the swale and potentially clogging debris is washed or blown off the pavement.

POROUS CONCRETE ON THE WEST COAST

In 1999 the West Coast acquired its first known installation of porous concrete (Tomoseen 1999; Craig Tomoseen, personal communication 2003). It is a 1500 foot long sidewalk along North Street in Olympia, Washington. The Washington State Aggregates and Concrete Association had a made test panels of porous concrete in

1994 (Chattin, 1999). To prepare for the Olympia project the city hired an experienced porous concrete installer to demonstrate placement of a further test panel to local contractors.

The sidewalk's aggregate is rounded gravel, selected following a trial which had shown that angular crushed rock produced higher strength, but round gravel produced a smoother surface. It was 1/16 inch to 3/8 inch in size, and washed.

Proportioning of the concrete mixture was based on laboratory tests with the local aggregate. The mixture was zero-slump. The aggregate-to-cement ratio was 4.5:1 by weight. The mixture contained polypropylene fibers, air entrainment, and water-reducing and retarding admixtures. The finished material had a porosity of 12 percent and compressive strength of 2400 psi. The zero-slump mixture was raked into the forms, hand-screeded to level, and then compacted with a 30 pounds-per-linear-foot roller. Expansion joints were placed every 200 feet and control joints every 20 feet.

Olympia's porous concrete material cost $5 more per square yard than dense concrete would have, and $10,000 more in engineering. However, because there had been no sidewalk in this area previously, this was a new development requiring rigorous stormwater control. The porous concrete saved $110,000 for off-pavement land acquisition and detention pond construction, making a large net saving. The Puget Sound Action Team (Wulkan et al., 2003) listed subsequent porous concrete installations in the Pacific Northwest region.

Large areas of California and adjacent states provide a theoretically ideal climate for porous concrete, having little frost and gentle, moderate rainfall. In 2001, northern California acquired its first installation in the parking lot of Bannister Park in Fair Oaks, near Sacramento (Andy Youngs, personal communication 2002; *Concrete Resources* 2001). The park is maintained by the Fair Oaks Recreation and Park District (www.fairoakspark.org); it supports field sports and recreational access to the American River. The district was motivated to use porous concrete for its presumed low contribution to the urban heat island, its control of oil pollution, and the low life cycle cost of concrete. Cooling of the pavement surface is by a combination of light surface color, porous material, and tree shading. Figure 11.17 shows the parking lot two years after construction.

Bannister Park's parking lot has 47 stalls, two of which are handicap spaces. The stalls are around the perimeter and in the center. The surface slopes at 2 or 3 percent. At low edges raised curbs collect any overflow drainage into inlets; elsewhere on the perimeter there are only wheel stops. Two years after installation the surface infiltration rate was high except in a small area which appeared smeared-over with paste.

Before placement of Bannister Park's pavement the subgrade was compacted. The wet concrete was leveled with a vibratory screed, then compacted with a roller. The joints were formed with a special roller. The contractors reported that after they became familiar with the material it required no more time or labor than dense concrete.

The parking lot's shade trees are planted in 8 feet × 8 feet soil areas among the central parking spaces. Space for them was claimed by shortening each facing stall by 4 feet, marking these short narrow stalls for compact cars, and giving them wheel stops to keep cars out of the soil areas. The other central stalls are 9 feet wide and have no wheel stops. The trees are spaced 26 feet apart, which is close enough for

FIGURE 11.17 Porous concrete parking lot at Bannister Park in Fair Oaks, California.

them to shade the central part of the pavement rather thoroughly upon maturity. The porous pavement surface will vitally assist root growth outside the tree pit. Similar trees are planted around the perimeter. Two years after installation the perimeter trees were larger than those in the center because they had the benefit of the park's irrigation system.

Southern California's first installation was being planned as of 2002. Overflow parking of 90,000 square feet was planned for the Cerritos Automall at State Road and Del Amo Boulevard in Cerritos, California (Andy Youngs, personal communication 2002).

TOWARD POROUS CONCRETE IN COLD CLIMATES

At the time of this writing (2004) few porous concrete pavements have been installed in areas colder than North Carolina or the Puget Sound area. Several questions deserve to be conclusively answered to assist porous concrete's further application in cold climates.

When water freezes into ice it expands by 9 percent. Growing ice crystals displace unfrozen water. At low water content no hydraulic pressure develops. But if the micropores in the cement binder are saturated or nearly saturated when freezing begins then hydraulic pressure builds up as freezing progresses (Kosmatka and Paranese, 1988, pp. 47–62). If water even in the aggregate particles begins to freeze, it expels water into the surrounding paste. Porous concrete's open voids offer a large internal surface area where water or deicing chemicals can enter the cement binder's micropores. The open voids that give porous concrete its porosity relieve the stress only at the microscopic surface of the cement matrix; freezing in micropores deeper within the binder must be relieved or resisted by other means. A standard method for testing the effect of freeze-thaw cycles on concrete is in ASTM C 666.

One of the very few known direct observations of porous concrete's behavior upon freezing was laboratory experiment by the U.S. Army's Cold Regions Research and Engineering Laboratory (Korhonen and Bayer, 1989, pp. 9–15). Samples of porous concrete without air entrainment, reinforcement, or other treatment for frost damage protection were repeatedly frozen and thawed. At intervals during the testing sequence samples were removed from the freezing cycle and put under compressive force to test their loss of breaking strength. Those that had been frozen in dry or damp (wetted, then drained) conditions showed little loss of strength over 160 freeze-thaw cycles.

However, similar samples that had been frozen in water-filled containers progressively deteriorated. Prior to the 45th cycle they withstood forces of 2000 psi or more, and when they broke under greater force they broke into a few large pieces. At the 45th cycle they had lost 11 to 21 percent of their strength and they broke into numerous small pieces. By the 80th cycle they showed numerous small internal cracks, they had lost 37 to 38 percent of their strength, and under breaking pressure they crumbled almost into powder. In comparison, dense concrete during 80 cycles lost only 7 percent of its strength and broke only into a few large pieces.

Meininger (1988) carried out a similar experiment and produced a similar result. He repetitively froze porous concrete samples with no admixtures. All failed fairly quickly under compressive load. Upon failure the samples cracked and split open, indicating that they had been weakened by internal freezing pressure.

In cold regions air entrainment is routinely added to concrete to protect it from frost damage (AASHTO, 1993, p. I-21). Experience primarily in building construction suggests that air entrainment improves the resistance of porous concrete to damage from freeze-thaw cycles, as it does for dense concrete (Kosmatka and Paranese, 1988, pp. 47–62; Monahan, 1981, p. 5).

Liquid polymer and latex additives could help by sealing the cement binder's micropores and preventing the entry of water. Pozzolans, polymer fibers, and liquid polymers can enhance a concrete's strength and thereby its resistance to freeze-thaw conditions and deicing chemicals (King, 2000; Pindado et al., 1999).

Resistance to abrasion from snowplowing also needs to be proven. Perhaps porous concrete is able to resist abrasion despite its rough-textured surface of bound aggregate particles. If not, then plow blades should be lifted above a concrete surface as they are for grass-based and aggregate-based pavements, as described in Chapter 8.

NRMCA (2004) speculated that dangerous freezing of a fully saturated concrete layer is most possible in locations where the air temperature is below freezing for long periods while precipitation continues. In this most severe climate, NRMCA recommends, in order beginning with the most important feature, 1) that the concrete slab be placed on an aggregate base 8 to 24 inches thick to drain water out of the concrete layer, 2) that the cement contain entrained air, and 3) that the base be positively drained by a perforated pipe or other means at an elevation well below the concrete layer. Not every situation warrants all three safeguards.

In State College, Pennsylvania, a porous concrete sidewalk at the Centre County-Penn State Visitor Center showed no appreciable freeze-thaw damage after five winters (NRMCA, 2004). It carries very heavy foot traffic on football game days. The slab is 4 inches thick; it is placed on an aggregate base 8 inches thick over uncompacted subgrade. The cement contains a cold-weather additive.

At Villanova University near Philadelphia, a porous concrete walkway constructed in 2003 is to be monitored for hydrologic and functional performance; results may be posted on the web site of the Villanova Urban Stormwater Partnership, www.villanova.edu/vusp (Traver et al., 2004). The walkway is at the center of campus and is very heavily used by pedestrians.

Several installations have reportedly been installed in Gallup, New Mexico, under the direction of engineer Frank Kozeliski (NRMCA, 2004). Gallup's climate is cold because the elevation is over 6000 feet, but annual precipitation is only about 12 inches. The dry winters do not present the most severe freeze-thaw damage hazard; in the dry, sunny air snowfall seldom stays on the ground for more than two or three days. A small parking lot at 102 West Hill Avenue was built in 1991. After 13 years it showed no appreciable structural distress. Its principal shortcoming was clogging by mud that dropped from parked four-wheel-drive vehicles; it has never been cleaned or maintained. Rainwater has infiltrated without surface ponding even during a 4-inch rainstorm. Snow is pushed off the surface, after which, on sunny days, there has been no ice build-up from surface meltwater. The cement contains no entrained air.

Research to prove conclusively the efficacy of known technologies in protecting porous concrete pavements in cold climates is called for.

TRADEMARKS

Table 11.7 lists the holders of registered trademarks mentioned in this chapter.

ACKNOWLEDGMENTS

The following persons and organizations generously provided information used in this chapter: Leigh Askew of the Georgia Department of Community Affairs; Bob Banka of Concrete Solutions, Woodburn, Oregon; Clifford Betts of Betts Engineering Associates, Chattanooga, Tennessee; Jim Black of Strader's Greenhouse, Grove City, Ohio; Dan Brown, formerly of the Georgia Concrete and Products Association (www.gcpa.org)

TABLE 11.7
Holders of Registered Trademarks Mentioned in This Chapter.

Registered Trademark	Holder	Headquarters Location
EcoCreto	EcoCreto of Texas	Austin, Texas
Perco-Bond	Michiels International	Kenmore, Washington
Perco-Crete	Michiels International	Kenmore, Washington
StoneyCrete	Stoney Creek Materials	Austin, Texas

and now of Holcim US (www.holcim.com/USA); Shelly Cannady and Linda Velasquez, at the time landscape architectural students at the University of Georgia, now in practice in Georgia; Bill Davenport of the American Concrete Pavement Association; Karen Enyedy of EcoCreto (www.ecocreto.com); Nancy Garza of The Aberdeen Group, publishers of *Concrete Construction* magazine; Bruce Glasby of Advanced Site and Paving, Longwood, Florida; John Gnoffo of Cerulea, Inc., Alpharetta, Georgia; Jeffrey Gremaud of Stoney Creek Materials (www.stoneycreekmaterials.com); Roy Keck of the Georgia Concrete and Products Association (www.gcpa.org); Charles Korhonen and librarian Elizabeth Smallidge of the Cold Regions Research and Engineering Laboratory (www.crrel.usace.army.mil); Jerry McBride of Ewell Industries, Tampa, Florida; Doug Imanishi and Frank Michiels of Michiels International (www.percocrete.com); Karthik Obla of the National Ready Mixed Concrete Association (www.nrmca.org); Chere Peterson of Petrus UTR (www.petrusutr.com); Andy Reese of AMEC (www.amec.com); Wilton "Bud" Roberts of the Oregon Department of Transportation; Betty Rushton of the Southwest Florida Water Management District (www.swfwmd.state.fl.us); Roy Settle of the Appalachian Resource Conservation and Development Council (Johnson City, Tennessee); staff of Signal Mountain Cement of Chattanooga, Tennessee; Craig Tomoseen of the City of Olympia, Washington; Lee Tribble of the Jones Ecological Research Center (www.jonesctr.org); Diep Tu of the Florida Concrete and Products Association (www.fcpa.org); Don Wade of Parkon; Joseph Whitaker of Whitaker Labs, Savannah Georgia; Andy Youngs of the California Cement Promotion Council.

The following persons constructively critiqued early drafts of this chapter: Dan Brown of Holcim US (www.holcim.com/USA), Chere Peterson of Petrus UTR (www.petrusutr.com), and Craig Tomoseen of the City of Olympia, Washington (www.ci.olympia.wa.us).

REFERENCES

American Association of State Highway and Transportation Officials (AASHTO) (1993). *AASHTO Guide for Design of Pavement Structures 1993*, Washington: American Association of State Highway and Transportation Officials.
American Concrete Institute (ACI) (1984). *Guide to Residential Cast-in-Place Concrete Construction*, 332R-84, Farmington Hills, MI: American Concrete Institute.
American Concrete Institute (in press). *Pervious Concrete*, Committee 522, Farmington Hills, MI: American Concrete Institute.
American Concrete Pavement Association (1996). *Construction of Portland Cement Concrete Pavements*, FHWA HI-96-027, Washington: U.S. Federal Highway Administration.
Asaeda, Takashi, and Vu Thanh Ca (2000). Characteristics of Permeable Pavement During Hot Summer Weather and Impact on the Thermal Environment, *Building and Environment* 35, 363–375.
Brady, Nyle C. (1974). *The Nature and Properties of Soils*, 8th ed., New York: Macmillan.
Chattin, Bruce T. (1999). Salmon Listing Requires New Stormwater Management Strategy, *Seattle Daily Journal of Commerce* (www.djc.com and www.washingtonconcrete.org/pervious.htm).
Concrete Resources (2001). Pervious Concrete Can Eliminate Traditional Stormwater Runoff, *Concrete Resources* (Southern California Ready Mixed Concrete Association) Vol. 1, 6.

Croney, David, and Paul Croney (1998). *Design and Performance of Road Pavements*, 3rd ed., New York: McGraw-Hill.
Eyman, Krystian H. (1963). Method of Proportioning Normal and No-Fines Concrete Mixtures, *Journal of the American Concrete Institute* July, 927–943.
Florida Concrete and Products Association (no date). *Construction of Portland Cement Pervious Pavement*, Orlando: Florida Concrete and Products Association.
Furukawa, K., and M. Fujita (1993). Advanced Treatment and Food-Production by Hydroponic Type Waste-Water Treatment-Plant, *Water Science and Technology* 28, 218–228.
Georgia Concrete and Products Association (no date). *Recommended Specifications for Portland Cement Pervious Pavement*, Tucker, Georgia: Georgia Concrete and Products Association.
Hanson, Dick, and Angelo Simeoni (1992). Concrete, in *Materials, Volume 4 of Handbook of Landscape Architectural Construction*, pp. 29–40, Scott S. Weinberg and Gregg A. Coyle, Eds., Washington: Landscape Architecture Foundation.
Harris, Charles W., and Kyle D. Brown (1998). Concrete, *Time-Saver Standards for Landscape Architecture*, section 830, 2nd ed., Charles W. Harris and Nicholas T. Dines, Eds., New York: McGraw-Hill.
Jacklet, Ben (2004). City Tests Its Paving Options, *Portland Tribune* February 5 (www.portlandtribune.com).
King, Bruce (2000). A Brief Introduction to Pozzolans, in *Alternative Construction: Contemporary Natural Building Methods*, Lynne Elizabeth and Cassandra Adams, Eds., New York: John Wiley and Sons.
Korhonen, Charles J., and John J. Bayer (1989). *Porous Portland Cement Concrete as an Airport Runway Overlay*, Special Report 89-12, Hanover, New Hampshire: U.S. Army Cold Regions Research and Engineering Laboratory.
Kosmatka, Steven H., and William C. Paranese (1988). *Design and Control of Concrete Mixtures*, 13th ed., Skokie, IL: Portland Cement Association.
Lin, Sheng H. (1992). Effective Diffusion Coefficient of Chloride in Porous Concrete, *Journal of Chemical Technology and Biotechnology* 54, 145–149.
Malhotra, V. M. (1976). No-Fines Concrete — Its Properties and Applications, *Journal of the American Concrete Institute* November, 628–644.
Maynard, D.P. (1970). A Fine No-Fines Road, *Concrete Construction* 15, 116–117.
Meininger, Richard C. (1988). No-Fines Pervious Concrete for Paving, *Concrete International* 10, 20–27.
Monahan, Alfred (1981). *Porous Portland Cement Concrete: The State of the Art*, Report No. FAA-RD-80-110, Vicksburg: U.S. Army Engineer Waterways Experiment Station.
National Ready Mixed Concrete Association (NRMCA) (2004). *Freeze-Thaw Resistance of Pervious Concrete*, Silver Spring, MD: National Ready Mixed Concrete Association.
Paine, John E. (1990). *Stormwater Design Guide, Portland Cement Pervious Pavement*, Orlando: Florida Concrete and Products Association.
Paine, John E. (1992). Portland Cement Pervious Pavement Construction, *Concrete Construction* September, 1992, pp. 655–659.
Pindado, Miguel Angel, Antonio Aguado, and Alejandro Josa (1999). Fatigue Behavior of Polymer-Modified Porous Concretes, *Cement and Concrete Research* 29, 1077–1083.
Portland Cement Association (PCA) (no date). *Concrete Basics: Chemical Admixtures*, www.portcement.org.
Portland Cement Association (PCA) (1991). *Fiber Reinforced Concrete*, Skokie, IL: Portland Cement Association.

Portland Cement Pervious Association (PCPA) (2003). *Standards and Recommendations for Design Criteria, Testing, Inspections, Certifications, and Specifications*, Savannah, GA: Portland Cement Pervious Association.

Rollings, Marion P., and Raymond S. Rollings, Jr. (1996), *Geotechnical Materials in Construction*, New York: McGraw-Hill.

Rushton, Betty, and Rebecca Hastings (2001). *Florida Aquarium Parking Lot, A Treatment Train Approach to Stormwater Management, Final Report*, Brooksville, FL: Southwest Florida Water Management District.

Shell, Scott (2001). High Volume Flyash: A Concrete Solution, *Environmental Design & Construction* May/June (www.edcmag.com).

Sorvig, Kim (1993). Porous Paving, *Landscape Architecture* 83, 66–69.

Stark, Richard E. (1986). Road Surface's Reflectance Influences Lighting Design, *Lighting Design + Application* April.

Stoney Creek Materials (no date). *StoneyCrete Specification Manual*, Austin, TX: Stoney Creek Materials (www.stoneycreekmaterials.com).

Tosomeen, Craig (1999). *Summary of Porous Concrete Sidewalk*, Olympia, WA: City of Olympia Engineering Department.

Tosomeen, Craig (2003). *Report on Cleaning of Porous Concrete Sidewalk*, Olympia, WA: City of Olympia Engineering Department.

Traver, Robert, Andrea Welker, Clay Emerson, Michael Kwiatkowski, Tyler Ladd, and Leo Kob (2004). Porous Concrete: Technical Lessons Learned in Material Handling at a Demonstration Site, *Stormwater* 5, 30–45.

U.S. Departments of the Army and the Air Force (1992). *Pavement Design for Roads, Streets, Walks, and Open Storage Areas*, Army TM 5-822-5 and Air Force AFM 88-7, Chapter 1, Washington: Headquarters, Departments of the Army and the Air Force; downloadable from www.usace.army.mil/inet/usace-docs/armytm/.

What Works (1998). The Parking Lot That Drinks, *What Works, Local Concrete Promotion Newsletter* March, 29.

Whitaker, Joseph (1999). *Report on Pervious Concrete Design Mixes*, Report to Georgia Concrete and Products Association, Savannah: Whitaker Laboratory.

Wingerter, Roger, and John E. Paine (1989). *Field Performance Investigation, Portland Cement Pervious Pavement*, Orlando: Florida Concrete and Products Association.

World Highways (1999). Reducing the Noise Factor, *World Highways* 8, 56–61.

Wulkan, Bruce, Steve Tilley, and Toni Droscher (2003). *Natural Approaches to Stormwater Management: Low Impact Development in Puget Sound*, Olympia, WA: Puget Sound Action Team (www.psat.wa.gov).

Yang, Jing, and Guoliang Jiang (2003). Experimental Study on Properties of Pervious Concrete Pavement Materials, *Cement and Concrete Research* 33, 381–386.

12 Porous Asphalt

CONTENTS

Components of Asphalt Mixtures	460
Asphalt Cement	460
Aggregate	461
Proportioning of Mixture	462
Early Porous Asphalt Mixtures	463
Development by the Franklin Institute, 1970–1972	463
The Problem of Asphalt Drain-Down	464
The University of Delaware Prototype, 1973	466
Contemporary Changes in Porous Asphalt Mixtures	468
Construction of Porous Asphalt Pavement	471
Required Thickness	473
Porous Asphalt Construction Cost	475
Cost without Stormwater Management	475
Cost with Stormwater Management	476
Cost Tradeoffs in Practice	477
Porous Asphalt Hydrology	479
Decline of Infiltration Rate under Traffic Load	480
Decline of Infiltration Rate under Winter Sanding	481
Water-Quality Effects	482
Porous Asphalt Maintenance and Rehabilitation	483
Porous Asphalt Pavements in Practice	484
Quiet Waters Park, Annapolis, Maryland	484
Siemens Office, Great Valley, Pennsylvania	486
Walden Pond State Reservation, Concord, Massachusetts	488
Morris Arboretum, Philadelphia, Pennsylvania	493
Residential Street, Luleå, Sweden	494
Rt. 87, Chandler, Arizona	495
Porous Asphalt In Overlays	496
The Development of Overlay Technology	496
Effects of Porous Asphalt Overlays on Safety	498
Effects of Porous Overlays on Noise	499
Effects of Porous Overlays on Hydrology	500
Cost of Porous Asphalt Overlays	502

    Recent Experience in Oregon 502
    Recent Experience in Georgia 504
Acknowledgments 507
References 507

"Asphalt" is an informal term for aggregate bound with asphalt cement, the technical term for which is asphalt concrete. This book follows the common informal convention of using the terms "asphalt" and "asphalt concrete" interchangeably. The aggregate can constitute 60 to 90 percent or more of the mixture's volume; the asphalt cement is a semifluid binder between aggregate particles.

The first asphalt pavements were installed in Europe in the mid-nineteenth century (Croney and Croney, 1998, pp. 10–11). In the United States, the first successful installations were in the District of Columbia in the 1870s (McShane, 1979). Many of the early installations had a surface of dense asphalt supported by an aggregate base. When automobile traffic began, the smooth surface of asphalt facilitated rapid travel, enlarging the market areas of local businesses and enhancing local economies. Soon most of America's cobbled, graveled, or unpaved city streets were covered over with smooth surfaces of asphalt or Portland cement. Today, asphalt covers more than three quarters of all the paved surface area in North America. Its marked advantage compared with other firm, heavy-duty surfaces is its low initial cost. Table 12.1 lists some terms with distinctive application in asphalt construction.

The asphalt binder never really hardens, so a pavement made with it is a flexible structure. An asphalt surface course and an aggregate base interact to form a single flexible structure. Flexible binding by the cement makes an asphalt surface able to withstand the abrasion of most kinds of traffic. However, a flexible asphalt pavement is vulnerable to kinds of deformation that do not occur in rigid materials. "Rutting" is the formation of a track or channel where the repeated passage of wheels has pushed the material aside. "Shoving" is the formation of surface waves or bulges where traffic has pushed the material forward. Weakness in the subgrade or base material can propagate upward into surface depressions or potholes. To resist these tendencies, asphalt concrete derives its flexible stability from the binder's adhesion and the interlock and friction of aggregate particles.

Porous asphalt differs from dense asphalt in its use of open-graded aggregate. Because no fine aggregate fills the voids between the single-sized particles, the material is porous and permeable. Porous asphalt has been used since the mid-twentieth century in highway and airfield overlays to improve drainage and safety. In the 1950s it was used for drainage of horse stables (A.E. Bye and Jack Daft, personal communications).

A landmark in the development of porous asphalt technology was an investigation in the early 1970s by Edmund Thelen and his colleagues at the Franklin Institute in Philadelphia. They conceived porous asphalt for stormwater control in 1968, and then developed it with research supported by the U.S.Environmental Protection Agency (EPA). EPA published their results in a detailed research report (Thelen et al., 1972). Subsequently Thelen and Howe (1978) published a concise summary of the results, to which they appended experiences with early field installations and

TABLE 12.1
Some Terms with Distinctive Application in Asphalt Construction

Term	Definition
Asphalt	Dark-colored bituminous cement
Asphalt concrete (bituminous concrete)	Aggregate bound with asphalt cement; informally called "asphalt"
Asphaltic content	Amount of asphalt binder, in percent by weight of total asphalt mixture (aggregate + binder)
Bitumen	Any of various mixtures of hydrocarbons derived from petroleum or related substances, of which asphalt is one
Blister	Pothole-like area where aggregate particles have raveled away
Cement (binder)	Any substance that promotes adhesion
Concrete	Aggregate bound with any binder
Cutback asphalt	Asphalt cement that has been liquefied by blending with petroleum solvents; upon exposure to the atmosphere the solvents evaporate
Drain-down	Downward migration of asphalt binder through the pores of porous asphalt, especially when warm, either during production and installation of the material or after installation
Flow	Deformation of an asphalt material under specified pressure
Hard asphalt (solid asphalt)	Relatively solid asphalt binder, specifically binder with penetration of less than 10
Marshall stability	Resistance to compressive load in pounds, used to evaluate alternative mixtures of asphalt and aggregate for use in a specific project
Neat	Without admixture
Penetration	Consistency of asphalt binder expressed as the distance a needle penetrates a sample under standard conditions
Prime coat	Application of bituminous material to coat and bind mineral particles preparatory to placing an overlying course
Raveling	Loss of pavement surface material as the binder degrades and strips away and loose particles are dislodged
Seal coat	Thin application of bituminous material
Semisolid asphalt	Asphalt binder with penetration between 10 and 300
Slurry seal	Seal coat consisting of asphaltic emulsion and fine aggregate
Spalling	Breaking of pavement or loss of material around cracks
Stability	Marshall stability
Stripping	Loss of adhesion between the binder and the aggregate, and the consequent separation of the film of binder from the particle surfaces, most specifically when moisture replaces the asphalt film coating on aggregate particles; the term has sometimes been used vaguely to lump together several mechanisms of water damage to asphalt concrete, not all of which involve loss of asphalt coating
Tack coat	Application of bituminous material for adhesion between successive layers of asphalt concrete
Tender asphalt	Asphalt concrete mixture that deforms and cracks under compaction during construction

Partly after Webster 1997, ASTM E 1778, and ASTM D 6433.

their recommended design guidelines. EPA supported demonstration projects through the early 1980s (Field et al., 1982). The Franklin Institute guidelines informed numerous installations and continue to be used to some extent to this day.

Porous asphalt has been installed in the field as much as any other type of porous material. Over the years, some of the installations have suffered from clogging and decline of infiltration rate, due to mediocre construction or the immaturity of the technology. Today, upgraded technology is being developed to improve porous asphalt's hydrologic and structural durability and reliability. This chapter describes both the experiences of the past and the direction of technology for the future.

The American asphalt paving industry is a large industry with much technical support. Unfortunately, it has only just begun to support the development and use of porous asphalt for environmental amelioration. The organizations listed in Table 12.2 provide much information about conventional industry technology and terminology, and may in the future provide updates on porous asphalt technology.

COMPONENTS OF ASPHALT MIXTURES

Asphalt's basic constituents are well known. Table 12.3 lists applicable AASHTO and ASTM standards.

Asphalt Cement

Asphalt cement is an oil-based substance. It is found in nature in a few places, but today it is obtained almost exclusively from refinement of petroleum.

Asphalt cement is a flexible binder. During production, an asphalt mixture is routinely heated to soften the binder and make the mixture amenable to handling and placement. The cement's consistency is often expressed in terms of viscosity, as it is for example in ASTM D 3381. Another way to express consistency is in terms of "penetration": the distance that a standard needle penetrates a sample of material under specified conditions of loading, time, and temperature (ASTM D 946). Penetration is reported in tenths of a millimeter ("pen"); thus, a 40 to 50 pen asphalt is harder than a 60 to 70 pen asphalt. Relatively hard (low-penetration) binders can be used in warm climates to inhibit bleeding away of the binder in warm weather; soft (high-penetration) binders are used in cool climates to avoid brittleness.

TABLE 12.2
Organizations and Sources of Information in the Asphalt Paving Industry

Organization	Contact Information
American Association of State Highway and Transportation Officials (AASHTO)	www.aashto.org
Asphalt Contractor magazine	www.asphalt.com
Asphalt Institute	www.asphaltinstitute.org
ASTM International	www.astm.org
National Asphalt Pavement Association	www.hotmix.org
National Center for Asphalt Technology	www.eng.auburn.edu/center/ncat
U.S. Federal Highway Administration	www.fhwa.dot.gov/pavement

TABLE 12.3
AASHTO and ASTM Specifications of Basic Constituents in Asphalt

Organization	Number	Title
AASHTO	M 20	Standard Specification for Penetration-Graded Asphalt Cement
AASHTO	M 226	Standard Specification for Viscosity-Graded Asphalt Cement
AASHTO	M P1	Standard Specification for Performance-Graded Asphalt Binder
AASHTO	R 12	Standard Specifications for Bituminous Mixture Design Using the Marshall and Hveem Procedures
ASTM	C 1097	Standard Specification for Hydrated Lime for Use in Asphalt-Concrete Mixtures
ASTM	D 692	Standard Specification for Coarse Aggregate for Bituminous Paving Mixtures
ASTM	D 946	Standard Specification for Penetration-Graded Asphalt Cement for Use in Pavement Construction
ASTM	D 1073	Standard Specification for Fine Aggregate for Bituminous Paving Mixtures
ASTM	D 3381	Standard Specification for Viscosity-Graded Asphalt Cement for Use in Pavement Construction
ASTM	D 3515	Standard Specification for Hot-Mixed, Hot-Laid Bituminous Paving Mixtures
ASTM	D 3515	Standard Specification for Hot-Mixed, Hot-Laid Bituminous Paving Mixtures
ASTM	D 5106	Standard Specification for Steel Slag Aggregates for Bituminous Paving Mixtures
ASTM	D 6114	Standard Specification for Asphalt-Rubber Binder
ASTM	D 6115	Standard Specification for Nontraditional Coarse Aggregates for Bituminous Paving Mixtures
ASTM	D 6373	Standard Specification for Performance-Graded Asphalt Binder

The amount of asphalt binder in a mixture is designated in terms of the percent of the total mixture's weight. In porous asphalt, enough binder must be included to form a film of asphalt around all the aggregate particles, but not so much that it drains quickly down through the pores or detracts excessively from the mixture's porosity.

With time, asphalt binder that is exposed to air oxidizes, forming a brittle crust around the remaining binder. If the asphalt film is thick, then the oxidized crust protects the remaining inner binder from oxidation.

Recycled rubber is a component of some asphalt binders. The asphalt-rubber mixture is standardized in ASTM D 6114. Information on this use of rubber is available from the Rubber Pavements Association, www.rubberpavements.com.

Aggregate

The aggregate in porous asphalt mixtures must meet the same standards of angularity, durability, and stability as the aggregate used in dense asphalt. It must be resistant to degradation such as freeze-thaw damage and disintegration in water. It should be resistant to stripping and development of tender mixtures. During production, aggregate is routinely heated to drive off moisture, which permits secure asphalt bonding.

The use of open-graded aggregate produces an aggregate "skeleton" which is vital to the material's porosity and permeability. A small amount of fine aggregate is added to some porous asphalt mixtures to form a matrix with the asphalt binder that controls the mixture's cohesiveness.

"Stripping" occurs when the asphalt cement loses its adhesion with the aggregate particles, most specifically when moisture intervenes. When a binder loses its adhesion, the pavement surface can ravel, which is the breaking up of the surface as traffic scatters the loose pieces about. A blister is a pothole-like area where the aggregate has raveled. Stripping is accelerated by prolonged contact with water and facilitated by the brittleness that comes with aging and oxidation. Various tests can be used to examine an aggregate's stripping potential. Most of them involve saturating a sample of asphalt concrete material with water, then determining the loss of asphalt coating or the change in the asphalt concrete's strength. Because of the severity of the problem should it happen, some agencies routinely specify anti-stripping agents such as lime or proprietary anti-stripping compounds.

Many aggregates can absorb water into the particles' micropores. Although some practitioners see absorption as a precursor to frost damage, the same pores that absorb water also act as pressure relief for expanding ice crystals, and many aggregates that have high absorption also have high durability (Rollings and Rollings, 1996, pp. 166–167). In the ASTM C 127 absorption test, the amount of water that enters an aggregate's micropores during 24 hours of soaking is expressed as a percentage of the aggregate's oven-dry weight.

Proportioning of Mixture

The required proportions of binder, aggregate, and additives in an asphalt mixture are project-specific. Correct proportioning balances competing requirements such as strength, porosity, and durability. Although experience shows that the proportions end up within a certain range, the quantitative proportions for a specific project depend on the character of the locally available aggregate and other variable conditions.

In a specific project proportioning may be determined by the design specifier based on knowledge of local materials, by a testing lab analyzing several alternative combinations on behalf of the specifier or the installer, or by the installer at the project site based on experience with local materials and porous concrete mixtures.

The Marshall, Hveem, and "Superpave" tests all operate in fundamentally similar ways, and seek similar characteristics in a proportioned mixture (Croney and Croney, 1998, p. 197; Rollings and Rollings, 1996, pp. 288–289). The Marshall test, for example, determines the effect of asphalt content on the material's load-bearing value (resistance to compressive load). The test is standardized in AASHTO T245. A machine places pressure on a sample of the material, and the maximum pressure required to make the material deform is recorded. The procedure measures the compressive load (Marshall stability in pounds) and corresponding deformation (flow in hundredths of an inch). A very high value of flow indicates a plastic mix that will deform under traffic; a very low value indicates a "lean" mixture that may ravel or be brittle. A common procedure is to perform the Marshall test on samples containing the aggregate to be used in a project, and different contents of asphalt binder; the amount of asphalt that produces the most desirable combination of stability and flow is selected. Although porous asphalt is adequate in Marshall properties for many applications, it does not always meet the criteria that would be used for dense asphalt (Thelen and Howe, 1978, pp. 13–14; Goforth et al., 1983, pp. 40, 43, 46).

After the components are mixed, the mixture is routinely kept heated in order to soften the binder and make the mixture workable.

EARLY POROUS ASPHALT MIXTURES

DEVELOPMENT BY THE FRANKLIN INSTITUTE, 1970–1972

The Franklin Institute began its landmark investigation in the early 1970s by identifying open-graded aggregate gradations previously used for asphalt highway overlays in Britain, California, and elsewhere. Table 12.4 lists the most important gradations in the researchers' tests and their resulting recommendations.

The California gradation was the most open-graded of the tested gradations, so it gave asphalt mixtures the highest porosity and permeability. The institute adopted it as the basis for its further evaluation of porous asphalt paving material. Mixtures based on it were strong enough to bear medium traffic loads according to the Marshall test, were not susceptible to frost damage, and might adequately resist aging and embrittlement as long as asphalt content was sufficient to maintain a film over the aggregate particles (Thelen et al., 1972, pp. 4, 29, 36). Thelen and Howe (1978, p. 64) subsequently published rounded central values which are referred to here as the "Thelen and Howe" gradation.

The original California gradation and the Thelen and Howe central values became the standards among most practitioners in the following decades (Brown, 1996b). In the institute's recommended mixture (Thelen et al., 1972, p. 70), the asphalt's penetration is 85 to 100; its content is 5.75 to 6 percent of the mixture's total weight.

TABLE 12.4
Aggregate Gradations in the Franklin Institute's Tests and Recommendations

	Percent (By Weight) Passing		
Sieve Size	California Specification	As Tested by Franklin Institute	Specified by Thelen and Howe
1/2 inch	100	100	100
3/8 inch	90 to 100	97	95
No. 4	35 to 50	34	35
No. 8	15 to 32	16	15
No. 16	0 to 15	13	10
No. 200	0 to 3	2	2

Data from Thelen et al., 1972, pp. 4 and 41; and Thelen and Howe, 1978, p. 64.

Anti-stripping agents were added if indicated by the ASTM test for asphalt stripping. The mixture produced porosity of 16 percent and infiltration rate of 170 inches per hour immediately after installation (Thelen and Howe, 1978, p. 13). EPA sponsored a further design manual expanding on the institute's results (Diniz, 1980).

Although the most open-graded of the institute's tested mixtures was the most favorable, that result did not cause researchers to consider further open-graded gradations, or gradations with larger particle sizes. As the institute did not continue its tests for more than a year or two, they did not discover the problem presented over time by asphalt drain-down.

THE PROBLEM OF ASPHALT DRAIN-DOWN

The hypothesis of gradual asphalt drain-down originated in 1996 during the examination of a porous asphalt residential driveway near Macon, Georgia. The driveway is used only for access to a workshed and for the washing of private cars. It is surrounded by grass and trees that supply abundant organic debris. The pavement had been built in 1990. Over the years the residents had observed the driveway's relative infiltration rate while washing their cars. For the pavement's first three or four years it had absorbed all rainfall and other water, without runoff. Then some visible runoff began to be generated.

In 1996, six years after construction, the residents gave an informal demonstration of the driveway's hydrologic performance as shown in Figure 12.1. Only some of the water from a garden hose infiltrated immediately into the surface; the remainder ran some distance across the surface before infiltrating.

FIGURE 12.1   A porous asphalt residential driveway near Macon, Georgia, in 1996.

On close examination, binder was partly missing from the surfaces of the uppermost aggregate particles, and a black layer was visible in the pavement's pores about 1/2 inch below the surface. Evidently the black material was asphalt binder that had drained down during hot summer days, mixed with tightly packed ground-up organic matter. Organic dust and hot asphalt made a slurry that flowed together to a lower, cooler level in the pavement, where they congealed into a clogging matrix.

In 2000, the residents reported that the surface was highly clogged and that it produced abundant runoff.

The hypothesis is that, after a porous asphalt mixture is installed, the asphalt binder can migrate downward through the pores under the influence only of heat and gravity. On hot summer days, the surface of an asphalt pavement can be notably hot. The heat-softened binder flows, very gradually, down from the surface until it meets a cooler level in the interior, where it slows down or stops, filling the pores. It drags with it any dust or debris from the surface, incorporating it in the clogging matrix.

The hypothesis has been reinforced during numerous field observations of porous asphalt parking lots. In 6- to 22-year-old pavements, the surfaces of the uppermost stone particles have been partly bare of asphalt binder. About 1/2 inch below the surface is a partially or wholly clogged layer, black in color. Probing into the clogging layer has revealed tightly packed dust or ground-up organic matter. The migration of the binder reduces the pavement surface's structural integrity, as the binder strips away from surface aggregate particles. The then-loose particles are scattered about, and the surface may become ruggedly raveled and blistered.

Observations in parking lots indicate that traffic intensity contributes to stripping and drain-down. Traffic abrasion helps separate the binder from the surface particles, freeing it to drain down and clog the pores. Oxidized, brittle outer films of binder may be particularly susceptible to being knocked away; the abraded cement particles become dust which joins the drained-down binder in the pores. The fastest drain-down and clogging occur under the greatest amount of traffic. For example, in a parking lot, the most-clogged and most-raveled places develop in the traveling lanes and especially close to the entrance where the traffic is greatest. In contrast, the little-trafficked parking spaces, and especially the areas where wheels seldom touch, keep their black color, open texture, and high infiltration rate relatively long.

The hypothesis was independently confirmed in field experience with highway overlays (Huber, 2000, pp. 11–12, 19). The asphalt paving industry had become aware of the phenomenon during porous asphalt production and installation, and gave it the term "drain-down." It was discovered that during production, storage and transport, while the mixture is still hot and soft, drain-down results in uneven distribution of the binder as the mixture is laid into place. Some parts of the pavement end up impervious with too much binder, and in other parts the aggregate particles quickly separate because there is too little binder.

The National Center for Asphalt Technology developed a test to determine a mixture's relative potential for drain-down (Mallick et al., 2000; Georgia Department of Transportation, no date). An open-graded asphalt mixture is placed in a wire basket having 1/4 inch mesh openings, and heated for one hour at 25°F above normal production temperature. The amount of asphalt cement that drains from the mixture is measured.

Systematic research to conclusively test the gradual migration of asphalt binder in installed material is called for. At intervals of time, sections need to be cut out of porous asphalt surfaces and examined microscopically to observe the changing concentrations of binder through the section. If the hypothesis is correct, then such an examination would show decreasing amounts of binder at the surface, and increasing amounts at some depth below the surface.

The reduction in porous asphalt's infiltration rate over time has contributed to a speculation that porous asphalt pavements require frequent vacuuming or other maintenance to remain permeable. In fact, the requirement for maintenance may be much lower and more site-specific than rumored, and vacuuming may not counteract this underlying cause of clogging. To preserve rapid infiltration and other desired surface properties, porous asphalt material must first be made to resist drain-down.

THE UNIVERSITY OF DELAWARE PROTOTYPE, 1973

The first major installation of the Franklin Institute mixture was a porous asphalt parking lot at the University of Delaware Visitors Center in Newark, in 1973. The project was well documented, and inspired subsequent installations (Bachtle, 1974; *Engineering News Record*, 1973; Robinette, 1976, pp. 62–68; Thelen and Howe, 1978, pp. 23–24, 82).

The parking lot held 74 parking spaces in two double-loaded bays. The underlying soil was moderately permeable silt loam and sandy loam, with increasing coarseness and permeability with depth.

Figure 12.2 shows the construction. Beneath the surface course of Franklin Institute porous asphalt was a 10-inch base reservoir on undisturbed subgrade. In the middle of the base course, a layer of relatively small aggregate was inserted to bind and stabilize the adjacent large stone (subsequent research would show that the intermediate layer was not necessary to increase stability). The uppermost layer of base aggregate was a "choker course" to provide a smooth platform for asphalt placement. The edge was restrained with a metal strip held by stakes.

The construction equipment and process were similar to those of dense asphalt paving. Total installation cost was 20 percent less than that of a dense asphalt pavement due to the reduction in curbing and stormwater conveyances (Bachtle, 1974; Robinette, 1976, p. 63; Thelen and Howe, 1978, pp. 23–24).

At the low corner of the parking lot, a trench was excavated 15 feet deep to allow excess water to infiltrate a deep, highly permeable soil layer. The trench was 3 feet wide and filled with 3/4 inch aggregate (data from drawings by Edward R. Bachtle and Associates in Robinette, 1976, p. 64). Because of the great hydraulic capacity of the base reservoir and the additional trench and the permeability of the subgrade soil, the newly installed pavement was able to infiltrate essentially all rain water into the subgrade.

Throughout the pavement's 27-year life, it was vacuumed about three times per year, as were all other parking lots on the campus. During the winter it was snow-plowed as necessary. Its only special care was the use of as little deicing salt as possible, and no sand for traction (Tom Taylor, personal communication 1993). It never required any structural repair.

Figure 12.3 shows the pavement in 1996, when it was 23 years old. The structural condition was superb for an asphalt pavement this old. Only one long crack was

Porous Asphalt

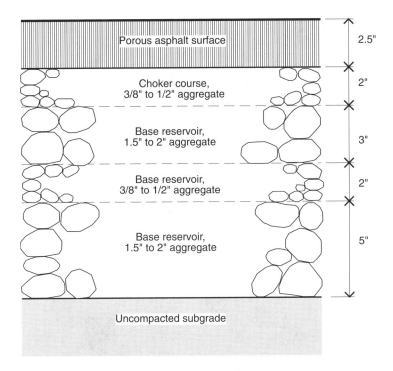

FIGURE 12.2 Porous asphalt pavement at the University of Delaware Visitors Center (after Bachtle, 1974).

visible on the surface. A few spalled-off potholes (apparently the result of stripping and raveling), about 1/2 inch deep, had little call for patching.

However, the pavement surface failed hydrologically due to gradual asphalt drain-down. The pavement lightened in color as the binder flowed away from the surface. The lightest portions were in the central turning lanes where traffic was heaviest and the lack of shade allowed the surface temperature to rise the highest. As of 1996, an asphalt clogging layer was visible in the pores about 1/2 inch below the surface, and the surface infiltration rate was low.

The University of Delaware experiment came to an end in 2000, when the parking lot was resurfaced with a thin layer of dense asphalt. The new layer hid the raveled blisters and reduced the infiltration rate to zero. All drainage was across the surface, and into a drop inlet and storm sewer. During its lifetime the porous surface had testified to porous asphalt's structural durability under moderate traffic load, and simultaneously to the need for technological refinements to maintain its infiltration rate.

The old reservoir course, permeable soil, and infiltration trench still underlie the now-dense surface. They could be recovered and their function restored by milling off all asphalt surface layers and replacing them with an equal thickness of modern porous asphalt or other reliable porous surfacing.

Subsequent installations that followed the Franklin Institute's findings and the University of Delaware prototype were listed by Thelen and Howe (1978, pp. 23–28). Some were initiated as practical solutions to stormwater control requirements; others

FIGURE 12.3 The University of Delaware porous asphalt parking lot in 1996.

were supported by government agencies for research and demonstration. Almost all were in Delaware, Maryland, and southeastern Pennsylvania. A few were reportedly clogged shortly after installation.

CONTEMPORARY CHANGES IN POROUS ASPHALT MIXTURES

At the time of this writing, changes are occurring in porous asphalt technology (Huber, 2000, p. 1, 10). The improvement in performance over older mixtures has been dramatic, although not all problems have necessarily been completely solved, and adjustments continue to be made.

The changes are founded on work in Europe, where porous asphalt began to be used in highway overlays in the 1980s. From the beginning of its experience, Europe has been a productive incubator for new porous asphalt technologies (Kuennen, 1996).

Porous Asphalt

Compared with the U.S., Europe's highway construction companies are fewer and larger, and they have close working relationships with their government highway agencies. Many European contractors develop their own proprietary mixtures, which they install and warrant for the agencies. In 1990 the U.S. National Academy of Sciences' Transportation Research Board brought the European developments to North America by dedicating an entire issue of *Transportation Research Record* (no. 1265) to the research and highway applications being done in many European countries. Ten years later the same agency published Huber's (2000) *Performance Survey on Open-Graded Friction Course Mixes*, which vitally summarized and updated the European developments, and the adaptation of those developments in American departments of transportation.

In the new mixtures, liquid polymer additives make the asphalt binder less susceptible to drain-down both during production and after installation (Huber, 2000, pp. 11–12, 19). The asphalt film viscosity is commonly four to six times that in dense asphalt material (Kuennen, 2003). Compared with unmodified asphalt it is stiffer at high temperatures, and more flexible at low temperatures (Georgia Department of Transportation, no date). Because the binder adheres to aggregate particles in thick, stable films, the interior of the film is protected from oxidation and embrittlement. The aggregate particles remain glued together, so the mixture resists rutting and raveling (Huber, 2000, pp. 11–12, 20).

Mineral or cellulose fibers have the same types of stabilizing effects as liquid polymer additives. Some models are almost microscopically thin, and only 1/4 inch long. The fibers, constituting perhaps 0.3 to 0.4 percent of the asphalt by volume, disperse evenly and, despite their tiny size, overlap and form a network (Kuennen, 2003). They interlock through the binder matrix as shown in Figure 12.4. They

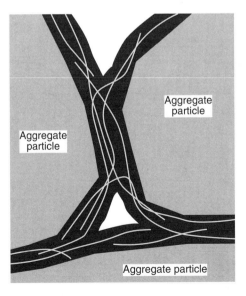

FIGURE 12.4 Concept of interlocking fibers in the asphalt coating of aggregate particles (after Georgia Department of Transportation, no date).

inhibit binder drain-down and prolong bonding (Georgia Department of Transportation, no date; Huber, 2000, p. 11). They permit the amount of binder in a mixture to be increased by 30 to 40 percent, which assures bonding and protects the interior of the film from oxidation.

During a mixture's production, both fibers and polymer modifiers permit heating of the mixture to high temperatures (Huber, 2000, p. 12). The heat expels moisture from the aggregate, so the bond between binder and particles is tight, and potential stripping and raveling are reduced.

Under the drain-down test procedure of the National Center for Asphalt Technology, drain-down must not exceed 0.3 percent of the total weight of the mixture to assure the mixture's integrity. Unmodified porous asphalt mixtures have not met this requirement. New mixtures which contain fiber, polymers or a combination of both have met it (Georgia Department of Transportation, no date).

Maximum aggregate sizes are being increased, from the Franklin Institute's 3/8 inch to 1/2 inch or more. Figure 12.5 contrasts a contemporary gradation with the old Thelen and Howe standard. The Oregon 19 mm (3/4 inch) Open gradation was developed about 2002 by the Oregon Department of Transportation. Compared with the old Thelen and Howe standard, the maximum particle size has doubled, and the quantity of very small particles has declined. New mixtures like Oregon's have 15 percent or more void space, as did the Franklin Institute mixture, but the larger single-sized particles produce larger individual voids and a higher, more durable infiltration rate (Huber, 2000, p. 2; Kuennen, 2003).

Any asphalt plant capable of producing high-quality dense asphalt can produce today's high-quality open-graded asphalt (Huber, 2000, p. 21).

The new changes have increased porous asphalt's average structural lifetime in highway overlays from about 7 to about 11 years. Highway agencies benefit greatly from this 50 percent increase in lifetime before replacement is necessary. But

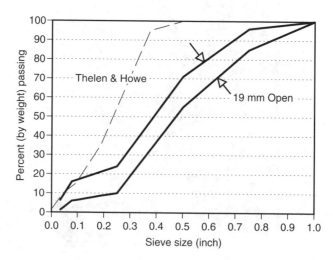

FIGURE 12.5 Aggregate gradations in evolving porous asphalt mixtures (data from Thelen and Howe, 1978, p. 64; and www.odot.state.or.us).

whether the changes extend the permeability of porous asphalt parking lots from 7 years to 30 years has not yet been demonstrated. The large pores in the coarse new gradations have not consistently resisted clogging compared with small pores, perhaps because the sticky binder tends to capture debris particles no matter how large the voids (Huber, 2000, p. 23). The larger voids have proven less susceptible to clogging only where they are applied in high-traffic highways, where, in wet weather, the rapid passage of tires pushes water through the pores in forceful bursts, loosening and moving debris in the pores.

To date, although the new American mixtures tend not to be as porous as European ones, their infiltration rate has been much improved by moving in the direction of European mixtures. New mixtures using polymer modifiers, high asphalt content, fibers, and open aggregate gradations have solved problems of rapid raveling and loss of permeability (Kandhal, 2002).

In taking advantage of the European developments, American agencies are struggling with several issues (Huber, 2000, pp. 1, 19, 21). Large aggregate particles require highway overlays to be correspondingly thick, and are thus correspondingly expensive. The high porosity requires high standards for aggregate's resistance to breaking. Nevertheless, contemporary materials are being written into the standard specifications of some state highway agencies. Among the best documented are those of Georgia and Oregon (Huber, 2000; Kuennen, 1996). One can find updated specifications on the web sites of those and other state agencies.

Outside of the highway agencies, the "old" mixtures ended their domination in about 2000. Practice is now rapidly being updated with the benefit of the new technology. Cahill Associates, in their numerous parking-lot applications, are now routinely specifying styrene-butylene-styrene (SBS) additive to inhibit drain-down (Thomas Cahill, personal communication 2004). As proven new technology becomes more available to general practitioners, porous asphalt is becoming a durable alternative among firm, heavy-duty porous paving materials.

CONSTRUCTION OF POROUS ASPHALT PAVEMENT

In order to minimize cost, most asphalt pavements have a bituminous wearing course only at the surface, and a base course of less expensive aggregate material below. After each layer of material is placed, it is compacted. Open-graded asphalt is compacted with rollers of only moderate weight, to avoid crushing aggregate particles and reducing the material's permeability.

If the base-course aggregate is much larger than the aggregate in the asphalt surface, an aggregate filter layer or "choker course" may be added between the two main courses. It prevents the aggregate in the surface course from collapsing into the base's large voids and makes a stable working platform for an asphalt paving machine (Jackson, 2003, p. 5).

Figure 12.6 shows the porous asphalt construction at the Hockessin (Delaware) Library, constructed in the early 1990s. This is an example of the practice of Cahill Associates, which has specified many porous asphalt installations in the mid-Atlantic region and elsewhere (www.thcahill.com). The pavement includes a thick base reservoir, a choker course, and the asphalt surface. The porous asphalt surface

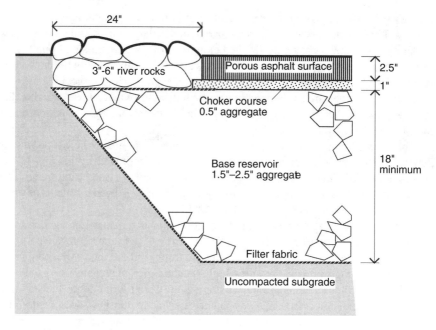

FIGURE 12.6 Construction of porous asphalt parking lot at the Hockessin Library in Hockessin, Delaware (after Brown, 1996b and Cahill, 1993, p. 11).

is based on the Franklin Institute specifications. The choker course is of 0.5 inch aggregate; it is lightly compacted to set the particles into the top of the base reservoir and to stabilize the surface to receive the porous surface course. The reservoir course is of coarse open-graded aggregate. At the floor of the reservoir a filter fabric prevents merging of soil fines with the base-course voids, while allowing water to infiltrate the subgrade. An extension of the base reservoir is covered with large river rocks, forming an extremely permeable perimeter "apron" that captures any runoff from the pavement surface in the event of pavement clogging or rainfall intensity that exceeds the surface's infiltration rate; any overflow from the reservoir discharges through the same apron.

Figure 12.7 shows the construction at Sweet Apple Park in Roswell, Georgia, constructed in 1997. This is an example of the practice of Cerulea, which has specified a number of installations in the Southeast. The pavement includes the same types of layers as those in the previous illustration, but the base reservoir is thinner because there is a firmly compacted subgrade, little frost penetration, and little demand for stormwater storage. The porous asphalt surface is based on a standard Georgia DOT specification of the time for "Type C" open-graded asphalt, which the DOT had developed for highway overlays. The aggregate in the Georgia mixture is roughly equivalent to ASTM No. 6 or 67. The availability of the standard specification allowed construction without a contractor's surcharge for a custom mixture. The choker course is of No. 810 crushed aggregate. Overflow water exits through the porous asphalt surface at low points in the pavement's edge.

Porous Asphalt

In the two examples there are no structural edge restraints because wheel stops prevent vehicles from crossing the pavement edge. In many other installations, where traffic may cross the edge, edge restraints like those described in Chapter 3 are required.

Successful porous asphalt installations have been characterized by diligent communication before, during, and after construction. The procedures necessary to assure a successful installation are not complicated, but most asphalt paving contractors are not accustomed to porous asphalt's special objectives. Thorough communication with contractors and testing labs and vigilant observation during installation are vital to implementing the objectives and specifications on which the installation's success depends.

REQUIRED THICKNESS

Adequate thickness of a pavement and each of the layers within it is vital for pavement stability. Because porous asphalt is not packed with fines, it has less shear strength and deflects more than dense asphalt (Adams, 2003; Wyant, 1992); so thickness guidelines for dense asphalt do not necessarily apply to porous asphalt.

One type of indicator of pavement thickness requirements comes from empirical experiences. Figure 12.8 shows a porous asphalt pavement of minimum thickness,

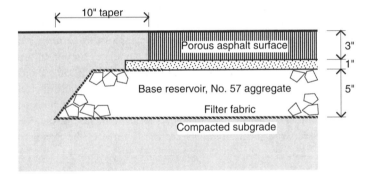

FIGURE 12.7 Construction of porous asphalt parking lot at Sweet Apple Park in Roswell, Georgia (after specifications by Cerulea).

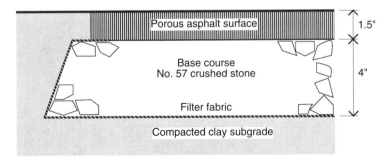

FIGURE 12.8 Construction of a porous asphalt residential driveway near Macon, Georgia (after specifications by Adele George).

built for a residential driveway near Macon, Georgia in 1990 (the hydrologic performance of this pavement was described earlier in this chapter). The 1.5 inch surface course and 4 inch base course add up to the minimum thickness of asphalt pavement, whether porous or dense, that could be built for any traffic load on almost any subgrade. The subgrade is compacted white clay of the Helena series. Frost penetration in this location is negligible; the only structural concern was light traffic load on occasionally wet clay. Stormwater storage was not required. The porous surface course followed Franklin Institute specifications. Ten years after installation the pavement was structurally in good condition, without rutting, shoving, or objectionable distortion. This experience indicates that, under a very light traffic load and on compacted subgrade, the thickness of porous asphalt pavement can be reduced to the same low-cost minimum to which dense asphalt could be taken.

A more robust empirical demonstration of pavement thickness comes from the many commercial and office parking lots overseen by Cahill Associates in the mid-Atlantic area (Adams, 2003; www.thcahill.com). The construction at the Hockessin Library shown above is typical of these projects. All of them have been structurally durable without objectionable or even noticeable deformation. The minimum base course thickness has been 18 inches, resulting in a total pavement thickness of over 20 inches. This substantial thickness has been sufficient to bear the parking lots' moderate traffic loads and to compensate for Cahill's routine specification against subgrade compaction while holding ponded water that saturates the uncompacted subgrade.

Another type of indication of required pavement thickness comes from structural theory. The Franklin Institute researchers calculated guidelines for porous asphalt thickness on a "worst-case" subgrade with California Bearing Ratio (CBR) of only 2, which represents a soft, uncompacted soil in wet conditions such as those that could occur under permeable pavement. They used Asphalt Institute design procedures which at that time specified minimum surface course thickness of 4 inches. Table 12.5 lists the resulting thicknesses, with traffic load designated in terms of the average number of

TABLE 12.5

Minimum Porous Asphalt Pavement Thickness Required to Bear Structural Load on Poor Subgrade with CBR 2

Traffic Category	Average ESAL per Day	Minimum Pavement Thickness (inches)		
		Porous Asphalt Surface Course	Aggregate Base Course	Total
Light (parking lots, residential streets)	1	4	6	10
	10	4	12	16
Medium (city business streets)	20	4.5	13	17.5
	50	5	14	19
	100	5	16	21
Heavy (highways)	1000	6	20	26
	5000	7	22	29

Thelen et al. 1972, pp. 5, 28, 31–32.

TABLE 12.6
Minimum Total Porous Asphalt Pavement Thickness (aggregate base course + porous asphalt surface course) Required to Bear Structural Load on Various Subgrades

	Minimum Total Pavement Thickness (inches)		
Traffic Load	Subgrade CBR 6 to 9	Subgrade CBR 10 to 14	Subgrade CBR 15 or more
Light (ESAL 5 or less per day)	9	7	5
Medium light (1,000 vpd max., ESAL 6 to 20 per day)	11	8	6
Medium (3,000 vpd max., ESAL 21 to 75 per day)	12	9	7

Diniz, 1980, Table 6; vpd is vehicles per day; ESAL is 18,000 pounds.

equivalent 18,000 pound single-axle loads (ESAL) per day during the pavement's intended life. The results validate Cahill's experience that parking lot pavements at least 20 inches thick are adequate or more than adequate on wet, uncompacted soils.

Diniz (1980, pp. 33–35) calculated minimum thicknesses for other subgrade CBR values. For subgrades with CBR of 5 or less, he recommended improving the subgrade to CBR 6 by adding angular aggregate. Table 12.6 shows his results for other CBR values. All the listed thicknesses are less than those calculated by the Franklin Institute because they apply to stronger, drier, or more compacted subgrade soils.

Further research is called for to quantify the strength of porous asphalt layers, the possible change over time after installation, and ways to increase strength if necessary.

POROUS ASPHALT CONSTRUCTION COST

The initial cost of porous asphalt in the 1970s was sometimes 35 to 50 percent higher than the cost of an equal amount of dense material. The major reason was the unfamiliar technology involved in porous material production, including the gradation requirements, and the narrow limits on the mixture's asphalt cement content (Diniz, 1980, p. 12). However, a porous pavement is not just a pavement; it is also part of a site's drainage and stormwater management system. In order to ascertain the total effect of porous pavement on site development costs, the costs of both pavement and drainage must be taken into account.

Cost without Stormwater Management

The Franklin Institute researchers (Thelen et al., 1972) calculated contrasting costs of porous and dense asphalt pavements, assuming both were designed to Asphalt Institute specifications of the day. In their calculations they gave the pavements drainage to meet the standards of the day, but no additional stormwater treatment or management. They assumed that dense pavements would have conventional drainage inlets and pipes, and counted them in the total pavement cost. Dense pavements' thicknesses varied to carry different traffic loads. For all porous pavements, they gave a minimum base reservoir thickness of 16 inches, which could retain the water from a 5.4-inch rain and so was assumed to eliminate the cost of drainage inlets and pipes. Figure 12.9 shows the results for four pavement categories. The "low standard" and "high standard" street categories reflect variations in municipal standards for residential street width, pavement thickness, and drainage. According to these results, the relative cost advantage of porous asphalt and dense asphalt depend on the pavement category. In light-duty applications porous pavement costs more because, although its 16-inch-thick base course traps stormwater, it is much thicker than a dense pavement, which needs only to carry traffic. In heavy-duty applications, dense pavement costs more because it requires both a thick pavement structure and thorough off-pavement surface drainage.

COST WITH STORMWATER MANAGEMENT

Goforth et al. (1983, p. 6) added stormwater detention to the cost comparison. Their calculations showed that for parking lots, using detention technology of the time, the total costs of porous asphalt and dense asphalt were comparable. At that time a porous asphalt parking lot with a reservoir base could be constructed (including engineering, inspection and testing) for about $10 per square yard. If a contrasting dense asphalt parking lot detained runoff on the pavement surface with a 6-inch curb and a restricted outlet, the cost (including engineering, inspection and testing) was again about $10 per square yard. Compared with the Franklin Institute results where no parking lot stormwater management had been assumed, this was a slight improvement in the cost comparison for porous pavements.

Today, more demanding stormwater management is routinely required, and the simple on-pavement detention assumed by Goforth et al. is seldom adequate. Cahill's work has shown that for parking lots meeting today's stormwater management

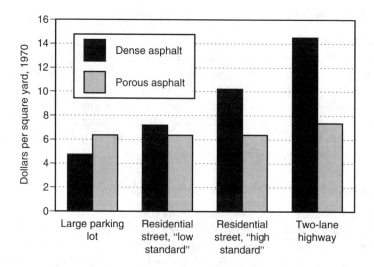

FIGURE 12.9 Comparative construction costs of porous asphalt and dense asphalt pavements (data from Field et al., 1982, and Thelen et al., 1972, pp. 7, 116 and 118).

Porous Asphalt

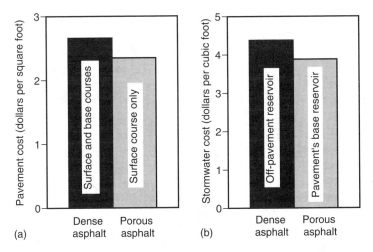

FIGURE 12.10 Comparative pavement and stormwater management construction costs for commercial parking lots in the mid-Atlantic area about 1990 (data from Brown, 1996b and Cahill, 1993, pp. 39–40).

standards, porous asphalt can actually cost less than dense asphalt. Figure 12.10 compares construction costs based on several of Cahill's projects in the mid-Atlantic area in the 1990s, in jurisdictions where rigorous stormwater management was required. The separate charts show the costs of the pavement structure (in dollars per square foot) and its stormwater management function (in dollars per cubic foot of hydraulic storage capacity); the total cost for a given project is the sum of the two costs. The estimate assumes that dense asphalt requires an off-pavement detention basin which brings in the costs of land, conveyance pipes, and outlet structures. For porous asphalt, the entire cost of the thick base reservoir is attributed only to stormwater storage; it assumed that the subgrade is permeable enough to infiltrate all required water, so no off-pavement drainage structures are required. The porous pavement cost includes special field testing to determine soil infiltration rate, the contractor's premium for porous asphalt's custom mixture, a filter fabric around the reservoir course, and special engineering design and inspection. These results show that as the demands for stormwater management go up, the relative cost savings for porous asphalt pavements become increasingly attractive. Site-specific project costs can vary, including the cost of land for off-pavement stormwater facilities.

COST TRADEOFFS IN PRACTICE

A parking lot at Sweet Apple Park in Roswell, Georgia, exemplifies the use of porous asphalt to reduce the size and cost of off-pavement stormwater management reservoirs. Its construction was illustrated earlier in this chapter. Figure 12.11 shows the 100-car parking lot soon after construction in 1997. In hydrologic modeling during site design, the pavement was given no "credit" for infiltration and runoff volume reduction during a storm, because its subgrade was slowly permeable compacted clay and the base course had little hydraulic storage capacity. However,

FIGURE 12.11 The porous asphalt parking lot in Roswell, Georgia's Sweet Apple Park.

the porous surface was given a runoff coefficient equivalent to that for turf on the same soil. This coefficient was much lower than would have been given to a dense pavement, and produced lower calculated runoff. Consequently, the park's stormwater detention facilities were made smaller and more economical. This outcome extends Cahill's conclusions about the combination of stormwater management and pavement costs. The subgrade need not be capable of infiltrating the entire volume of runoff from a large design storm for porous asphalt to have a cost advantage compared with dense pavement.

A porous asphalt parking lot at the DuPont Agricultural Chemicals office in Barley Mills, Delaware solved still larger issues of drainage, land allocation, and economy (Brown, 1996b; *Highway & Heavy Construction,* 1988; *Land and Water,* 1988). It was designed by Andropogon Associates and Cahill Associates, with construction similar to that at the Hockessin Library described earlier. It has eight double-loaded parking bays, each of which is 60 feet × 175 feet in area. The base reservoirs in adjacent bays are connected by pipes; the final bay overflows via a discharge pipe, and is undersized to assure detention in the base reservoirs. The base reservoirs receive runoff from large office-building roofs, in addition to the rainfall directly on the pavement. The reservoirs' purpose is to prevent excess runoff from entering a nearby floodplain lowland. Topographic mapping and hydrologic calculations had shown that, if a dam were placed across the floodplain to use it for stormwater detention, rising of the stored runoff during storms would inundate adjacent properties. As an alternative, modeling studies showed that the discharge from the pavement's base reservoirs duplicated the predevelopment hydrograph without the assistance of any off-pavement detention basin. This outcome used the porous asphalt pavement to meet stormwater control requirements without the cost of a dam, the destruction of riparian forests, or the liability of off-property flooding.

Orange County (Florida) has made an unusual cost tradeoff in using porous asphalt to pave previously unpaved low-traffic county roads (Donald Jacobovitz, personal communication 1998). The self-draining porous asphalt material reduces the cost of grading and drainage improvements. It is considered an economical temporary road upgrade, to be replaced some time later by more costly and complete street improvements. The first installation of this type was on Fourth Street in 1995. The subgrade was stabilized with soil cement; a porous asphalt surface course 2.5 inches thick was placed over a limerock base. Several months after installation the surface infiltration rate was 1.29 inches per hour. After installing 8 miles of porous asphalt roads in the mid-1990s, the average installed cost was $6.61 per square yard, which was one fifth the cost of other upgrade alternatives. The expected lifetime until resurfacing will be required is five to eight years; at that time the county has the option of upgrading the grading and drainage to meet higher design standards and to carry growing amounts of traffic.

POROUS ASPHALT HYDROLOGY

Newly installed porous asphalt can have porosity of 15 percent to over 20 percent, depending on the aggregate gradation and variations in components. Increasing void content is associated with a higher infiltration rate and less clogging potential (Tan et al., 2000). Table 12.7 lists measured infiltration rates in relatively freshly installed material constructed under the Franklin Institute specifications.

TABLE 12.7
Infiltration Rate of Porous Asphalt within 18 Months after Construction

Location	Infiltration Rate (inches per hour)	Reference
The Woodlands, Texas	11 to 33	Thelen et al., 1972, p. 30
Franklin Institute (lab sample)	160 to 176	Thelen et al., 1972, p. 30
Austin, Texas	152 to 5,290	Goforth et al., 1983, p. 56

DECLINE OF INFILTRATION RATE UNDER TRAFFIC LOAD

Following installation, porous asphalt's porosity and infiltration rate decline under the influence of traffic. This happens with both "old" asphalt mixtures and new polymer-modified mixtures (Huber, 2000, p. 12).

In parking lots, porosity can decline at discrete locations where vehicles brake or turn repeatedly. The friction of the wheels drags the plastic asphalt across the pores. As the porosity declines, infiltration declines with it. The decline is likely to be permanent, but its occurrence is limited to only a few square feet under each wheel (Goforth et al., 1983, p. 38).

In residential streets and commercial parking lots, field tests in France found that porous asphalt's infiltration rate declines rapidly in the first two or three years (Balades et al., 1995). The initial infiltration rate remained constant for one year and then decreased during the second year by 35 to 50 percent. After the first couple of years any further decline was much slower. Although the decline was observed in 18 sites, its cause was not explained.

The heavy traffic load of highways and busy city streets can reduce porous asphalt's porosity and infiltration rate soon after construction. Figure 12.12 shows the decline in void space in porous asphalt overlays on Britain's heavily traveled A45 trunk road. The two curves represent two different sizes of aggregate in the porous asphalt mixture. In 22 months, the void space in both mixtures declined to about two thirds of the initial values.

Figure 12.13 shows the decline in infiltration rate accompanying A45's decline in porosity. After the initial decline, infiltration in the smaller aggregate layer continued to decline gradually; that in the larger aggregate with larger, more open voids declined relatively little. After 22 months infiltration declined to about 0.2 to 0.5 of the initial values. On heavily trafficked high-speed roadways such as interstate highways,

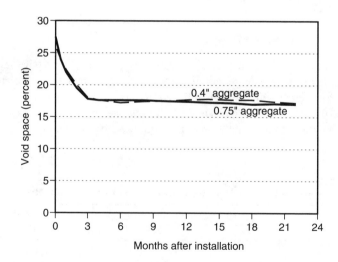

FIGURE 12.12 Decline in void space in porous asphalt overlays on Britain's A45 trunk road (after Croney and Croney, 1998, p. 497).

Porous Asphalt

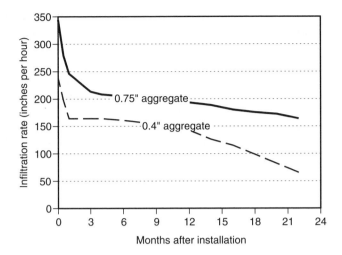

FIGURE 12.13 Decline in infiltration in porous asphalt overlays on Britain's A45 trunk road (after Croney and Croney, 1998, p. 497).

maintenance for infiltration rate has not been required because the suction produced by numerous, fast-moving tires tends to pull sediment out of the pores (Kuennen, 2003).

Similar declines of porosity and infiltration rate were seen on a highway in Denmark (Bendtsen, 1997, cited in Huber, 2000, pp. 6–7). In two years void space decreased from 21 or 22 percent of the pavement volume to 18 percent. Surface infiltration rate declined by 70 to 80 percent. Below the surface, the voids remained relatively open and interconnected.

In France, on heavily trafficked highways, in the first year infiltration rate declined 30 to 50 percent in rural areas, 40 to 70 percent in cities, and 60 to 90 percent in "very polluted areas." After this initial decline, the pavement stabilized and infiltration declined much less rapidly (Balades et al., 1995).

DECLINE OF INFILTRATION RATE UNDER WINTER SANDING

St. John and Horner (1997) observed the role of winter sanding in clogging porous asphalt. Their monitoring site was a road shoulder in Redmond, Washington. The porous asphalt surface was based on an Arizona mixture (Hossain and Scofield, 1991); its aggregate was crushed gravel 1/2 inch to 3/4 inch in size. The 3.5-inch-thick surface was applied directly on the sandy subgrade without a base course.

Prior to the sanding experiment, the porous asphalt shoulder had been in place for two years, receiving routine sand loads equivalent to 0.09 inch of sand per year (derived from St. John, 1997, p. 11). The sand was up to 3/8 inch in size and contained abundant fine particles. Then the researchers applied an additional amount of the same type of sand equivalent to four years' routine application (0.36 inch deep) and worked it into the pores with a broom.

The sand caused the infiltration rate to decline from 1750 inches per hour to 1.44 inches per hour (St. John and Horner, 1997, pp. 14–16). The sand particles were embedded only in the upper 1/2 inch of the asphalt; there was no indication that the subgrade soil surface was also becoming clogged (St. John, 1997, p. 20).

The declining infiltration rate was not accompanied by a corresponding increase in runoff coefficient (of 0.20 to 0.25) because during runoff measurements the experimental inflow rate of water to the porous asphalt was 0.23 inch per hour, which was below the pavement's infiltration rate even after the sand application. However, projecting the rate of infiltration decline into the future, it would decline to a rate less than the inflow rate in five to seven years, at which point the researcher considered that the pavement would begin to be functionally clogged (St. John, 1997, p. 18).

Sweeping of the surface after sanding maintained the infiltration rate to some degree. Vacuuming was not attempted.

WATER-QUALITY EFFECTS

The surface of porous asphalt captures and stores particles so they can be recovered without washing off downstream. Colandini et al. (1995) observed the composition of trapped particles by collecting the material vacuumed from the surface pores of porous asphalt freeways and parking lots during cleaning. The trapped surface material was made mainly of sand and contained lead, copper, zinc and cadmium. Metallic pollutants were concentrated in the fine particles. Figure 12.14 shows that the concentrations of pollutants increased with traffic intensity. The researchers recommended that porous asphalt in polluted settings such as heavily trafficked urban streets be cleaned regularly to evacuate metal-contaminated particles and prevent runoff pollution. Upon disposal, some clogging particles may have to be treated as potentially toxic waste because they contain heavy metals which could be leached out under environmental conditions different from those in the pavement where they were trapped.

Below a porous asphalt surface, the pores support an ecosystem of microbes. Aerobic bacteria can digest organic contaminants such as the hydrocarbons from

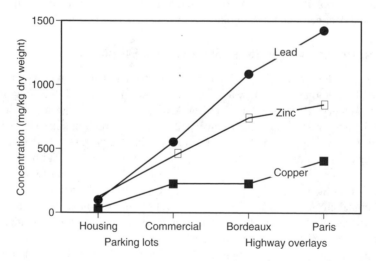

FIGURE 12.14 Concentration of metals in sediment trapped in the surface pores of porous asphalt (traffic intensity increases from left to right; data from Colandini et al., 1995).

automobile oil. The completeness of purification depends on the supply of air and water, the temperature, the nature of the pollutant, and the pollutants' residence time in the pavement. The existence of these microbes was first noticed by the Franklin Institute. Their biochemical activity was confirmed in a temporary porous asphalt parking lot in The Woodlands, Texas. In The Woodlands' percolate, total organic carbon and chemical oxygen demand were low and nitrogen forms had been aerobically transformed, all apparently because of bacterial activity within the pavement (Diniz and Espey, 1979, pp. 55; Thelen and Howe, 1978, pp. 5, 11–12, 17).

POROUS ASPHALT MAINTENANCE AND REHABILITATION

The maintenance of porous asphalt is different from that of dense asphalt. Seal coats must not be applied as they would be on dense pavements. Topcoating of any kind should be practiced only if a fully porous coating material is available. The pavement must not be sanded in the winter because sand clogs the pores.

A porous asphalt pavement that becomes partially clogged by sediment can be rehabilitated by vacuuming and washing, the equipment for which is widely available. In an early experiment, researchers artificially clogged a newly constructed pavement, reducing its infiltration rate from 21 inches per hour to 6 inches per hour; subsequent brushing, vacuuming, and high-pressure washing restored the pavement to full function (Thelen and Howe, 1978). The required frequency of vacuuming depends on the site-specific rate of clogging, for example by wind-borne dust particles, washed-on sediment, or traffic's dust and oil. If dirt is limited to a specific pavement area, such as that tracked on by dirty tires, a vacuuming effort can be economically focused on the limited part of the surface where dirt is present.

Figure 12.15 shows the effects of different cleaning methods found in an experiment in France (Balades et al., 1995). High-pressure washing with simultaneous vacuuming was consistently the most effective at restoring infiltration after

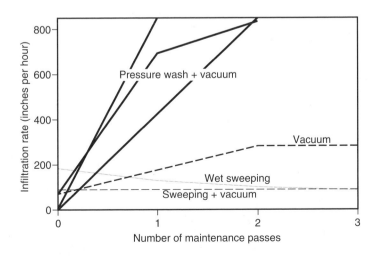

FIGURE 12.15 Change in infiltration rate resulting from alternative porous asphalt maintenance procedures (data from Balades et al., 1995).

clogging. Vacuuming alone and sweeping followed by suction had relatively little effect on tightly clogged surfaces. Wet sweeping (moistening followed by sweeping) actually reduced infiltration by lodging sediment particles more tightly in the surface pores.

All types of cleaning are most effective when they are done before clogging is complete. If cleaning is delayed and dirt is allowed to be ground into the surface by rain and traffic, it is harder to vacuum out later. Theoretically, a surface that is completely clogged can be restored by milling off the top one or two inches, where clogging is usually concentrated, and replacing it with an equal thickness of porous asphalt or another porous paving material.

When a porous asphalt surface layer reaches the end of its lifetime and requires rehabilitation, it is possible to mill it, heat it, and recycle it in place. Recycled porous asphalt can have the same infiltration rate as new porous asphalt (Huber, 2000, p. 25).

POROUS ASPHALT PAVEMENTS IN PRACTICE

Experiences in the field point to many features of porous asphalt design and performance that can inform further installations in the future. Certain installations have retained significant infiltration for many years; their distinctive features hint at factors that could be used to assure successful installations in the future, even before proven new technologies become generally available.

The mid-Atlantic region has acquired the greatest concentration of porous asphalt pavements in North America. This is the region closest to the Franklin Institute and the prototype installation at the University of Delaware. The region's design practitioners have historically been aware of the availability of porous asphalt in general, and some have tenaciously advocated appropriate application. Other porous asphalt parking lots described in this book are in State College, Pennsylvania (Chapter 2) and Warrenton, Virginia (Chapter 4). Further installations are listed by Thelen and Howe (1978) and updated on Cahill's website (www.thcahill.com). Overseas, numerous installations of porous asphalt parking lots, streets and highways cab be found in France, Germany, Britain, Japan, Singapore, and other countries.

QUIET WATERS PARK, ANNAPOLIS, MARYLAND

Quiet Waters Park in Annapolis, Maryland, was built in 1990 by the Anne Arundel Department of Recreation and Parks. A porous asphalt parking lot close to the visitors' center retained significant infiltration as of 2000, ten years after it was built.

When water was poured on the surface as shown in Figure 12.16, some of it infiltrated in place leaving a round stain; the rest made a wide, slow, shallow stream about 3 feet long, then infiltrated without running further. After a few minutes the entire contact area was drying.

The parking lot had retained its infiltration despite evident drain-down of the asphalt binder. There was almost no binder on top of the surface particles. One half inch below the surface, the pores were largely filled with black asphalt binder, perhaps mixed with dust and debris.

Porous Asphalt 485

FIGURE 12.16 Partial surface infiltration in a ten-year-old porous asphalt parking lot at Quiet Waters Park, Annapolis, Maryland, in 2000.

The parking lot's near absence of overhanging vegetation may have contributed to its durable infiltration. There is little source of organic debris to thicken the draining-down binder.

Another factor that could have contributed to durable infiltration is the aggregate's great void space and superb structural quality. The aggregate is crushed hard limestone with a markedly uniform size of 3/8 inch. The particles are sharply angular with planar, even conchoidal facets. Adjacent particles have maintained their positions by locking together along planar facets. Highly angular aggregate such as this has very high porosity of 40 percent and more, as explained in Chapter 6. This aggregate's high porosity has been maintained over the years by the hard, angular particles' resistance to settling, compaction, and fragmentation. The abundant void space has been a buffer that has compensated to a degree for the accumulation of draining binder in the pores.

Siemens Office, Great Valley, Pennsylvania

A porous asphalt parking lot at the Siemens corporation office in Great Valley (Malvern), Pennsylvania, demonstrates the same effect of good aggregate as does the one at Quiet Waters, in an even older installation (Adams, 2003; Brown, 1996b; Cahill, 1993). It does so in a large and carefully detailed installation. The parking lot was built in 1983. Figure 12.17 shows it in 2001, 18 years after construction.

Andropogon Associates and Cahill Associates laid out the 600 spaces in eight double-loaded bays forming terraces on a long slope. In each terrace both the bottom of the base and the pavement surface are level. Between the terraces are grass banks that take up the difference in elevation. The various base reservoirs are

FIGURE 12.17 Parking lot at the Siemens corporation office, Great Valley (Malvern), Pennsylvania, in 2001.

Porous Asphalt

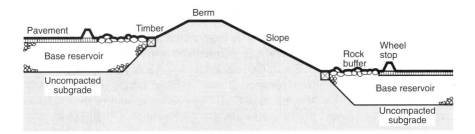

FIGURE 12.18 Arrangement of parking terraces at Siemens.

connected by overflow pipes; the lowest one discharges overflows to a final retention basin. Figure 12.18 shows a typical terrace arrangement. The berm rises about 12 inches above the upper terrace level.

As of 2001 the pavement surface was strikingly white. The white color was that of the asphalt's limestone aggregate: after 18 years the asphalt binder had drained off the surface aggregate particles. Below the surface particles were clogged-looking pockets of binder. Small, loose aggregate particles had washed up against the wheel stops in little banks, and between the wheel stops in alluvial fans.

The asphalt's infiltration rate is moderate. Some infiltration had been retained because of the hard, highly angular aggregate, as at Quiet Waters. Reportedly no maintenance had been applied during the life of the parking area, including any type of vacuuming or flushing. Beginning in about 1991, small (2 feet diameter) shallow pools of standing water were observed after heavy rains (Cahill, 1993, pp. 44–45).

The pavement surface showed very few cracks or distortions. There was no sign that the surface had been repaired or sealed over the years. This is evidence of the durability of porous asphalt in a climate with moderate freezing in the winter, and the structural stability of a thick aggregate base reservoir.

Around the edge of the pavement is a 24-inch-wide buffer of 3- to 4-inch river rock, edged by 4 inch×4 inch timbers. This "backup" system keeps the installation working hydrologically. Even if the pavement surface is partially clogged, the base reservoir still receives and stores water, and the subgrade still infiltrates rainwater. There is no sign of overflow beyond the rock buffer. If any base reservoir were to overflow, the raised grassed berm wold holds on the terrace until it infiltrates or discharges to the next terrace downslope.

The ten-year frost depth in this location is about 24 inches; in most years the penetration is less than that. The pavement's total thickness is about 20 inches, so in all years frost penetrates part of the pavement thickness, and in some years it penetrates to the base of the reservoir and into the subgrade. Cahill has specified bases of this thickness in many porous asphalt projects in the mid-Atlantic region, to hold the water from large storms while it infiltrates the subgrade. The great thickness keeps ponded water below the frost penetration depth in some winters, limits the intensity of the reservoir's freezing temperature in most winters, and supplies weight that counteracts the pressure of growing ice at all times.

WALDEN POND STATE RESERVATION, CONCORD, MASSACHUSETTS

The main parking lot at Walden Pond State Reservation in Concord, Massachusetts, is one of the northernmost of porous asphalt installations in North America, and one of the oldest extant. The U.S. Environmental Protection Agency funded the installation as a "technology transfer" project; it was installed during 1977 and 1978. Northeastern University's Department of Civil Engineering monitored the results for several years after construction. The considerable documentation illuminates a number of useful points for future porous pavement design (Wei, 1986; Eaton and Marzbanian, 1980; Keating, 2001; www.millermicro.com/porpave.html). The site's nominal frost depth for design is 48 inches (Wei, 1986, pp. 2–6); the great depth of winter freezing was illustrated in Chapter 3. Walden's experience demonstrates the ability of porous paving material to endure deep freezing over many winter seasons. The observed hydrology of this installation was described in Chapter 4. The historic park receives 600,000 visitors per year. Figure 12.19 shows the installation in 1999, 21 years after construction.

The subgrade soil is deep, very well-drained gravelly sand in the Hinckley series (Koteff, 1964; Bruce Thompson, personal communication 1999). It is ideal as a pavement structural foundation with Unified classification of GW and little susceptibility to frost damage. It is ideal also as a medium for infiltrating water, having a rapid infiltration rate of 40 inches per hour (Wei, 1986, pp. 2–3).

The 3.5-acre pavement was constructed on a preexisting gravel parking lot in a curved, broken layout among preexisting woodland trees. The surface is essentially level. There are no curbs or other structural edgings. At curves and intersections

FIGURE 12.19 Porous asphalt parking lot at Walden Pond State Reservation, Massachusetts, in 1999.

bollards inhibit veering off the pavement edge. At perimeter parking spaces there is a wooden bumper fence, which was illustrated in Chapter 2.

A number of asphalt surface mixtures were installed for comparative testing. Two of the mixtures ("K" and "J3") were installed within the Franklin Institute specifications. These were the most open-graded mixtures, with porosity of 16 to 23 percent (Wei, 1986, pp. 6–24). Other, more intermediate-graded "porous" mixtures were also attempted.

The installation involved a problem with the paving contractor that confirms that caution must be taken when installing porous asphalt (or many other kinds of porous pavement). Although the paving contractor was a large, well-known company, the contractor had considerable difficulty meeting, or paying attention to, the porous asphalt specifications (Wei, 1986, pp. 3-5–3-8). Almost all samples taken from the several installed surface courses had aggregate more dense-graded than specified. When this was discovered, an inspector was placed at the asphalt plant to turn away batches that would not meet specifications. But even truckloads that had been rejected at the plant continued to show up at the site for installation. Before construction the contractor had proposed an alternative mixture that would have been valuable for the research of the time because, on paper, it was coarser and more open-graded than previously used gradations, and it would have been easier to supply from the contractor's stated viewpoint. The mixture was accepted as one of several to be installed in sample areas, but the contractor failed to comply even with his own specification. After further testing of installed pavement samples and negotiation with the contractor, some areas were reconstructed in 1978. Even the reconstructed areas were not completely within specifications.

Figure 12.20 shows the locations of the different surface types in the parking lot. They are arrayed in various areas for comparative research and are further fragmented by the partial 1978 reconstruction. The mixtures within the Franklin Institute specifications are labeled "Porous." The thickness of the asphalt surface courses varied from 1.5 inch to 4 inches.

Various base layer sequences and thicknesses were installed to test their performance. Figure 12.21 shows the different gradations of aggregate used in the base and subbase courses. "Gravel borrow" was unsorted natural material that had surfaced the preexisting gravel parking lot, and that remained for the new asphalt pavements as subbase. "Processed gravel" was sorted natural material. Both the processed and borrowed gravels had moderate infiltration rates of 12 inches per hour (Wei, pp. 2-4 and 2-11). "Type p" and "Type q" were open-graded crushed stone with very high infiltration rates over 1000 inches per hour.

Figure 12.22 shows in section view how those materials were assembled into various combinations of base and subbase courses. All sections had a considerable thickness of gravel borrow, mostly remaining from the preexisting gravel parking lot, that could be considered subbase for the new asphalt pavements. The "gravel" base had a relatively thin layer of processed gravel above the subbase and immediately under the surface course. The "thin" type had a thin base of 2 inch stone and a choker course of 3/8-inch stone supporting the asphalt surface course. The "thick" type had a more substantial base of 2-inch stone. The "thick-layered" base was as deep as the "thick" base, and was divided into alternating layers of small and large stone, analogous to

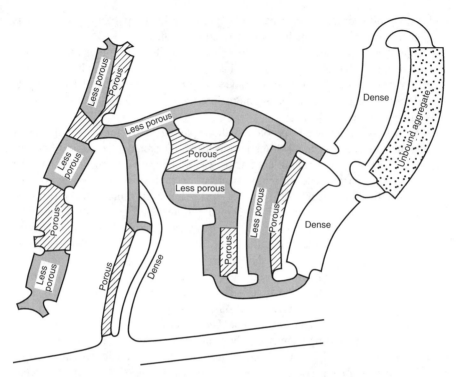

FIGURE 12.20 Pavement surface types at Walden Pond (after Wei, 1986, pp. 2-13a and 3-10a).

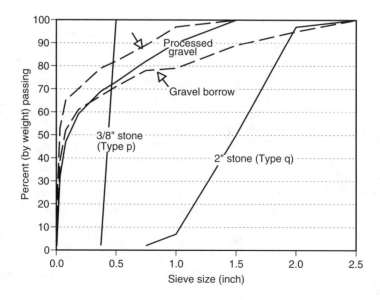

FIGURE 12.21 Aggregate gradations used in Walden Pond's base and subbase courses (data from Wei, 1986, pp. 2-4a and 2-11).

Porous Asphalt 491

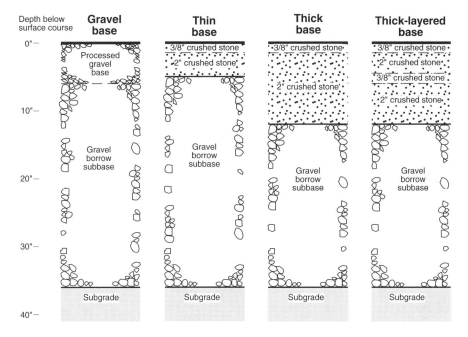

FIGURE 12.22 Base types at Walden Pond (data from Wei, 1986, pp. 2-10–2-11a, 3-2, 9-12a, and 9-34).

those at the University of Delaware described earlier in this chapter. All types were overlain by various mixtures and thicknesses of surface courses. With the surface courses, bases and subbases, most pavements had total thicknesses of 36 inches. Filter fabric was not used. Figure 12.23 shows where the base types were located in the parking lot.

The Army Corps of Engineers (Eaton and Marzbanian, 1980; Wei, 1986, pp. 9-37–9-39) tested the strength of the various pavement types by measuring the deflection when a heavily loaded plate was pressed into the surface. They found (pp. 12 and 20) that porous asphalt was not as strong as dense asphalt with a comparable base course, and did not distribute traffic load as widely. Nevertheless, a surface layer of porous asphalt 2.5 inches thick was suitable for parking lots and other light and moderate traffic loads; greater thickness produced greater strength. Layering of an open-graded base course, like that done in Walden's "thick-layered" base and previously at the University of Delaware, did not add to strength.

In 1999, the entire parking lot was in superb structural condition despite 22 years of traffic and deep winter freezing. There were no noticeable cracks, surface irregularities, or raveled areas.

This experience confirms that well-drained porous asphalt pavement material is not damaged by significant frost. Ordinary pavement construction practice in the Concord area provides total non-frost-susceptible pavement thickness of 12 to 14 inches; evidently, that thickness produces sufficient dead weight to counteract the heaving force of frozen soil. By that measure, Walden's base and subbase layers

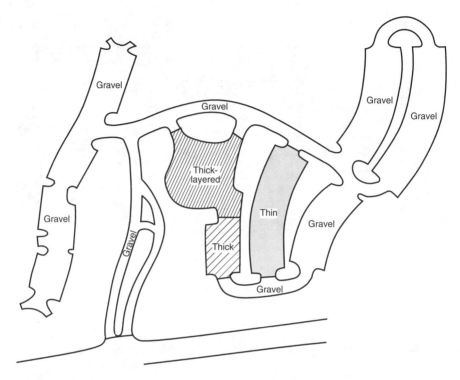

FIGURE 12.23 Locations of base types at Walden Pond (after Wei, 1986, pp. 2-13a, 2-18a, 3-11d, and 7-5a).

provided more than sufficient thickness of non-frost-susceptible material. The Walden pavement has been further protected by the subgrade's high permeability, thorough drainage, and low frost susceptibility.

The surface infiltration rate declined within a few years. Informal demonstrations in 1999, 22 years after installation, suggested that infiltration had further declined to very low levels, certainly less than 1 inch per hour. Asphalt binder was nearly absent from the upper surfaces of the aggregate particles. In the pores just below the surface, a continuous black layer was visible, visibly clogging the voids with drained-down asphalt binder perhaps mixed with organic matter from overhanging trees.

In addition, in 1999, sand was visible on the porous asphalt's surface and in its pores; in places it was abundant enough to be picked up with the fingers. Sand can contribute to pavement clogging, especially where trapped in a matrix of sticky asphalt. The sand had several apparent sources. Pedestrians returning to their cars from the park's beaches and trails visibly track the local Hinckley sand onto the pavement. Vehicles that stray off the edge of the pavement at curves and intersections track further sand. In the winter, park staff apply sand to nearby walkways for traction; pedestrians then track that sand onto the parking lot. There is nothing to prevent enthusiastic staff from spreading the same sand onto portions of the parking lot. A sign identifying the pavement, which was shown in Chapter 2, does not remind

Porous Asphalt

park staff of any special maintenance concerns such as keeping sand off. According to park staff, the parking lot is not vacuumed. This experience argues for signage that advocates appropriate maintenance for park staff, in addition to identifying the pavement for the public.

MORRIS ARBORETUM, PHILADELPHIA, PENNSYLVANIA

A 1987 parking lot at the Morris Arboretum in Philadelphia illustrates the beauty of which porous asphalt pavements are capable with careful detailing. The installation has been widely documented and is still extant as of this writing (Adams, 2003; Hughes, 1990; Sorvig, 1993; Steiner and Johnson, 1990; Strom and Nathan, 1998, pp. 275–278). It is located at the arboretum's Public Garden and Visitor Center on Northwestern Avenue. Andropogon Associates laid it out elegantly with granite curbs and bands consistent in character with the arboretum's nearby historic architecture. Figure 12.24 shows the parking lot in 2000. The parking bays are in two topographic terraces, separated by a vegetated slope. Surrounding trees and plants have grown up attractively under the arboretum's superb landscape maintenance. Grass slopes drain across the granite curbs onto the pavement, but the thickness of the grass prevents erosion that would discharge sediment onto the pavement.

Figure 12.25 shows the construction in one of the parking terraces. The central traveling lane is of dense asphalt; the parking stalls on each side are of porous asphalt. The top size of the aggregate in the porous asphalt is 3/8 inch to 1/2 inch. The base reservoir extends across the entire terrace, under porous and dense surfaces alike, turning the entire parking lot into an infiltration bed. The base is 14 inches

FIGURE 12.24 Porous asphalt parking lot at the Morris Arboretum in Philadelphia, Pennsylvania.

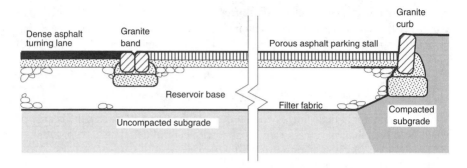

FIGURE 12.25 Construction of the Morris Arboretum visitors' parking lot (adapted from Strom and Nathan, 1998, p. 278).

thick to hold the precipitation from the 100-year storm (Colin Franklin, personal communication 1996). The subgrade is made up of sandy and gravelly Coastal Plain sediments.

A surface inlet for any overflow drainage is in the low corner of each parking bay. The inlets drain to a perforated pipe buried off-pavement, where water has a second chance to infiltrate the site's soil before discharging onto the vegetated ground surface.

Maintenance such as vacuuming has not been done. During visits in 1996, 2000, and 2001, respectively 9 to 14 years after installation, the porous asphalt showed no sign of cracks, deformation, raveling, or other structural concerns. However, the infiltration rate was low; a continuous black clogging layer was visible in the pores just below the surface particles.

Aggregate particles became loose on the surface, freed by gradual asphalt drain-down. Below the surface a mass of black binder gradually developed, jammed with small aggregate particles. The drain-down and the decline in infiltration were greatest in the most-trafficked areas. The asphalt's clogging may have been abetted by the aggregate, which came from the Brunswick Shale. Many particles are platy in shape, and may be less resistant to compaction and crushing than other aggregate materials.

Residential Street, Luleå, Sweden

Stenmark (1995) reported a porous asphalt installation in the extremely cold subarctic climate of Luleå, Sweden. The pavement's construction and structural performance were introduced in Chapter 3. Luleå's greatest occurrences of liquid stormwater tend to be at times of snowmelt, or precipitation on melting snow.

The pavement's installation was the reconstruction of a preexisting residential street that had suffered from excess runoff and standing water. The preexisting street was of dense asphalt, drained by side ditches with culverts at driveway intersections. During snowmelt periods, surface water had backed up on driveways and streets due to inadequately sloping impervious surfaces and ice blockages in culverts. The dense asphalt was in bad condition due to frost heave and settling, especially near manholes and shallow buried pipes that created local differences in heat properties and soil hydrology.

The porous pavement was built in 1993–1994 (Backstrom, 1999). It was intended to reduce runoff on and around the street by increasing permeable cover, hydraulic storage volume, and the draining away of excess water. The new street area, including vegetated roadside swales, is 1.21 acres. The drainage area is 8.15 acres; it includes two other streets and adjacent houses.

The porous asphalt surface is narrower than the previous street surface, to increase the site's vegetated area. It slopes toward roadside swales more decisively than did the previous surface. In the swales, culverts under driveways were replaced with prefabricated trenches covered with metal grids to simplify access for breaking up ice formations.

Water that infiltrates through the porous surface into the base reservoir either percolates into the subgrade, or discharges through a perforated pipe to a ditch off the road. According to infiltration tests performed during excavation, the soil's infiltration rate varies from 0.1 to 5.4 inches per hour.

The asphalt's infiltration during meltwater conditions was studied in the laboratory using samples cut from the installed pavement (Backstrom, 1999; Backstrom and Bergstrom, 2000). Infiltration of 685 inches per hour at room temperature declined to 307 inches per hour at just below freezing (29 to 30°F) due to ice formation in the pores. When the temperature fluctuated between freezing and thawing over a period of days as it could during snowmelt conditions, during which the asphalt remained generally cold and wet, infiltration declined farther to 10 percent of its value at room temperature (about 68 inches per hour). Thawing of porous asphalt was rapid because of meltwater infiltration early in each snowmelt event. The researchers concluded that freshly installed porous asphalt has sufficient permeability during snowmelt, as long as the street is not completely covered with ice. Unfortunately, the change in the surface's permeability over a period of years was not studied.

In the field, during spring snowmelt, ice and snow on the porous surface melted a couple of days before pipe discharge began, indicating some combination of storage and infiltration of the first meltwater. For the overall winter–spring snow and snowmelt seasons, only 30 to 40 percent of the total precipitation discharged as meltwater runoff; the remainder was disposed of by infiltration and evaporation (Stenmark 1995).

Infiltration kept the porous pavement's subgrade continuously supplied with moisture. In late autumn and early winter, the water table (which is 4.3 to 9.5 feet deep in the area) fell relatively little under the porous asphalt compared with that under nearby dense asphalt. During spring snowmelt, it rose one month earlier (Backstrom, 2000). As described in Chapter 3, the pavement's high moisture content insulated it from freezing and frost damage.

Rt. 87, Chandler, Arizona

Porous asphalt was reportedly successful in its first five years in the desert environment of Chandler, Arizona (Hossain and Scofield, 1991; Hossain et al., 1992). The infiltration rate declined during those years, but not below the level needed to absorb the area's precipitation. The installation is a 3500-foot-long section of Rt. 87 between

Elliot and Warner Roads, bearing 30,000 vehicles per day. It opened to traffic in 1986. The northbound lanes were paved in porous asphalt; the southbound lanes were paved in dense asphalt for comparison. The porous asphalt has a 6-inch surface course, a 6-inch base of open-graded aggregate treated with asphalt for stability, and an 8-inch open-graded subbase reservoir. The combined reservoir capacity of the pavement is sufficient to hold the ten-year storm. At the edge of the pavement, the subbase drains into an aggregate-filled trench filled for infiltration into the soil.

In wet weather, the pavement's painted stripes have been much more visible than those on the adjacent dense surface. Minor raveling occurred. No cracking or significant deformation occurred.

The initial infiltration rate of the porous asphalt surface was 77 inches per hour. Four years after installation the infiltration rate had declined to 38 to 40 inches per hour, with the lower values in the wheel tracks. Although this was only half of the initial value, it was well above the 13 inches per hour required for performance in a design storm. The installation was still extant as of 2003 (Jackson, 2003, p. 4).

POROUS ASPHALT IN OVERLAYS

Highway agencies have installed porous asphalt overlays on highway pavements to improve driving safety, traffic flow, and the highway environment. The overlays' removal of water from the driving surface essentially eliminates tire spray and hydroplaning, and reduces glare on wet nights. The water drains onto highway shoulders or through openings in the sides of grate inlets. Porous overlays also reduce pavement noise by attenuating the sound energy generated at the interface of pavement and tires. Because of their good traction they are often called "open-graded friction courses" (OGFCs).

Highway overlays have the longest history of all porous asphalt applications. Experience with them began with the "old" porous asphalt mixtures. Although the experience of various agencies with OGFCs has differed widely, today overlays are the most widespread use of porous asphalt in the world, and highway personnel are centrally involved in the development of new porous asphalt technologies. Approximately 20 U.S. state departments of transportation have a standard specification for some type of OGFC material, which could be referenced in porous asphalt applications of many kinds (Kandhal and Mallick, 1998). Contemporary developments in this field are vital to future porous asphalt applications of all types.

THE DEVELOPMENT OF OVERLAY TECHNOLOGY

The Oregon highway department began experimenting with open-graded asphalt overlays as early as the 1930s; their tests showed high skid resistance and decreased glare from headlights. By the late 1940s and 1950s, other states were using porous highway overlays to some degree (Brown, 1996b; Huber, 2000, p. 3; Thelen et al., 1972, p. 25). Although the early mixtures had void space hardly over 10 percent, they were often called "popcorn" mixtures because of their granular appearance.

Safety improvement has similarly motivated porous overlays on airfields. In 1962 Britain's Royal Air Force used a porous asphalt overlay on an airfield, after which American military engineers cautiously brought the concept to North American airfields (Forrest, 1975). One of the early applications was at the Dallas (Texas) Naval Air Station. Aircraft landing on Dallas' flat, poorly drained, dense asphalt runway had been hydroplaning, with alarmingly poor steering and braking; incoming flights had to be diverted to other landing sites. Immediately after the addition of a porous overlay, all reports of ponded water and poor braking ceased. Figure 12.26 shows porous asphalt's attractive cost advantage in early airfield installations, compared with the principal alternative treatment of grooving the preexisting dense pavement. At Dallas, the cost of the porous overlay was barely half the cost of cutting traction grooves.

The Federal Highway Administration encouraged more widespread testing and trial use of OGFCs in the 1960s and especially in the 1970s following the Franklin Institute's hopeful early results (Smith et al., 1974, cited in Goforth et al., p. 36).

But under highway traffic loads the limiting factor in porous overlay durability has been structural integrity (Huber, 2000, pp. 9, 12, 23; Kuennen, 1996). Heavy traffic wears out the material even while the hydraulic effect of fast-moving vehicles in wet weather prolongs environmental performance by flushing the voids of debris. As asphalt drains down through the material's pores, the thin films remaining on surface particles oxidize and become brittle. The binder's brittleness allows traffic to dislodge aggregate particles, and the surface ravels. Porous asphalt overlays were failing only six to nine years after installation. If the binder had oxidized sufficiently, raveling would progress very rapidly; an entire overlay could deteriorate in a matter of months. Many highway agencies are not organized to respond to such an abrupt call for repair. The combination of short life and catastrophic failure mechanism discouraged many highway agencies from continuing tests of the material.

But the next stage of porous asphalt development was in Europe. Porous highway overlays came to Britain in 1967, and to continental Europe in the 1980s following

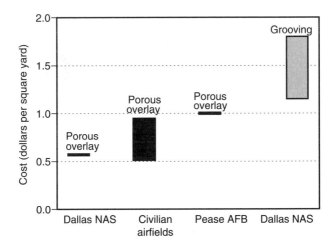

FIGURE 12.26 Costs of alternative pavement treatments for airfields in the early 1970s (NAS is Naval Air Station; AFB is Air Force Base; data from Forrest, 1975 and Jones, 1973).

the experience in airfield applications (Croney and Croney, 1998, pp. 494–495; Stotz and Krauth, 1994). The European highway overlays were motivated by driving safety as they were in the U.S. An additional motivation was the reduction of noise where heavily traveled streets are in the midst of densely inhabited urban areas. Research by large European contracting companies has been competitive and innovative.

The European mixtures typically contained large, highly durable aggregate, producing rigorous stone-to-stone contact that gave open-graded material its strength. They routinely produced porosity of 20 percent and consequently high infiltration and noise attenuation. The coarse surface texture produced high traction. Large quantities of asphalt containing new additives produced thick, durable binder films.

The European developments were imported back to North America through initiatives such as those of the Transportation Research Board described earlier in this chapter. Particularly good American results have come from mixtures using polymer-modified asphalt binder and relatively coarse open-graded aggregates. Good drainage of the porous layer (by open voids and a drainage outlet) makes the layer structurally durable by reducing the penetration of moisture between the aggregate particles and their films of asphalt binder. High binder content makes thick films that protect inner asphalt from oxidation; the unoxidized internal binder retains its adhesion to the aggregate particles. The new, stiff mixtures can be produced and placed at notably high temperatures without immediate drain-down, so moisture is driven off and the binder clings firmly to the aggregate particles. Solid bonding of asphalt onto the particles reduces drain-down.

Effects of Porous Asphalt Overlays on Safety

By their skid resistance properties, porous overlays enhance highway safety and traffic flow. Figure 12.27 shows the degree to which porous asphalt maintains tire friction in dry and wet weather conditions. In dry conditions all the materials in the

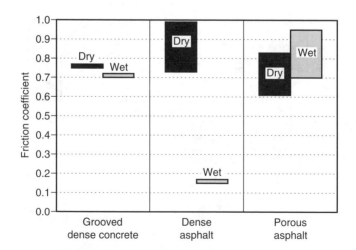

FIGURE 12.27 Friction coefficients of recently installed pavement surfaces in dry and wet conditions (data from Diniz, 1980, p. 27).

figure have essentially similar friction. In wet conditions porous asphalt keeps its friction value as does grooved concrete, while dense asphalt loses its friction, so vehicles slip and hydroplane.

Porous asphalt's high friction in wet weather results from a combination of factors (Thelen et al., 1972, p. 50; Thelen and Howe, 1978, pp. 13–14). The surface texture is coarse. The porous drainage eliminates surface sheets of water that would separate tires from the pavement. Where a tire does encounter surface water, tire pressure pushes the water away through the voids, keeping the tire in contact with the pavement.

The effect of porous asphalt on the formation and removal of surface ice has been a controversial question. Informal observations suggest that in marginally freezing temperatures less ice forms on porous asphalt than on dense asphalt because freezing water and melting snow drain directly off the surface (Bachtle, 1974; *Highway & Heavy Construction,* 1988). On the other hand, it has been speculated that porous material's low thermal conductivity may allow formation of ice sooner on porous asphalt than on dense material (Huber, 2000, p. 9). It has also been speculated that when deicing salt is applied to a porous surface it may have less deicing effect than on a dense surface (Huber, 2000, p. 24). Salt has to dissolve into melting ice in order to have effect. On a porous surface, when some ice has dissolved the salt and melted, the brine drains away through the surface pores; the remaining patches of ice may require a further application of salt. In contrast, on a dense pavement the brine can stay on the surface and melt more ice.

Porous overlays contribute further to safety by improving pavement visibility. On wet nights, porous asphalt produces little headlight glare compared with that from dense asphalt, because water does not pool on the surface. Painted markings such as those at road edges and centerlines are more visible because they are not covered by a glaring film of water.

Also important is backspray kicked up from wet pavements by moving vehicles. Traffic mist restricts visibility more than natural fog, because its large droplets make the mist denser. Directly behind a large truck a driver might see no farther than 15 to 30 feet. Freshly installed porous asphalt reduces the amount of spray to less than 10 percent of that from a dense surface. However, the spray-reducing capability declines over time alongside infiltration: in a British study, porous overlays lost their spray reduction capabilities in five to eight years, the same time frame as porous overlays' other structural and functional lifetimes (Huber, 2000, pp. 6, 23).

EFFECTS OF POROUS OVERLAYS ON NOISE

At moderate to high speed, a rolling tire generates noise as the tread rolls into contact with the pavement surface, trapping compressed air, and then leaves it, sucking the air away from the pavement. Noise reduction has been of great interest in Europe, where heavily traveled streets and highways intrude into densely populated urban areas. The technical terminology of sound was introduced in Chapter 1.

Recently installed porous asphalt reduces noise compared with dense asphalt by 3 dBA or more (Bendtsen, 1997, cited in Huber 2000, pp. 6–7; Bendtsen and Larsen, 1999; Kuennen 1996), which is comparable to the noise reduction that would result from halving the amount of traffic on the road, or doubling the distance from the noise source. A porous surface absorbs sound energy, in contrast with the reflection from a dense surface. A porous surface also allows some of the air around tires to be pressed into the voids, dissipating the pressure before the noise is generated (Thelen et al., 1972, p. 50).

Porous asphalt particularly reduces the higher frequencies of sound, leaving residual noise that is both lower in pitch and lower in volume than that from dense asphalt. A lower pitch could multiply the impression of less noise. Absorption of low-pitched noise increases with porous layer thickness (Bendtsen and Larsen 1999; Kuennen 1996).

Figure 12.28 shows that porous asphalt's noise reduction effect has been more durable on high-speed highways than on urban streets. At high highway speeds, tires on wet pavement create hydraulic forces that flush the voids and reduce clogging; the open voids reduce noise. In contrast, on urban streets, the lower-speed traffic produces less noise, but the noise-absorbing effect of the porous surface declines as the voids become clogged.

Effects of Porous Overlays on Hydrology

The University of Stuttgart provided a thorough study of the hydrology of a porous asphalt overlay in Germany (Stotz and Krauth, 1994). The installation's water-quality effects were described in Chapter 4. It was installed in late 1988 on federal highway A6 in Weinsberg. A gutter along the shoulder collected the overlay's discharge water; the gutter was sealed to prevent contamination by other sources of water. The highway carried 35,000 vehicles per day, of which 25 percent were trucks. The climate in this region has gentle rainfall and mild, wet winters. Monitoring began after the pavement had been in operation for two-and-a-half years, and continued for 12 months. The overlay was 1.5 inches thick and had a 1.5 percent slope. It was made with 5/8 inch aggregate and had 19.1 percent porosity.

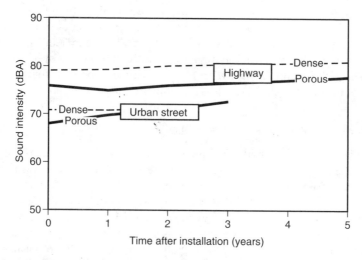

FIGURE 12.28 Traffic noise on asphalt pavements in Denmark (average of data from Bendtsen, 1997, cited in Huber, 2000, p. 7).

Porous Asphalt

Figure 12.29 shows the surface infiltration rates in different segments of the Weinsberg overlay's surface. All the values were significantly lower than those typically given for newly built porous asphalt. Infiltration in the right (slow) lane was particularly slow; this lane had a distinctively high load of heavy vehicles and frequent stop-and-go driving.

Table 12.8 lists the precipitation conditions that produced discharge. The greater readiness to produce lateral discharge in the winter apparently resulted from the cold, damp winter weather, which kept the porous asphalt moist for long periods. During small rainfall events, the discharge from the porous overlay began only after a delay, and then lasted for a long time; peak flows were low. During intense or prolonged events the peak discharge followed relatively closely after the peak rainfall. In all events the total discharge volumes were significantly smaller than the runoff volumes from dense asphalt.

Table 12.9 lists the total seasonal and annual outcomes. Almost all rainwater evaporated. Only 4.4 percent of the year's precipitation discharged laterally. The particularly high evaporation in the summer was due to prolonged dryness of the asphalt in combination with high temperature.

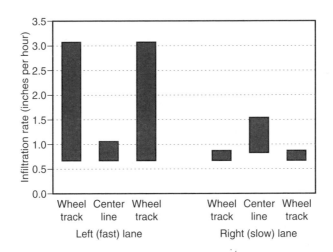

FIGURE 12.29 Range of surface infiltration rates in a porous asphalt overlay in Weinsberg, Germany 2.5 to 3.5 years after construction (data from Stotz and Krauth, 1994, Table 3).

TABLE 12.8
Conditions to Initiate Discharge from the Porous Asphalt Overlay in Weinsberg, Germany

Minimum Precipitation to Produce Discharge	Summer	Winter
Depth (inches)	0.06	0.02
Intensity (inches per hour)	0.02	0.01

Data from Stotz and Krauth, 1994.

TABLE 12.9
Seasonal and Annual Hydrology of the Porous Overlay in Weinsberg, Germany during a Year of Monitoring

	Summer	Winter	Year
Total precipitation (inches)	12.2	9.3	21.5
Total lateral discharge (inches)	0.45	0.49	0.94
Total lateral discharge (% of precipitation)	3.7	5.3	4.4
Evaporation (% of precipitation)	96.3	94.7	95.6

Data from Stotz and Krauth, 1994.

Pagotto et al. (2000) studied overlay hydrology from a slightly different angle. They compared the discharge from an overlay in Nantes, France with that from same segment of highway in the previous year, when the road had been surfaced with dense asphalt. The overlay was 1 inch thick and carried traffic of 12,000 vehicles per day on each side. In the six months from June to November similar depths of rain had fallen during the two monitoring periods. With the porous overlay response times (delay of runoff after beginning of rainfall) were two times longer and the discharge lasted 15 percent longer, due to storage in the overlay's pores. Peak discharge rates were 11 percent lower. For total precipitation, the dense pavement's runoff coefficient had been 0.71. Surprisingly, the corresponding coefficient for the porous overlay, the ratio of lateral discharge to precipitation, was greater, at 0.86. The increase was attributed to the porous overlay's reduction in splashing, and consequent increase in surface infiltration and lateral discharge.

COST OF POROUS ASPHALT OVERLAYS

Porous asphalt material for highway overlays costs more to produce than does dense asphalt. Figure 12.30 contrasts the initial (construction) costs of two kinds of porous asphalt with that of dense asphalt on Arizona highways in the late 1990s. The Arizona highway agency's new "modified" binder is of rubber–asphalt. Both neat (unmodified) and modified porous asphalt materials cost more per ton than did the dense material.

Figure 12.31 contrasts the lifetimes of similar materials as they have been experienced on Georgia highways. The average life of today's modified material is greater than that of both unmodified porous asphalt and dense asphalt. The longer life equalizes the average annual life-cycle costs of dense and porous asphalt overlays.

RECENT EXPERIENCE IN OREGON

Oregon was one of the few American states that continued trials with porous asphalt overlays during the period when many American agencies had lost interest because of the failures they had experienced, and while Europe was beginning a whole new generation. Tenacious states such as Oregon were in a position to benefit quickly from the European experience, and are now sources of the latest information in porous asphalt material.

Porous Asphalt

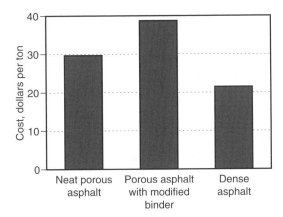

FIGURE 12.30 Initial (construction) costs of pavement surface materials on Arizona highways in the 1990s (data from Huber, 2000, pp. 27–28).

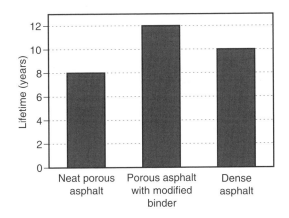

FIGURE 12.31 Life of pavement surface materials on Georgia highways in the late 1990s (data from Huber, 2000, pp. 27-28).

In the 1970s, the Oregon Department of Transportation (DOT) had been using "popcorn" mixtures in overlays up to 1.5 inches thick. The aggregate's top size was only 3/8 or 1/2 inch. As in other states, Oregon's early experience was a combination of encouraging performance of freshly installed overlays, but disappointing durability (Brown, 1996a). As early as 1979, Oregon expanded its use of porous overlays as it began experimenting with a coarser aggregate mixture and correspondingly thicker overlay placement (Kuennen, 1996).

As of 2003, Oregon referred to its standard open-graded asphalt mixture as "19 mm Open" (19 mm refers to the nominal top aggregate size of 0.75 inch. The specifications are at the DOT web site, www.odot.state.or.us. The DOT has specified this material for Interstate highway surfaces throughout the state. The polymer-modified material reduces road noise, resists deformation and oxidation, and reduces backspray.

The widespread use of the material has led some of Oregon's county highway agencies to adopt it for their own overlays (Kuennen, 1996; Schlect, 1994).

The aggregate gradation in 19 mm Open was shown in a chart earlier in this chapter. It is roughly comparable to ASTM No. 67. The large 19 mm particles interlock to produce the material's structural stability. The void space is 15 to 18 percent, with large individual voids that produce free drainage into and through the material. The material is placed in layers 2 inches thick. The thick layer drains large quantities of water below the surface, effectively reducing splash and spray and increasing friction.

Table 12.10 lists additional specifications for the mixture as of 2003. To meet the tensile strength requirement, anti-stripping additives are added (Oregon DOT, 1999, p. 7). Oregon does not specify a standard asphalt cement content for the mixture; the quantity is left up to project-specific testing to meet other specifications.

On Oregon's heavily trafficked highways, the lifetime of the 19 mm mixture seems to be limited primarily by wearing from snow tires, to which the 19 mm mixture is slightly more sensitive than dense asphalt mixtures (Bud Roberts, personal communication 2003).

The Oregon DOT (2002, p. 9-1) is cautious about using open-graded asphalt in complex urban settings where significant amounts of handwork and feathering would be required around curbs, utility appurtenances, and driveways. It is also cautious in areas of heavy snowplow activity, although the experience that led to that attitude has not been documented. In determining required pavement thickness, ODOT gives an open-graded asphalt layer the same structural credit as a dense-graded mixture.

RECENT EXPERIENCE IN GEORGIA

Another state department of transportation with unbroken experience in porous asphalt overlays is that of Georgia (Georgia Department of Transportation, no date). When Georgia started using open-graded overlays in 1970, drivers appreciated their drainage performance but the structural durability was unsatisfactory. After the extremely hot summer of 1980, the DOT surveyed damaged installations and found significant cases of stripping and raveling; the porous layers were visibly wearing downward from the surface (Kuennen 1996). Georgia immediately suspended new installations but continued research into the material.

The DOT began new trial installations in 1985, and in 1986 it reinstituted standard use of OGFCs with new requirements under mixture designation "Type D."

TABLE 12.10
Requirements for Oregon's 19 mm Open OGFC Mixture

Item	Requirement
Air voids, %	13.5–16.0
Drain-down, %	70–80
Tensile strength ratio, minimum	80
Coating, %, minimum	—
Voids filled with asphalt (VFA), %	40–50

Oregon DOT 1999, p. 10.

Porous Asphalt

It used this mixture on many heavily trafficked highways to improve traction and visibility. In 1991, Georgia made further tests involving a mixture which it called Porous European Mix, modeled on the European developments. The performance of the mixture was very favorable, but the large size of the mixture's aggregate required a correspondingly thick pavement layer. The cost of such a great amount of material discouraged direct use of the European model.

Consequently Georgia experimented with a compromise design, using a gradation coarser than the old "Type D" but not as coarse as the European one. Figure 12.32 compares the evolving aggregate gradations. The new mixture's aggregate is roughly equivalent to ASTM No. 7. Georgia calls its new mixture "12.5-mm OGFC"; this is the mixture that Georgia is using as of this writing (Huber, 2000, pp. 11–12; Georgia DOT, no date). The complete specifications are not yet part of Georgia's standard manuals; they are available on request from the DOT.

In the new 12.5 mm mixture the asphalt cement is "harder," at viscosity grade 30, to reduce stripping and drain-down. Hydrated lime is added to deter stripping; alternative deterrents are used where required for compatibility with local aggregate. Polymer modifiers and mineral fibers further reduce the cement's migration. The binder is added in greater amounts for thick coating on the aggregate particles.

Georgia DOT measured the alternative mixtures' infiltration rates by applying water at the surface of recently installed overlays (Georgia DOT, no date; Jared et al., 1998). The resulting infiltration values reflected a composite of the actual hydraulic conductivity of the material, and the thickness of the layer through which the water drained away laterally. By this measure, the infiltration rate of the old Type D in-place mixture was 67 inches per hour, and that of the modified 12.5 mm mixture (in a thicker layer) was 120 inches per hour.

The new material's rapid drainage reduces hydroplaning and improves night-time visibility. Drivers have commented on the safe condition of the newly overlaid highways during rain, mentioning the absence of standing water and the great reduction

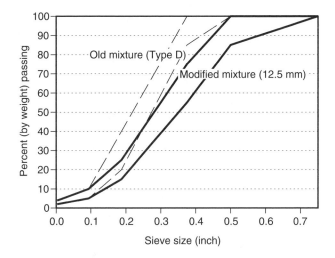

FIGURE 12.32 Aggregate gradations in two open-graded asphalt mixtures considered by the Georgia Department of Transportation for its highway overlays (data from Georgia DOT, no date).

in back-spray (Kuennen, 1996). The improvements in driving conditions and traffic flow in effect raise the service level of the highways during wet weather without the expense of widening the road.

Figure 12.33 compares the costs of the old and new materials for a typical section of highway overlay in the late 1990s. The old Type D mixture was placed at 60 pounds per square yard; the modified 12.5 mm material is placed in a thicker layer at 75 pounds per square yard. The initial cost for the modified material is 34 percent higher than that of the old mixture as a result of the extra components in the mixture, the thickness of the overlay, and certain production factors.

However the new material's interval between rehabilitations is greater, so its annualized cost is lower. The old Type D mixture's service life was eight years; that of the modified 12.5 mm mixture is 10 to 12 years. The new material's less frequent rehabilitation entails correspondingly less disruption of traffic. In terms of average annual cost, the modified material becomes an economically attractive alternative if it lasts only 11 months longer than the old mixture. Figure 12.34 shows the annualized cost

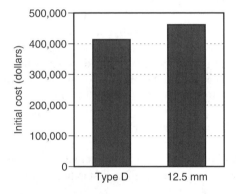

FIGURE 12.33 Initial costs of alternative porous asphalt overlays on a typical section of Georgia highway in the late 1990s (data from Georgia DOT, no date).

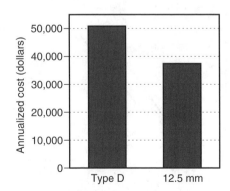

FIGURE 12.34 Annualized costs of alternative porous asphalt overlays on a typical section of Georgia highway in the late 1990s (data from Georgia DOT, no date).

based on a 12-year life for the new material; at this rate the return on investment in the modified material comes quickly.

ACKNOWLEDGMENTS

Near the beginning of this project, Shelly Cannady, then a graduate student of landscape architecture at the University of Georgia, completed a meticulously thorough and documented telephone survey of porous asphalt experiences in North America. She is now a landscape architect practicing in Georgia.

The following persons generously provided information used in this chapter: Robert E. Boyer of the Asphalt Institute; Nancy Brickman of the Oregon State University Transportation Research Institute; Leland H. Bull of GWSM (Pittsburgh); Steve Burges and Richard Horner, University of Washington; A.E. Bye, landscape architect of Ridgefield, Connecticut; Thomas Cahill and Michelle C. Adams, Cahill Associates (www.thcahill.com); Jack Daft, landscape architect of New Bern, North Carolina; Karen Enyedy, Austin, Texas; Colin Franklin of Andropogon Associates (www.andropogon.com); Adele George, landscape architect of Grey, Georgia; John Gnoffo of Cerulea, Inc. (Alpharetta, Georgia); P. Rexford Gonnsen of Beall, Gonnsen and Company (Athens, Georgia); Kent Hanson of the National Asphalt Pavement Association; L. Fielding Howe, landscape architect of Merion Station, Pennsylvania; Donald Jacobovitz of the Orange County (Florida) Roads and Drainage Department; Peter Kumble of Land Ethics (Ann Arbor, Michigan); Kit Leshk, Wilton "Bud" Roberts and Kevin Thiel of the Oregon Department of Transportation; Beverly Norwood of Norwood Design (Raleigh, North Carolina); Gary Robinette of the University of Texas at Arlington; Leslie Sauer, formerly of Andropogon Associates; Mary Silva of the Arizona Transportation Research Center; Gary Smith, formerly of the University of Delaware Department of Plant Sciences; Tom Taylor, University of Delaware; Ed Thelen, formerly of the Franklin Institute; Bruce Thompson, NRCS (Amherst, Massachusetts); the staff of the Virginia Transportation Research Council; Linda Velasquez of Greenroofs.com (www.greenroofs.com); Robert "Alfie" Vick of the University of Georgia and ECOS Environmental Design (www.ecosedi.com); Donald Watson, Georgia DOT Office of Materials and Research; and Irvine Wei of Northeastern University.

The following persons constructively critiqued an early draft of this chapter: Thomas Cahill of Cahill Associates (www.thcahill.com), Wilton "Bud" Roberts of the Oregon Department of Transportation, and Ali Soltani of the Pennsylvania Transportation Institute at Pennsylvania State University.

REFERENCES

Adams, Michelle C. (2003). Porous Asphalt Pavement with Recharge Beds: 20 Years & Still Working, *Stormwater* 4, 24–32.
Anderton, Gary L. (1992). *Evaluation of Asphalt Rubber Binders in Porous Friction Courses*, CPAR-GL-92-1, Vicksburg: U.S. Army Corps of Engineers Waterways Experiment Station.

Ann, Tan Siew, Fwa Tien Fang, and Chuai Chip Tiong (1997). A New Apparatus for Measuring the Drainage Properties of Porous Asphalt Mixes, *Journal of Testing and Evaluation* 25, 370–377.

Ann, Tan Siew, Fwa Tien Fang, and Chuai Chip Tiong (1999). Automatic Field Permeameter for Drainage Properties of Porous Asphalt Mixes, *Journal of Testing and Evaluation* 27, 57–62.

Bachtle, Edward R. (1974). The Rise of Porous Paving, *Landscape Architecture* Vol. 65, October, 385–387.

Backstrom, Magnus (1999). *Porous Pavement in a Cold Climate*, Licentiate Thesis, Luleå, Sweden: Luleå University of Technology (http://epubl.luth.se).

Backstrom, Magnus (2000). Ground Temperature in Porous Pavement During Freezing and Thawing, *Journal of Transportation Engineering* 126, 375–381.

Backstrom, Magnus, and A. Bergstrom (2000). Draining Function of Porous Asphalt During Snowmelt and Temporary Freezing, *Canadian Journal of Civil Engineering* 27, 594–598.

Balades, J.-D., M. Legret and H. Madiec (1995). Permeable Pavements: Pollution Management Tools, *Water Science and Technology* 32, 49–56.

Batiato, G., M. Donada, and P. Grandesso (1996). DDL (Double Draining Layer): A New Generation of Porous Asphalt Pavement Developed by Autovie Venete, in *Proceedings of Eurasphalt and Eurobitume Congress*, Vol. 3, Strasbourg, France, May 7-10, 1996.

Bendtsen, Hans Christian (1997). Noise Reduction by Drainage Asphalt, *Nordic Road and Transport Research* 1, 6–8.

Bendtsen, Hans Christian, and Lars Ellebjerg Larsen (1999). Noise-Reducing Pavements for Urban Roads, *Nordic Road and Transport Research* 3, 14–16.

Berengier, M., J. F. Hamet, and P. Bar (1990). Acoustical Properties of Porous Asphalts: Theoretical and Environmental Aspects, *Transportation Research Record* 1265, 9–24.

Brown, Daniel C. (1996a). Oregon Likes Porous Pavements, *Asphalt Contractor* January, 70–76.

Brown, Daniel C. (1996b). Porous Asphalt Pavement Rescues Parking Lots, *Asphalt Contractor* January, 70–77.

Brown, Daniel C. (2003). Managing Stormwater with Porous Asphalt, *Public Works* 134, 1–16.

Cahill, Thomas (1993). *Porous Pavement With Underground Recharge Beds, Engineering Design Manual*, West Chester Pennsylvania: Cahill Associates.

Cahill, Thomas (1994). A Second Look at Porous Pavement/Underground Recharge, *Watershed Protection Techniques* 1, 76–78.

Cahill, Thomas, Michelle Adems, and Courtney Marm (2003). Porous Asphalt: The Right Choice for Porous Pavements, *Hot Mix Asphalt Technology* September–October.

Camomilla, Gabriele, Mauro Malgarini, and Sandro Gervasio (1990). Sound Absorption and Winter Performance of Porous Asphalt Pavement, *Transportation Research Record* 1265, 1–8.

Colandini, Valérie, Michel Legret, Yves Brosseaud and Jean-Daniel Balades (1995). Metallic Pollution in Clogging Materials of Urban Porous Pavements, *Water Science and Technology* 32, 57–62.

Cooley, L. Allen, R. Ray Brown, and Donald E. Watson (2000). *Evaluation of OGFC Mixtures Containing Cellulose Fibers*, Report No. 2000-05, Auburn, Alabama: Auburn University National Center for Asphalt Technology.

Croney, David, and Paul Croney (1998). *Design and Performance of Road Pavements*, 3rd ed. New York: McGraw-Hill.

Decoene, Y. (1990). Contribution of Cellulose Fibers to the Performance of Porous Asphalts, *Transportation Research Record* 1265, 82–86.

Diniz, Elvidio V. (1976). Quantifying the Effects of Porous Pavements on Urban Runoff, 63-70, and Author's Response, 45-43, in *National Symposium on Urban Hydrology, Hydraulics, and Sediment Control*, University of Kentucky, Lexington, Kentucky, July 26–29, 1976.

Diniz, Elvidio V. (1980). *Porous Pavement, Phase 1 — Design and Operational Criteria*, EPA-600/2-80-135, Cincinnati: U.S. Environmental Protection Agency Municipal Environmental Research Laboratory (www.epa.gov/ednnrmrl/repository/abstrac2/abstra2.htm).

Diniz, Elvidio V., and William H. Espey, Jr. (1979). *Maximum Utilization of Water Resources in a Planned Community — Application of the Storm Water Management Model, Volume I*, EPA-600/2-70-050c, Cincinnati: U.S. Environmental Protection Agency Municipal Environmental Research Laboratory; National Technical Information Service (NTIS) No. PB80121437.

Eaton, Robert A., and Peter C. Marzbanian (1980). *Structural Evaluation of Porous Pavement Test Sections at Walden Pond State Reservation, Concord, Massachusetts*, Special Report 80-29, Hanover, NH: U.S. Army Corps of Engineers Cold Regions Research and Engineering Laboratory.

Engineering News Record, 1973, When It Rains, It Pours Through the Pavement, *Engineering News Record* October 11, 38.

Field, Richard, Hugh Masters, and Melvin Singer (1982). Porous Pavement: Research, Development, and Demonstration, *Transportation Engineering Journal* 108, 244–258.

Forrest, James B. (1975). Porous Asphalt Airfield Surface, *Military Engineer* 167,78–79.

Galloway, B.M., and J.A. Epps (1974). *Mixture Design Concepts, Laboratory Tests and Construction Guides for Open-Graded Bituminous Overlays*, Research Report TTI-2-10-74-36-1F, College Station: Texas A&M University, Texas Transportation Institute.

Georgia Department of Transportation, Office of Materials and Research (no date). *Georgia's Modified Open-Graded Friction Course*, downloadable from web page for "Office of Materials and Research" at www.dot.state.ga.us.

Goforth, Gary F., Elvidio V. Diniz, and J. Brent Rauhut (1983). *Stormwater Hydrological Characteristics of Porous and Conventional Paving Systems*, EPA-600/2-83-106, Cincinnati: U.S.E.P.A. Municipal Environmental Research Laboratory, National Technical Information Service (NTIS) # PB-84-123728.

Highway & Heavy Construction (1987). Porous Asphaltic Concrete Pavement Test, *Highway & Heavy Construction* 130, December, 82–83.

Highway & Heavy Construction (1988). Storm Water Recharge Beds Replace Retention Basins, *Highway & Heavy Construction* 131, July, 50–52.

Hogland, W., M. Larson, and R. Berndisson (1990). The Pollutant Build-Up in Pervious Road Construction, in *5th International Conference on Urban Storm Drainage*, Osaka, pp. 845–852.

Hossain, Mustaque, and Larry A. Scofield (1991). *Porous Pavement for Control of Highway Runoff*, Publication AZ-352, Phoenix: Arizona Department of Transportation, Transportation Research Center.

Hossain, Mustaque, Larry A. Scofield, and W. R. Meier, Jr. (1992). Porous Pavement for Control of Highway Runoff in Arizona: Performance to Date, *Transportation Research Record* 1354, 45–54.

Huber, Gerald (2000). *Performance Survey on Open-Graded Friction Course Mixes*, National Cooperative Highway Research Program Synthesis of Highway Practice 284, Washington: Transportation Research Board.

Huet, M., A. de Boissoudy, J.-C. Gramsammer, A. Bauduin, and J. Samanos (1990). Experiments with Porous Asphalt on the Nantes Fatigue Test Track, *Transportation Research Record* 1265, 54–58.

Hughes, Agatha H. (1990). Making a Parking Lot Into an Exhibit, *The Public Garden* January, 14–17.

Isenring, Thomas, Harld Koster, and Ivan Scazziga (1990). Experiences with Porous Asphalt in Switzerland, *Transportation Research Record* 1265, 41–53.

Jackson, Newt (2003). *Porous Asphalt Pavements*, Information Series 131, Lanham, Maryland: National Asphalt Pavement Association.

Jared, David, Andrew Johnson and Donald Watson (1998) Georgia Department of Transportation's Progress in Open-Graded Friction Course Development, *Transportation Research Record* 1616, 30–33.

Jones, Michael P. (1973). Friction Overlay Improves Runway Skid Resistance, *Civil Engineering* 43, 45–48.

Kandhal, Prithvi S. (2002). *Design, Construction, and Maintenance of Open-Graded Asphalt Friction Courses*, Information Series 115, Lanham, Maryland: National Asphalt Pavement Association.

Kandhal, Prithvi S., and Rajib B. Mallick (1998). *Open-Graded Asphalt Friction Course: State of the Practice*, Report No. 98-7, Auburn, Alabama: Auburn University National Center for Asphalt Technology.

Kandhal, Prithvi S., and Rajib B. Mallick (1999). *Design of New-Generation Open-Graded Friction Courses*, Report No. 99-2, Auburn, Alabama: Auburn University National Center for Asphalt Technology.

Keating, Janis (2001). Porous Pavement, *Stormwater* 2, 30–37.

Koteff, Carl (1964). *Surficial Geology of the Concord Quadrangle, Massachusetts*, Geologic Quadrangle 331, Washington: U.S. Geological Survey.

Kuennen, Tom (1996). Open-Graded Mixes: Better the Second Time Around, *American City & County* August, 40–53.

Kuennen, Tom (2003). A New Era for Permeable Pavements, *Better Roads* April (www.betterroads.com).

Land and Water (1988). Recharge Beds Save Space, Preserve Environment, *Land and Water* January, 14.

Legret, M., V. Colandini, and C. Le Marc (1996). Effects of a Porous Pavement with Reservoir Structure on the Quality of Runoff Water and Soil, *The Science of the Total Environment* 190, 335–340.

Livet, J. (1994). The Specific Winter Behavior of Porous Asphalt: the Situation as It Stands in France, in *Proceedings of the 9th PIARC International Winter Road Congress*, Seefield, Austria, March 1994.

Mallick, Rajib B., Prithvi S. Kandhal, L. Allen Cooley Jr., and Donald E. Watson (2000). *Design, Construction, and Performance of New-Generation Open-Graded Friction Courses*, Report No. 2000-01, Auburn, Alabama: Auburn University National Center for Asphalt Technology.

McShane, Clay (1979). Transforming the Use of Urban Space: A Look at the Revolution in Street Pavements, 1880-1924, *Journal of Urban History* 5, 279–307.

Meiarashi, S., M. Ishida, T. Fujiwara, M. Hasebe, and T. Nakatsuji (1996). Noise-Reduction Characteristics of Porous Elastic Road Surfaces, *Applied Acoustics* 47, 239–250.

Nelson, P.M., and P.G. Abbott, (1988). Acoustical Performance of Pervious Macadam Surfaces for High-Speed Roads, *Transportation Research Record* 1265, 25–33.

Nicholls, J.C. (1997). *Review of UK Porous Asphalt Trials*, Report 264, London: Transport Research Laboratory.

Niemczynowicz, J., W. Hogland and T. Wahlman (1985). Consequence Analysis of the Unit Superstructure, *VATTEN* 41, 250–258 [in Swedish].

Oregon Department of Transportation (1999). *Supplemental Standard Specifications, April 1999, Section 00745 QA — Hot Mixed Asphalt Concrete (HMAC)*, Salem, OR: Oregon Department of Transportation (www.odot.state.or.us).

Oregon Department of Transportation (2002). *ODOT Pavement Design Guide*, Salem, OR: Oregon Department of Transportation (www.odot.state.or.us).

Pagotto, C., M. Legret, and P. Le Cloirec (2000). Comparison of the Hydraulic Behaviour and the Quality of Highway Runoff Water According to the Type of Pavement, *Water Resources* 34, 4446–4454.

Parker, Frazier, Jr., Robert C. Gunkel and Thomas D. White (1977). *Observations of Portland Cement Concrete and Porous Friction Course Pavement Construction*, Miscellaneous Paper S-77-26, Vicksburg: U.S. Army Engineer Waterways Experiment Station.

Pérez-Jiménez, F. E., and J. Gordillo (1990). Optimization of Porous Mixes through the Use of Special Binders, *Transportation Research Record* 1265, 59–68.

Roberts, Wilton A. (2003). *FY 2004 Research Statement: Pervious Pavement Full-Depth Design for Highways*, Salem, Oregon: Oregon Department of Transportation (www.odot.state.or.us).

Robinette, Gary O. (1976). *Parking Lot Landscape Development*, McLean, Virginia: Environmental Design Press (republished 1993 by Agora Communications, Plano, Texas).

Rollings, Marian P., and Raymond S. Rollings, Jr. (1996). *Geotechnical Materials in Construction*, New York: McGraw-Hill.

Ruiz, Aurelio, Roberto Alberola, Félix Pérez, and Bartolomé Sanchez (1990). Porous Asphalt Mixtures in Spain, *Transportation Research Record* 1265, 87–94.

Sainton, Alain (1990). Advantages of Asphalt Rubber Binder for Porous Asphalt Concrete, *Transportation Research Record* 1265, 69–81.

Schlect, Ed (1994). Open-Graded Asphalt Surfaces Offer Safety and Environmental Advantages, *Stone Review* April, 10–11.

Smith, R.W., J.M. Rice, and S. R. Spelman (1974). *Design of Open-Graded Asphalt Friction Courses*, Report No. FHWA-RD-74-2, Washington: U.S. Federal Highway Administration.

Sorvig, Kim (1993). Porous Paving, *Landscape Architecture* 83, 66–69.

St. John, Matthias S. (1997). *Porous Asphalt Road Shoulders: Effect of Road Sanding Operations and Their Projected Life Span*, Renton: King County Department of Transportation.

St. John, Matthias S., and Richard R. Horner (1997). *Effect of Road Shoulder Treatments on Highway Runoff Quality and Quantity*, WA-RD 429.1, Olympia: Washington State Department of Transportation.

Steiner, Frederick, and Todd Johnson (1990). Fitness Adaptability, Delight: Three Firms Explore Ecological Design in Practice, *Landscape Architecture* 80, 96–101.

Stenmark, Christer (1995). An Alternative Road Construction for Stormwater Management in Cold Climates, *Water Science and Technology* 32, 79–84.

Stotz, G., and K. Krauth (1994). The Pollution of Effluents from Pervious Pavements of an Experimental Highway Section: First Results, *The Science of the Total Environment* 146/147, 465–470.

Strom, Steven, and Kurt Nathan (1998). *Site Engineering for Landscape Architects*, 3rd ed. New York: John Wiley and Sons.

Tan, Siew Ann, Tien Fang Fwa, and Vincent Y.K. Guwe (2000). Laboratory Measurements and Analysis of Clogging Mechanism of Porous Asphalt Mixes, *Journal of Testing and Evaluation* 28, 207–216.

Tappeiner, Walter J. (1993). *Open-Graded Asphalt Friction Course*, Information Series 115, Lanham, MD: National Asphalt Pavement Association.

Thelen, Edmund, and L. Fielding Howe (1978). *Porous Pavement*, Philadelphia: Franklin Institute Press.

Thelen, Edmund, Wilford C. Grover, Arnold J. Holberg, and Thomas I. Haigh (1972). *Investigation of Porous Pavements for Urban Runoff Control*, 11034 DUY, Washington: U.S. Environmental Protection Agency.

van der Zwan, J. Th., Th. Goeman, H.J.A. Gruis, J.H. Swart, and R.H. Oldenburger (1990). Porous Asphalt Wearing Courses in the Netherlands: State of the Art Review, *Transportation Research Record* 1265, 95–110.

van Heystraeten, G., and C. Moraux (1990). Ten Years' Experience of Porous Asphalt in Belgium, *Transportation Research Record* 1265, 34–40.

Webster, L. F. (1997). *The Wiley Dictionary of Civil Engineering and Construction*, New York: John Wiley and Sons.

Wei, Irvine W. (1986). *Installation and Evaluation of Permeable Pavement at Walden Pond State Reservation*, Report to the Commonwealth of Massachusetts Division of Water Pollution Control, Boston: Northeastern University Department of Civil Engineering.

Woelfl, G.A., I.W. Wei, C.N. Faulstich, and H.S. Litwack (1981). Laboratory Testing of Asphalt Concrete for Porous Pavements, *Journal of Testing and Evaluation* 9, 175–181.

Wyant, David C. (1992). *Final Report, Field Performance of a Porous Asphaltic Pavement*, VTRC 92-R10, Charlottesville: Virginia Transportation Research Council.

13 Soft Porous Surfacing

Kim Sorvig

CONTENTS

The Theory and Practice of Softness	514
Where to Use SPS	516
Safety and Accessibility of Loose-Fill Surfaces	517
SPS Installation	518
Considerations for Use of Organic Materials	519
Wood and Bark Materials	520
Examples of Wood and Bark Mulch Installations	523
Engineered Wood Fiber	525
Organic Crop By-Products	528
Mollusk Shells	529
Granular Recycled Rubber	529
Trademarks	535
Acknowledgments	535
References	535

"Paved" usually means hard-surfaced — yet soft alternatives also exist. Soft porous paving materials are so unobtrusive and simple that they go virtually unrecorded in the literature of site construction. Soft porous paths can be found at the Rockefellers' Kykuit estate in Pocantico, New York, between public allotments at Melvin Hazen Community Garden in Washington DC, and on trails in Alaska's Sitka National Forest.

Soft Porous Surfacing (SPS) stabilizes and protects the earth's surface with granular or fibrous particles of organic or recycled materials. SPS is a simple approach with often-overlooked potential. For many paving applications SPS offers adequate surface stability with porosity and low cost. It has much to recommend it in parks, gardens, trails, and low-traffic parking areas. This class of porous paving materials can set a restrained tone, save money, protect soil, conserve resources, and blend gracefully with a number of landscape design styles.

SPS is similar to stone aggregate in that both have granular structure and loose installation. The concepts of particle size and shape, gradation, void space, porosity, permeability, and particle interlock introduced for aggregate in Chapter 6 apply also to SPS materials. However, unlike stone, SPS materials are organic or recycled. Many are biodegradable. Most are far lighter and, literally, softer than stone. These characteristics make the production, transport, use, and maintenance of SPS materials quite different from those of aggregate.

By another name, SPS materials are mulch, and as such can contribute to soil structure or fertility. Biodegradability and soft granular structure are the sources both of SPS's advantages and of its limitations.

This chapter describes five types of SPS materials: chipped wood and bark mulches, "engineered wood fiber," organic crop by-products, mollusk shells, and granular recycled rubber. The chapter describes where and how to use soft surfacing— and its history, strengths and weaknesses— and then describes individual SPS materials.

Table 13.1 lists examples of information sites where suppliers of many kinds of SPS materials can be identified. Calkins (2002a) describes further directories for sources of recycled materials.

THE THEORY AND PRACTICE OF SOFTNESS

Many soft surfacing materials have their origins in historic gardens and streetscapes. Historically, wood and crop by-products were readily available and could be worked with simple technologies. Today, soft materials have similar advantages and serve the same functions as they did in the past.

SPS materials are sometimes specified precisely because of their historic associations. Public historic landscapes use soft materials for authenticity. Private estates use them to convey a sense of age and conservative respectability.

Like many other traditional materials, soft surfacing materials are useful for "green building," which seeks improved resource and energy efficiency in construction. Preindustrial building materials were obtained locally and installed with simple low-tech methods. Many of those traditional materials and methods are more resource-efficient than modern mechanized approaches. SPS materials fit the profile of "green" materials: they are by-products or reused materials; little fuel energy is involved in producing and installing them; and they are minimally processed and in that sense "natural" (Thompson and Sorvig, 2000, pp. 4–6). One green development, DeWees Island, near Charleston, South Carolina, includes in its site-protective covenants the requirement that all paths and maintenance roads, if not left as native sand, must be SPS-surfaced with shells or wood chips (Wilson, 1997).

Soft surfacing materials achieve permanence by regular maintenance of materials which are easy and inexpensive to work. This approach can be called the Taos

TABLE 13.1
Examples of Information Sites for Identifying Suppliers of all Types of Soft Paving Materials

Name	Contact Information
Business and Industry Resource Venture	www.resourceventure.org
California Integrated Waste Management Board	www.ciwmb.ca.gov
Landscape Link	www.elandscapelink.com
Recycler's World	www.recycle.net
RecycleXchange	www.recyclexchange.com

Pueblo strategy, in honor of North America's oldest continuously used building. Taos Pueblo is made of mud and has been replastered every other year for over a millennium. This approach is the opposite of that of hard materials, which rigidly resist weathering and wear, but which are energy-intensive and costly to install and repair. Because extracting, installing, and maintaining soft materials is low-tech, even frequent replenishment over a lifetime of use can be more resource-efficient than doing the same job with hard materials.

Behind this approach is an ethic articulated by Seattle landscape architect Richard Haag in an interview as the "Theory of Softness" (J. William Thompson, personal communication). The theory asks that no surface be any harder than its function truly demands. Specifying hard materials like asphalt, concrete or blocks is wasteful overdesign for little-used pedestrian pathways. Soft surfaces offer nonintrusive construction methods. They are a welcome relief from the hard surfaces of the modern landscape. They can be elegant in their simplicity. The soft alternatives discussed in this chapter embody Haag's concept.

The roots of living trees can thrive under an SPS surface. A covering of soft porous material protects the rooting zone from compaction and erosion. Granular and fibrous materials adapt to expansive subsurface root growth because raking and replenishment of material is a built-in aspect of their maintenance whether or not root growth occurs.

SPS creates a yielding surface comfortable for walkers, joggers, and equestrians, and is especially valued by elderly pedestrians (Flink et al., 2001, p. 73). Where pedestrian traffic mingles with that of vehicles such as mountain bikes, soft surfaces can calm traffic by controlling the speed of travel, increasing safety and enjoyment for all who share the trail.

The initial installation cost of most SPS materials is low. Rails-to-Trails (1999) rated the per-mile cost of wood chip trails at 1/20th of that for concrete. The only cheaper alternatives for new trail surfaces were native soil and grass at 1/25th the cost of concrete. The purchase cost for the SPS materials is low; in some locales, some types such as municipal tree chippings are free. By default, transportation to deliver them can form a high percentage of their total cost. Labor cost for most installations is significantly less than that for any hard material. An additional indirect or societal cost reduction when using recycled SPS materials is that they keep bulky materials out of landfills.

In the long term, the life-cycle cost of SPS can be low compared to that of hard surfaces. Replenishing, leveling, and weeding of SPS surfaces require only simple tools; laborers without special skills can achieve good results; replacement materials are inexpensive (Michigan State University Extension, 1996). By contrast, although repair of hard surfaces is infrequent, when it is required it is intensive. Concrete and asphalt must be sawed or jackhammered; patched surfaces can be unsightly and structurally inferior.

SPS materials compare favorably with hard surfaces also in terms of energy cost. Embodied energy is the total amount of energy used in producing, delivering, and installing a product (Calkins, 2002b; Thompson and Sorvig 2000, pp. 229–258). Embodied-energy statistics available today for most materials are provisional. Nevertheless, the energy required to install SPS is negligible when compared to that for the machinery that mixes, places, and cuts most hard surfacing. Many SPS materials are by-products or recycled products, which are considered "free" of energy costs except for transportation and installation (the energy to produce them is assigned to the main product). Some energy accounting even considers such materials "credits" which offset the energy used to produce the main product; for example the use of bark or wood waste as SPS offsets the energy cost of producing lumber.

WHERE TO USE SPS

Table 13.2 summarizes some considerations for choosing between soft and hard surfacing for specific applications. They include concerns of traffic, cost, maintenance, slope, and perceptual "fit."

SPS is suited for settings where traffic is too high to leave a surface in grass or native soil, but too low to require hard surfacing. Most SPS is best for moderate-use pedestrian and equestrian paths, or for lightly used pedestrian plazas. Some SPS materials are suited for playground surfacing, where impact attenuation is a primary goal. Very few SPS materials should be used for automobile traffic.

On steep slopes (over 5 percent), SPS particles can be displaced by traffic. Where surface drainage is concentrated they can be washed out. Some kinds of SPS

TABLE 13.2

Considerations for Choosing between SPS and Hard Surfaces for Specific Applications

Use Soft Porous Surfacing	Use Hard Surfacing
Where traffic is lightweight: pedestrian, equestrian, nonmotorized bikes, possibly some low-ground-pressure vehicles	Where traffic includes heavy weights such as trucks or busses
Where volume is low: occasional-use parking, private driveways	Where traffic volume is high: public streets, frequently used parking, trails carrying large numbers of bikes or horses
Where soft contact with feet is desirable, or to cushion impact in case of falls: comfortable paths, safe playgrounds	Where traffic could abrade surface: tracked vehicles, motorized bikes, and high ground-pressure vehicles
Where low life-cycle cost is required	Where high installation and maintenance costs would be offset by other considerations
For "fit" in historic or natural areas and in naturalistic or soft design styles	For hard, tidy appearance, or where hard surfaces were historically used
Where the surface must call little attention to itself	Where lane marking or directional signage must be applied to the surface
Where slope < 5% or cross-slope < 2%, and runoff from adjacent surfaces can be kept from flowing across the surface	Where slope > 5% or cross-slope > 2%
Where simple installation and maintenance are desirable	Where long-term routine maintenance must be minimized, and skilled labor is available when necessary for repairs

resist displacement by the interlocking of particles or entanglement of fibers, and can be used on steeper slopes than those with particles that do not interlock.

Table 13.3 lists specific traffic levels at which degradation of soil and vegetated surfaces makes the surfaces unusable, and a covering of SPS necessary. The thresholds in the table assume rates of soil and vegetation loss that occur in the climate of northeastern U.S.; they could occur at lower traffic levels in more severe climates (Flink et al., 2001, pp. 62–63).

SPS allows completely unhindered access to underground utilities and makes a seamless "patch" when replenished afterwards. It can be easily removed where pavement may have only a limited useful lifetime or may have to be relocated. Organic SPS granules which cannot be raked or shoveled away can be left to decompose in the soil. Geotextile beneath an SPS surface can ensure complete removal if necessary.

SAFETY AND ACCESSIBILITY OF LOOSE-FILL SURFACES

A combination of ASTM (www.astm.org) standards guides SPS loose-fill materials to be at once soft enough to limit injury from falls in playgrounds, and maneuverable enough for wheelchair accessibility (U.S. Architectural and Transportation Barriers Compliance Board, 2000).

ASTM F 1292, *Standard Specification for Impact Attenuation of Surface Systems Under and Around Playground Equipment*, defines conditions for safety in the event of a fall. It establishes criteria for what the Consumer Products Safety Commission called "critical height" (King, 1999), which was introduced in Chapter 6. A fall from a height less than a material's critical height would not be expected to produce a life-threatening head injury. Safety increases with increasing critical height.

ASTM F 1951, *Standard Specification for Determination of Accessibility of Surface Systems Under and Around Playground Equipment*, defines accessibility in terms of the work necessary to propel a wheelchair across the surface, in comparison

TABLE 13.3
Traffic Levels at Which Degradation of Soil and Vegetated Surfaces Occurs

	Assumed Conditions			
Type of Traffic	Maximum Slope	Maximum Cross-slope	Average Speed	Frequency of Use above which Soil and Grass Trails become Unstable
Pedestrians	Any	4%	3 to 7 mph	> 100 users per week
Universally accessible	3 to 8%	2%	2 to 3 mph	Not applicable
Bicycles (non-motorized)	3 to 8%	2 to 4%	15 to 30 mph	> 10 users per week
Motorized bikes	Any	Any	> 30 mph	1 user per week
Horseback riding	10%	4%	4 to 6 mph	1 user per week
Skiers	5%	4%	2 to 8 mph	Compacted snow thaws late, possibly suppressing vegetation
Inline skaters	Unknown	Unknown	2 to 10 mph	Require paved surface

Based on Flink et al., 2001, pp. 62–63, and Irving 1985, pp. 96–98.

with that on a 7.1 percent grade on a smooth, hard surface. Alternative measures of accessibility continue to be proposed; Laufenberg and Winandy (2003) measured it in terms of surface "firmness" and "stability" under the weight and movement of wheelchair casters.

SPS INSTALLATION

Figure 13.1 shows an example of SPS installation. The SPS granules are laid at least a few inches thick (for example 3 to 4 inches), up to 12 inches if impact attenuation is a primary goal.

Specifications for SPS material can stipulate source, composition, particle size, and color. The material should not contain any substance which could harm plants, animals, or water quality, or produce an objectionable odor. It must not contain any substance harmful to humans through inhalation or skin contact, or extraneous matter hard or sharp enough to injure persons working on or using the surface. It should be free of viable plant seeds. Samples may be required before installation. Before placement, the material should be stored in such a way as to ensure that it remains free of contaminants.

Upon placement the surface can be leveled and lightly compacted. Compaction must not be so great as to prevent free drainage through the surface. A common practice with organic materials is to place the material at 1.5 times the intended thickness, and then to compact it to its finished elevation (Laufenberg and Winandy, 2003).

In many settings edge restraint is required to contain the particles and to define the perimeter. Several types of edge restraint were described in Chapter 3. The edging should ordinarily extend from the finished surface elevation to the bottom of the SPS layer, and be placed or fastened to resist settlement and lateral movement. If the setting presents a steep slope, edging can be used to form steps or terraced surfaces that hold SPS particles. A geotextile may be used to separate the SPS material from the underlying soil.

Drainage in and around the SPS material should prevent concentrated runoff from flowing across or through an SPS surface and saturating it, or floating or washing it away. On poorly drained soil elevating the path a few inches might protect the material. A bed of open-graded aggregate below the SPS layer might improve drainage, especially if underdrains are added, but underdrains are costly. Impervious

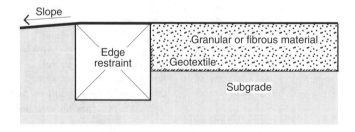

FIGURE 13.1 Example of SPS installation.

weed-barrier sheeting must not be placed under SPS because it would impede drainage. In settings where wet conditions are unavoidable, the choice of SPS materials is limited to those that are not floatable or biodegradable.

CONSIDERATIONS FOR USE OF ORGANIC MATERIALS

Organic SPS materials such as wood and bark mulches and organic crop by-products biodegrade and influence the ecology of the rooting-zone soil, special considerations must be taken into account while planning for their use and maintenance. The materials and their breakdown products are substrates for organic growth having nutrients, organic matter, pH, and cation exchange capacity.

Decomposing organic matter can support weeds. However, a thick (perhaps 8 inches or more) layer of mulch naturally suppresses growth of buried weed seedlings by blocking sunlight (Martin and Gershuny, 1992, pp. 158–159; Stout, 1961, pp. 6–19). Fabric weed barriers are probably not needed at the base of such a thick layer. Weeds that root in the layer itself are fairly easy to uproot and may be naturally suppressed by traffic if the surface is regularly used. If herbicides are used for weeding where tree roots are present in the soil below, the herbicides should be of a type that selectively targets non-woody plants or that is neutralized by soil contact.

Fresh organic matter such as wood and bark chips is highly carbonaceous and low in nitrogen. Table 13.4 shows that the material's ratio of carbon to nitrogen (C/N) is far higher than that of typical native soil. As it decomposes, decay microorganisms call on nitrogen from adjacent mulch or soil for growth, and can rob nitrogen from soil with which they are in contact. However, a deeper, older, more decomposed layer of mulch separates the nitrogen-demanding fresh material from the underlying soil (Martin and Gershuny 1992, pp. 157–159; Stout, 1961 pp. 6–19). Substantial decomposition of a layer of organic surfacing takes six months to two years (Martin and Gershuny 1992, pp. 137–139). In that time, an uncompacted 8-inch layer subsides to a mass 2 or 3 inches thick. In a maturely established mulch, the lowest layer, in contact with the soil, has the C/N ratio most in equilibrium with natural soil. The finished material adds organic matter and cation exchange capacity to the soil (Brady, 1974, pp. 151–154).

A deep installation of mulch, continuously replenished with fresh material on the top, imitates the healthy continuous natural addition of organic matter to native forest and prairie soils. Although the decomposed material is fine-textured, it retains

TABLE 13.4
Carbon/Nitrogen (C/N) Ratios in Soil and Mulch Materials

Material	Carbon:Nitrogen	Reference
Wood and bark chips	115:1 to 615:1	Craul, 1992, pp. 364–366
Decay micro-organisms	4:1 to 9:1	Brady, 1974, p. 151
Finished (decomposed) compost	10:1	Bradley and Ellis, 1992, p. 140
Typical native soil	8:1 to 15:1	Brady, 1974, pp. 151–154

its porosity and permeability if it is not compacted. In general, in using organic SPS materials over the rooting zone of plants, it is prudent not to import materials whose breakdown products differ greatly from those that created the native soil (Sauer 1998, p. 156).

WOOD AND BARK MATERIALS

Wood and bark are made into mulch by shredding. The resulting chips are informal, even rustic in appearance. They make a yielding, comfortable surface for walking or riding, and are adequate for trail-bike use. Some suppliers distinguish fibrous "shredded" products from chunky "chips."

Table 13.5 lists examples of commercial suppliers of wood and bark mulch materials. Home-and-garden centers stock materials from these and other suppliers.

Table 13.6 lists examples of information sites for identifying additional sources of material. Further suppliers can be identified at the web sites listed in Table 13.1, and at those of various state and regional "green industry" and nurserymen's associations.

As an alternative to commercial sources like those listed in the table, wood and bark chips can be obtained directly from local sawmills, tree-care companies, utility

TABLE 13.5
Examples of Suppliers of Wood and Bark Mulch Materials

Supplier	Contact Information
Area Mulch and Soils	www.areamulchandsoils.com
Beaver Brook Landscape Supply	www.beaverbrooksupply.com
Boss Mulch	www.bossmulch.com
Carolina Mulch Plus	www.carolinamulchplus.com
Consolidated Resource Recovery	www.resourcerecovery.com
Corbitt Manufacturing	www.cypress-mulch.com
Earth Products	www.earthproducts.net
Fertile Garden Supply	www.fertilegarden.com
Garick	www.garick.com
GSO America	www.gsoamerica.com
J&J Materials	www.jjmaterials.com
Mallard Creek	www.mallardcreekinc.com
Monty's Mulch	www.montysmulch.com
Mulch Masters	www.mulchmasters.com
Mulch Mountain	www.mulchmountain.com
Ohio Mulch	www.ohiomulch.us
Pacific Topsoils	www.pacifictopsoils.com
Parsons Farm	www.parsonsfarm.com
Rivendell	www.rivendelldistribution.com
S & J Exco	www.sjexcoinc.com
SJAP NaturaLink	www.naturalink.com
Tioga County Waste Wood Recyclers	www.woodrecyclers.com
Zeager Bros.	www.zeager.com

TABLE 13.6
Examples of Information Sites for Identifying Additional Suppliers of Wood and Bark Mulch Materials

Name	Contact Information
American Forest & Paper Association	www.afandpa.org
Composting Council of Canada	www.compost.org
Green Industry Yellow Pages	www.giyp.com
Green Media Online	www.greenmediaonline.com
Mulch and Soil Council	www.mulchandsoilcouncil.org
Pacific Recycle	http://pacific.recycle.net
Recycler's World	www.recycle.net
U. S. Composting Council	www.compostingcouncil.org

companies, and municipal yardwaste-collection agencies. Chips from these sources can be among the least costly SPS materials because they are the result of direct, local recycling. Sorting by particle size may or may not be available. Arborists' and municipal chippings may contain a wide variety of species from the pruning of diverse yard and roadside trees; those from a sawmill or wood processing plant may come from a narrow range of species or from recycled products such as shipping pallets. Chipped preservative-treated wood should ordinarily be avoided because it is likely to be capable of leaching toxic chemicals.

On wooded sites, material and transportation costs can be eliminated by harvesting and processing wood and bark materials on-site. Some suppliers provide these on-site services. Clearing woods to develop a site might generate enough chips to surface the project's footpaths. When the initial surface wears down, replenishment chips may have to be transported from off-site.

The following sample specification indicates some of the characteristics that a user may wish to stipulate in wood and bark surfacing materials. The language was developed by the Environmental Purchasing Program of King County, Washington (www.metrokc.gov). It assumes that the species are those available in the King County area; in other regions different species are available. It does not distinguish between wood and bark materials. Other designers should modify and expand the language to suit specific needs.

> Material shall be shredded bark or wood and shall be ground so that a minimum of 95 percent of the material will pass through a 5-inch sieve and no more than 45 percent will pass through a 3/4-inch sieve. The material shall contain a minimum of extraneous material. It shall contain no material that would be deleterious to equipment and shall not contain resin, tannin, or other compounds in quantities that would be detrimental to animals, plant-life or water quality.

Small twigs among wood and bark fragments bind the particles together (Walter Cook, personal communication 2001). The material generated during trimming of street tree branches for utility lines can contain a large quantity of twigs, as long as the chipping machine does not break them up. The fresh material fluffs up with some depth as the twigs hold the fragments apart. With traffic and decomposition it settles into a soft, stable mass that does not easily get kicked away or eroded. In contrast chipped wood that does not contain a large amount of twigs is easily kicked out of place; an example is the chipped fragments of logs generated at some stages in the wood pulp industry.

The color of natural wood chips may fade to gray before any functional deterioration begins. Raking might counteract fading, delaying any perceived need for replenishment. Dyed chips may retain their artificial colors for some time.

Table 13.7 shows the critical heights of three products for safety in playground falls. It shows that wood and bark mulches are capable of producing substantial impact attenuation, especially in a thick layer of randomly sized particles. The cushioning effect can decline over time with decomposition, pulverization, compaction, and displacement.

Fresh green wood and bark products may harbor pests or diseases of living trees or plants. In areas where plant diseases have been reported, it would be prudent not to place fresh local chippings near vulnerable plants. Dry, dead wood seldom harbors the same organisms that infect living plants.

Decaying hardwood mulch can harbor one microorganism that can cause property damage: the so-called Artillery Fungus, *Sphaerobolus stellatus* (Kuhns, 1997). The fungus ejects its spores explosively, shooting them up to 20 feet. The sticky spores adhere to any surface they hit. The fungus "aims" its artillery barrages at the lightest surfaces nearby, including polished surfaces like cars and windows where people are least happy to find sticky black specks. The hazard of this fungus may make wood-based SPS inappropriate near some structures. Research on the fungus' life cycle, diagnosis, and prevention is continuing.

Wood-based SPS could theoretically harbor termites and carpenter ants. Actual problems of this type have not been reported. In Iowa, Jauron (1995, pp. 99–100) stated that "Sawdust, wood chips, and bark mulches can be safely used around the home. These materials will not 'attract' termites to the house." Nevertheless, it may be prudent to consider the possibility in regions where these insects are present. Keeping SPS a certain distance from certain wood structures may be considered.

Table 13.8 lists standard wood and bark products defined by the the Mulch and Soil Council (2001).

TABLE 13.7
Critical Height of Three Wood and Bark Materials Placed in a Compacted Layer 9 in. Thick

Tested Material	Description	Critical Height (ft)
Wood mulch	Randomly sized chips, twigs and leaves, chipped from tree limbs and brush	10
Double-shredded bark mulch	Type commonly used to mulch planting areas	7
Uniform wood chips	Relatively uniformly sized shredded hardwood fibers, containing no bark or leaves	6

Consumer Products Safety Commission, no date.

TABLE 13.8
Standard Wood and Bark Products Defined by the Mulch and Soil Council (2001)

Product Name	Source Genus	Particle Size
Wood products (wood > 15 percent of total weight):		
Cedar mulch	Thuja or Juniperus	< 3 inches
Cypress mulch	Taxodium	< 3 inches
Hardwood mulch	Deciduous hardwood trees	< 3 inches
Hemlock mulch	Tsuga	< 2 inches
Pine mulch	Pinus	< 1.5 inches
Western mulch	Conifers common in western North America	< 1 inch
Bark products (wood < 15 percent of total weight):		
Cypress bark mulch	Taxodium	< 3 inches
Hardwood bark mulch	Deciduous hardwood trees	< 3 inches
Hemlock bark mulch	Tsuga	< 2 inches
Pine bark mini-nuggets	Pinus	0.5 to 1.5 inches
Pine bark mulch	Pinus	< 1.5 inches
Pine bark nuggets	Pinus	1.25 to 3.5 inches
Western large bark	Conifers common in western North America	1.75 to 3.0 inches
Western medium bark	Conifers common in western North America	1 to 2 inches
Western pathway bark	Conifers common in western North America	0.25 to 1.5 inch
Western small bark	Conifers common in western North America	0.5 to 1.0 inch

Table 13.9 lists additional characteristics of some of the standard products. The listed colors may vary with local tree species. A product's "longevity" is how long it substantially maintains its original appearance in a temperate climate with moderate sun exposure. If appearance is not taken into account and compaction is negligible, some types of chips require replenishment only every two or three years (Michigan State University Extension, 1996). Some kinds of bark last longer than wood (Craul, 1992, p. 365; Mulch and Soil Council, 2001); conifer barks in large granule sizes are rated longest-lasting. The moisture retention of most wood mulches is higher than that of bark.

The particles of many types of bark are thin and light. They are poor choices for windy sites. Pine bark in particular floats in surface water and is easily knocked out of place. Hardwood bark is more fibrous and holds its place more durably on steep slopes, although it has the disadvantage of decomposing faster (Michigan State University Extension, 1996). A material that combines stable fibrous structure with slow decomposition is cedar bark; it is a by-product of industries that process cedar posts by peeling bark from the logs (Walter Cook, personal communication 2001).

EXAMPLES OF WOOD AND BARK MULCH INSTALLATIONS

The following examples illustrate the appropriate and effective placement of wood and bark mulches in a variety of specific settings.

Figure 13.2 shows a footpath of chipped hardwood in the English Woodland Garden portion of the Missouri Botanical Garden in Saint Louis (www.mobot.org).

TABLE 13.9
Characteristics of Wood Mulches

Product Name	Color	Longevity (Seasons)	pH
Wood products (wood > 15 percent of total weight):			
Cypress Grade B	Golden brown	Single	Neutral
Pine Mulch	Light brown	Single	Slightly acidic
Western Mulch	Reddish brown	Single	Slightly acidic
Bark products (wood < 15 percent of total weight):			
Cypress Grade A	Golden brown	Single	Neutral
Hardwood Mulch	Dark brown to black	Single	Slightly acidic
Pine Mini-nuggets	Dark brown	Multiple	Neutral
Pine Nuggets	Reddish brown	Multiple	Neutral
Western Large	Reddish brown	Multiple	Neutral
Western Medium	Reddish brown	Multiple	Neutral
Western Pathway	Reddish brown	Single	Slightly acidic

Data from Mulch and Soil Council Mulch Selection Table, www.nbspa.org.

The surface permeability is high. An edging of cut tree limbs confines the material and distinguishes the path from the adjacent woodland soil. The source of both the chips and the limbs is in the routine pruning and maintenance operations of the garden complex. As the mulch decomposes, it loses volume, permeability, and appearance. It is replenished approximately annually, but not removed and replaced. It rots in place and becomes part of the rooting medium under the path, continuous with the surrounding planting beds.

Figure 13.3 shows a path of cypress mulch in the Bok Tower Garden in Lake Wales, Florida (www.boksanctuary.org). The garden was established with a design by Frederick Law Olmsted Jr. in the 1920s and is now fully grown and well maintained. The sandy substrate is dry except in irrigated planting areas. The cypress material is replenished annually. It comes in long, fibrous, golden-brown shreds. Fresh material is strongly aromatic. Thin edging strips divide the path from adjacent planting areas.

Figure 13.4 shows a path of chipped redwood in Armstrong Woods State Reserve in Guerneville, California (www.parks.ca.gov). The reserve's old-growth redwood forest has many ancient trees and huge downed logs. The trail gives day visitors access to the colossal forest and its facilities. Walkers in the characteristic gentle rain are enveloped in the redwood trees, the redwood path, and growing moss and ferns, all blended into one by the soundless mist. It is a strongly naturalistic sense of place.

Figure 2.15 showed wood-mulch paths in the Melvin Hazen Community Garden on Sedgwick Street in Washington DC. This is one of several community garden areas overseen by the National Park Service's National Capital Region office. The garden is divided into 16-foot- wide plots which are allocated to city residents. In return for an annual fee each gardener has gardening space, irrigation water, and wood chips from a stockpile. Paths 3 to 4 ft wide separate the plots. The garden's rules require that each plot holder keep the paths on two sides weed-free and covered

Soft Porous Surfacing

FIGURE 13.2 Chipped hardwood path in the English Woodland Garden portion of the Missouri Botanical Garden, Saint Louis.

with chips. After several generations of use, the path's edges are wavy, and plot edgings are delightfully diverse brick, stone, 4inch × 4inch timber, etc. Some large plants overhang into the paths, requiring careful treading to traverse.

ENGINEERED WOOD FIBER

ASTM F 2075 defines "engineered wood fiber" (EWF) in order to enable specification of wood and bark material which can be designated suitable for safe playground surfacing. According to the standard, EWF is a special variety of wood and bark mulch. It is "processed wood that is ground to a fibrous consistency, randomly sized, approximately ten times longer than wide with a maximum length of 2 inches, free

FIGURE 13.3 Path of cypress mulch in Bok Tower Garden, Lake Wales, Florida.

of hazardous substances." In order to meet the requirements of the specification, the material must have the correct particle size, be free of metal particles and heavy-metal contaminants, and meet the requirements of F 1292 for impact attenuation. Species are not stipulated. Wood and bark products are not distinguished.

Figure 13.5 shows EWF's required particle-size gradation. The specified sizes are the small dimensions of the particles, not their lengths. The specified gradation is believed to provide sufficient porosity for drainage, and enough large particles to limit compaction and maintain resilience while limiting over-sized solid particles that would cause injury.

Table 13.10 lists examples of products designated as engineered wood fiber.

Laufenberg and Winandy (2003) tested a 12-inch- thick layer of EWF for performance in playgrounds. The material was adequately stable for wheelchair maneuverability, and adequately firm for the short wheelchair travel distances expected on

Soft Porous Surfacing

FIGURE 13.4 Chipped redwood path in Armstrong Woods State Reserve, Guerneville, California.

playgrounds. It met ASTM criteria for impact attenuation with a 10-foot fall. Its impact attenuation was relatively durable after six months of exposure to weather. The porous, permeable structure allowed material deep in the layer to stay relatively dry.

The same researchers also tested 12-inch- thick installations of which the top 1.5 inches to 2.5 inches was stabilized with polyurethane or latex binder. The installations met all criteria for playground accessibility and impact attenuation as did the unbound EWF. However, during six months of exposure to weather, the bound-surface installations had lost some impact attenuation relative to the unbound material. The material under the bound surfaces was persistently wet; evidently the bound surfaces impeded the drying of subsurface material. The researchers attributed the decline of impact attenuation to the subsurface moisture. They speculated that the moisture could hasten biodegradation and reduce impact attenuation in freezing conditions.

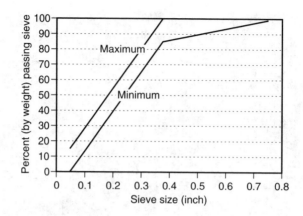

FIGURE 13.5 Particle-size gradation in engineered wood fiber (data from ASTM F 2075).

TABLE 13.10
Examples of Products Designated as Engineered Wood Fiber

Product	Company	Contact Information
Bodyguard	CWLM	www.cwlm.com
Engineered wood fiber	B.Y.O. Products	www.byoplayground.com
Engineered wood fiber	Fibar	www.fibar.com
Engineered wood fiber	Park Structures	www.parkstructures.com
Engineered wood fiber	ABCreative	www.abcreative.net
GTImpax Fiber	GameTime	www.gametime.com
Kids Karpet	Garick	www.kidskarpet.com
Sof'Fall	Sof'Fall	www.sof-fall.com
Woodcarpet	Zeager Bros.	www.zeager.com

ORGANIC CROP BY-PRODUCTS

Organic crop by-products such as nut shells and the hulls removed from other crop products are available for use as SPS. Most are seasonal and regional in their availability, being driven by the production of a main crop or product (for example shell by-products from nut production).

Some nut shells are more durable than wood, and some have intense, long-lasting colors. However, they can be hard and brittle. Pecan shells have sharp edges, so they should not be used where walkers may be in bare feet (Walter Cook, personal communication June 2001).

Many other crop by-products are softer than shells. The hulls of buckwheat, peanuts, and rice may be too fragile and lightweight to serve as SPS under traffic.

Licorice by-products may be available in bags from garden centers. They are used for their aroma. However, they are likely to be expensive, and less durable than wood or bark.

Cocoa hulls are available from food processing factories. They are valued for their aroma. However, they can contain theobromine, which is toxic to cats, dogs and horses (Fitzgerald, 2000). Cocoa shell mulch "has a high potassium content which may injure some plants. Young maples, lilacs, rhododendrons, and azaleas have been found to be susceptible to damage" (Michigan State University Extension, 1996).

Some by-products may contain food residues (such as pecan nut dust) which can attract animals that scatter the material, for example, dogs or ground-feeding birds. Carefully selected and specified material may escape animal disturbance.

MOLLUSK SHELLS

Shells of oysters, clams, and other mollusks are locally plentiful near natural deposits and shellfish processing plants, and in bags from commercial distributors. Shells have been used as sparkling white or pink surfacing in coastal areas around the world. They are particularly common in coastal parts of the southeastern states, where their historical use is well known.

Shells from processing plants should be cleaned and, to bear traffic comfortably, crushed to pass a given sieve size. In contrast, shells taken from natural deposits tend to be naturally broken, bleached, and smoothed from exposure.

The availability of shells from processing plants is affected by fluctuations in shellfish harvest and by requirements that harvesters return a portion of shells to the shellfish beds to ensure a next generation of shellfish.

Figure 13.6 shows a shell drive at Joan M. Durante Park in Longboat Key, Florida, a town park used for small performances and gatherings. (www.longboatkey.org). The white shell is easily obtained and replenished from local natural deposits in the form of broken, weathered fragments. Perceptually and ecologically it is highly compatible with the park's barrier-island setting. The shell fragments are in effect highly angular aggregate particles with a maximum size of about half an inch and abundant small particles. They are not easily displaced. However, they are not durable; the material crushes and compacts under the weight of vehicles.

The shell covers the park's entry drive and all 17 parking stalls, including one stall marked for handicap parking. The stalls are in groups of two or three tucked between preexisting trees and shrubs. Individual stalls are marked by timber wheel stops. Eight years after installation the shell on the drive was compacted with low permeability; that in the parking stalls remained open-textured and of moderate permeability. Perhaps raking would restore permeability where necessary.

The park's numerous walkways are of the same shell material, edged by wooden strips. The walkways wind around restored tidal wetlands for use by walkers, bird watchers, and school groups. Their surfaces have been crusted as much by raindrop impact as by pedestrian traffic, and now have moderate permeability.

GRANULAR RECYCLED RUBBER

Granular rubber is made by chipping or shredding recycled tires. Technically, modern tires are only 50 percent true natural or synthetic rubber; the remainder is a

FIGURE 13.6 Shell drive at Durante Park in Longboat Key, Florida.

combination of carbon black, metal, fiber, and other components (ISRI, no date). Nevertheless the term "rubber" is used generically to refer to essentially all the elastic material in recycled tires. The U.S. generates 280 million scrap tires per year (EPA, 2003). Grants are available from environmental agencies to those who keep scrap tires from landfills by putting the tires to productive use (Prince, 1998; *Recycling Today,* 2003). In recycling, tire material may be mixed with rubber scraps from other industrial sources. Unbound granular rubber has been used on walkways, playgrounds, equestrian areas, and reputedly as an alternative to unbound aggregate in low-traffic drives (ISRI, no date).

Common sizes for loose-fill chips such as those for playgrounds are 1/2 inch to 3/4 inch (Rubber Manufacturers Association, no date). Rubber chips tend to resist degradation and compaction. They retain their particle size, so their porosity and

Soft Porous Surfacing

permeability are relatively durable. They are considerably denser than bark, so they do not float in water.

Rubber does not leach toxic chemicals, but used tires may have surface contaminants which must be washed off before processing (Pliny Fisk, cited in Thompson and Sorvig, 2000, p. 334). Rubber is less likely than organic materials to give weeds, insects, or microorganisms a foothold, or to attract animals. It absorbs little water. It is free of allergenic molds (Newland, 1999).

Some granular rubber products can absorb more playground impact than wood chips. The critical height of tire chips in a 6-inch uncompacted layer was 12 feet, compared with 7 feet for an equal depth of wood mulch (Rubber Manufacturers Association, no date). After one parents' group had experienced an actively used rubber-chip school playground, their only suggestions for improvement related to granules' dispersing and

TABLE 13.11
Examples of Associations in the Tire Recycling Industry

Name	Contact Information
Institute of Scrap Recycling Industries	www.isri.org
Rubber Association of Canada	www.rubberassociation.ca
Rubber Manufacturers Association	www.rma.org
Tire Industry Association	www.tireindustry.org

TABLE 13.12
Examples of Suppliers of Granular Recycled Rubber

Supplier	Contact Information
American Rubber Technologies	www.americantire.com
Davis Rubber	www.davisrubbercompany.com
Emanuel Tire	www.emanueltire.com
Future Mulch	www.futuremulch.com
Global Tire Recycling	www.gtrcrumbrubber.com
JAI Tire	www.jaitire.com
Rainbow Turf	www.rainbowturfproducts.com
Recovery Technologies Group	www.rtginc.com
Recycled Rubber Resources	www.rubber.com/rubber/trade
Recylage Granutech	www.rubber.com/rubber
Rubber Recycle	www.rubberecycle.com
Rubber Recycling and Manufacturing	http://crumb.rubber.com/trade
Rubber Resources	www.stopmulching.com
Rubberific	www.rubberificmulch.com
Tierra Verde Industries	www.grubblellc.com
Tire Environmental Systems	http://www.recycle.net/trade
Tire Grinders Transporters	www.grn.com/trade
Tire Recycling Atlantic Canada	www.traccnb.ca
Tire Recycling Systems	www.recycle.net/trade
Tire Shredders Unlimited	www.rubber.com/rubber/trade

TABLE 13.13
Examples of Information Sites for Identifying Further Suppliers of Granular Recycled Rubber

Name	Contact Information
Pacific Recycle	http://pacific.recycle.net
Recycler's World	www.recycle.net
Scrap Tire News	www.scraptirenews.com
The Rubber Room	www.rubber.com

TABLE 13.14
Common ASTM Grades of Recycled Particulate Rubber

Grade	Characteristics
Grade 1	The material from whole car, truck and bus tires from which the fiber and metal have been removed
Grade 2	The material from the treads of car, truck and bus tires
Grade 3	The material from "retread buffings," from the tread and shoulder area of car, truck and bus tires

ASTM D 5603.

TABLE 13.15
Examples of Product Grades in the Shredded Tires Category at The Rubber Room

Grade	Characteristics
No.1 Shredded Tires	"Shredded debeaded tires sized to minus 3 inches"
No.2 Shredded Tires	"Shredded whole tires sized to minus 3 inches"
No.1 Tread Chips	"Uniform chipped tire treads, sized to minus 2.5 inches plus 10 mesh. Must be bead free and black only."
No.1 Rubber Chips	"Uniform chipped bias tire or metal free sidewalls, sized to minus 2.5 inches plus 10 mesh. Must be guaranteed metal free (magnetically separated material is not acceptable) and black only."
No.2 Rubber Chips	"Uniform chipped bias tire or metal free sidewalls, sized to minus 2.5 inches plus 10 mesh. Must be guaranteed metal free (magnetically separated material is not acceptable) and may contain black & white."

www.rubber.com/rubber.

sticking to shoes (Newland, 1999) were: rake the material daily to maintain uniform depth; have kids empty their shoes of particles before leaving the playground; install mats at playground exits to scrape excess rubber bits from shoes.

The tire-recycling industry is supported by associations such as those listed in Table 13.11.

Table 13.12 lists examples of suppliers of granular recycled rubber. Home and garden centers stock the products of these and other suppliers. Some processors are

Soft Porous Surfacing 533

TABLE 13.16
Holders of Registered Trademarks Mentioned in This Chapter

Registered Trademark	Holder	Headquarters Location
ABCreative	ABCreative	DeSoto, Kansas
American Rubber Technologies	American Rubber Technologies	Jacksonville, Florida
Area Mulch and Soils	Area Mulch and Soils	Raleigh, North Carolina
B.Y.O. Products	B.Y.O. Products	Dallas, Georgia
Beaver Brook Landscape Supply	Beaver Brook Supply	Danbury, Connecticut
Bodyguard	Tierra Verde Industries	Irvine, California
Boss Mulch	A. C. Hesse	Toms River, New Jersey
Carolina Mulch Plus	Carolina Mulch Plus	Pisgah Forest, North Carolina
Consolidated Resource Recovery	Consolidated Resource Recovery	Sarasota, Florida
Corbitt Manufacturing	Corbitt Manufacturing	Lake City, Florida
Davis Rubber	Davis Rubber	Little Rock, Arkansas
Earth Products	Earth Products	Marietta, Georgia
Emanuel Tire	Emanuel Tire	Baltimore, Maryland
Fertile Garden Supply	Fertile Garden Supply	San Antonio, Texas
Fibar	Fibar Systems	Armonk, New York
Future Mulch	Future Mulch	Fairfax, Virginia
Garick	Garick Corporation	Cleveland, Ohio
Global Tire Recycling	Global Tire Recycling	Wildwood, Florida
GSO America	GSO America	Columbus, Ohio
GTImpax Fiber	GameTime	Fort Payne, Alabama
J&J Materials	J&J Materials	Rehoboth, Massachusetts
JAI Tire	JAI Tire	Denver, Colorado
Kids Karpet	Garick Corporation	Cleveland, Ohio
Mallard Creek	Mallard Creek	Rocklin, California
Monty's Mulch	Monty's Mulch	Douglassville, California
Mulch Masters	Mulch Masters	Jacksonville, Florida
Mulch Mountain	Mulch Mountain	Elizabethtown, Pennsylvania
Ohio Mulch	Ohio Mulch	Columbus, Ohio
Pacific Topsoils	Pacific Topsoils	Everett, Washington
Park Structures	Park Structures	Coral Springs, Florida
Parsons Farm	Parsons Farm	Independent Hill, Virginia
Rainbow Turf	Rainbow Turf Products	Saint Cloud, Florida
Recovery Technologies Group	Recovery Technologies Group	East Guttenberg, New Jersey
Recycled Rubber Resources	Recycled Rubber Resources	Macon, Missouri
Recylage Granutech	Recylage Granutech	Plessisville, Quebec
Rivendell	Rivendell Distribution	Glenwood Springs, Colorado
Rubber Recycle	Rubber Recycle	Lakewood, New Jersey
Rubber Recycling and Manufacturing	Rubber Recycling and Manufacturing	Caguas, Puerto Rico
Rubber Resources	Rubber Resources	Hudson, Florida

TABLE 13.16 Continued

Registered Trademark	Holder	Headquarters Location
Rubberific	Rubberific	Saint Louis, Missouri
S & J Exco	S & J Exco	South Dennis, Massachusetts
SJAP NaturaLink	SJAP NaturaLink	Elmer, New Jersey
Sof'Fall	Sof'Fall	Draper, Utah
Tierra Verde Industries	Tierra Verde Industries	Irvine, California
Tioga County Waste Wood Recyclers	Tioga County Waste Wood Recyclers	Owego, New York
Tire Environmental Systems	Tire Environmental Systems	Muscatine, Iowa
Tire Grinders Transporters	Tire Grinders Transporters	Aurora, Illinois
Tire Recycling Atlantic Canada	Tire Recycling Atlantic Canada	Minto, New Brunswick
Tire Recycling Systems	Tire Recycling Systems	London, Ontario
Tire Shredders Unlimited	Tire Shredders Unlimited	Fenton, Missouri
Woodcarpet	Zeager Bros.	Middletown, Pennsylvania
Zeager Bros.	Zeager Bros.	Middletown, Pennsylvania

capable of accepting scrap tires and returning them shredded into the desired particle size.

Further suppliers can be identified from the member lists of the industrial associations, and from information sites such as those listed in Tables 13.1 and 13.13.

ASTM D 5603, *Standard Classification for Rubber Compounding Materials — Recycled Vulcanizate Particulate Rubber* defines grades of recycled particulate rubber based on particle-size distribution and types of rubber material. The grades described in Table 13.14 stipulate the source of recycled-tire rubber materials. The material must be substantially free of foreign contaminants. A specification of maximum particle size completes the specification. A size designation consists of the nominal maximum size (sieve size that a value such as 85 to 95 percent of the material passes) and a sieve size that 100 percent of the material passes. A particular size specification can be agreed upon between the vendor and the user.

Many suppliers of particulate rubber surfacing materials classify their products in terms other than those of the ASTM standard. Examples of conventions established by individual suppliers are those for playgrounds, walkways, and equestrian areas of American Rubber Technologies, Emanuel Tire, and Rubber Recycle. In facilitating exchange of information about available products, The Rubber Room (www.rubber.com) uses the grades of materials listed in Table 13.15. They are distinguished by the material's detailed origin, particle size, color, and presence of metal.

Specifications for individual projects can stipulate particle sizes, color, and freedom from residual metal particles. Refinement and adoption of industry-wide standards that would facilitate practical surfacing applications is called for.

Granular rubber is made into surfacing mats with the addition of binders (Nichols, 1993). Perm-A-Mulch (www.permamulch.com) is one of the few examples of currently available manufactured mats which may be significantly porous and permeable.

TRADEMARKS

Table 13.16 lists the holders of registered trademarks mentioned in this chapter.

ACKNOWLEDGMENTS

The following persons generously contributed information used in this chapter: Jim Anderson of Recovery Technologies Group (www.rtginc.com); Michael Blumenthal of the Rubber Manufacturers Association (www.rma.org); Steven M. Bopp of Austin Tao & Associates (http://austintao.com); Walter Cook, trail designer of Athens, Georgia; Tamas Deak of Koonce Pfeffer Bettis (www.kpb-architects.com); C. Hedlund; Lynn Lewis of the National Trust for Historic Preservation; Stefan Luger of Eximlink U.K. Ltd (www.eximlink.com); Aaron Pryor of the Institute of Scrap Recycling Industries (www.isri.org); J. William Thompson of *Landscape Architecture*; S. Titco; Robert LaGasse of the Mulch and Soil Council (www.mulchandsoilcouncil.org).

Walter Cook, trail designer of Athens, Georgia, constructively critiqued an early draft of this chapter.

REFERENCES

Agate, Elizabeth (1982). *Footpaths*, Wallingford, England: British Trust for Conservation Volunteers.

Birchard, William, Jr., and Robert D. Proudman (2000). *Trail Design, Construction, and Maintenance*, Harpers Ferry, WV: Appalachian Trail Conference.

Birchard, William, Jr., and Robert D. Proudman (1982). *Appalachian Trail Fieldbook, A Self-Help Guide for Trail Maintenance*, 2nd ed., Harpers Ferry, WV: Appalachian Trail Conference.

Blomgren, Paige Gilchrist (1999). *Making Paths and Walkways, Creative Ideas and Simple Techniques*, Asheville, NC: Lark Books.

Bradley, Fern Marshall, and Barbara W. Ellis (1992). *Rodale's All-New Encyclopedia of Organic Gardening*, Emmaus, PA: Rodale Press.

Brady, Nyle C. (1974). *The Nature and Properties of Soils*, 8th ed., New York: Macmillan Publishing.

Calkins, Meg (2002a). Closing the Loop, *Landscape Architecture*. 92, 42–48 and 102–105.

Calkins, Meg (2002b). Green Specs, *Landscape Architecture* 92, 40–45 and 96–97.

Consumer Products Safety Commission (no date). *Playground Surfacing Materials*, Document No. 1005, Washington: Consumer Products Safety Commission (downloadable at www.kidsource.com).

Craul, Phillip J. 1992. *Urban Soil in Landscape Design*, New York: John Wiley and Sons.

Duffy, H. (1992). *Surface Materials for Multiple-Use Pathways*, Lakewood, CO: National Park Service, Rocky Mountain Region.

EDAW (1981). *Trail Construction Guidelines*, Denver: Colorado Division of Parks and Outdoor Recreation.

Fitzgerald, D. K. (2000). *Chocolate Poisoning in Dogs*, Apogee Communications Group (viewable at www.apogeecomgrp.com).

Flink, Charles A., Kristine Olka and Robert M. Searns (2001). *Trails for the Twenty-First Century: Planning, Design, and Management Manual for Multi-Use Trails*, 2nd ed., Washington: Island Press.

Institute of Scrap Recycling Industries,(ISRI) (no date). Rubber Recycling Rolls Along, *Industry Information*, www.isri.org.

Irving, J. A. (1985). *The Public in Your Woods*, Chichester, UK: Packard Publishing.

Jauron, R. (1995). Organic Mulches for Gardens and Landscape Plantings, *Horticulture & Home Pest News*, Iowa State University Extension.

King, Steve (1999). Creating Grounds for Play, *Landscape Design & Build*, supplement to *Landscape & Irrigation* 12, S14–S15.

Kuhns, L. (1997). Artillery Fungus Threatens Homeowners, *Mulch Industry* (http://aginfo.psu.edu).

Lancaster, R. (1987). *Recreation, Park and Open Space Standards and Guidelines*, Ashburn, VA: National Recreation and Park Association.

Laufenberg, Theodore, and Jerrold Winandy (2003). *Improved Engineered Wood Fiber (EWF) Surfaces for Accessible Playgrounds, Final Report: Phase II, Field Performance Testing*, Washington: U.S. Access Board (www.access-board.gov).

Martin, Deborah L., and Grace Gershuny, editors (1992). *The Rodale Book of Composting*, Emmaus, PA: Rodale Press.

McCoy, M. and M. A. Stoner (1992). *Mountain Bike Trails: Techniques for Design, Construction and Maintenance*, Missoula, MT: Bikecentennial.

Michigan State University Extension (1996). *Mulches in the Landscape*, Woody Ornamental Series (viewable at www.msue.msu.edu/genesee/hort/mulches.htm).

Mulch and Soil Council [formerly National Bark & Soil Producers Association] (2001). *Voluntary Uniform Product Guidelines for Horticultural Mulches, Growing Media and Landscape Soils*, Manassas, VA: Mulch and Soil Council (www.mulchandsoilcouncil.org or www.nbspa.org).

Newland, Nancy (1999). A Playground Odyssey, *Scrap Tire News* October (www.scraptirenews.com).

Nichols, David (1993). Fresh Paving Ideas for Special Challenges, *Landscape Design* 6, 17–20.

Prince, John (1998). *News Release: Recycled Waste Tires for Playgrounds*, Cardwell, MO: Southland School District, http://rebel.southland.k12.mo.us.

Proudman, Robert D., and Reuben Rajala (1981). *Trail Building and Maintenance*, Boston: Appalachian Mountain Club.

Rails-to-Trails (1999). *Surfacing Your Trail*, www.railtrails.org.

Recycling Today (2003). Michigan Agency Awards Scrap Tire Grants, in *Recycling Today* June 16, ww.recyclingtoday.com.

Rubber Manufacturers Association (no date). Scrap Tire Markets, *Scrap Tires*, Rubber Manufacturers Association, www.rma.org.

Sauer, Leslie (1998). *The Once and Future Forest*, Washington: Island Press.

Sorvig, Kim (1994). The Path Less Traveled, *Landscape Architecture* 84, 30–33.

Stout, Ruth (1961). *Gardening Without Work*, New York: Devin-Adair.

Thompson, J. William, and Kim Sorvig (2000). *Sustainable Landscape Construction: A Guide to Green Building Outdoors*, Washington: Island Press.

U.S. Architectural and Transportation Barriers Compliance Board (2000). Americans with Disabilities Act (ADA) Accessibility Guidelines for Buildings and Facilities; Play Areas; Final Rule, *Federal Register* October 18, 2000, 36 CFR Part 1191.

U.S. Environmental Protection Agency (EPA) (2003). In the News — Scrap Tires, U.S. EPA Region 6 (www.epa.gov/region06).
U.S. Forest Service (1984). *Standard Specifications for Construction of Trails*, Washington: U.S. Forest Service.
Vogel, Charles (1982). *Trails Manual*, Sylmar, CA: Equestrian Trails, Inc.
Wilson, Alex (1997). Dewees Island: More Than Just a Green Development, *Environmental Building News* 6, 5–7.

14 Decks

Gregg A. Coyle

CONTENTS

Deck Components and Effects 542
    Decking and Its Effects 542
Natural Wood in Deck Construction 544
    Preservative Treatment of Decay-Susceptible Species 545
Plastic and Composite Lumber in Deck Construction 546
On-Grade Decks 547
    Carpinteria Marsh, Carpinteria, California 548
    1010 On the Green, Anchorage, Alaska 549
Elevated Decks 550
    Foundations for Elevated Decks 551
    Curbs and Railings 551
    Prefabricated Bridges 554
    Elm Brook Crossing, Minuteman National Park, Massachusetts 554
    Arcadia Sanctuary, Easthampton, Massachusetts 556
Decks in Practice 557
    Minnesota Landscape Arboretum, Chanhassen, Minnesota 557
    Columbus Zoo, Columbus, Ohio 557
    Riverwalk, Jacksonville, Florida 558
    The Landings, Skidaway Island, Georgia 559
    Puncheon Boardwalks in Alaska 560
    New River Gorge, West Virginia 561
    Mar-Vista Restaurant, Longboat Key, Florida 561
    Japanese Garden, Missouri Botanical Garden, Saint Louis 561
Trademarks 562
Acknowledgments 565
References 565

Decks are surrogates for pavement that are porous and permeable as long as their surfacing components are spaced or perforated. They differ from pavements in that they are not coverings of the earth surface; they are bridge-like structures that span over the soil without bearing continuously on it. They are built either with a ramping surface to conform to a slope, or level with connecting stairs as necessary. Decks are rapidly drained, easily cleaned surfaces, potentially accessible to all but those wearing the narrowest spike heels. They can integrate fixed seating, shelters, planters, arbors, tables, storage bins, and lights. They can be configured to fit a huge variety of site-specific conditions and requirements.

Compared with ground-contact pavements, the construction of decks leaves underlying soil relatively undisturbed. Decks contact the subgrade only at footings or along "sleeper" beams, leaving the remaining soil undisturbed and uncompacted for infiltration of water and air and rooting of trees (Watson and Himelick, 1997, p. 44). Decks can straddle obstacles such as utility lines, tree roots, and boulders. With appropriate adaptation of their footings decks can be stable amid swelling and frost-susceptible subgrades.

The potential loads on a deck pose structural requirements for types and sizes of deck materials, foundations, fasteners, and bracing. The load includes the "dead" load of the structure, the "live" load of traffic upon it, and snow. The consequences of structural failure could be severe. Substantial loads must be anticipated where people may crowd together for parties or performances. In places subject to inundation decks are also subject to lateral loads from waves and flood flows, especially where debris gets trapped and catches the force of flowing water. Rising water can lift unanchored wood off its foundations.

Because decks are not ground-contact pavements, detailed technical coverage of their design and construction is outside the scope of this chapter. Instead, this chapter introduces decks as substitutes for pavements by surveying typical deck components and materials and some of the variations of configuration that are possible to fit differing site conditions and deck purposes.

Detailed guidelines for deck design and construction are available elsewhere. Examples of technical references are listed at the end of this chapter. Local building codes may specify certain features. Associations and manufacturers mentioned later in this chapter provide detailed guidelines for use of their specific products.

Table 14.1 lists terms with distinctive application in deck construction.

Early deck-like structures were wooden bridges (Ritter and Stanfill-McMillan, 1995). In the nineteenth century, wooden bridges in America were roofed to protect the wood from moisture and decay. Beginning in 1865, the pressure treating of wood with preservative (initially creosote) allowed bridges to be built without roofs. Iron and reinforced concrete quickly joined wood in bridge construction.

In 1870, the city of Atlantic City, New Jersey, built the first section of its seaside boardwalk, originally to reduce the amount of sand tracked into railroad cars and newly developed hotels (Library of Congress, no date). The 8-foot-wide wooden structure connected the center of town to the beach. At first, it was intended to be dismantled each winter, but it was soon extended along enormous amusement piers. The only vehicles allowed on the walk were rolling chairs, introduced in 1884. In subsequent decades, similar boardwalks were constructed in other resort cities and extended and reconstructed in stages with wooden planks on piers of wood and concrete.

More versatile use of deck structures began in California after World War II. Landscape architect Thomas Church (1955) took advantage of California's fine climate to build outdoor "rooms" as extensions of indoor living spaces. His decks

TABLE 14.1
Terms with Distinctive Application in Deck Construction

Term	Definition
Beam	Long, heavy piece of timber
Board	Long, thin piece of sawed timber
Boardwalk	Deck providing a walkway
Bulkhead	Upright structure or partition that resists pressure or separates compartments
Cantilever	Member supported sufficiently over part of its length to project upward or outward at the end
Composite lumber	Lumber made of plastic and other components blended together
Curb (kerb)	Siding built along an edge
Deck	Flat floored open structure
Decking	The surfacing material of a deck
Footing	Bottom unit of a structure, in contact with the subgrade
Foundation	The supporting part of a structure at or below ground level including footings
Grate	Frame of parallel bars or crossbars forming an open lattice-work
Joist	Small beam supporting the planking of a floor
Lumber	Timber ready for use, or other structural material prepared in a similar form
Pier (post)	Upright structural support
Pile	Long, slender support driven into the ground
Plank	Heavy, thick board
Plastic lumber	Lumber made at least partly from plastic (ASTM D 6662 limits plastic lumber specifically to that containing more than 50 percent plastic resin)
Puncheon	Split log or heavy slab with the face smoothed, or any roughly dressed timber
Rail	Bar extending from one post or support to another as a guard or barrier
Railing	Barrier consisting of a rail and its supports
Sleeper	Beam placed horizontally on or near the ground to support a structure
Spindle	Long slender piece
Stanchion	Post, especially a free-standing post
Stringer (string)	Long horizontal timber used to connect uprights or to support a floor, or one of the inclined sides of a stair supporting the treads and risers
Tie (cross-tie)	Structural element that links other elements
Timber	Wood used for building, or a large piece of wood forming part of a structure

exploited the region's naturally decay-resistant redwood timber to utilize steep slopes and sandy coastal settings.

Today, many companies supply materials for deck construction and many contractors specialize in their installation. The alternatives in deck materials and configurations are numerous and rapidly evolving. The Deck Industry Association (www.deckindustry.org) advocates product standards and provides information and education. It was established in 2001 and is rapidly growing. The association's member list identifies deck designers, builders, and product suppliers. It supports the magazine *Professional Deck Builder*. Table 14.2 lists additional sources of information about contemporary deck materials, design, and construction.

TABLE 14.2
Examples of Sources of Information about Deck Materials, Design and Construction

Name	Contact Information
ASTM	www.astm.org
BestDeckSite	www.bestdecksite.com
Better Homes and Gardens	www.bhg.com
Boardwalk Design	www.boardwalkdesign.com
Deck Industry Association	www.deckindustry.org
Decks USA	www.decksusa.com
Professional Deck Builder Magazine	www.deckmagazine.com
Western Trailbuilders Association	www.trailbuilders.org

DECK COMPONENTS AND EFFECTS

Figure 14.1 illustrates the major components of deck construction, in some of their possible arrangements. Beams rest at ground level or are elevated by piers. Where joists are used, they are distinguished from beams by their relatively small size and their direct support of the decking. Joists can either rest on top of the beams or hang from the sides. The choice of arrangement is a project-specific question of cost and the relationship of the deck to its setting and purpose.

DECKING AND ITS EFFECTS

Lumber is the most common material for decking and many other deck components. For decking it is used in the form of planks. The various kinds of lumber are described in special sections later in this chapter.

As alternatives to lumber decking planks, some companies manufacture decking panels; examples are listed in Table 14.3. Some can span from beam to beam; others require placement directly on earth or aggregate for continuous support. Some are made of selected woods such as cedar and teak; others are perforated plastic tiles; some others have thin veneers of decorative material laminated over structural slabs.

Tree grates and trench grates are in effect manufactured decking panels. Examples of grate products were listed in Chapter 5. Many are made of metal such as iron, bronze, or aluminum; others are made of plastic or reinforced concrete. Many stock designs are available from the companies' catalogs; it is also possible to specify custom designs.

For porosity and permeability of the surface the decking planks or panels must be spaced to leave open joints; tongue-and-groove decking is out of the question. A common spacing is 1/8 to 1/4 inch; this is sufficiently wide for relatively free drainage and sufficiently narrow for accessibility and a perception of secure solidity. Wider joints allow free drainage, discharge of leaves and debris, snow removal, and penetration of light to support vegetation below, but might give an impression of little support. Narrower joints would give an impression of great security, but could inhibit drainage. A level deck surface or one with joints running perpendicular to the slope promotes dispersion of water through the joints.

Decks

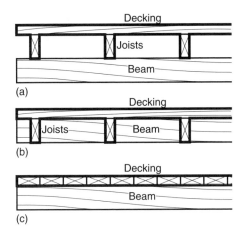

FIGURE 14.1 Components and alternative arrangements of deck framing.

TABLE 14.3
Examples of Manufactured Decking Panels

Product	Company	Contact Information
EZ Deck	EZ Deck	www.ez-deck.com
Mister Boardwalk	Mister Boardwalk	www.misterboardwalk.com
Neoterra	Mateflex	www.mateflex.com
Stone Deck	Deck Technologies	www.stonedeck.biz
Superdeck	Superdeck	www.superdecksystems.com
ThruFlow	Otron Tech	www.thruflowpanel.com
Unnamed	Smith & Hawken	www.smithandhawkentrade.com
Unnamed	Walpole Woodworkers	www.walpolewoodworkers.com

For accessibility it is believed that joints or openings in a deck should be no wider than 1/2 inch, and oriented with the long dimension perpendicular to the direction of traffic movement (O'Connell, 1999). These provisions are to prevent wheelchair and shopping-cart casters, and the points of canes, crutches, and walkers from falling through or getting stuck in the openings.

In hydrologic models decks are sometimes assumed to be perfectly permeable, as long as the decking surface has numerous open joints or perforations. If this assumption is correct, then the runoff coefficient of a deck area could be considered equal to that of the underlying ground surface. In a more complex model the deck surface could be treated as a separate layer, where rain water is attenuated to some degree before it falls through the joints. The fall from the joints, concentrated along the joint lines, would then be the inflow to the soil layer. The character of the earth surface below the deck determines the remaining hydrology of infiltration and runoff. A covering over the soil of aggregate or vegetation protects the surface from crusting or erosion by water falling from the deck. The permeability and runoff coefficients of deck surfaces have not been studied quantitatively; research on this subject is called for if decks are to be used for quantitative control of infiltration and runoff.

NATURAL WOOD IN DECK CONSTRUCTION

Natural wood has been the dominant material in deck construction. It is a strong, easily worked, renewable, and widely available material. It is dimensionally stable when its moisture content is at equilibrium with the atmosphere; for this purpose much commercially sold wood has been kiln-dried. Almost any wood exposed outdoors weathers naturally to a silver-gray color unless stained for color.

Wood industry associations advocate uniform standards and provide technical guidelines for use of their products. Each species group has an association or inspection bureau that classifies and grades the wood for specific applications (McDonald et al., 1996, p. 8). Table 14.4 lists some of the many associations in the wood industry and some additional sources of information about wood material.

Recycled wood from demolished structures is available from sources such as those listed in Table 14.5. Further sources can be identified at Recycler's World

TABLE 14.4
Examples of Associations and Information Sources in the Wood Industry

Name	Contact Information
American Lumber Standard Committee	www.alsc.org
American Wood Council	www.awc.org
American Wood Preservers Institute	www.preservedwod.com
American Wood-Preservers' Association	www.awpa.com
California Redwood Association	www.calredwood.org
National Hardwood Lumber Association	www.natlhardwod.org
Southern Pine Council	www.southernpine.com
U.S. Forest Service, Forest Products Laboratory	www.fpl.fs.fed.us
Western Red Cedar Lumber Association	www.wrcla.org
Western Wood Products Association	www.wwpa.org
Wood World	www.woodfibre.com

TABLE 14.5
Examples of Suppliers of Recycled Timber from Demolished Structures

Supplier	Contact Information
Barco Products	www.barcoproducts.com
Chestnut Woodworking & Antique Flooring	www.chestnutwoodworking.com
Mountain Lumber	www.mountainlumber.com
Pioneer Millworks	www.pioneermillworks.com
TerraMai	www.terramai.com
Trestlewood	www.trestlewood.com
Vintage Log and Lumber	www.vintagelog.com
Vintage Timberworks	http://vintagetimber.com
Will Branch Antique Lumber	www.willbranch.net
Wind River Collections	www.windrivercollections.com

TABLE 14.6
Examples of Wood Species with Natural Resistance to Decay

Group	Species
Domestic softwoods	Redwood, cypress, several species of cedar
Domestic hardwoods	Black cherry, chestnut, black walnut, several species in white-oak family
Tropical hardwoods	Teak, ipé
Other examples	Black locust, Osage orange, red mulberry

Data from Wilson 1999b.

(www.recycle.net). Some recycled wood is suitable for outdoor use as a result of its species or prior preservative treatment.

Wood species that are naturally decay-resistant are listed in Table 14.6. These species can be used outdoors with little or no treatment to protect them from damage by insects or decay.

PRESERVATIVE TREATMENT OF DECAY-SUSCEPTIBLE SPECIES

Many other species of wood are subject to degradation and decay in outdoor weather (McDonald et al., 1996, pp. 13–17). Susceptible wood is subject to attack by insects and microorganisms and to decay caused by fungi, especially in warm humid climates. Moisture accelerates degradation because it raises the grain and harbors fungi. Any place where wood contacts the soil, and is thus surrounded by moisture, presents a special decay hazard (Illman and Ferge, 1998).

The service life of decay-susceptible species can be greatly extended by forcing preservative chemicals into the wood under pressure. Preservative chemicals tend to be broad-spectrum pesticides whose toxicity eliminates wood as a potential food source for decay fungi and insects (Lebow, 2001). Pressure-treated wood is common in deck construction. It is strong and durable enough to be used in almost any deck component, and economical compared with some alternative materials. However, it tends to be subject to cracking or splintering as the lumber wets and dries, so in some decks its use has been limited to structural components such as beams that do not have routine human contact. Wood given great amounts of preservative is used for applications in contact with water or soil. Wood treated with toxic preservative chemicals is surrounded by precautions of use, handling, exposure, and disposal.

A common preservative chemical since the 1930s has been chromated copper arsenate (CCA). To some degree, CCA's chemical reactions within wood bond the chemicals with the wood and make the toxic ingredients insoluble in water (Lebow, 2001). Nevertheless, some toxins gradually leach out of the wood over time, particularly from small construction debris such as sawdust which has a high surface-area-to-volume ratio. A wave of concern over the use of CCA-treated wood began in the late 1990s when high levels of carcinogenic arsenic were found on a playground in Florida where the timber play equipment had been treated with CCA. In addition, the copper in CCA is toxic to aquatic organisms if it leaches out of treated wood (Wilson 1999a, 2002a). Consequently, beginning in 2004, wood intended for residential use is no longer treated with CCA (West, 2003).

One type of alternative is wood treated with nonarsenic preservatives. Examples are alkaline copper quaternary (ACQ) and copper azole (CBA). Other, nonarsenic preservatives are being developed and used, such those based on copper and borate, copper and azole, and copper and quat ("quat" is an abbreviated name for fungicidal quaternary ammonium compounds).

PLASTIC AND COMPOSITE LUMBER IN DECK CONSTRUCTION

Another type of alternative is lumber manufactured wholly or partly of plastic. The manufacture of plastic and composite lumber began in the 1980s as a way to recycle large amounts of waste plastic (Wilson, 1999a; Winterbottom, 1995a). The most common type of plastic in these materials is high-density polyethylene (HDPE) recycled from plastic products such as milk jugs. Among other types of plastic used in lumber are polyvinyl chloride (PVC) and vinyl.

Plastic and composite lumber is commonly manufactured in dimensions matching those of timber and installed with tools and fasteners similar to those used with natural wood. It is free of splinters, knots, and other types of imperfections that may occur in natural wood. It does not rot and is not damaged by termites. It produces no leachable toxins. It is impervious to water and requires no painting, staining, or sealing. Table 14.7 lists examples of plastic and composite lumber products. Some of the companies provide installation and maintenance guidelines for their specific products.

The production of plastic and composite lumber is growing and developing rapidly in response to the demand created by the decline of CCA-treated wood and rapidly evolving industrial technologies. The industry is supported by the Plastic Lumber Trade Association, which promotes uniform standards for plastic lumber products (www.plasticlumber.org; Robbins 2002). The American Plastics Council makes general information on plastic applications available at its web site (www.plasticsresource.com).

ASTM D 6662, *Standard Specification for Polyolefin-Based Plastic Lumber Decking Boards*, defines plastic lumber narrowly as lumber containing more than 50 percent plastic resin ("polyolefin" includes essentially all the plastics now being used in plastic lumber). It sets standards such as finish, appearance, and outdoor weathering for plastic lumber decking boards. It defines structural properties and permits project-specific specification of required strength and stiffness.

In lumber made of pure plastic the surfaces are smooth, unlike those of natural wood. At the current state of development the strength and stiffness of some of the pure plastic products are less than those of wood, so they tend to be used mostly for decking and railing, and less for structural applications such as beams and posts. In heat the material softens, expands, and sags (Wilson, 1999a).

Composite lumber combines plastic with other materials, most commonly organic materials such as wood fiber, flax or rice hulls. The organic components tend to be recycled industrial by-products. Many composites are outside the ASTM D 6662 definition of plastic lumber because they contain less than 50 percent plastic; standards for them fall under other ASTM committees. The surface texture of some products is reminiscent of sand. Compared with pure plastic lumber, composites can be stronger and stiffer, and less expansive with heat (Wilson, 1999a). As the technology improves,

TABLE 14.7
Examples of Plastic and Composite Lumber Products for Deck Construction

Product	Company	Contact Information
Bedford Plastic Timbers	Bedford Technology	www.plasticboards.com
Boardwalk	CertainTeed	www.ctboardwalk.com
Bufftech	CertainTeed	www.bufftech.com
Carefree Xteriors	U.S. Plastic Lumber	www.carefreexteriors.com
ChoiceDek	Advanced Environmental Recycling Technologies	www.choicedek.com
Correct Deck	Correct Building Products	www.correctdeck.com
Deck Lok	Royal Crown	www.royalcrownltd.com
Dream Deck	Thermal Industries	www.thermalindustries.com
E-Z Deck	Pultronex	www.ezdeck.com
Ecoboard	American Ecoboard	www.recycledplastic.com
Evergrain	Epoch	www.evergrain.com
Fiberon	Fiberon	www.fibercomposites.com
Geodeck	Kadant Composites	www.geodeck.com
Lumber	EPS	www.epsplasticlumber.com
Mikronwood Decking	Mikron Industries	http://mikronwood.com
Monarch	Green Tree Composites	www.monarchdeck.com
Nexwood	Nexwood Industries	www.nexwood.com
Polywood	Polywood	www.polywood.com
Rhino Deck	Master Mark Plastics	www.rhinodeck.com
TenduraPlank	Tendura	www.tendura.com
Timberlast	Kroy Building Products	www.kroybp.com
TimberTech	Crane Plastics	www.timbertech.com
Trex	Trex	www.trex.com
Unnamed	The Plastic Lumber Company	www.plasticlumber.com
Unnamed	Plastic Recycling of Iowa Falls	www.hammersplastic.com
Unnamed	Renew Plastics	www.renewplastics.com
WeatherBest	Louisiana Pacific	www.weatherbest.lpcorp.com

structural capabilities increase. New composite materials are being rapidly introduced, and even the use of existing composite materials is expanding rapidly.

ON-GRADE DECKS

On-grade decks are characterized by sleeper beams resting directly and continuously on subgrade soil or aggregate bedding. The sleepers are at once the foundation of the deck and the ties that hold the joists or the decking in position. They are commonly of lumber or concrete. They need not be large because they have no span; they are fully supported at grade. Because the elevation above the ground is slight, on-grade decks seldom require curbs or railings for safety.

Figure 14.2 illustrates some possible configurations of on-grade construction. Bedding the sleepers in open-graded aggregate can keep them dry and resistant to lateral shifting. For a deck surface flush with the surrounding ground, a block, band

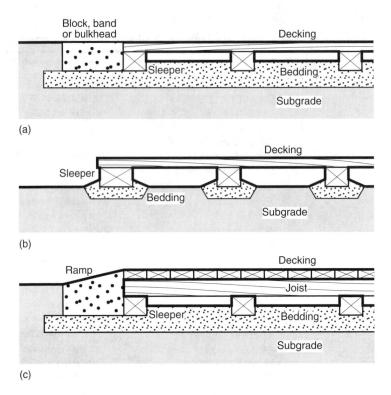

FIGURE 14.2 Some possible configurations of on-grade decks.

or bulkhead of durable material (for example, concrete, stone or decay-resistant wood) can separate soil and moisture from potentially decayable deck components. For a deck surface raised one step above the surrounding ground, a particularly firm sleeper may stabilize the edge. A ramp of aggregate, lumber or concrete might assist access.

Carpinteria Marsh, Carpinteria, California

A boardwalk in Carpinteria, California, illustrates an on-grade deck giving access through a fragile and dynamic setting. Figure 14.3 shows the boardwalk crossing coastal dunes in the city's Carpinteria Salt Marsh Nature Park toward overlook points in the interior of the marsh preserve.

The walk surface is only a few inches above the sand so it has no curbs or railing. The decking planks are composite lumber nominally 2 inches × 6 inches in size. The planks rest on 4-inch-wide composite sleepers, where they are fastened with flathead screws. The screws are countersunk 1/8 inch below the top of the boards. Drifting sand has filled in above the metal heads, hiding them from view.

The sleepers rest directly on the sand, and so may shift if the sand beneath them shifts. However, they keep the planks in general alignment with each other, so if the walk shifts slightly it could maintain some integrity for walking.

Decks 549

FIGURE 14.3 Boardwalk at Carpinteria Marsh Nature Park in Carpinteria, California.

The joint width between planks varies from 0 to 1/2 inch as the boards' orientation follows the walk's curving alignment. The open joints between boards allow both rainwater and drifting sand to flow through, finding their natural places in the dune formations. Native dune vegetation sprouts up through the joints.

1010 ON THE GREEN, ANCHORAGE, ALASKA

A boardwalk in Anchorage, Alaska, illustrates an on-grade deck in a very different type of setting. Figure 14.4 shows the boardwalk giving access to a building called 1010 On the Green which houses offices and bed-and-breakfast rooms. In Anchorage entry boardwalks are common at residences and other small buildings because they are easy to clean of snow (Tamas Deak, personal communication 2003).

FIGURE 14.4 On-grade boardwalk in Anchorage, Alaska.

This one is of native wood. The wood might not be pressure-treated; decay in Alaska's cold climate is very slow. The deck rises to the building in five levels with a wooden riser at each step. The planks rest on longitudinal sleepers. At the sides wooden bulkheads separate out adjacent soil. Moderate frost-heaving could be tolerable, as at the various levels the sleepers are tied flexibly together.

ELEVATED DECKS

Elevated decks are distinguished by posts or foundations which raise the structure above the soil. Beams span from post to post to support the decking. Curbs or railings are required where the deck surface is high above the ground. The surface connects to the ground via stairs, ramps, or an alignment that intersects elevated ground or the floor level of a building.

Elevated decks can take people into sensitive and unruly environments, because their posts suspend them above the earth without the need to level the subgrade. Their construction can require markedly little disturbance of soil or tree roots. Points for seating the foundations can be found in tangled ground such as mangrove roots and rock fields. They are uniquely adaptable to steep slopes. Their heights can be adjusted to rise above flood levels or to allow movement of small wildlife below.

FOUNDATIONS FOR ELEVATED DECKS

Figure 14.5 shows some types of foundations for elevated decks. Various configurations of foundations and posts are put together to keep decay-susceptible materials well drained, assure stability in given soil conditions, and minimize cost.

Posts are commonly made of lumber, masonry, concrete, or metal. Tall posts require bracing to inhibit bending or leaning. Posts set into the ground can be backfilled or collared with concrete, packed aggregate, or packed soil. Concrete can be poured directly into an excavated hole or into tube-like forms. In firm rock substrate, narrow holes can be drilled to accept metal supports. Distinct footings can be added to spread out deck load on soils with limited bearing capacity.

Piles driven into the earth support a deck by transferring the load to a strong soil layer below the surface or to the surrounding soil by friction along their length. Many piles are of wood, steel, or structural fiberglass. For example, piles of 2-inch-diameter steel pipe were used to support pedestrian boardwalks on wet, soft soil at Juanita Bay Park in Kirkland, Washington (www.kirkland.wa.us; Thompson and Sorvig, 2000, p. 143). Because the piles were of small diameter, heavy equipment was avoided; they were installed by one laborer with an air hammer and scaffolding.

Pin and screw foundations are special types of piling. Table 14.8 lists some products of this type. Small ones can be driven or screwed into the soil with little disruption (Chambers, 2001). For example, the Diamond Pier uses pins of galvanized pipe or tubing from 2 to 10 feet in length; two to four pins are hammered at angles into the soil through a special metal or concrete bracket which holds a post or a beam. Helical piers such as those manufactured by A.B. Chance are installed with a rotary torque motor. The lead section of the pier has one or more helical bearing plates that drive the pier into the soil; extensions can be added until the pier hits the required bearing resistance.

The bases of stairs and ramps require their own foundations. Beams or stringers can be laid on concrete foundations or aggregate beds. Figure 14.6 shows a concrete foundation combined with the foot of a ramp to give access to a deck structure.

Where the subgrade is susceptible to swelling, footings can be seated below the zone of seasonal moisture change. Where the subgrade is susceptible to frost action, footings can be seated below frost depth or far enough into the soil such that the weight of the structure concentrated on the footing resists the remaining uplift pressure.

CURBS AND RAILINGS

Curbs are added to some elevated decks to guide and confine users, to restrain wheeled vehicles such as golf carts and wheelchairs, and to prevent stepping off the edge below an open railing.

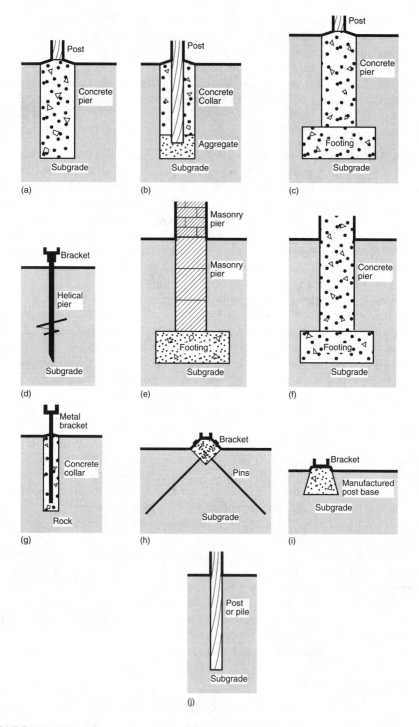

FIGURE 14.5 Examples of foundations for elevated decks.

Decks

TABLE 14.8
Examples of Pin and Screw Foundation Products

Product	Company	Contact Information
Atlas-Helical	Atlas Systems	www.atlassys.com
Diamond Pier and others	Pin Foundations	www.pinfoundations.com
Eco-Tech	Eco-Tech Foundations	www.eco-techfoundations.com
Fasteel	Fasteel Systems	www.fasteel.com
Helical Pier	A.B. Chance	www.abchance.com
Helix Pier	Dixie Anchoring Systems	www.dixieanchor.com
Precision Pier	MC Group	www.mcgroup.us

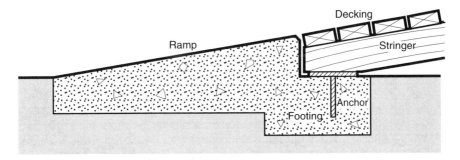

FIGURE 14.6 Example of ramp foundation (adapted from Breeden, 1976, Figure 51).

Railings, more obtrusive than curbs, are added to some decks to prevent falls, to guide partially sighted or nighttime users, to support the handicapped and users at rest, to confine visitors so they do not damage the surroundings, and to increase the perception of safety. Local building codes may specify the conditions where a railing is required. For example, a railing may be required where the deck floor is more than 18 or 24 inches above the ground (McDonald et al., 1996, p. 57). There is a special need for handrails at steps and steep ramps. The potential drawbacks of adding a handrail are the disruption of views for seated persons, perceptual confinement, visual intrusion into the deck's setting, and the catching of flood debris and flood flow forces.

The configuration of a railing is a design question almost independent of the construction of the rest of a deck. Handrails require particularly careful detailing because they are near the eye and under the hands of users. Railings should be impenetrable to small children where children may be present and a fall would be hazardous. Local building codes may specify some features of railings such as railing height and the width of openings.

Potential railing materials are enormously diverse. Rails and their supports are commonly of wood, metal, or plastic. Panels between supports can be of glass, bamboo, stainless steel wires, perforated metal sheets, trellis, prefabricated fencing, spindles, rails, ropes, chains, nets, or fabrics.

TABLE 14.9
Examples of Special Products for Railing Assemblies

Product	Company	Contact Information
Armor-Rail	Shakespeare Composite Structures	www.armor-rail.com
Bufftech	Certainteed	www.bufftech.com
Cable-Rail	Feeney Wire Rope & Rigging	www.cablerail.com
ChoiceDek	Advanced Environmental Recycling Technologies	www.choicedek.com
Clearview	Clearview Railing Systems	www.cvrail.com
Deckorators	Deckorators	www.deckorators.com
Endurance Railing	Railing Dynamics	www.rdirail.com
Fortress Iron Railing	Fortress Iron	www.fortressiron.com
Poly Rail	Digger Specialties	www.polyrail.com
Precision Rail	Mid Atlantic Vinyl Products	www.mvp97.com
Ultra-Tec	The Cable Connection	www.thecableconnection.com
WeatherBest	Louisiana Pacific	www.weatherbest.com

Table 14.9 lists some special products for railing components. They are made of wood, glass, plastic, and metal such as aluminum with baked-on finishes. All can be fitted to specific decks using couplers, sleeves, and other fasteners.

Where an actual falling hazard is not present but may be perceived, increasing the width of a deck might make it "feel" as safe as if it had a handrail (Pierce and Associates, 1998), especially if the extended width is beyond an edge indication such as a row of stanchions.

PREFABRICATED BRIDGES

Table 14.10 lists some sources of deck-like prefabricated bridges. The bridges are constructed of timber, steel, or aluminum. Some are capable of long spans and of carrying substantial loads. They come in standard configurations, or can be customized for specific applications (Winterbottom, 1995b). They must be supported by adequate site-specific foundations.

ELM BROOK CROSSING, MINUTEMAN NATIONAL PARK, MASSACHUSETTS

A boardwalk in Minuteman National Park in Massachusetts illustrates many features of elevated deck construction as they were resolved for one project. The park's Battle Road Trail gives walkers and bikers access to historic sites across fields, forests, and wetlands (www.nps.gov/mima; Hammatt, 2002). Figure 14.7 shows how the trail becomes a boardwalk where it crosses the Elm Brook floodplain wetland. Most of the boardwalk is level; it is considered universally accessible. It winds around the wetland's few trees in a series of straight segments.

The deck surface is a couple of feet above the wetland soil surface. This is high enough not to obstruct flood flows or migration of small wildlife, but low enough that a safety rail is not required (Hammatt, 2002). Instead, the deck has 4 inch × 4 inch curbs, held above the decking by segments of 2-inch-thick lumber. The curbs and

TABLE 14.10
Examples of Suppliers of Prefabricated Bridges

Company	Contact Information
Continental Bridge	www.continentalbridge.com
Echo Bridge	www.echobridgeinc.com
EnWood Structures	www.enwood.com
Excel Bridge Manufacturing	www.excelbridge.com
Hughes Brothers	www.hughesbrothers.com
Marine Bridge & Iron	www.marineworks.com
North American Bridge	www.nabcousa.com
PermaPost	www.permapost.com
Redd Team	www.reddteam.com
Steadfast Bridges	www.steadfastbridge.com
Western Wood Structures	www.westernwoodstructures.com
Wheeler Consolidated	www.wheeler-con.com

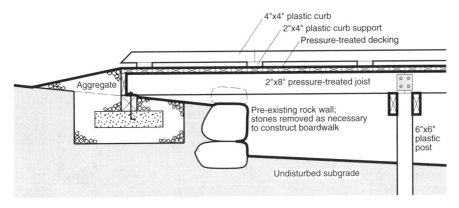

FIGURE 14.7 Boardwalk crossing the Elm Brook floodplain on the Battle Road Trail, Minute Man National Park, Massachusetts (after drawing by Carol R. Johnson Associates in Hammatt, 2002).

their supports are of gray plastic lumber. The omission of handrails gives the walk an open character with unobstructed views of the surrounding wetlands.

Post holes 12 inches in diameter were dug 4.5 feet into the wetland soil. Posts were dropped into the holes and the holes were backfilled around the posts with native soil. The posts are of plastic lumber to avoid contact of wood preservatives with the wetland. Pressure-treated beams support longitudinal joists. At the edge, the decking cantilevers 6 inches beyond the outermost joists so the deck appears to "float" above the wetland vegetation.

The trail's transition onto the boardwalk is at a preexisting stone wall which makes the change in elevation from trail above to the wetland below. A 6 inch × 6 inch plastic beam on a footing holds the ends of the joists. The aggregate trail is firmly compacted so it will not subside relative to the boardwalk structure.

Arcadia Sanctuary, Easthampton, Massachusetts

Deck construction was resolved in different ways for a boardwalk at Arcadia Sanctuary in Easthampton, Massachusetts (www.massaudubon.org; Chambers, 2001). Figure 14.8 shows how the boardwalk gives visitors and schoolchildren access to the wooded floodplain of the Mill and Connecticut Rivers. It is considered fully accessible. It is accessed at floor level from the sanctuary's visitor center; outward from there its surface is level or very gently sloping. It follows the edge of a pond, winding around preexisting trees and taking advantage of wetland views.

The foundations are helical piers. The piers were driven 5 feet into the wetland soil with essentially no disturbance of the soil, its rooting zones, or its flows of air and water (Steven Walker, personal communication 2003). The deck elevation is set so the bottom of the beams is just above the 100-year flood level.

Wooden posts support handrails. PVC-coated wire-mesh panels stapled to the railing frame provide safety for little children while allowing an open view.

The deck's lumber is a combination of materials selected for finish and cost. The decking and cap rail are of composite lumber or naturally decay-resistant red cedar to minimize splinters. The other components are of decay-resistant black locust or pressure-treated wood for strength. Only the galvanized steel shafts of the helical piers are in contact with wetland soil and water.

The end of the boardwalk adjoins an open field. On the meadow side the deck surface is only 6 to 12 inches above the ground, so the railing is omitted. In its place is a 2 inch × 4 inch wooden curb separated from the decking by small wooden blocks.

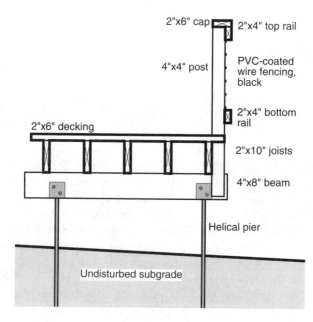

FIGURE 14.8 Boardwalk at Arcadia Sanctuary in Easthampton, Massachusetts (after drawing by Walter Cudnohufsky Associates, in Chambers, 2001).

Decks 557

The end of the boardwalk is supported by a sleeper resting on well-drained aggregate. Further aggregate ramps up to meet the decking without the need for a step.

DECKS IN PRACTICE

The following case studies illustrate the adaptability of decks to diverse settings and purposes.

MINNESOTA LANDSCAPE ARBORETUM, CHANHASSEN, MINNESOTA

Figure 14.9 shows a deck constructed around a landmark tree in the Rock Garden portion of the Minnesota Landscape Arboretum in Chanhassen, Minnesota (www.arboretum.umn.edu). Wooden posts support the deck a foot or two above the ground surface. The deck's pressure-treated wood is stained silver-gray. The deck and its attached bench wrap around the preexisting tree with no curb or railing. The tree's roots grow through the soil under the deck essentially undisturbed by the structure.

COLUMBUS ZOO, COLUMBUS, OHIO

Boardwalks at the Columbus Zoo in Columbus, Ohio (www.colszoo.org) illustrate a variety of construction treatments corresponding to the various kinds of animal habitats. The one in Figure 14.10 makes a gently sloping ramp as it enters the area housing African lions. The decking planks are set perpendicular to the slope for traction

FIGURE 14.9 Deck in the Rock Garden portion of the Minnesota Landscape Arboretum in Chanhassen, Minnesota.

FIGURE 14.10 Deck in the lion area of the Columbus Zoo, Columbus, Ohio.

and accessibility. Posts set the deck's surface elevations without disturbing the surrounding vegetated slope. The railing is a palisade of slender logs of different lengths and diameters, set in random order, simulating construction with rough-hewn timber in a safari-like setting. Elsewhere in the zoo are boardwalks featuring sleek wire-mesh panels, smoothly sawn timber stained silver-gray, diagonally patterned decking boards, glass panels, and wide sloping handrails, all corresponding perceptually to the types of animals and landscapes around them.

RIVERWALK, JACKSONVILLE, FLORIDA

Figure 14.11 shows a deck supporting heavy public traffic along the Riverwalk boardwalk in Jacksonville, Florida. The city developed the Riverwalk to connect

Decks

FIGURE 14.11 The Riverwalk boardwalk at Jacksonville, Florida.

museums, hotels, public plazas, commercial areas and water-taxi landings along the St. Johns River waterfront. The deck is supported in some places by concrete seawalls and at others by large wooden piles reminiscent of old wharves. The water side is continuously secured by railings of various sturdy designs.

THE LANDINGS, SKIDAWAY ISLAND, GEORGIA

Figure 14.12 shows a deck used for golf-cart travel. It is part of an extensive system of trails connecting residential neighborhoods and commercial and recreational centers in a planned community called The Landings on Skidaway Island, Georgia (www.thelandings.com). In this setting golf carts partly replace autos. The boardwalk curves artfully as it follows pockets of stable soil across a tidal marsh.

FIGURE 14.12 Golf cart trail at The Landings, Skidaway Island, Georgia.

Posts support decking a couple of feet above the wetland surface. A curb at the side restrains the wheels of golf carts; it is raised from the walk surface for debris discharge.

PUNCHEON BOARDWALKS IN ALASKA

In Alaska, puncheon boardwalks are used to stabilize rural trails for pedestrians and off-highway vehicles across wet and muddy soil (Meyer, 2002, pp. 35–38). Where convenient the lumber is harvested locally from beetle-killed spruce. The U.S. Forest Service has a standard construction technique of this type which it calls Puncheon with Decking; Figure 14.13 illustrates the construction. Logs 8 inches in diameter are used for sleepers and stringers. Larger logs are split into decking slabs 2 inches thick and 6 feet long. Very small logs are used directly for further decking. Planking cut from tapered logs conveniently accommodates curves in the trail alignment. The timbers are fastened to each other with pins and spikes. In Alaska's climate, installations of this type have been in place for 15 years and more with only occasional replacement of surface planks.

Similar wooden boardwalks are common in Inuit villages in Alaska's Yukon-Kuskokwim delta region (Tamas Deak, personal communication March, 2000). This area is a flat, poorly drained coastal plain with numerous lakes, bogs, and sluggish streams. The alluvial soil is shallow over permafrost and constantly wet (Gallant et al., 1995, pp. 32–34). The villages are built on wet tundra with limited high ground. Extensive boardwalk systems connect the various parts of each village.

Decks

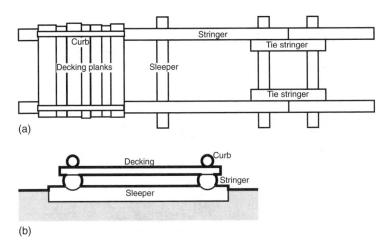

FIGURE 14.13 Plan and section of puncheon boardwalk (after U.S. Forest Service Standard Drawing 93-2, in Meyer, 2002, p. 37).

NEW RIVER GORGE, WEST VIRGINIA

Figure 14.14 shows a deck negotiating an extraordinarily steep slope at the New River Gorge National River in West Virginia (www.nps.gov/neri). It gives pedestrians from the Canyon Rim Visitor Center access to observation decks overlooking the deep river gorge and its landmark bridge. The natural wood material is intended to harmonize with the rugged native setting. Numerous flights of stairs connect level segments of boardwalk. Posts of various lengths support both beams and railings. The roots of native trees overlap freely through the largely undisturbed soil. Enormous earthwork would have been required for a ground-contact pavement.

MAR-VISTA RESTAURANT, LONGBOAT KEY, FLORIDA

Figure 14.15 shows a deck used for outdoor dining at the Mar-Vista Restaurant on Longboat Key, Florida. The restaurant is located beside a marina on Sarasota Bay. The deck is built on-grade, of pressure-treated lumber. It is accessed from a shell walkway by a short wooden ramp. The surface is only a few inches above the surrounding ground, and level with the floor in the rest of the restaurant. Wheelchairs vibrate only slightly as they are pushed across the surface. The legs of plastic chairs and tables rest stably on the surface, spanning over the narrow joints. Wooden bulkheads segregate the deck material from adjacent soil. The deck's edge is marked by a row of plantings and a thick braided rope draped between wooden posts.

JAPANESE GARDEN, MISSOURI BOTANICAL GARDEN, SAINT LOUIS

Figure 14.16 shows a deck intended for aesthetic experience in the Japanese Garden portion of the Missouri Botanical Garden in Saint Louis. The deck is a series of stepped platforms with no curb. Wooden piles support sturdy planks about 18 inches

FIGURE 14.14 Boardwalk at New River Gorge National River, West Virginia.

above the ground and water. Each plank span is in one direction in the zig-zagging alignment, and supports the planks in the next direction. In Japanese tradition following the crooked route is supposed to free one from evil spirits (Holborn, 1978, pp. 90–92). Walking on the narrow plank surface requires concentration. The winding alignment encourages one to linger and to be drawn to the beauty of the irises in the artificial marsh below. Each turn in the walk brings a different vista of the flowers.

TRADEMARKS

Table 14.11 lists the holders of registered trademarks mentioned in this chapter.

FIGURE 14.15 Dining deck at Mar-Vista Restaurant on Longboat Key, Florida.

FIGURE 14.16 Boardwalk in the Japanese Garden section of the Missouri Botanical Garden in Saint Louis.

TABLE 14.11
Holders of Registered Trademarks Mentioned in This Chapter

Registered Trademark	Holder	Headquarters Location
Armor-Rail	Shakespeare Composite Structures	Newberry, South Carolina
Atlas-Helical	Atlas Systems	Independence, Missouri
Bedford Plastic Timbers	Bedford Technology	Worthington, Minnesota
Boardwalk	CertainTeed	Valley Forge, Pennsylvania
Bufftech	Certainteed	Valley Forge, Pennsylvania
Cable-Rail	Feeney Wire Rope & Rigging	Oakland, California
Carefree Xteriors	U.S. Plastic Lumber	Boca Raton, Florida
ChoiceDek	Advanced Environmental Recycling Technologies	Junction, Texas
Clearview	Clearview Railing Systems	Orangeville, Ontario
CorrectDeck	CorrectDeck	Biddeford, Maine
Deckorators	Deckorators	Saint Louis, Missouri
Diamond Pier	Pin Foundations	Gig Harbor, Washington
Dream Deck	Thermal Industries	Pittsburgh, Pennsylvania
Eco-Tech	Eco-Tech Foundations	Saint Marys, West Virginia
Endurance Railing	Railing Dynamics	Egg Harbor Township, New Jersey
EverGrain	Epoch Composite Products	Lamar, Missouri
EZ Deck	PlayPower	Saint Louis, Missouri
Fasteel	Fasteel Systems	Crestwood, Missouri
Fiberon	Fiberon	New London, North Carolina
Fortress Iron Railing	Fortress Iron	Richardson, Texas
Geodeck	Kadant Composites	Bedford, Massachusetts
Helical Pier	A.B. Chance	Centralia, Missouri
Helix Pier	Dixie Anchoring Systems	Memphis, Tennessee
Mister Boardwalk	Mister Boardwalk	Point Pleasant, New Jersey
Monarch	Green Tree Composites	Saint Clair, Michigan
Neoterra	Mateflex	Utica, New York
Nexwood	Nexwood Industries	Fallbrook, California
Polywood	Polywood	Edison, New Jersey
Precision Pier	MC Group	Peachtree City, New Jersey
Rhino Deck	Master Mark Plastics	Albany, Minnesota
Stone Deck	Deck Technologies	Woodbury, Minnesota
Superdeck	Superdeck Marketing	Minneapolis, Minnesota
TenduraPlank	Tendura	Troy, Alabama
Timberlast	Kroy Building Producttts	York, Nebraska
TimberTech	TimberTech	Wilmington, Ohio
Trex	Trex	Winchester, Virginia
WeatherBest	Louisiana Pacific	Portland, Oregon
Continental Bridge	Continental Bridge	Alexandria, Minnesota
Echo Bridge	Echo Bridge	Elmira, New York
EnWood Structures	EnWood Structures	Morrisville, North Carolina
Excel Bridge Manufacturing	Excel Bridge Manufacturing	Santa Fe Springs, California
Hughes Brothers	Hughes Brothers	Seward, Nebraska
North American Bridge	North American Bridge	Tampa, Florida
PermaPost	PermaPost	Hillsboro, Oregon

Decks

TABLE 14.11 Continued

Registered Trademark	Holder	Headquarters Location
Redd Team	Redd Team	Keystone Heights, Florida
Steadfast Bridges	Steadfast Bridges	Fort Payne, Alabama
ThruFlow	Otron Tech	Wallaceburg, Ontario
Western Wood Structures	Western Wood Structures	Tualitin, Oregon
Wheeler Consolidated	Wheeler Consolidated	Bloomington, Minnesota

ACKNOWLEDGMENTS

The following persons generously provided information used in this chapter: Jim Bernier of Consumers Energy Hydro Operations (http://cmsenergy.com); Steven M. Bopp of Austin Tao & Associates (http://austintao.com/); R. Chowdhary of QuatChem; Tamas Deak of Koonce Pfeffer Bettis (www.kpb-architects.com); Dave Fowlie of Chemspec; Doug Johnson of Superdeck Marketing (www.superdecksystems.com); Robin Lindsay of Wagga Iron (www.wagga-iron.com.au); Kenneth L. Mitchell of the California Department of Parks and Recreation; Mike and Jan Riter of Trail Design Specialists (http://traildesign.tripod.com); and Steven Walker of the Massachusetts Audubon Society.

The following persons constructively critiqued early drafts of this chapter: José Buitrago of the University of Georgia, and Clayton Eads of the Deck Industry Association (www.deckindustry.org).

REFERENCES

Breeden, J. Brooks (1976). Decks and Overhead Structures, in *Handbook of Landscape Architectural Construction*, pp. 511–531, Jot D. Carpenter, ed., Washington: Landscape Architecture Foundation.

Chambers, Joseph (2001). Light on their Feet, *Landscape Architecture* 91, 38–42 and 94–99.

Church, Thomas Dolliver (1955). *Gardens Are for People, How to Plan for Outdoor Living*, New York: Reinhold.

Fishbeck, Gary M., and Jeffrey D. Blankenship (1998). Wood Decks and Boardwalks, in *Time-Saver Standards for Landscape Architecture*, section 460, 2nd ed., Charles W. Harris and Nicholas T. Dines, Eds., New York: McGraw-Hill.

Gallant, Alisa L., Emily FR. Binnian, James M. Omernik, and Mark B. Shasby (1995). *Ecoregions of Alaska*, Professional Paper 1567, Washington: U.S. Geological Survey.

Gibbs, Harold R. (1992). Wood and Wood Structures, in *Handbook of Landscape Architectural Construction, Materials, Volume 4* pp. 231–288, Scott S. Weinberg and Gregg A. Coyle, Eds., Washington: Landscape Architecture Foundation.

Hammatt, Heather (2002). Retreating through History, *Landscape Architecture* 92, 64–71 and 89–90.

Harris, Charles W., Jefrey D. Blankenship and Tess Canfield (1998). Pedestrian Bridges, in *Time-Saver Standards for Landscape Architecture*, section 470, 2nd ed., Charles W. Harris and Nicholas T. Dines, Eds., New York: McGraw-Hill.

Holborn, Mark (1978). *The Ocean in the Sand, Japan: From Landscape to Garden*, Boulder: Shambhala Publications.

Illman, Barbara, and Les Ferge (1998). *Effect of Climate on Durability of Wood* and *On-Site Protection of Structures*, Madison, WI: U.S. Forest Service, Forest Products Laboratory.

Lebow, Stan (2001). *Environmental Impact of Preservative-Treated Wood*, Madison, WI: U.S. Forest Service, Forest Products Laboratory.

Library of Congress (no date). *Today in History, June 26: On the Boardwalk*, at memory.loc.gov.

McDonald, Kent A., Robert H. Falk, R. Sam Williams, and Jerrold E. Winandy (1996). *Wood Decks: Materials, Construction, and Finishing*, Madison, WI: Forest Products Society.

Meyer, Kevin G. (2002). *Managing Degraded Off-Highway Vehicle Trails in Wet, Unstable, and Sensitive Environments*, 2E22A68-NPS OHV Management, Missoula, MT: U.S. Forest Service Technology and Development Program.

O'Connell, Kim A. (1999). Designing ADA-Compliant Tree Grates, *Landscape Architecture* 89, 22.

Pierce, James, and Associates (1998). *Boardwalk Design Guide* and *Boardwalk Engineering Guide*, Gatton, QL, Australia: Gatton Sawmilling (www.outdoorstructures.com).

Ritter, Michael A., and Kim Stanfill-McMillan (1995). Wood Bridges in New England, in *Restructuring: America and Beyond, Procedures of Structures Congress 13*, pp. 1081–1084, Masoud Sanayei, Ed., New York: American Society of Civil Engineers.

Robbins, Alan E. (2002). *2001–2002 State of the Recycled Plastic Lumber Industry,* Plastic Lumber Trade Association (downloadable from www.plasticlumber.org).

Thompson, J. William, and Kim Sorvig, (2000). *Sustainable Landscape Construction, A Guide to Green Building Outdoors*, Washington: Island Press.

Watson, Gary W., and E. B. Himelick (1997). *Principles and Practice of Planting Trees and Shrubs*, Savoy, IL: International Society of Arboriculture.

West, Anne W. (2003). The CCA Story, Fact vs. Fiction, *Professional Deck Builder* 2, 10–16, 58, and 93.

Wilson, Alex (1999a). Recycled Plastic Lumber in the Landscape, *Landscape Architecture* 89, 56–62 and 139.

Wilson, Alex (1999b). Using Wood Outdoors, *Landscape Architecture* 89, 38–43.

Wilson, Alex (2002a). CCA to Be Phased Out, *Environmental Building News* 11, 4–5.

Wilson, Alex (2002b). Treated Wood: An Update, *Landscape Architecture* 92, 46–50.

Winterbottom, Daniel (1995a). Plastic Lumber, *Landscape Architecture* 85, 34–36.

Winterbottom, Daniel (1995b). Wooden Wonders, *Landscape Architecture* 85, 30–33.

Index

AASHTO, 64, 72
  see also Standards
Accessibility
  block pavement, 329
  decks, 542
  porous concrete, 421
  soft porous surfacing, 517–518
Admixtures, 429–430
  definition, 419
Adsorption, definition, 153
ADT (Average Daily Traffic), definition, 71
ADV (Average Daily Volume), definition, 71
Aesthetics, 26–29; *see also* Appearance
Aggregate, 42–43, 52–58, 199–240; *see also*
    Flexible pavement
  California Bearing Ratio, 206–207, 234
  cost, 22
  definition, 43, 200, 201
  displacement by traffic, 215–216, 226
  dust generation, 216–217
  ESCS, *see* Expanded slate, clay and shale
  gradation, *see* Gradation
  in base reservoir, 233–235
  in block joints, 332–333, 338–339, 345, 349–356
  in freezing climate, 93, 209–210, 215
  in geocells, 57–58, 287, 309–314
  in open-celled grids, 385
  in porous asphalt, 461–462, 468–471, 505
  in porous concrete, 421, 422, 427–428
  in structural soil, 180–181
  infiltration rate, 124, 206
  maintenance, 219–220
  on swelling soil, 105, 215, 231–234
  on uncompacted subgrade, 215
  open-graded, *see* Open-graded
  porosity, 190–191, 206, 233–234
  recycled, 214–215, 227–228, 396
  root heaving, 179, 215
  runoff coefficient, 126
  stability, 75, 200–202

  stabilizing binders, 217–219, 222–223
  standards for, 200
  water storage capacity, 132
Air-entrained soil, 184–187
  see also Arnold, ESCS, Structural soil
Air entrainment, 429–430, 451
  definition, 419
Alabama, Mobile, 57–58, 287–289
Alaska
  Anchorage, 291, 549–550
  Palmer Hay Flats, 291–293
  puncheon boardwalks, 560–561
  White Mountains National Recreation Area, 293
Ambergrid concrete grid, 388–390
Amsterdam sand, *see* Amsterdam tree soil
Amsterdam tree soil, 187–189
Andropogon Associates
  Dupont Agricultural Chemicals, 478–479
  Morris Arboretum, 493–494
  Siemens office, 136, 486–487
Antecedent moisture, definition, 123
Appearance, 26–29
Aquifer, definition, 143
  see also Groundwater, Subsurface disposition
Arizona
  Chandler, 495–496
  Flagstaff, 310–312
  Phoenix, 266–267
Arnold, Henry, 55–56, 184–187, 191, 347–349
  see also Air-entrained soil, Metrotech, Pier A Park
Asphalt, *see also* Flexible pavement, Porous asphalt
  components and proportions, 458–463
  concrete, definition, 458, 459
  definition, 43, 458, 459
  recycling, 484
ASTM, 64, 71, 72, 200, 326, 382, 460
  see also Standards
Atterberg limits, 71, 77
Australia, Sydney, 171–172

567

Base course, 36–37
  extension at edge, 106–107
  reservoir, *see* Reservoir
  rooting media, *see* Structural soil
Base flow, definition, 143
Base flow, infiltration effect on, 150–152
Base reservoir, *see* Reservoir
Bassett, James, 271–272
Beam, definition, 541
Bearing value, *see* California Bearing Ratio
Beauty, *see* Aesthetics, Appearance
Bedding (bed, bedding layer), definition, 325, 382
Best management practice, definition, 120
Biodegradation, definition, 153
Biodegradation, of oil, 159–160, 482–483
Biological oxygen demand (BOD), definition, 153
Bitumen, definition, 459
Bituminous concrete, definition, 459
Black-layer hypoxia, 244, 262
Blister, asphalt, definition, 459
Block, *see also* Flexible pavement, Open-jointed block
  definition, 43, 325, 382
  natural stone, 368–371
  open laying patterns, 372–373
  porous concrete, 366
  porous stone, 366
  root heaving, 179
  spacers for, 372–373
  wooden, 366
Boardwalk, *see also* Deck
  definition, 541
BOD, *see* Biological oxygen demand
Bollards, 109–113, 192
Borgwardt, Soenke, 334, 336, 360, 366, 373
Brick, 370–372
  crushed, 214, 220–222, 396
  definition, 325
  open-celled, 410–411, 413
Bridge, prefabricated, 554–555
Brownfield, 6, 152, 345–347
Building, infiltration near, 152
Built cover, *see* Impervious cover
Bulk density, definition, 71, 201
Bulk volume, definition, 201
Bunch grass, definition, 244

Cadmium, *see* Metals
Cahill, Thomas (Cahill Associates)
  base compaction, 234
  contemporary asphalt technology, 471–472
  edge drain, 41–42, 471
  geotextile installation, 88–89
  hydrologic performance, 146
  infiltration rate protection, 92, 145
  pavement thickness experience, 103, 474
  porous asphalt pavement construction, 472
  porous asphalt cost, 476–479
  Siemens office, 134–135, 486–487
Calcined, definition, 201
California
  Avila Beach, 404–405
  Carpinteria, 548–549
  Central Valley, 160
  Cerritos, 450
  Davis, 267–268
  Fair Oaks, 449–450
  Guerneville, 524, 527
  La Palma, 305–306
  Livermore, 360–362
  Long Beach, 304–305
  Los Angeles, 162–163, 222–223, 394–395
  Pasadena, 290–291
  Pomona, 364–366
  Santa Clarita, 111, 355–356
  Santa Monica, 308–309
California Bearing Ratio, 76–77, 79–80, 206–207, 234
  definition, 71, 173
Caliper, definition, 173
Canopy, tree, definition, 173
Cantilever, definition, 541
Capillary, definition, 173
Cast in place, definition, 382
Castellation, definition, 382
Catchment, definition, 123
Cation exchange capacity (CEC), definition, 153, 161, 173–175
CBR, *see* California Bearing Ratio
CEC, *see* Cation exchange capacity
CeePy concrete block, 129–131
cfs, definition, 123
Chamfer, definition, 325
Checkerblock concrete grid, 388–390
Chemical oxygen demand (COD), definition, 153
Chloride, 154–155; *see also* Deicing salt
Choke layer (choker course), 234–235, 466, 471
Church, Thomas, 540–541
Coarse-graded (coarse-grained, coarse-textured), definition, 75, 204
Cobblestone, 368
COD, *see* Chemical oxygen demand
Coefficient of uniformity, aggregate, 203–204
Cohesion, 76
  definition, 71
Cold climate, *see also* Freezing, Frost, Ice formation, Winter maintenance
  aggregate, 93, 209–210, 215
  geocell, 93, 286, 310–312
  open-jointed block, 99–102

Index

Cold climate (*continued*)
    porous asphalt, 491–492, 494–495
    porous concrete, 450–452
    turf, 262, 277–280
Compaction, 89–93, 431–432
    aggregate, 217
    base course, 234–235
    curve, 90
    effect on soil infiltration rate, 144–145
    effect on tree rooting, 192
    requirement, 91–92
    test, *see* Proctor
Composite lumber, 546–547
    definition, 541
Composite pavement, definition, 83
Computer, *see* Software
Concrete, *see also* Porous concrete
    definition, 43, 418, 419, 459
    rubble, 214, 364–366
    slabs, manufactured, 363–364
Connecticut, Farmington, 261–262, 298–301
Connecticut, Waterford, 353–355
Construction
    open-jointed block, 329–332
    geocell, 286–287
    open-celled grid, 384–386
    porous asphalt, 471–475
    porous concrete, 430–435
    protection during, 61–63, 353–355
    turf, 250–257
Construction Specifications Institute, 64
Contractor, communication with, 65, 473
Contractor, qualifications and certification, 65, 489
Copper, *see* Metals
Cornell structural soil, 180–184
Cost
    aggregate, 22
    geocell maintenance, 287–289
    porous asphalt overlay, 497, 503, 506–506
    porous asphalt, 475–479
    porous concrete, 22–23, 421–422, 444, 449
    soft porous surfacing, 515
    stormwater control, 10, 22–24, 475–479
Course, definition, 36
Critical height, 229–233, 522
    see also Playground safety
Crushed gravel, definition, 201
Crushed stone, definition, 201
Crusher run, 204, 211
Cultivar, 245
    definition, 244
Curb, pavement, 83, 107–111, 194, 330–331
Curing, definition, 418, 419
Curve number, 123, 126
Cutback asphalt, definition, 459

Darcy's law, 123
dbh, definition, 173
Decibel, definition, 19
Deck, 24–25, 51, 55, 539–566
    components, 542–544
    construction, elevated, 550–557
    construction, on-grade, 547–550
    curb, 541, 550–552
    definition, 539, 541
    environmental effects, 542–544
    history, 540–541
    on swelling soil, 105
    railing, 550–552
Decking, definition, 541
Decomposed granite, 201, 211, 222–223, 310–312
Deicing salt, 154, 262, 407
    see also Chloride
Delaware
    Barley Mills, 478–479
    Hockessin, 471–472
    Newark, 234–235, 466–468
Denmark
    Copenhagen, 189
    porous asphalt, 481, 500
    water quality observation, 160
Dense pavement, definition, 1
Dense pavement, infiltration and runoff, 125–126
Dense-graded, definition, 75, 204
Depression storage, 121, 123
Design storm, definition, 120
    see also Precipitation frequency
Development intensity, 5–6
Direct runoff, definition, 123
District of Columbia, Washington, 50, 184, 524–525
Drainage
    area, definition, 123
    at pavement edge, 59–63
    below pavement surface, 129–131
    layer, 38; *see also* Reservoir
    outlet, *see* Pipe outlet
Drain-down, asphalt, 48, 464–466, 470
    definition, 459
Drainstone concrete block, 339–344
Dripline, definition, 173
Durability, definition, 201
D_x, definition, 204

ECO I Paver concrete block, 363
EcoCreto porous concrete, 430
EcoGrid concrete grid, 410
Ecoloc concrete block, 342–347
Eco-Logic concrete block, 55–56, 347–358

Ecosystem preservation
  deck, 24–25, 550–551, 554–557, 559–560
  porous concrete, 25–27, 443–444
Edge treatments, 59–63, 106–112, 331, 448
Egra Stone concrete block, 363
Energy cost, soft porous surfacing, 515
Energy use in buildings, 16
Engineered wood fiber, 525–528
England
  A45 Trunk Road, 480–481
  Nottingham, 140–142, 157–160
  Wheatley, 142
Equivalent Single-Axle Load (ESAL), 70, 71
Equivalent Wheel Load, (EWL), definition, 71
ESAL, see Equivalent Single-Axle Load
ESCS, see Expanded slate, clay and shale
Evaporation, definition, 120
Evapotranspiration
  cooling effect, 17
  definition, 120
  infiltration effect on, 150–152
EWL, see Equivalent Wheel Load
Exfiltration, definition, 143
Expanded slate, clay and shale
  aggregate, 211–214
  CEC, 184–185, 187
  definition, 173, 201
  durability, 213–214
  standard gradations, 213
  tree rooting media, 184–187
  turf rooting media, 255–256
  water-holding capacity, 190–191

Fence, edging, 110, 112
Fibers, polymer, in porous asphalt, 469
Field stone, 368
Fill, infiltration into, 152
Filter criteria, aggregate, 236–237
Filter criteria, geotextile, 237–238
Filter layer, 36, 42, 86–87, 235–238
Fine-grained (fine-textured), definition, 75, 204
Fines, definition, 71, 201
First flush, definition, 153
Flag (flagstone), definition, 325
Flexible pavement, 82–83
Florida
  Brandon, 44
  Cape Coral, 420
  Fort Myers, 420
  Hillsborough County, 268–269
  Hobe Sound, 441–442
  Jacksonville, 558–559
  Lake Wales, 524, 526
  Longboat Key, 529, 561, 563
  Miami, 29, 297–298
  Naples, 420
  North Fort Myers, 420
  Orange County, 479
  Sanford, 422
  Stuart, 439–443
  Tampa, 11, 422–426
  Wilton Manors, 350–353
Flow, asphalt, definition, 459
Fly ash, definition, 419
Fly ash, in porous concrete, 428
Footing, definition, 541
Forest, urban, see Trees, urban
Formpave concrete block, 142
Foundation, deck, 541, 551–553
Fraction, definition, 71, 204
France
  Nantes, 502
  porous asphalt in, 480–481
  Rezé, 156–157
Franklin Institute, 458, 463–464
Freezing, see also Cold climate, Frost
  pavement adaptation to, 93, 96–102
Friction, porous asphalt, 498–499
Frost, see also Cold climate, Freezing
  depth, 93–96

Gage number, see Sieve number
Gap-graded, definition, 173, 204
Geoblock geocell, 289–295
Geocell, 25, 29, 43, 45, 285–322
  base course, 314–317
  in cold climate, 93, 286, 310–312
  installation, 286–287
  maintenance, 287–289
  on swelling soil, 105, 286, 301–302
  on uncompacted soil, 286
  root heaving, 179
  with aggregate, see Aggregate
  with turf, see Turf
Geomembrane, 36, 42
Georgia
  Alpharetta, 422, 447–448
  Athens, 25, 220–222, 228–232
  Atlanta, 45, 162–163, 340–344, 410–411, 444–447
  Buford, 269–270
  Guyton, 431
  Macon, 464–465, 473–474
  Newton, 27, 443–445
  Roswell, 472, 477–478
  Savannah, 24
  Skidaway Island, 559–560
  Stone Mountain State Park, 318–319
Georgia Department of Transportation, 471, 472, 504–507

Index

Geotechnical survey, 79–81
Geotextile, 36, 42, 86–89, 235, 237–238
   conditions requiring, 88
Geoweb, 318–319
Germany, Duisburg, 367–368
Germany, Weinsberg, 158, 500–502
Glenn Rehbein Company, 277–280
Gradation, 202–205, 207–209, 216–217
   definition, 71, 202, 204
Graded aggregate base, 211
Graded, definition, 204
Grass Grid concrete grid, 403
Grass, *see also* Turf
   appearance, 26–29
   cooling effect, 17
   cool-season, 244, 248–250
   crown, definition, 244
   Manning's roughness, 128
   warm-season, 244, 246–248
Grasscrete concrete grid slab, 393–398
Grasspave^2 geocell, 57–58, 287–289, 295–303
Grassroad Pavers geocell, 303–306
Grasstone concrete grid, 410
Grassy Pavers geocell, 306–309
Grate, definition, 541
Gravel, 201, 211
   in playgrounds, 229–234
Gravelpave^2 geocell, 57–58, 287–289, 309–314
Grids, *see* Open-celled grids
Groundwater, 8, 10, 143, 152
   recharge, 143, 148–150

Haag, Richard, 515
Hard asphalt, definition, 459
Head, hydraulic, definition, 123
Heat island, 15–18
   see also Turf temperature
Heavy metals, *see* Metals
History
   decks, 540–541
   geocells, 286
   open-celled grids, 382–384
   open-jointed block, 324–325
   porous asphalt, 438–460, 463–464, 466–468, 496–498, 502–507
   porous concrete, 418–420
   soft porous surfacing, 514
   turf, 243, 250–252
Hoboken South Waterfront, *see* Pier A Park
Horizons, soil, 71, 81, 143
Hydration, definition, 418, 419
Hydraulic conductivity, *see* Saturated hydraulic conductivity, Infiltration rate
Hydrocarbons, *see* Biodegradation, Oils

Hydrogel, in structural soil, 173, 181–182
Hydrology, 119–170
   decks, 542
   see also Groundwater, Infiltration, Runoff, Stormwater, Watershed, Water quality

Ice formation, 499
Illinois, Schaumburg, 15
Impervious cover, *see also* Watershed
   effects, 9
   extent, 1–5
InfiltraStone concrete block, 363
Infiltration, *see also* Drainage below pavement surface
   cumulative, 161–164
   ponding time, *see* Reservoir ponding time
   soil, *see* Subsurface disposition, Subgrade infiltration
Infiltration rate
   block pavements, 332–333, 338–339
   pavement surface types, 122–125
   porous asphalt, 479–484, 501, 505
   porous concrete, 422
   sand, 124
   soil, 143–145
   turf, 262–264, 276
Installer, *see* Contractor
Interlock, in block pavements, 337–339
Internal friction, definition, 71
Iron, in runoff, 426
   see also Metals
Irrigation, tree, 187
Irrigation, turf, *see* Turf maintenance

James, William, 121, 325, 334, 349
Joist, definition, 541
Joints, between blocks, definition, 325
Joints, in concrete, 419, 432–435
Jordan Cove research project, 353–355

Kansas, Overland Park, 369
kip, definition, 71
Kozeliski, Frank, 452

Langsdorff, *see* von Langsdorff
Lateral outlet, *see* Pipe outlet
Lawn Sav'r geocell, 317
Lawn, *see* Turf
Laying bed, definition, 325, 382
Lead, *see* Metals
Lift, in fill and aggregate placement, 217
   definition, 173
Limerock, 201, 211, 427
Liner, *see* Geomembrane
Liquid limit, definition, 71

Load, of water constituent, definition, 153
Loam, definition, 75
Los Angeles abrasion test, 209, 214
Lumber, *see* Composite lumber, Plastic lumber, Wood
Lyle, John, 364–366

Macropores, 173, 190–191
Magnesium sulfate test, 209–210, 214
Maintenance, *see also* Irrigation, Winter maintenance
  aggregate, 219–220
  dense asphalt, 287–289
  geocell, 287–289
  open-jointed block, 334–335
  porous asphalt, 466, 483–484
  porous concrete, 426, 440–441
  porous pavement, 65–66
  turf, 243, 257–262
Manning's equation, 123, 127
Manning's roughness coefficient, 127–128
Marshall stability, 459, 462–463
Maryland
  Annapolis, 225–226, 408–410, 484–486
  Baltimore, 401–402
  Frostburg, 312–314
Massachusetts
  Boston, 162–163
  Easthampton, 556–557
  Minuteman National Historic Park, 217–219, 554–555
  Walden Pond State Reservation, 66, 94, 127, 235, 488–493
McHarg, Ian, 52
meq, definition, 173
Metal plates, paving, 367–368
Metals, in stormwater, 155–158, 425–426, 482
Metrotech, 14–15, 185–186
Michigan, Southfield, 302–303
Micropores, 173, 190–191
Minnesota
  Chanhassen, 51, 537
  Lino Lakes, 277–278
  Minneapolis, 278–280
Mississippi, Clinton, 301–302
Mississippi, Jackson, 307–308
Missouri, Saint Louis, 232–235, 388–390
  Missouri Botanical Garden, 265, 369–370, 397–398, 523–525, 561–563
Models, *see also* Software
  hydrologic, 121–122
Moisture content, in soil or aggregate, definition, 71

Mollusk shell surfacing, 529–530
Monoslab concrete grid, 386–388, 398–403
Morris Arboretum, 493–494
Mulch, *see* Engineered wood fiber, Soft porous surfacing, Wood and bark

Nature preservation, *see* Ecosystem preservation
Neat (asphalt), definition, 459
Netherlands, 158–159, 187–189
Netpave geocell, 293, 314–317
New Jersey
  Cherry Hill, 220
  Hoboken, 55–56, 185–187, 348–349
  Medford, 52–55, 223–224
  Newark, 363
New Mexico, Gallup, 452
New York
  Brooklyn, 14–15, 185–186
  Cortland, 345–347
  Garden City, 224–225
  Ithaca, 182–183
  Long Island, 160
  Sands Point, 391–393
Nitrogen, in stormwater, 426
  see also Nutrients
Noise
  block pavement, 329
  perception, 19
  porous asphalt, 499–500
  porous concrete, 421
  street, 18–20
Nonpoint pollution, definition, 153
North Carolina
  Boone, 431
  Charlotte, 395–397
  Henderson, 309–310
  Wilmington, 362–262
NPDES, 153
NRCS, 71
Nutrient-holding capacity, *see* Cation exchange capacity
Nutrients, in stormwater, 154
  see also Phosphorus, Nitrogen

Ohio
  Cincinnati, 179
  Columbus, 227–228, 557–558
  Dayton, 407–409
  Dublin, 271–272
  Gahanna, 399–400
  Grove City, 229–231
  Saint Clairsville, 364
Oils, in stormwater, 154
  see also Biodegradation

Index

Ontario
  Guelph 99–101
  Richmond Hill, 100–102
  Toronto, 46, 356–358
Open-celled grids, 23–24, 28, 43, 47, 324, 381–415
  bedding, 384–386
  definition, 381
  industry, 382–384
  infiltration rate, 124
  joints, 382
  root heaving, 179
  runoff coefficient, 126
  runoff observations, 128–129
  see also Flexible pavement
  with aggregate, *see* Aggregate
  with turf, 242–243
Open-graded, definition, 75, 173, 204
Open-jointed block, 45–46, 323–379; *see also* Block
  clogging and restoration, 334–336
  configurations and uses, 326–329
  construction, 329–332
  definition, 324–328
  hydrologic observations, 129–131, 140–142
  in cold climate, 99–102
  industry, 324–326
  infiltration rate, 124
  joint-fill materials, 332–337
  Manning's roughness, 128
  runoff coefficient, 126
  stability, 337–339
  surface drainage, 350–353
  water quality treatment, 157–160
Optimum moisture content, 71, 90
Orange Bowl, 297–298
Oregon Department of Transportation, 470–471, 502–504
Organic compound, definition, 153
Organic crop by-products, 528–529
Organic debris, clogging, 426
Organic surfacing materials, considerations for use, 519–520
Outlet, *see* Pipe outlet
Overlay, porous asphalt, 21, 36–38, 470–471, 496–507
  durability, 502–503
  water quality treatment, 158–159
Overseeding, definition, 244
Oxidation, of asphalt cement, 461

Paine, John (Jack), 419
Particulates, 154
  definition, 153

Patterson, James, 184, 256
Pavement
  area in America, 1–5
  components, 36–42
  definition, 1, 36
  effects, 1–33
  structure, *see* Structure, pavement
Paver, definition, 325, 382
Pavior, definition, 325
Pea gravel, 211
Penetration, asphalt, definition, 459
Pennsylvania
  Allison Park, 44
  Exton, 22
  Great Valley (Malvern), 134–135, 486–487
  Philadelphia, 493–494
  Pittsburgh, 7–8
  State College, 49, 452
  Swarthmore, 265–266
  Villanova, 452
  Willow Grove, 148–150
Perco-Bond concrete additive, 430
Perco-Crete porous concrete, 430
PermaTurf geocell, 317
Permeability, *see also* Infiltration rate
  definition, 120
pH, definition, 153
Phosphorus, in runoff, 425–426
  see also Nutrients
Pier A Park, 55–56, 185–187, 348–349
Pile foundation, definition, 541
Pipe outlet, 39–42
  discharge from, 136–143, 146–148
Pit run, 21, 204
Planting pit, 173
Plastic geocell, *see* Geocell
Plastic limit, definition, 71
Plastic lumber, 546–547
  definition, 541
Plastic web, 318–319
Plasticity, 71, 76
Playground
  accessibility, 526–528
  aggregate, 229–234
  engineered wood fiber, 525–528
  granular rubber surfacing, 530–532
  safety, 229–233, 517–518, 526–528; *see also* Critical height
Plug (of grass), definition, 244
Pollutants, *see also* Solids, Metals, Nutrients, Oils, Water quality
  leaching, 42, 152
Poorly graded, definition, 75
Porosity, definition, 120
Porous aggregate, *see* Aggregate

Porous asphalt, 22, 48–49, 457–512; *see also* Asphalt
  areas where used, 484
  construction, 471–475
  contemporary technology, 468–471, 497–498, 502–503
  drain-down, *see* Drain-down
  history, 438–460, 463–464, 468–469, 496–498
  hydrology, 479–483
  in cold climate, 101–102, 488–495
  maintenance and rehabilitation, 483–484
  noise, 20, 499–500
  overlay, *see* Overlay
  pavement thickness, 473–475
  recycling, 484
  root heaving, 179
  runoff coefficient, 126
  safety, 21
  standard specification, 503
  strength, 491
  water quality observations, 156–157, 159
Porous concrete, 25–27, 47–48, 417–455
  advantages and disadvantages, 420–422
  clogging and rehabilitation, 426, 440–441
  color, 421
  components, 418–419, 426–430, 436–437
  cracking, 432
  history, 418–420
  hydrology, 422
  in cold climate, 450–452
  infiltration rate, 124
  installation, 435–439
  joints, 419, 432–435
  root heaving, 179
  testing for quality control, 438–439
Porous pavement
  alternative materials, 42–51, 58–59
  definition, 1
  effects, 1–33
  provisions for all applications, 58–66
Portland cement, 217, 419, 428
  concrete, 419; *see also* Concrete, Porous concrete
Portland Cement Pervious Association, 420
Pozzolan, 419, 428
Pratt, Chris, 140–142, 157–160
Precast, definition, 325, 382
Precipitation
  affected by heat island, 16
  frequency, 161; *see also* Design storm
  small storms, 161–164
Prime coat, definition, 459
Proctor compaction test, 71, 89–92
Profile (soil), definition, 71
Puncheon, 541, 560–561

Rada thickness method, 85–86
Railing, 541, 550–552
Rainfall, *see* Precipitation
Raveling, asphalt, 459
Recharge, *see* Groundwater, Subsurface disposition
Recycled materials
  aggregate, 214–215
  asphalt, *see* Asphalt recycling
  brick, crushed, 214, 220–222, 396
  bricks, 370–371
  concrete, *see* Concrete rubble
  metal plates, 367–368
  plastic, in deck lumber, 546–547
  plastic, in geocells, 290, 314
  rubber, granular, 529–535
  rubber, in asphalt, 461
  scoria and slag, 215
  stone, 368
  vitrified clay, crushed, 214, 227–228
  wood, 544–545
Regulation, of development and stormwater, 23–24
Reinforcement, concrete, 429
Reinforcement, turf, *see* Turf reinforcement
Research needed
  decks, 543–544
  ESCS, 187
  frost damage and protection, 102, 452
  geocell standards, 286
  hydrology, 136, 426, 543–544
  open-jointed block structure, 339
  porous asphalt drain-down, 466
  porous concrete, 426, 430, 437, 452
  thermal effects, 18
  tree-rooting media, 187, 190, 191
Reservoir
  configuration, 36, 38–40, 132–135
  hydraulic storage, 131–136, 145–146
    conflict with tree rooting, 189
  in freezing climate, 98–99, 102–103
  outlet, 39–42; *see also* Pipe outlet
  ponding time and required thickness, 135–136, 145–146
  routing, 120
Rhizome, definition, 244
Rhode Island, Middletown, 314–317
Rigid pavement, 82–83
RKM, *see* Grassy Pavers geocell
Rollings and Rollings, 71, 236–237, 325, 349
Root base (root flare), *see* Tree base
Root heaving, 178–179
Rooting
  channels, 177, 192, 194
  media, *see* Tree rooting media, Turf construction

Index **575**

Rubber, granular recycled, 529–535
Rubber, in asphalt, 461
Rubble, *see* Concrete rubble
Runner (grass), definition, 244
Runoff, *see also* Runoff coefficient
    block pavements, 336–337
    definition, 123
    infiltration effect on, 150–152
    porous concrete, 422–426
    velocity and travel time, 127–128
Runoff coefficient, *see also* Curve number
    contrast with infiltration rate, 126–127
    definition, 123, 125
    open-celled grids, 408–409
    pavement surface types, 125–127
    porous asphalt overlay, 502
    tight-jointed blocks, 329
Rutting, definition, 458

Safety, driving, 20–21; *see also* Friction
    porous asphalt, 499, 505–506
    porous concrete, 421
Safety, playground, *see* Critical height, Playground safety
Salt, *see* Deicing salt
Sanding, and clogging, 481–482, 492–493
Saturated hydraulic conductivity, 122, 123, 253
    see also Infiltration rate
Scoria aggregate, 214
Screening, definition, 201, 409
Screenings, definition, 201, 204
SCS, *see* Soil Conservation Service
Seal coat, 459
Sediment, *see* Drainage at pavement edges, Construction, protection during, Solids
Segmental pavement, definition, 325, 326
Semisolid asphalt, definition, 459
Set time (setting time), 419
Setting bed, definition, 325, 382
SF Concrete Technology, 358
SF-Rima concrete block, 358–363
Shackel, Brian, 324, 325
Shell, *see* Mollusk shell, Organic crop by-product
Shoving, definition, 458
Shrinkage limit, definition, 71
Siemens office, 134–135, 486–487
Sieve number, definition, 71
Sieving, definition, 201
Signage, 65–66
Single-axle load, definition, 71
Single-sized, definition, 75, 173, 204
    see also Open-graded
Skeletal soil, 180
Skid resistance, *see* Friction

Slab, definition, 325
Slag aggregate, 140–142, 214, 428
Sleeper, definition, 541
Slope
    decks on, 561–562
    pavement surface, 38–40, 132–135
    reservoir, 132–135
    subgrade, 152
Slump test, 419, 436
Slurry seal, definition, 459
Smith, David, 325, 329
Snowmelt, definition, 120
Snow removal, *see also* Cold climate, Winter maintenance
    aggregate, 216, 312
    decks, 542
    geocells, 300, 312
    open-celled grids, 407
    open-jointed block, 354
    porous asphalt, 504
    porous concrete, 451
    turf, 262, 300
Sod, definition, 244
Soft porous surfacing, 49–51, 513–537
    accessibility, 517–518
    considerations for use, 515
    drainage, 518–519
    history, 514
    installation, 518–519
    safety, 517–518
    tree rooting, 179, 515
Software, hydrologic modeling, 121–122, 349
Software, structural design, 86, 339, 349
Soil Conservation Service (SCS), 71, 123
Soil, *see also* Subgrade
    classification, *see* Unified soil classification
    moisture, *see* Subsurface disposition
    particle size, 73–74
    structure, definition, 71
    surveys, 79–81
    swelling, 42, 72, 152, 231–234
    testing, 81
    texture, 74
Solid asphalt, definition, 459
Solids, capture and treatment, 155–159
    see also SS, TSS, Particulates
Sound, *see* Noise
South Carolina, DeWees Island, 514
Southface Institute, 45, 411, 444–447
Spalling, asphalt, 459
Specification, 63–64
    see also Contractor, communication with
Sprig (of grass), definition, 244
SS, definition, 153
Stability, *see also* Structure
    asphalt, *see* Marshall stability

Standards, AASHTO
  asphalt constituents, 461
  geotextile, 88
Standards, ASTM
  aggregate, 334, 427–428
  air entrainment, 430
  asphalt constituents, 461–462
  brick, 372
  concrete constituents, 426–428
  concrete grids, 383–384, 403
  concrete reinforcing, 429
  ESCS, 213
  gradation, 208, 213
  infiltration and permeability tests, 123
  particulate rubber, 532, 534
  plastic lumber, 546
  sieve sizes, 203
  turf maintenance, 257
  turf reinforcement mat, 274
  turf root zone tests, 254
Stolon, definition, 244
Stone, definition, 325
Stone dust, definition, 204
StoneyCrete porous concrete, 430
Storage, hydraulic, see Reservoir
Storm flow
  definition, 120
  duration, 146–148
  peak, 146–147
Stormwater, 7–10; see also Hydrology
  constituents, 154–155
  control, regulation, 22–24
  cost of control, 23–24
  definition, 120
  detention in reservoir, 146–148
Strength of porous pavement, 84
  see also Structure
Stringer, definition, 541
Stripping, asphalt, 459, 462
Structural soil, 14, 171–197, 287, 385
  see also Tree rooting
Structure, pavement, 69–117
  see also Flexible pavement, Rigid pavement
Subbase, 36, 419, 431
Subgrade
  bearing value, 75
  definition, 36, 71
  infiltration rate, preservation, 92–93, 145
  infiltration, 142–152; see also Subsurface disposition
  plastic, 105–106
  structural role, 71–81
  swelling, 103–106
  tree rooting, 192
  uncompacted, 92–93

water quality treatment, 160–161
  wet, 85–87, 103
Subsurface disposition of infiltrated water, 148–150
  hazards, 152
Surface course, 36–37
  role in tree rooting, 177–179
  structural role, 83–84
Sweden, Luleå, 101–102, 494–495

Tack coat, 459
Temperature, see Heat island, Water quality
Tender asphalt, definition, 459
Tennessee, Chattanooga, 48
Texas
  Arlington, 104, 231–234
  Austin, 128–129, 133–134, 479
  The Woodlands, 159, 479, 483
Texture, soil, 72
Thatch, turf, 244
Thelen, Edmund, 458
Thelen and Howe gradation, 463
Thermal effects, see Heat island, Water quality
Thickness
  block pavement, 337–339
  design, 84–86
  for frost, 93, 96–98
  geocell pavement, 315–316
  porous concrete pavement, 430–431
  structural role, 71–73
Throughflow, see Drainage below surface
Tie (cross-tie), definition, 541
Tight-jointed, definition, 327–328
Time of concentration, 123, 127
Top size, definition, 204
Topdressing, definition, 244
Toxic chemicals in soil, 42
  see also Brownfield
Trace metals, see Metals
Traffic calming, 329
Traffic load, 70–71
Trails, 291–295
Transpiration, definition, 120
Traprock, definition, 201
Tree
  base, 173, 192–194
  grates, 193–194, 541, 542
  preservation, 23–24, 177, 192, 360–362, 397–398
  rooting, 55–56, 443, 515
    media, 171–197; see also Structural soil
    volume, 12, 175–177, 190
    zone, 172–174, 177–179
  sand, see Amsterdam tree soil
  selection, 190

Index

Tree (*continued*)
   shade, cooling effect, 17
   urban, 10–15
      benefits, 11–12
Trunk flare, *see* Tree base
TSS, definition, 153
Tufftrack geocell, 303–304
Turf, 43–45, 55–56, 241–283
   accessibility, 243
   construction, 250–257
      base course, 255–257
      California profile, 251–252
      root zone, 252–254
      USGA profile, 250–251
   environmental effects, 243
   in freezing climate, 93
   in geocells, 57–58, 242–243, 287–289, 297–302
   in open-celled grids, 385–388, 391–393
   infiltration rate, 124, 262–264, 276
   maintenance, 243, 257–262
   on swelling soil, 105
   reinforced, 242–243, 277–280
   runoff coefficient, 126
   strength, 242–243
   temperature, 407–408
   tree-root heaving, 179
   turfgrass, *see also* Grass
      definition, 244
      installation, 254–255
      species and varieties, 245–250
      wear and compaction, 243–245, 386–388
Turfblock concrete grid, 403
Turfgrid concrete grid, 403, *see also* Turfstone
Turf Pavers concrete grid, 410
Turf-Slab concrete grid, 403, *see also* Turfstone
Turfstone concrete grid, 384, 386, 402–410
   hydrologic observations, 128–131

Unified soil classification, 72, 76–80
Uni-Green concrete grid, 410
Uni-Group USA, 121–122, 342, 349
USDA, 72

Virginia, Alexandria, 26
Virginia, Warrenton, 133–135
Vitrified clay, crushed, 214, 227–228

Void space, definition, 120
Void space, hydraulic storage in, 131–132
von Langsdorff Licensing, 342, 349
VSS, definition, 153

Washington, DC, *see* District of Columbia
Washington
   King County, 521
   Olympia, 47, 405–407, 426, 448–449
   Redmond, 481–482
   Renton, 130–131, 404
   Seattle, 162–163
Water harvesting, 6, 42, 152
Water holding capacity, 173–175
Water quality, 7–10; *see also* Pollutants
   in porous asphalt, 482–483
   in porous concrete, 422–426
   thermal, 16
   treatment, 152–161; *see also* Biodegradation
Watershed discharge of infiltrated water, 150–152
   see also Base flow, Storm flow, Subsurface disposition
Watershed, urban, 7–10, 123
   see also Impervious cover, Stormwater, Water quality
Water table, *see* Groundwater
Wearing course, *see* Surface course
Web, *see* Plastic web
Well graded, definition, 75, 204
West Virginia, New River Gorge, 561–562
Westfarms Mall, 261–262, 298–301
Wheel stops, 109–111
Widely spaced blocks, definition, 327–328
Winter maintenance, *see* Deicing salt, Sanding, Snow removal
Wisconsin, Hudson, 278–279
Wisconsin, Kettle Moraine State Forest, 294–295
Wood, *see also* Lumber
   in deck construction, 544–546
   preservative-treated, 545–546
   recycled, 544–545
Wood and bark surfacing, 520–525
   see also Engineered wood fiber
Workability, of porous concrete, 419, 436

Zinc, *see* Metals